Developing Windows-Based and Web-Enabled Information Systems

Developing Windows-Based and Web-Enabled Information Systems

Nong Ye
Teresa Wu

CRC Press
Taylor & Francis Group
Boca Raton London New York

CRC Press is an imprint of the
Taylor & Francis Group, an **informa** business

CRC Press
Taylor & Francis Group
6000 Broken Sound Parkway NW, Suite 300
Boca Raton, FL 33487-2742

First issued in paperback 2017

Version Date: 20140716

ISBN 13: 978-1-4398-6059-5 (hbk)
ISBN 13: 978-1-138-07377-7 (pbk)

Library of Congress Cataloging-in-Publication Data

Ye, Nong, 1964-
Developing Windows-based and Web-enabled information systems / authors, Nong Ye, Teresa Wu.
pages cm
Includes bibliographical references and index.
ISBN 978-1-4398-6059-5 (hardcover : alk. paper) 1. Database design. 2. Web site development. 3. Information technology. 4. Microsoft Windows (Computer file) I. Wu, Teresa, 1972- II. Title.

QA76.9.D26Y43 2015
005.74'3--dc23 2014023646

Visit the Taylor & Francis Web site at
http://www.taylorandfrancis.com

and the CRC Press Web site at
http://www.crcpress.com

Contents

Section II Database Design and Development

Section VI Case Studies

Preface

An information system is an integrated system that contains databases, programs, and graphical user interface to generate, send, receive, store, and process data into meaningful information and knowledge for supporting decision making. Two types of information systems are common: those for Windows applications used by individual users and those for Web applications accessible simultaneously by multiple users through the Internet. This book covers concepts, methods, programming languages, and software tools for designing and developing both Window-based and Web-enabled information systems. Specifically, data modeling and design methods, including entity–relationship modeling, relational modeling and normalization, and object-oriented data concepts are covered. The use of Windows application development tools and languages, including Microsoft Access, MySQL, structured query language (SQL), Visual Studio, Visual Basic for Applications (VBA), and extensible markup language (XML), is introduced and explained. Concepts, methods, and tools will be illustrated in a self-contained manner with easy-to-understand small examples and large-scale case studies. This book will enable readers in engineering, business, and science fields to gain knowledge and skills for developing Windows-based and Web-enabled information systems.

Acknowledgments

Nong Ye would like to thank her family, Baijun and Alice, for their love, understanding, and unconditional support. Teresa Wu wishes to express her love and gratitude to her beloved family, Aimin and Tristan, who support and encourage her in spite of all the time it took her away from them. Teresa Wu would also like to thank Eugene Rex Jalao, who provided support, offered comments, and assisted in the editing and proofreading of some book materials. We would like to thank Cindy Carelli, a senior editor at CRC Press. This book would not have been possible without her being responsive, helpful, understanding, and supportive for a few years since we started preparing this book. It has been a great pleasure working with her. We also thank other people at CRC Press who helped us in publishing this book.

Authors

Nong Ye is a professor at the School of Computing, Informatics, and Decision Systems Engineering, Arizona State University, Tempe, Arizona. She holds a PhD in Industrial Engineering from Purdue University, West Lafayette, Indiana, an M.S. degree in Computer Science from the Chinese Academy of Sciences, Beijing, People's Republic of China, and a B.S. degree in Computer Science from Peking University, Beijing, People's Republic of China. Her book publications include this book, *Data Mining: Theories, Algorithms, and Examples*, *The Handbook of Data Mining*, and *Secure Computer and Network Systems: Modeling, Analysis, and Design*. Her publications also include over 80 journal papers in the fields of data mining, statistical data analysis and modeling, computer and network security, quality-of-service optimization, quality control, human–computer interaction, and human factors.

Teresa Wu is an associate professor at the School of Computing, Informatics, and Decision Systems Engineering, Arizona State University, Tempe, Arizona. She holds a PhD in Industrial Engineering from the University of Iowa. Her book publications include this book and *Managing Supply Chain Risk and Vulnerability: Tools and Methods for Supply Chain Decision Makers*. Her publications also include over 60 journal papers in the fields of information systems, decision algorithms, data mining, and health informatics.

Overview

Topics for developing information system applications are organized into six sections as follows:

I Foundation of Information Systems
II Database Design and Development
III Windows Application Development
IV Web Application Development
V Design of Information Systems
VI Case Studies

Section I of this book describes computer hardware and software that provide the foundation for developing information system applications. Section I covers the following:

- Boolean algebra and digital logic circuits for computer hardware in Chapter 1
- Digital data representation in Chapter 2
- Computer and network system software in Chapter 3
- Computer and network applications for information systems in Chapter 4

Section II describes concepts, models, and tools related to database design and development. A database is an organized collection of data and is the core of an information system. Data in a database are organized using a model, such as a relational model, an object-oriented model, or an object-relational model. Data are then retrieved via data queries. Section II covers the following:

- Conceptual database design: entity–relational modeling in Chapter 5
- Logical database design: Relational modeling and normalization in Chapter 6
- Database implementation in Microsoft Access in Chapter 7
- Structured query language in Chapter 8
- MySQL: A relational database management system in Chapter 9
- Object-oriented database systems in Chapter 10

With the development of a database, applications are then developed to enable users to access and manipulate data in a database through a graphical user interface (GUI). Section III of this book covers the development of Windows-based applications with a database embedded. Section III includes the following:

- Windows forms and controls in Microsoft Visual Studio in Chapter 11
- Visual Basic programming in Microsoft Visual Studio in Chapter 12
- Database connection in Microsoft Visual Studio in Chapter 13

- Windows forms and controls with Microsoft VBA in Chapter 14
- Data connectivity with VBA embedded within Microsoft Excel in Chapter 15

Section IV of this book describes the development of Web-enabled applications. Some fundamental concepts and standards (including XML) are explained. The development of Web applications and Web services is illustrated. Section IV includes the following:

- Web applications in Microsoft Visual Studio in Chapter 16
- Working with XML (I) in Chapter 17
- Working with XML (II) in Chapter 18
- Web services in Chapter 19

With Section II to Section IV covering the database and GUI aspects of information systems, Section V addresses some other aspects of information systems: computing efficiency of algorithms, usability, security, extraction of useful information and knowledge from data through data mining, and knowledge-based reasoning to support decision making. Section V includes the following:

- Computing efficiency of algorithms in Chapter 20
- User interface design and usability in Chapter 21
- Computer and network security in Chapter 22
- Data mining in Chapter 23
- Expert systems in Chapter 24
- Decision support systems in Chapter 25

Section VI of this book demonstrates the development of information systems through three case studies. Section VI includes the following:

- Development of a healthcare information system using Microsoft Access and Visual Studio in Chapter 26
- Development of an online system for imaging device productivity evaluation in radiology practices in Chapter 27
- Development of radiology skin dose simulation tools using VBA and database in Chapter 28

Distinctive Features

Many existing textbooks are written for readers having a background in computer science and knowledge, experience, and skills in programming and working on developing large-scale computer and network software. However, many professionals in engineering, business, and science fields have needs to develop information systems for managing and using data to support decision making in their fields, although they are not professional programmers or software developers. This book aims to help students and professionals in engineering, business, and science fields, who are not professional programmers or software developers, learn concepts, methods, and software tools for developing Windows-based and Web-enabled information systems. This book takes readers through the entire process of developing a computer and network application for an information system, including the following:

1. Design and development of a database
2. Development of a Windows-based or Web-enabled application with a database embedded and with GUI to allow users to access and manipulate data in the database
3. Design of the information system for computing efficiency, usability, security, and capabilities of turning data into useful information and knowledge and performing knowledge-based reasoning for decision support

This book provides both concepts and operational details of developing a Windows-based or a Web-enabled application for an information system using Microsoft software development environments and tools. In each chapter, small data examples are used to manually walk through concepts and operational details. Hence, this book will enable readers in non–computer science fields to obtain the conceptual understanding and the practical skills of independently developing Windows-based or Web-enabled applications for their specialized information systems.

Teaching Support

Topics covered in this book involve different levels of difficulty. An instructor, who uses this book as a textbook for a course on information systems, may select book materials to meet their instructional needs based on the level of the course and the difficulty levels of book materials. The book materials in Chapters 5 to 7, 11 to 16, and 26, which cover database design and Windows-based and Web-enabled applications, are appropriate for both undergraduate-level and graduate-level courses. The case studies in Chapters 26 to 28 can be selectively used by the instructor to illustrate topics covered in Chapters 5 to 7, 11 to 16. The book materials in Chapters 17 to 19, which cover XML standards and Web services, may be appropriate for a graduate-level course only. The book materials in Section V, including Chapters 20 to 25, take the development of information systems from the basic components of a database and GUI to more advanced design issues. Hence, the book materials in Section V are appropriate for a graduate-level course. Concepts in Section I of this book, which explain the foundation of hardware and system software for computers and networks, are required to understand the materials on computer and network security in Chapter 22. Although the book materials in Section I of this book are not required for learning the concepts and skills of developing information systems in Section II to Section IV, readers who are interested in learning how Windows and Web applications are enabled by hardware and system software for computers and networks are encouraged to read the book materials in Section I to unveil the mystery of the digital revolution enabled by computing technologies and to obtain a complete picture of computer and network applications.

Exercises are provided at the end of each chapter. Additional teaching support materials can be obtained from the publisher and include the following:

- Solution manual
- Lecture notes, which include the outline of topics, figures, tables, and equations from this book to support teaching

Section I

Foundation of Information Systems

Information systems are computer and network applications running on networks of computers. Figure I.1 shows the computer architecture in a hierarchy from the hardware level of digital logic circuits to the software levels of machine instruction code, operating system, system software, and applications. There are many types of computer and network applications for scientific computing, data visualization, data management, decision support, and so on. Information systems are computer and network applications for managing data and information to provide decision support.

Section I of this book covers computer hardware and system software, which lay the foundation for the development of information systems that is covered after Section I. Computer hardware is made up of digital logic circuits that can be designed using Boolean logic. Boolean logic and digital logic circuits are covered in Chapter 1. Digital logic circuits directly support the representation and manipulation of binary data. Chapter 2 explains how non-binary data and information are represented using binary data. System software, including the operating system and network software, is described in Chapter 3. Chapter 4 gives an overview of various information systems.

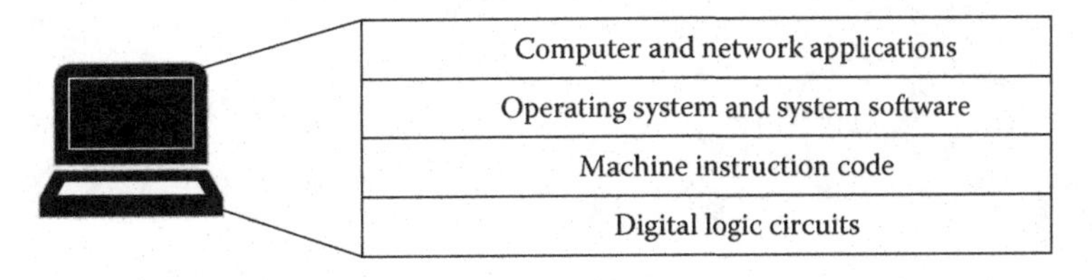

FIGURE I.1
Hierarchy of Computer Hardware and Software.

1

Boolean Algebra and Digital Logic Circuits for Computer Hardware

Computer hardware is made of electrical circuits. High and low voltages of electrical circuits are considered as true and false or 1 and 0, respectively, of binary data in Boolean logic, which is also known as "symbolic logic," and deals with true and false values in logical reasoning. The implementation of a digital circuit can be expressed as a Boolean expression. Hence, Boolean logic is used to design digital circuits (called "digital logic circuits"). This chapter first introduces Boolean logic and then gives the design of digital logic circuits for such computer components as the arithmetic logic unit (ALU) of the central processing unit (CPU) and memory to illustrate how computer hardware works to process and store binary data.

1.1 Boolean Logic

This section introduces binary values and Boolean operators, gives properties of Boolean algebra, and describes the method of converting a truth table for a function into a Boolean expression, which has one-to-one correspondence to the implementation of a digital logic circuit for the function.

1.1.1 Binary Values and Boolean Operators

Boolean logic represents two values: false, with 0 as the binary value representation, and true, with 1 as the binary value representation. There are three basic Boolean operators: AND, OR, and NOT, which are used to construct Boolean expressions. Tables 1.1, 1.2, and 1.3 give the truth table for the three Boolean operators. A truth table for a Boolean expression lists all the combinations of values for the input variables in the Boolean expression and gives the resulting values of the Boolean expression for input values. In Table 1.1, for the AND operator, there are two input variables, x and y, which take binary values, 0 and 1, and one binary output for the result of the AND operation, x AND y, which is represented by xy. For the AND operator, the output result is true (1) only when x takes the true value (1) and y takes the true value (1). In Table 1.2, the output result of the OR operator, x OR y, which is represented by $x + y$, is true (1) when either x or y takes the true value (1). In Table 1.3, the output of the NOT operator, NOT x, which is represented by $\overline{x}$, takes the opposite value of x.

TABLE 1.1
Truth Table for the Boolean Operator, AND

x	y	xy
0	0	0
0	1	0
1	0	0
1	1	1

TABLE 1.2
Truth Table for the Boolean Operator, OR

x	y	$x + y$
0	0	0
0	1	1
1	0	1
1	1	1

TABLE 1.3
Truth Table for the Boolean Operator, NOT

x	$\overline{x}$
0	1
1	0

1.1.2 Basic Properties of Boolean Algebra

Table 1.4 gives the basic properties of Boolean algebra, showing a basic set of identical Boolean expressions. These properties can be proven using the truth table. For example, Table 1.5 gives the truth table to prove the OR form of de Morgan's law. In Table 1.5, fourth column gives the values for the left side of the equation for the de Morgan's law, and the last column gives the values for the right side of the equation for the de Morgan's law. The

TABLE 1.4
Properties with Identical Expressions in Boolean Algebra

Name of Property	AND Form	OR Form
Identity properties	$1x = x1 = x$ $xx = x$	$0 + x = x + 0 = x$ $x + x = x$
Null property	$0x = x0 = 0$	$1 + x = x + 1 = 1$
Inverse property	$x\overline{x} = 0$	$x + \overline{x} = 1$
Commutative property	$xy = yx$	$x + y = y + x$
Associative property	$(xy)z = x(yz)$	$(x + y) + z = x + (y + z)$
Distributive property	$x + yz = (x + y)(x + z)$	$x(y + z) = xy + xz$
de Morgan's law	$\overline{xy} = \overline{x} + \overline{y}$	$\overline{x + y} = \overline{x}\overline{y}$

TABLE 1.5

Truth Table to Prove the OR Form of de Morgan's Law

x	y	$x+y$	$\overline{x+y}$	$\overline{x}$	$\overline{y}$	$\overline{x}\overline{y}$
0	0	0	1	1	1	1
0	1	1	0	1	0	0
1	0	1	0	0	1	0
1	1	1	0	0	0	0

basic properties of Boolean algebra can be used to prove more complex identical Boolean expressions and simplify a Boolean expression. Example 1.1 shows using the basic properties of Boolean algebra to simplify a Boolean expression. Example 1.2 proves two identical Boolean expressions.

Example 1.1

Simplify the following Boolean expression:

$$(x+y)(x+\overline{y})$$

$$(x+y)(x+\overline{y}) = x + y\overline{y} \quad \text{(AND form of distributive property)}$$
$$= x + 0 \quad \text{(AND form of inverse property)}$$
$$= x \quad \text{(OR form of identity property)}$$

Example 1.2

Use the basic properties of Boolean algebra to prove the following Boolean equation:

$$x(x+y) = x$$

$$x(x+y) = xx + xy \quad \text{(OR form of distributive property)}$$
$$= x + xy \quad \text{(AND form of identity property)}$$
$$= x1 + xy \quad \text{(AND form of identity property)}$$
$$= x(1+y) \quad \text{(OR form of distributive property)}$$
$$= x1 \quad \text{(OR form of null property)}$$
$$= x \quad \text{(AND form of identity property)}$$

1.1.3 Conversion of a Truth Table for a Function into a Boolean Expression

To implement a function using a digital logic circuit, we often use a truth table to express the function and then covert the truth table into the corresponding Boolean expression, which is used as the design of the digital logic circuit implementing the function. Table 1.6 gives the truth table for a parity generator, which produces a parity bit for data containing 3 bits of x, y, z. If the 3 bits of x, y, z have an odd number of value 1, the parity generator produces the parity bit of 1. If the input variables x, y, z have an even number of value 1, the parity generator produces the parity bit of 0. When the 3-bit data are transmitted, the

TABLE 1.6

Truth Table for the Parity Generator

x	y	z	Parity Bit
0	0	0	0
0	0	1	1
0	1	0	1
0	1	1	0
1	0	0	1
1	0	1	0
1	1	0	0
1	1	1	1

parity bit is transmitted along with the data. If the received data and the parity bit are not consistent, there is an error in transmitting the data and the parity bit. For example, if the data of 000 and its parity bit of 0 are transmitted but the received data are 100 and the received parity bit is 0, we know that a transmission error occurs because the data of 100 should have the parity bit of 1. The parity bit can be used for detecting only an odd number of bit transmission errors. For example, if the data of 000 and the corresponding parity bit of 0 are transmitted and there are 2-bit transmission errors to produce the received data of 101 and the received parity bit of 0, we cannot detect the transmission errors because the received parity bit of 0 is the correct parity for the received data of 101.

Table 1.7 gives the truth table for a 1-bit full adder that takes carry-in.

A function expressed in a truth table can be represented as a Boolean expression in the sum-of-product form in the following steps:

Step 1. Express the input variables in a product term for each row of the truth table that has the output variable taking the value of 1, by expressing an input variable x with the value of 0 as $\bar{x}$ and an input variable x with the value of 1 as x.

Step 2. Add all the product terms from Step 1 to produce the sum-of-product form of a Boolean expression.

Step 3. Simplify the Boolean expression in Step 2.

TABLE 1.7

Truth Table for the 1-Bit Adder

Inputs			Outputs	
x	y	z (Carry-In)	Sum	Carry-Out
0	0	0	0	0
0	0	1	1	0
0	1	0	1	0
0	1	1	0	1
1	0	0	1	0
1	0	1	0	1
1	1	0	0	1
1	1	1	1	1

These steps are illustrated in Examples 1.3 and 1.4.

Example 1.3

Produce a Boolean expression in the sum-of-product form for the truth table in Table 1.6.

In Step 1, Rows 2, 3, 5, and 8 have the output value of 1 among the eight rows of data. The product terms for these rows with the output value of 1 are

Row 2: $\overline{x}\overline{y}z$
Row 3: $\overline{x}y\overline{z}$
Row 5: $x\overline{y}\overline{z}$
Row 8: xyz

In Row 2, we have the input variables $x = 0$, $y = 0$, and $z = 1$. The only product term of the input variables producing the output value of 1 is $\overline{x}\overline{y}z$ for Row 2. Similarly, $\overline{x}y\overline{z}$, $x\overline{y}\overline{z}$, and xyz are the only product form of the input variables producing the output value of 1 for Rows 3, 5, and 8, respectively.

In Step 2, the product terms for Rows 2, 3, 5, and 8 are added together through the OR operator into the following Boolean expression in the sum-of-product form:

$$\overline{x}\overline{y}z + \overline{x}y\overline{z} + x\overline{y}\overline{z} + xyz.$$

This Boolean expression cannot be further simplified in Step 3.

In this example, the input values in Rows 2, 3, 5, or 8 make one of the product terms in the Boolean expression of the parity generator produce the output value of 1 and, thus, make the entire Boolean expression produce the output value of 1. Any other combinations of input values (those in the other rows of the truth table) are different from those that are represented by the product terms in the Boolean expression make each of the product terms produce the output value of 0 and thus make the entire Boolean expression produce the output value of 0. Hence, the Boolean expression constructed in the above three steps produces the desired output value for each row of input values.

Example 1.4

Produce a Boolean expression in the sum-of-product form for the output variable, carry-out, in the 1-bit full adder in Table 1.7.

In Step 1, Rows 4, 6, 7, and 8 have the output value of 1 among the eight rows of data. The product terms for these rows with the output value of 1 are

Row 4: $\overline{x}yz$
Row 6: $x\overline{y}z$
Row 7: $xy\overline{z}$
Row 8: xyz

In Step 2, the product terms for Rows 4, 6, 7, and 8 are added together through the OR operator into the following Boolean expression in the sum-of-product form, which is simplified in Step 3:

$$\overline{x}yz + x\overline{y}z + xy(\overline{z} + z) = \overline{x}yz + x\overline{y}z + xy.$$

Note that the output for the sum bit in Table 1.7 is the same as the output for the parity bit in Table 1.6. Hence, the Boolean expression from Example 1.3, $\overline{x}\overline{y}z + \overline{x}y\overline{z} + x\overline{y}\overline{z} + xyz$, can be used for the sum bit in Table 1.7, where z is carry-in.

1.2 Digital Logic Circuits

This section introduces digital logic gates for a set of basic Boolean operators and describes combinational circuits and sequential circuits that are used to build computer hardware components such as the ALU in the CPU and memory.

1.2.1 Digital Logic Gates

Digital logic gates are digital circuits built for a set of basic Boolean operators, including AND, OR, NOT, and other operators such as exclusive-OR (XOR) and NOT AND (NAND). Table 1.8 gives the truth table for XOR, which produces the output value of 1 if only one of the two input variables x and y has the value of 1, that is, if the input variables have the value of 1 exclusively. Table 1.9 gives the truth table for x NAND y, whose Boolean expression is $\overline{xy}$. Figure 1.1 shows the symbols used to represent the logic gates for AND, OR, NOT, XOR, and NAND.

TABLE 1.8
Truth Table for XOR

x	y	x **XOR** y
0	0	0
0	1	1
1	0	1
1	1	0

TABLE 1.9
Truth Table for NAND

x	y	x **NAND** y
0	0	1
0	1	1
1	0	1
1	1	0

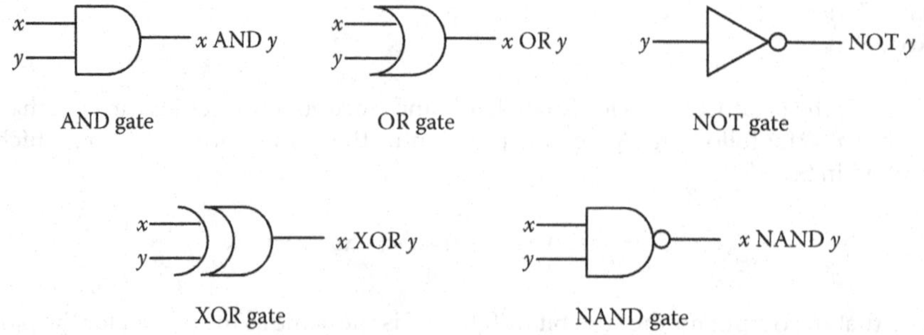

FIGURE 1.1
Symbols for logic gates.

XOR is useful for many functions. For example, XOR can be used to clear the content of x stored at a memory location on a computer by writing the result of x XOR $x = 0$ to the memory location. That is, $x = x$ XOR $x = 0$. XOR can also be used to exchange the values of x and y stored at two memory locations by performing the following operations in sequence:

$$x = x \text{ XOR } y$$

$$y = x \text{ XOR } y$$

$$x = x \text{ XOR } y.$$

Table 1.10 shows the change of values in x and y step by step in the above operations. The values of y in the second last column in Table 1.10 are the same as the original values of x in the first column. The values of x in the last column in Table 1.10 are the same as the original values of y in the second column.

The NAND gate can be used to construct a digital logic circuit for any function. For example, the AND, NOT, and OR gates can be constructed using only the NAND gates as follows:

$$\overline{\overline{xy}\;\overline{xy}} = \overline{\overline{xy}} = xy$$

$$\overline{xx} = \overline{x}$$

$$\overline{\overline{xx}\;\overline{yy}} = \overline{\overline{x}\;\overline{y}} = \overline{\overline{x}} + \overline{\overline{y}} = x + y$$

Figure 1.2 gives the logic diagrams showing the implementation of digital logic circuits based on the Boolean expressions of $\overline{\overline{xy}\;\overline{xy}}$, $\overline{xx}$, and $\overline{\overline{xx}\;\overline{yy}}$ for the AND, NOT, and OR gates, respectively.

TABLE 1.10

Illustration of Operations for Using XOR to Exchange Values of x and y

x	y	$x = x$ XOR y	$y = x$ XOR y	$x = x$ XOR y
0	0	0	0	0
0	1	1	0	1
1	0	1	1	0
1	1	0	1	1

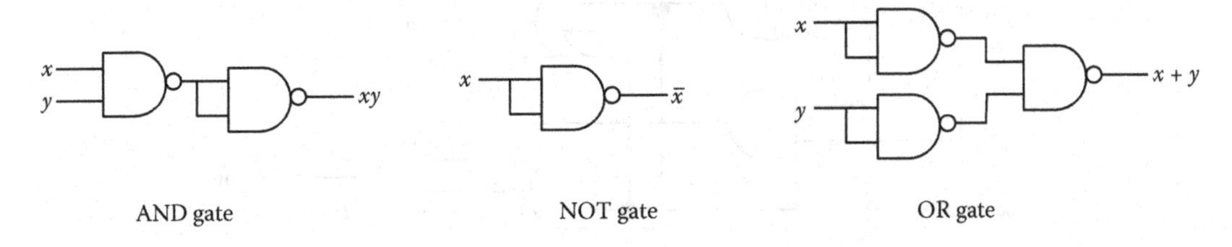

FIGURE 1.2
Construction of the AND, NOT, and OR gates, using only the NAND gates.

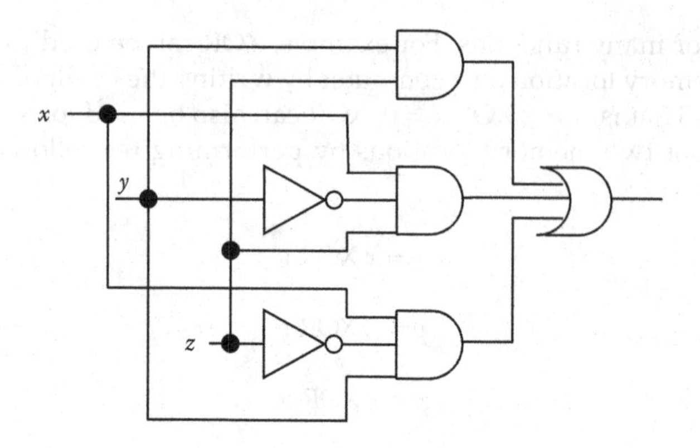

FIGURE 1.3
Logic diagram for $yz+x\overline{y}z+xy\overline{z}$.

Example 1.5

Use the logic diagram to show the implementation of the digital logic circuit for the Boolean expression $yz+x\overline{y}z+xy\overline{z}$ from Example 1.4.

Figure 1.3 gives the logic diagram for the Boolean expression.

1.2.2 Combinational Circuits

A combinational circuit produces an output based on entirely the current inputs without memory or storage of the past inputs. Figure 1.4 gives the combinational circuit that implements the 1-bit half adder without the carry-in bit. Figure 1.5 shows a combinational circuit of the 1-bit full adder in Table 1.7.

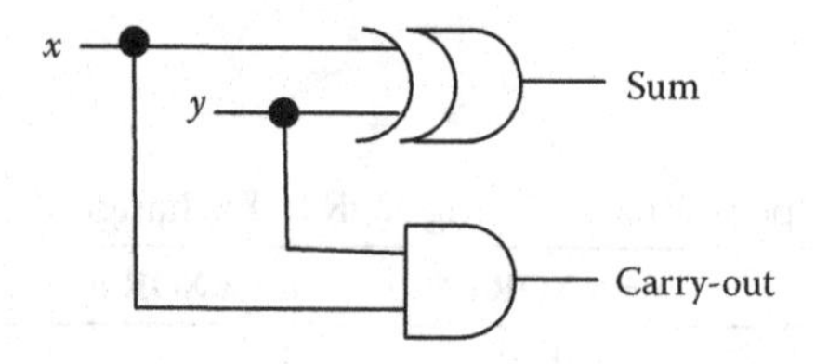

FIGURE 1.4
Logic diagram of the digital logic circuit for the 1-bit half adder.

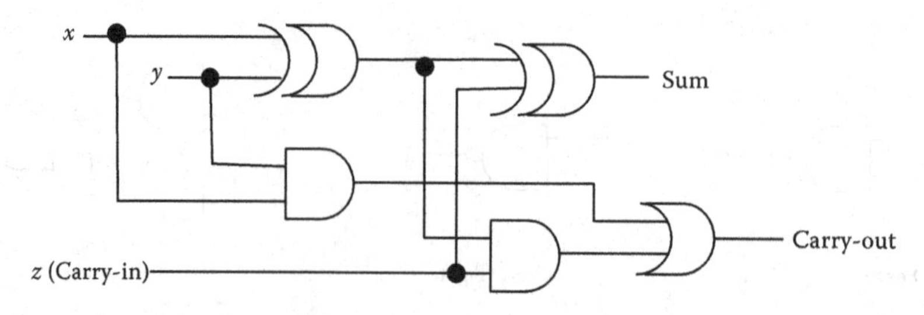

FIGURE 1.5
Logic diagram of the digital logic circuit for the 1-bit full adder.

The Boolean expression for the sum bit in the combinational circuit of the full adder in Figure 1.5 is Sum = z XOR (x XOR y). As shown below, this Boolean expression is equal to the Boolean expression for the sum bit in the sum-of-product form from Example 1.4.

$$\begin{aligned}\text{Sum} &= z \text{ XOR } (x \text{ XOR } y) = z \text{ XOR } (\bar{x}y + x\bar{y}) = \bar{z}(\bar{x}y + x\bar{y}) + z(\overline{\bar{x}y + x\bar{y}})\\ &= \bar{x}y\bar{z} + x\bar{y}\bar{z} + z\left(\overline{\bar{x}y}\right)\left(\overline{x\bar{y}}\right) = \bar{x}y\bar{z} + x\bar{y}\bar{z} + z(x+\bar{y})(\bar{x}+y)\\ &= \bar{x}y\bar{z} + x\bar{y}\bar{z} + x\bar{x}z + xyz + \bar{x}\bar{y}z + y\bar{y}z = \bar{x}y\bar{z} + x\bar{y}\bar{z} + xyz + \bar{x}\bar{y}z.\end{aligned}$$

As shown below, the Boolean expression for the carry-out bit in the combinational circuit in Figure 1.5 is equal to the Boolean expression for the carry-out bit from Example 1.4.

$$\text{Carry-out} = xy + (x \text{ XOR } y)z = xy + (\bar{x}y + x\bar{y})z = xy + \bar{x}yz + x\bar{y}z$$

The digital logic circuit for the 1-bit full adder in Figure 1.5 can be used to construct a ripple-carry adder for any bit length (e.g., 128 bits) by feeding the carry-out bit of the adder for 0 bit to the carry-in bit of the adder for the next bit.

Figure 1.6 gives the logic diagram of the digital logic circuit for the 2- to 4-bit decoder, which has a 2-bit input representing a memory address. Two bits can represent four different memory addresses since $2^2 = 4$. The decoder has four output bits, each of which corresponds to one of the four memory locations. When a memory address is given to the decoder as the input, the decoder produces the output value of 1 for the memory location represented by the input and produces the output value of 0 for the other memory locations. Hence, the decoder is to select a memory location specified by the memory address. For example, for the input of 00 ($x = 0$ and $y = 0$), the output line for $\bar{x}\bar{y}$ in Figure 1.6 has the value of 1, and the three other output lines have the value of 0. Hence, the memory address of 00 selects the output line for $\bar{x}\bar{y}$ in Figure 1.6. The n-to-2^n decoder with n input bits for a memory address and 2^n output lines for the memory locations can be designed similarly.

Figure 1.7 gives the logic diagram of the digital logic circuit for a multiplexer. The multiplexer in Figure 1.7 lets two control bits C_1 and C_0 determine which of the four input lines, I_3, I_2, I_1, and I_0 become the output line. Specifically, if the control bits C_1C_0 are 00, the multiplexer makes I_0 the output; if the control bits C_1C_0 are 01, the multiplexer makes I_1 the

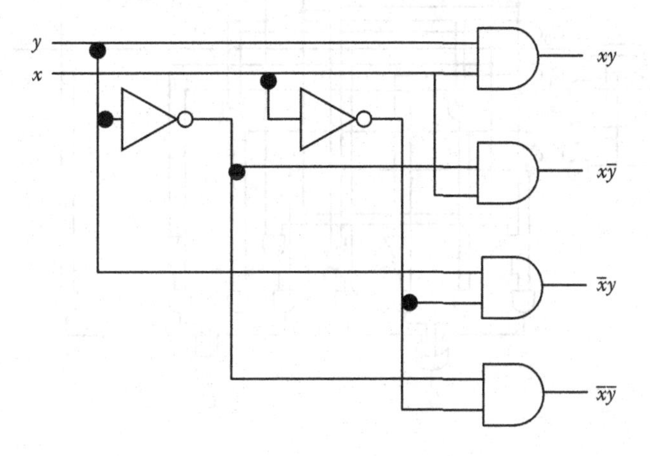

FIGURE 1.6
Logic diagram of the digital logic circuit for a 2- to 4-bit decoder.

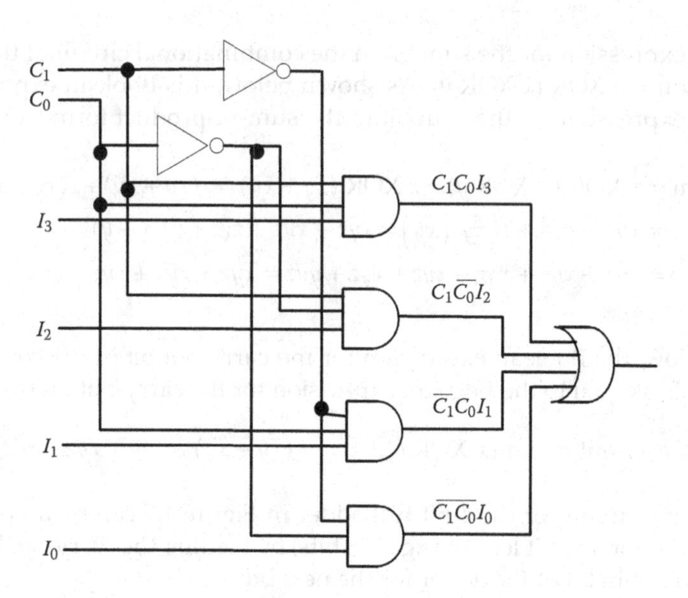

FIGURE 1.7
Logic diagram of the digital logic circuit for a multiplexer.

output; if the control bits C_1C_0 are 10, the multiplexer makes I_2 the output; and if the control bits C_1C_0 are 11, the multiplexer makes I_3 the output.

CPU is an important part of computer hardware. It consists of the ALU, registers including the program counter, and the control unit. Figure 1.8 shows the digital logic circuit for a simple 2-bit ALU, which performs four basic operations: AND, OR, NOT, and addition. The digital logic circuit for the ALU uses the digital logic circuits of the half

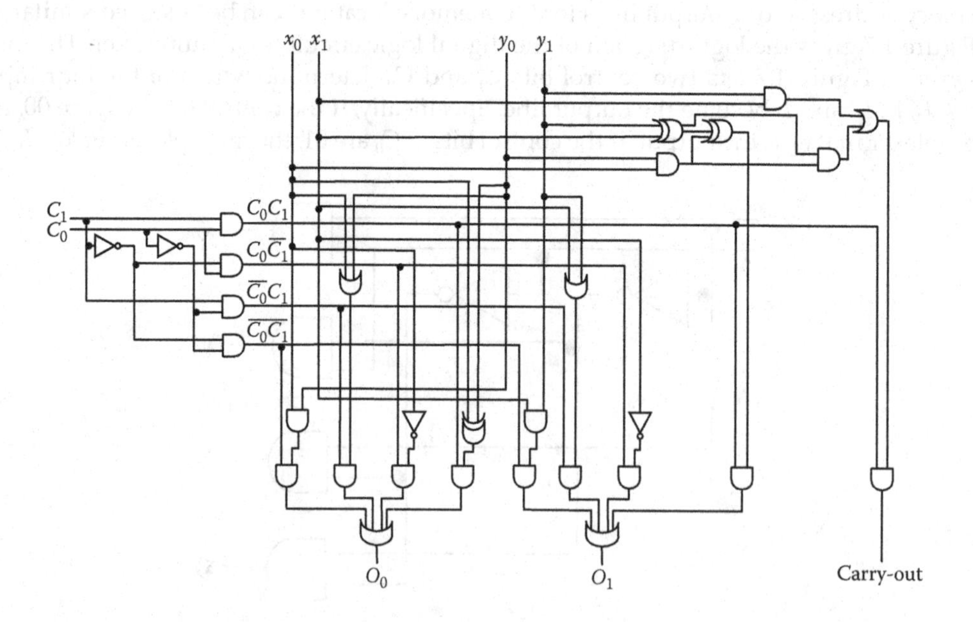

FIGURE 1.8
Logic diagram of the digital logic circuit for a simple ALU.

adder in Figure 1.4, the full adder in Figure 1.5, and the decoder in Figure 1.6. The two control bits C_1 and C_0 determine which operation is performed on two inputs, x_1x_0 and y_1y_0. If C_0C_1 are 00, x_1y_1 and x_0y_0 for AND is performed. If C_0C_1 are 01, $x_1 + y_1$ and $x_0 + y_0$ for OR is performed. If C_0C_1 are 10, $\overline{x_1}\overline{x_0}$ for NOT is performed. If C_0C_1 are 11, an addition of x_1x_0 and y_1y_0 is performed.

1.2.3 Sequential Circuits

The outputs of a sequential circuit depend not only on the current inputs, but also on the effect of the past inputs on the state of the sequential circuit. Table 1.11 gives the truth table of a sequential circuit called the set/reset (SR) flip-flop. Figure 1.9 shows the logic diagram of the SR flip-flop. When $S \neq 1$ or $R \neq 1$, the two outputs of the SR flip-flop take opposite values, Q and $\bar{Q}$. However, when $S = 1$ and $R = 1$, $Q(t+1) = \overline{1+\overline{Q(t)}} = 0$ and $\bar{Q}(t+1) = \overline{1+Q(t)} = 0$, regardless of what value $Q(t)$ takes. The same output value of 0 for Q and $\bar{Q}$ contradicts the expected opposite values for the two outputs. Hence, the outputs for $S = 1$ and $R = 1$ are considered undefined.

The SR flip-flop in Figure 1.9 is called an "asynchronous sequential circuit," since it updates its outputs when any input value changes. A synchronous sequential circuit uses a clock to synchronize the update of the outputs. A clock can be implemented by a circuit that emits a series of pulses with high and low levels, as shown in Figure 1.10. A level-triggered circuit updates its outputs when the clock signal changes its level to high or low. Figure 1.11 shows the digital logic circuit for the SR flip-flop that has the clock signal as an input and changes the output value when the level of the clock signal is high with the value of 1.

TABLE 1.11
Truth Table of the SR Flip-Flop

S	R	Present State $Q(t)$	Next State $Q(t + 1)$
0	0	0	0
0	0	1	1
0	1	0	0
0	1	1	0
1	0	0	1
1	0	1	1
1	1	0	Undefined
1	1	1	Undefined

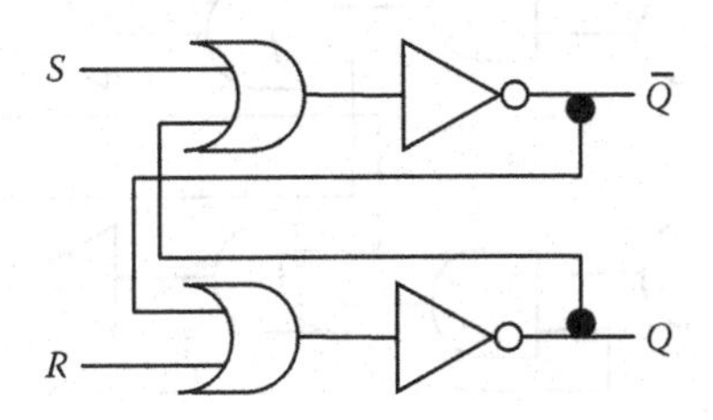

FIGURE 1.9
SR flip-flop without clock.

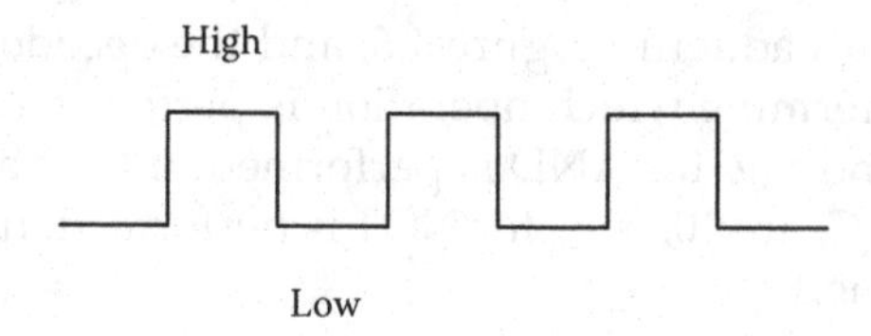

FIGURE 1.10
Clock signal with a series of pulses.

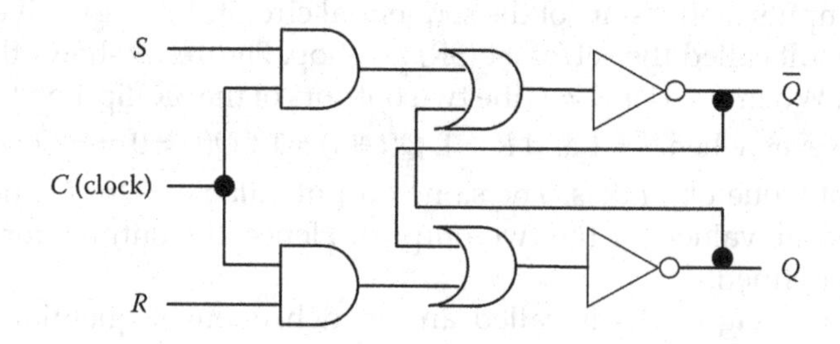

FIGURE 1.11
SR flip-flop with clock.

The JK flip-flop, which is named after its inventor, Jack Kilby, addresses the problem of the undefined outputs of the SR flip-flop for $S = 1$ and $R = 1$ using the digital logic circuit shown in Figure 1.12. Table 1.12 gives the truth table of the JK flip-flop. By having Q and $\bar{Q}$ join R and S inputs of the SR flip-flop, respectively, the JK flip-flop eliminates the input combination of $S = 1$ and $R = 1$ to the SR flip-flop. When $J = 1$ and $K = 1$, Q and $\bar{Q}$ change their state to the opposite value (Figure 1.12).

The data (D) flip-flop addresses the problem of undefined outputs in the SR flip-flop by eliminating the input combination of $S = 1$ and $R = 1$, and the input combination of $S = 0$ and $R = 0$ as shown in Figure 1.13. Table 1.13 gives the truth table for the D flip-flop. As seen in Table 1.13, the D flip-flop can be used as a memory unit since it sets the state and the output of the memory unit to the input value. Figure 1.14 gives the logic diagram symbols of the SR, JK, and D flip-flops.

Figure 1.15 shows a 4-bit register that is made of four D flip-flops. Figure 1.16 shows a memory that can store four words, with 2 bits in each word. In Figure 1.16, A_1A_0 are the inputs for a 2- to 4-bit decoder and give the memory address to select one of the four

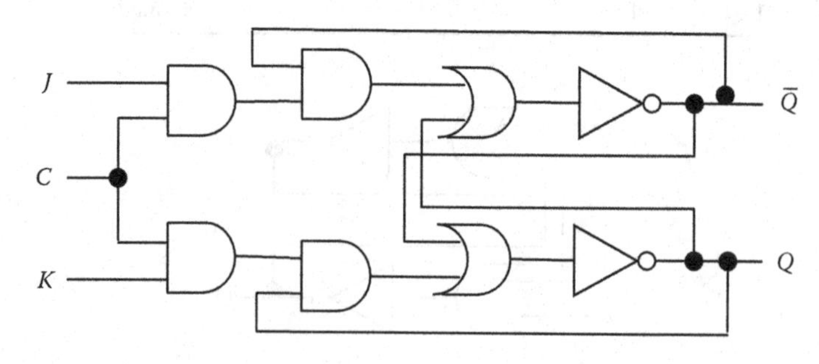

FIGURE 1.12
JK flip-flop with clock.

TABLE 1.12

Truth Table of the JK Flip-Flop

J	*K*	Present State *Q*(*t*)	Next State *Q*(*t* + 1)
0	0	0	0
0	0	1	1
0	1	0	0
0	1	1	0
1	0	0	1
1	0	1	1
1	1	0	1
1	1	1	0

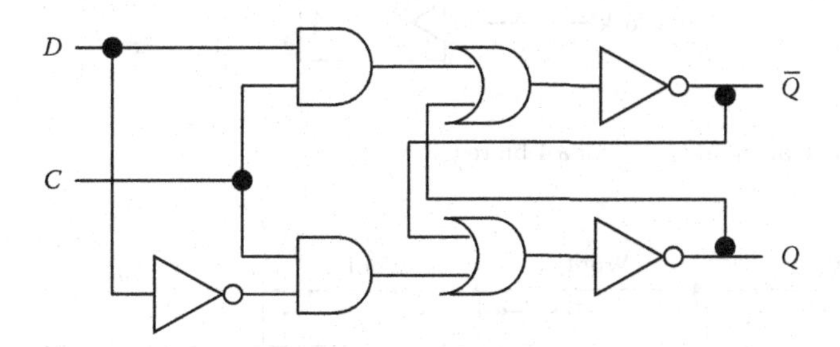

FIGURE 1.13
D flip-clop with clock.

TABLE 1.13

Truth Table for the D Flip-Flop

D	*Q*(*t*)	*Q*(*t* + 1)
0	0	0
0	1	0
1	0	1
1	1	1

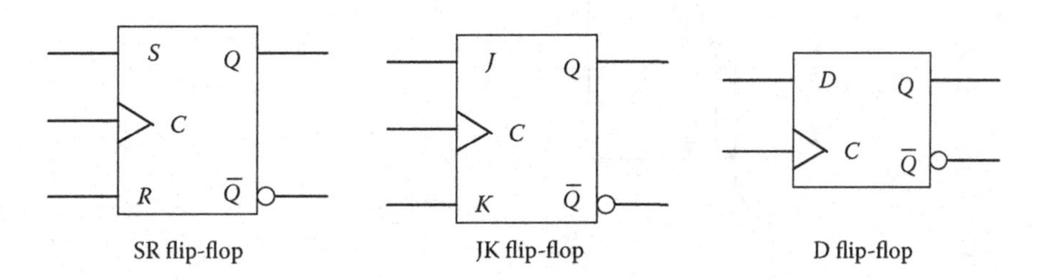

FIGURE 1.14
Logic diagram symbols for the SR, JK, and D flip-flops.

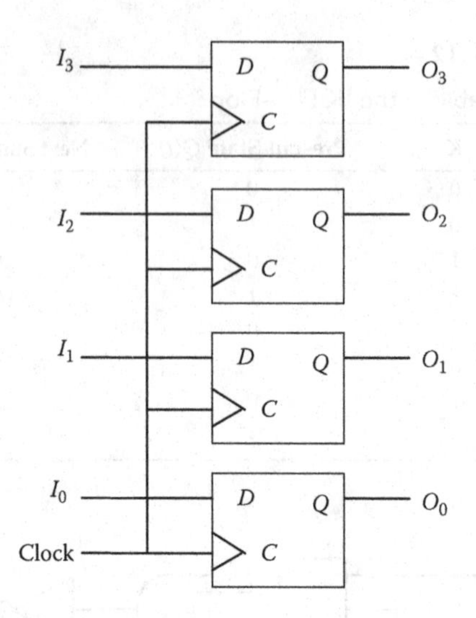

FIGURE 1.15
Logic diagram of the sequential circuit for a 4-bit register.

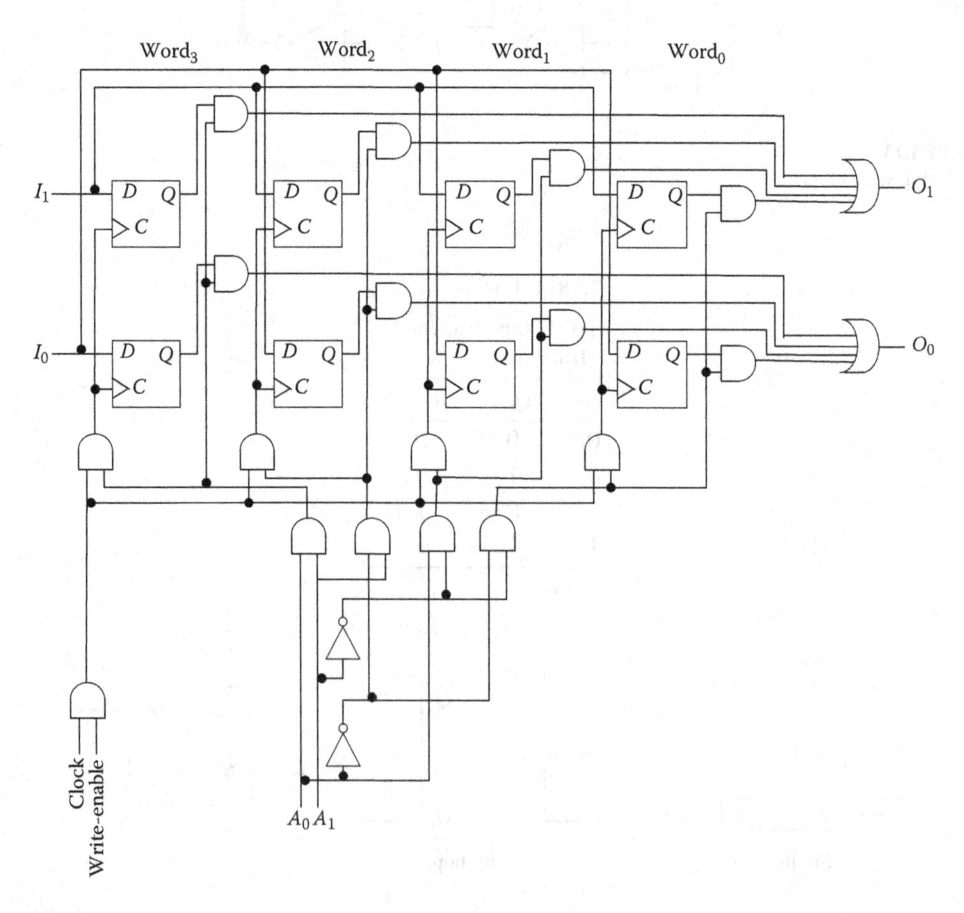

FIGURE 1.16
Logic diagram of the sequential circuit for a 2-bit memory to store four words.

words. If A_1A_0 are 00, $Word_0$ is selected. If A_1A_0 are 01, $Word_1$ is selected; if A_1A_0 are 10, $Word_2$ is selected; and if A_1A_0 are 11, $Word_3$ is selected. When the bit for Write-Enable in Figure 1.16 is set to 1, the D flip-flops for the selected word are allowed to update its outputs from its inputs, that is, to write the input data to the memory at the selected word location. When the bit for Write-Enable in Figure 1.16 is set to 0, the memory content at the selected word location is read since the outputs O_1O_0 present what are stored in the memory.

1.3 Summary

This chapter gives the basics of how computer hardware, including key components such as the ALU in the CPU and memory, is made of digital logic circuits. Boolean algebra, which is useful in designing digital logic circuits, is introduced. The method of designing a digital logic circuit for a function is described to include first using a truth table to express the function and then converting the truth table into a Boolean expression, which is directly implemented by a digital logic circuit. The combinational or sequential circuits for some key computer hardware components such as the ALU and memory are provided in this chapter. With an understanding of how electronic devices are used to construct computer hardware and how computer hardware stores and operates on binary data in this chapter, we explain how binary data are used to represent non-binary numbers, characters, and other forms of information on computers in the next chapter.

EXERCISES

1. Use a truth table to verify the AND and OR forms of the distributive property in Table 1.4.
2. Use a truth table to verify the AND form of de Morgan's law in Table 1.4.
3. Given three input variables with binary values, x, y, and z, a majority voting function produces the output of 1 if the majority of the input variables take the value of 1, and the output of 0, otherwise. Express the majority voting function using a truth table, and express this function using a Boolean expression in the sum-of-product form.
4. Use a truth table to express the 2-bit adder, which performs the addition of two input bits with sum and carry-out but no carry-in. Express this function in a Boolean expression in the sum-of-product form.
5. Table 1.14 expresses the function of the parity checker for examining whether the received parity bit is consistent with the received data to detect transmission errors. Express the parity checker using a Boolean expression in the sum-of-product form.
6. Develop a method of converting a truth table into a Boolean expression in the product-of-sum form.
7. Use only the NAND gates to construct a digital logic circuit for the XOR gate, and use the logic diagram to show the implementation of the digital logic circuit.

TABLE 1.14
Truth Table for the Parity Checker

x	*y*	*z*	Parity Bit	Parity Checker (Error)
0	0	0	0	0
0	0	0	1	1
0	0	1	0	1
0	0	1	1	0
0	1	0	0	1
0	1	0	1	0
0	1	1	0	0
0	1	1	1	1
1	0	0	0	1
1	0	0	1	0
1	0	1	0	0
1	0	1	1	1
1	1	0	0	0
1	1	0	1	1
1	1	1	0	1
1	1	1	1	0

8. Design a combinational circuit for a 3- to 8-bit decoder.
9. Design a sequential circuit for a memory that can store four words with 4 bits in each word.
10. Design a synchronous sequential circuit for a 4-bit synchronous counter, which goes through a predetermined sequence of states that reflect the binary number sequence 0000, 0001, 0010, 0011, …, 1111, and then back to 0000, as the clock pulses. The counter uses an input, Count-Enable, to enable counting when Count-Enable is set to 1. The sequential circuit does not change the state if Count-Enable is set to 0.

2

Digital Data Representation

This chapter describes how numbers, characters, and other forms of information are represented on computers using binary data. Some methods of detecting and correcting errors in data storage and transmission are also given in this chapter.

2.1 Representation of Numbers

In this section, we first introduce the conversion between decimal numbers and binary numbers for unsigned numbers. We then give the representation of signed numbers. An unsigned number is a number without the negative sign. A memory address is an unsigned integer on computers. A signed number has a positive or a negative sign. Data used in computation are typically declared as signed numbers on computers.

2.1.1 Conversion between Unsigned Binary and Decimal Numbers

A decimal number has a base of 10 and digit values of 0, 1, 2, ..., 9. A binary number has a base of 2 and digit values of 0 and 1. The representation of a decimal number and a binary number using their base is given as follows:

Decimal number: $123.45_{10} = 1 \times 10^2 + 2 \times 10^1 + 3 \times 10^0 + 4 \times 10^{-1} + 5 \times 10^{-2}$

Binary number: $101.01_2 = 1 \times 2^2 + 0 \times 2^1 + 1 \times 2^0 + 0 \times 2^{-1} + 1 \times 2^{-2}$

The conversion of an unsigned binary number to a decimal number is carried out by computing the sum using the base representation of the binary number, as shown:

$$101.01_2 = 1 \times 2^2 + 0 \times 2^1 + 1 \times 2^0 + 0 \times 2^{-1} + 1 \times 2^{-2} = 4_{10} + 0_{10} + 1_{10} + 0_{10} + 0.25_{10} = 5.25_{10}$$

We use the decimal number system in our daily activities. An unsigned decimal number must be represented on computers using a binary number that is supported by computer hardware. A binary integer with N binary digits (called "bits") can represent an unsigned decimal integer from 0 to 2^N-1. For example, the decimal integers from 0 to 15 can be represented by 4-bit binary numbers, as shown in Table 2.1.

One method of converting an unsigned decimal number to a binary number is the repeated subtraction method, which has the following steps:

1. Look for the highest power of 2 that is smaller than the decimal number.
2. Subtract that power of 2 from the decimal number.
3. Take the result of the subtraction in Step 2 as the decimal number.

TABLE 2.1

Decimal Integers Represented by 4-Bit Binary Numbers

Decimal Numbers	Binary Numbers
0	0000
1	0001
2	0010
3	0011
4	0100
5	0101
6	0110
7	0111
8	1000
9	1001
10	1010
11	1011
12	1100
13	1101
14	1110
15	1111

4. If the decimal number is 0 or meets another stopping criterion, go to Step 5; otherwise, go back to Step 1.
5. Construct the binary number by filling 1 in each bit position, where the power of 2 is used in the subtraction in Step 2 and 0 in each bit position, where the power of 2 is not used in the subtraction in Step 2.

A stopping criterion in Step 4 can be defined using the value after the decimal point, for example, the value after the decimal point <0.01, or the number of binary digits after the decimal point, for example, two binary digits after the decimal point. The repeated subtraction method requires the knowledge of decimal numbers for powers of 2, as shown in Table 2.2.

Example 2.1 illustrates the repeated subtraction method.

Example 2.1

Use the repeated subtraction method to convert the decimal number 123.45 into a binary number, with four binary digits after the decimal point.

In Step 1, we determine that the highest power of 2 smaller than 123.45 is $2^6 = 64$.

In Step 2, we perform the subtraction:

$$123.45 - 2^6 = 59.45 \text{ or } 123.45 = 1 \times 2^6 + 59.45$$

59.45 is taken as the next decimal number in Step 3. Since the decimal number is not 0, we go back to Step 1. The remaining steps are given as follows:

$59.45 - 2^5 = 27.45$ or $123.45 = 1 \times 2^6 + 1 \times 2^5 + 27.45$
$27.45 - 2^4 = 11.45$ or $123.45 = 1 \times 2^6 + 1 \times 2^5 + 1 \times 2^4 + 11.45$
$11.45 - 2^3 = 3.45$ or $123.45 = 1 \times 2^6 + 1 \times 2^5 + 1 \times 2^4 + 1 \times 2^3 + 3.45$
$3.45 - 2^1 = 1.45$ or $123.45 = 1 \times 2^6 + 1 \times 2^5 + 1 \times 2^4 + 1 \times 2^3 + 0 \times 2^2 + 1 \times 2^1 + 1.45$
$1.45 - 2^0 = 0.45$ or $123.45 = 1 \times 2^6 + 1 \times 2^5 + 1 \times 2^4 + 1 \times 2^3 + 0 \times 2^2 + 1 \times 2^1 + 1 \times 2^0 + 0.45$

TABLE 2.2

Decimal Numbers for Some Powers of 2

Power of 2	Decimal Number
2^{-4}	$\frac{1}{2^4} = 0.0625$
2^{-3}	$\frac{1}{2^3} = 0.125$
2^{-2}	$\frac{1}{2^2} = 0.25$
2^{-1}	$\frac{1}{2^1} = 0.5$
2^0	1
2^1	2
2^2	4
2^3	8
2^4	16
2^5	32
2^6	64
2^7	128
2^9	256
2^{10}	512

$0.45 - 2^{-2} = 0.20$ or $123.45 = 1 \times 2^6 + 1 \times 2^5 + 1 \times 2^4 + 1 \times 2^3 + 0 \times 2^2 + 1 \times 2^1 + 1 \times 2^0 + 0 \times 2^{-1} + 1 \times 2^{-2} + 0.20$

$0.20 - 2^{-3} = 0.075$ or $123.45 = 1 \times 2^6 + 1 \times 2^5 + 1 \times 2^4 + 1 \times 2^3 + 0 \times 2^2 + 1 \times 2^1 + 1 \times 2^0 + 0 \times 2^{-1} + 1 \times 2^{-2} + 1 \times 2^{-3} + 0.075$

$0.075 - 2^{-4} = 0.0125$ or $123.45 = 1 \times 2^6 + 1 \times 2^5 + 1 \times 2^4 + 1 \times 2^3 + 0 \times 2^2 + 1 \times 2^1 + 1 \times 2^0 + 0 \times 2^{-1} + 1 \times 2^{-2} + 1 \times 2^{-3} + 1 \times 2^{-4} + 0.0125$

We stop with four binary digits after the decimal point. In Step 5, we construct the binary number: 1111011.0111.

Another method of converting a decimal number into a binary number is the division remainder method, which has the following steps for the integer part of the decimal number:

1. Divide the decimal number by 2, obtain the remainder of the division, and take the quotient of the division as the next decimal number.
2. If the decimal number is 0, go to Step 3; otherwise, go back to Step 1.
3. Construct the binary number by reading the division remainders from the last step to the first step of Step 1 performed.

The division remainder method has the following steps for the fractional part of the decimal number:

1. Multiply the decimal number by 2, obtain the unit digit of the multiplication product, and take the fractional part of the multiplication product as the next decimal number.
2. If the decimal number is 0 or a required number of digits after the decimal point is obtained, go to Step 3; otherwise, go back to Step 1.
3. Construct the binary number by reading the unit digit of the multiplication product from the first step to the last step of Step 1 performed.

Example 2.2

Use the division remainder method to convert the decimal number 123.45 into a binary number, with four binary digits after the decimal point.

We first convert the integer part of the decimal number: 123. In Step 1, we divide 123 by 2, obtain 1 as the reminder of the division, and take the quotient, 61, as the next decimal number. In Step 2, because the decimal number is not 0, we go back to Step 1. The complete procedure is shown as follows.

$123 \div 2 = 61$ with the remainder of **1** or $123 = 61 \times 2^1 + \mathbf{1} \times 2^0$ (note that the remainder is highlighted in bold)

$61 \div 2 = 30$ with the remainder of **1** or $123 = (30 \times 2^1 + \mathbf{1}) \times 2^1 + 1 \times 2^0 = 30 \times 2^2 + \mathbf{1} \times 2^1 + 1 \times 2^0$

$30 \div 2 = 15$ with the remainder of **0** or $123 = 15 \times 2^3 + \mathbf{0} \times 2^2 + 1 \times 2^1 + 1 \times 2^0$

$15 \div 2 = 7$ with the remainder of **1** or $123 = 7 \times 2^4 + \mathbf{1} \times 2^3 + 0 \times 2^2 + 1 \times 2^1 + 1 \times 2^0$

$7 \div 2 = 3$ with the remainder of **1** or $123 = 3 \times 2^5 + \mathbf{1} \times 2^4 + 1 \times 2^3 + 0 \times 2^2 + 1 \times 2^1 + 1 \times 2^0$

$3 \div 2 = 1$ with the remainder of **1** or $123 = 1 \times 2^6 + \mathbf{1} \times 2^5 + 1 \times 2^4 + 1 \times 2^3 + 0 \times 2^2 + 1 \times 2^1 + 1 \times 2^0$

$1 \div 2 = 0$ with the remainder of **1** or $123 = (0 \times 2 + \mathbf{1}) \times 2^6 + 1 \times 2^5 + 1 \times 2^4 + 1 \times 2^3 + 0 \times 2^2 + 1 \times 2^1 + 1 \times 2^0 = \mathbf{1} \times 2^6 + 1 \times 2^5 + 1 \times 2^4 + 1 \times 2^3 + 0 \times 2^2 + 1 \times 2^1 + 1 \times 2^0$

As illustrated above, the last remainder is associated with the highest power of 2. Hence, we obtain the binary integer of 1111011 by reading the remainder from the last step to the first step of Step 1 performed.

We then convert the fractional part of the decimal value: 0.45. In Step 1, we multiply 0.45 by 2, obtain 0 as the unit digit of the multiplication product, and take the fractional part of the multiplication product, 0.90, as the next decimal value. In Step 2, because the decimal value is not 0, we go back to Step 1. The complete procedure is shown as follows:

$0.45 \times 2 = \mathbf{0}.90$ or $0.45 = (\mathbf{0} + 0.90) \times 2^{-1} = \mathbf{0} \times 2^{-1} + 0.90 \times 2^{-1}$ (note that the unit digit is highlighted in bold)

$0.90 \times 2 = \mathbf{1}.80$ or $0.45 = 0 \times 2^{-1} + 0.90 \times 2^{-1} = 0 \times 2^{-1} + (\mathbf{1} + 0.80) \times 2^{-1} \times 2^{-1} = 0 \times 2^{-1} + \mathbf{1} \times 2^{-2} + 0.80 \times 2^{-2}$

$0.80 \times 2 = \mathbf{1}.60$ or $0.45 = 0 \times 2^{-1} + 1 \times 2^{-2} + 0.80 \times 2^{-2} = 0 \times 2^{-1} + 1 \times 2^{-2} + (\mathbf{1} + 0.60) \times 2^{-1} \times 2^{-2} = 0 \times 2^{-1} + 1 \times 2^{-2} + 1 \times 2^{-3} + 0.60 \times 2^{-3}$

$0.60 \times 2 = \mathbf{1}.20$ or $0.45 = 0 \times 2^{-1} + 1 \times 2^{-2} + 1 \times 2^{-3} + 0.60 \times 2^{-2} = 0 \times 2^{-1} + 1 \times 2^{-2} + 1 \times 2^{-3} + (\mathbf{1} + 0.20) \times 2^{-1} \times 2^{-3} = 0 \times 2^{-1} + 1 \times 2^{-2} + 1 \times 2^{-3} + 1 \times 2^{-4} + 0.20 \times 2^{-4}$

As illustrated above, the unit digit from the first multiplication product is associated with the highest power of 2. Hence, we obtain the binary number of 0.0111 by reading the unit digit from the first step to the last step of Step 1 performed. Hence, the division remainder method produces the binary number 1111011.0111, which is the same as the result in Example 2.1.

2.1.2 Representation of Signed Integers

Using the methods in Section 2.1.1, we can convert an unsigned decimal number into an unsigned binary number that is directly stored and manipulated on computers. This section introduces the signed magnitude method, one's complement method and two's complement method of representing a signed integer. ALU described in Chapter 1 includes the addition operation. The addition of signed integers represented by the signed magnitude method, one's complement method and two's complement method is described in this section.

TABLE 2.3

Examples of Representing Signed Numbers Using the Signed Magnitude Method

Signed Decimal Numbers	8-Bit Binary Numbers
−123	11111011
123	01111011
0	00000000
−127	11111111
127	01111111

2.1.2.1 Signed Magnitude Method

The signed magnitude method uses the left-most bit of a binary integer to represent the sign, specifically, 0 for the positive sign and 1 for the negative sign. An N-bit binary integer can represent any value in the range $[-(2^{N-1} - 1), (2^{N-1} - 1)]$. For example, using 8 bits, we have the left-most bit representing the sign of the binary integer and the remaining 7 bits representing the magnitude (absolute value) of the binary integer. Hence, an 8-bit binary integer can represent any value in the range $[-(2^7 - 1), (2^7 - 1)]$ or [–127, 127]. Table 2.3 gives examples of using 8 bits and the signed magnitude method to represent signed numbers –123, 123, 0, –127, and 127. To perform the addition of two signed integers using the signed magnitude method, we need to process the sign bit separately from the number bits. For example, if we perform the addition of one positive integer and one negative integer, we need to determine which integer has a smaller magnitude (a smaller absolute value), perform the subtraction of the smaller integer from the larger integer using the number bits, and assign the sign of the larger integer as the sign of the subtraction result. Hence, it is not straightforward to perform the addition of signed integers in digital circuits.

2.1.2.2 One's Complement Method

One's complement method uses the diminished complement of an integer. Given an integer a with base b and N digits, the diminished complement of a is defined by $(b^N - 1) - a$. For example, given a decimal integer 43 with a base of 10 and 3 digits, the diminished complement of 43 is $(10^3 - 1) - 43 = 999 - 43 = 956$. To see how the diminished complement of an integer is used in the addition, we look at an example of performing 156 – 43 as follows:

$156 - 43 = x$

$156 - 43 + 999 = x + 999$

$156 + 956$ (the diminished complement of 43) $= x + 999$

$1112 = x + 999$

$1112 - 1000 + 1$ (end carry around: the carry-out beyond three digits is moved as the unit digit and added) $= x + 999 - 1000 + 1$

$112 + 1 = x$

$113 = x,$

TABLE 2.4

Representation of Signed Binary Integers with 4 Bits Using One's Complement Method

Signed Integer	Binary Integer
–7	1000
–6	1001
–5	1010
–4	1011
–3	1100
–2	1101
–1	1110
0	0000 or 1111
1	0001
2	0010
3	0011
4	0100
5	0101
6	0110
7	0111

which is the result of 156 – 43 = 113. As illustrated above, the addition with one negative integer can be performed by representing the negative integer (–43) using the diminished complement of that integer without the negative sign, adding the positive integer and the diminished complement, and performing the end carry-around.

Given a binary integer 0101 with a base of 2 and 4 digits, the diminished complement of 0101 is $(2^4 - 1) - 0101 = 1111 - 0101 = 1010$. Hence, obtaining the diminished complement of a binary integer is easy to implement in digital circuits by simply flipping each bit.

One's complement method uses the diminished complement to represent a negative binary integer. A binary integer with N bits can represent a signed integer in the range of $[-(2^{N-1} - 1), (2^{N-1} - 1)]$, with the left-most bit indicating the sign. For example, a binary integer with 4 bits can present an integer in the range of $[-(2^{4-1} - 1), (2^{4-1} - 1)]$ or [–7, 7], as shown in Table 2.4. Using one's complement method to represent a signed integer, we have the left-most bit of 0 for a positive integer and the left-most bit of 1 for a negative integer. Using one's complement method, we convert and represent a negative integer using its diminished complement and leave a positive integer unchanged. Note that both 0000 and 1111 represent 0 in one's complement method.

The addition of two binary integers is carried out in the following steps:

1. Represent the binary integers using one's complement method.
2. Perform the addition of the binary integers represented by one's complement method.
3. Perform the end carry-around to obtain the result.

Example 2.3

Perform the addition of one positive 4-bit binary integer and one negative 4-bit binary integer, 0100 – 0011 (4 – 3), using one's complement method.

In Step 1, we represent –0011 using its diminished complement 1100.
In Step 2, we perform the addition:

0100
+ 1100 (the deminished complement of 0011)
= 10000.

In Step 3, we perform the end carry-around by removing the carry-out bit 1 beyond the 4 bits and adding that bit at the unit position:

0000
+ 1
= 0001,

which is 1.

Example 2.4

Perform the addition of one positive 4-bit binary integer and one negative 4-bit binary integer, 0011 – 0100 (3 – 4), using one's complement method.
In Step 1, we represent –0100 using its diminished complement 1011.
In Step 2, we perform the addition:

0011
+ 1011 (the diminished complement of 0100)
= 1110.

In Step 3, we perform the end carry-around:

1110
+ 0
= 1110

which is –1.

2.1.2.3 Two's Complement Method

Two's complement method is the most common method used on computers. In Table 2.4, for one's complement method, we have 0000 and 1111 representing 0. Two's complement method addresses this problem by using the complement that is defined by $b^N - a$ for an integer a with a base of b and N digits. That is, two's complement is one's complement plus 1. For example, the complement of 0101 is 2^4 – 0101 = 10000 – 0101 = 1011. It is easy to obtain the complement of a binary integer in digital circuits by flipping each bit and then adding 1. Table 2.5 shows the representation of signed binary integers using two's complement method and 4 bits. Using N bits, two's complement method can represent signed integers in the range $[-2^{N-1}, 2^{N-1} - 1]$.

TABLE 2.5

Representation of Signed Binary Integers with 4 Bits Using Two's Complement Method

Signed Integer	Binary Integer
−8	1000
−7	1001
−6	1010
−5	1011
−4	1100
−3	1101
−2	1110
−1	1111
0	0000
1	0001
2	0010
3	0011
4	0100
5	0101
6	0110
7	0111

The addition of two signed integers with N bits represented by two's complement method is performed in the following steps:

1. Represent the binary integers using two's complement method.
2. Perform the addition of the binary integers represented by two's complement method.
3. Discard the carry-out beyond N bits.

Example 2.5

Perform the addition of one positive 4-bit integer and one negative 4-bit integer, 0100 − 0011 (4 − 3), using two's complement method.

In Step 1, we represent −0011 by its complement 1101.

In Step 2, we perform the addition:

  0100
+ 1101 (the complement of 0011)
= 10001.

In Step 3, we drop the carry-out bit 1 beyond 4 bits and obtain 0001 as the result.

Example 2.6

Perform the addition of one positive 4-bit integer and one negative 4-bit integer, 0011 − 0100 (3 − 4), using two's complement method.

TABLE 2.6

Representation of a Floating Point Number

Sign	Exponent	Significand
0	00011	1000100000

In Step 1, we represent –0100 by its complement 1100.
In Step 2, we perform the addition:

0011
+ 1100 (the complement of 0100)
= 1111.

Since there is no carry-out, Step 3 produces the result 1111, which is –1.

2.1.3 Representation of Signed Floating Point Values

A floating point value is represented using the sign bit (0 for the positive sign and 1 for the negative sign), bits for the exponent, and bits for the fractional part of the floating point value, which is called "significand." For example, a floating point value, 100.01, is equal to 0.10001×2^3, whose representation is shown in Table 2.6 using 16 bits, with 1 bit for the sign, 5 bits for the exponent, and 10 bits for the significand. The number of bits used for the exponent determines the range of values that can be represented. The number of bits used for the significand determines the precision of the value that is represented.

An exponent may be a positive integer or a negative integer. To represent a signed exponent, a bias value is used. For example, 5 bits for an exponent can represent a value in the range [0, 31]. A value near the middle of the range, for example, 16, is used as the bias value for a 5-bit exponent. The biased exponent is obtained by adding the bias value to an exponent. For example, the exponent value of 3 is represented by the biased exponent, which is 16 + 3 = 19 or 10000 + 11 = 10011. The exponent value of –1 is represented by the biased exponent, which is 16 – 1 = 15 or 10000 – 1 = 1111. Hence, in the biased exponent representation, a biased exponent larger than the bias value represents a positive exponent, and a biased exponent smaller than the bias value represents a negative exponent. The exponent is obtained by subtracting the bias value from the biased exponent.

A floating point number, 100.01, is equal to 0.10001×2^3, 0.010001×2^4, 0.0010001×2^5, and so on, which correspond to different representations of the same floating point number. To produce a unique representation of a floating point number, normalization is necessary to require that the left-most bit of the significand must be 1. With normalization, 100.01 can be represented only in the form in Table 2.6 for 0.10001×2^3.

2.2 Representation of Alphabet and Control Characters

Text and control information is represented on computers by using codes for alphabet, number, control, and other characters. Table 2.7 gives American Standard Code for Information Interchange (ASCII) character codes that use 8 bits and have code values from 0 to 127 ($2^7 - 1$). A code itself uses 7 bits, and the left-most bit (the eighth bit) is used as the

TABLE 2.7
ASCII Character Codes

Character	Code	Character	Code	Character	Code	Character	Code
Null	0	Space	32	@	64	`	96
Start of heading	1	!	33	A	65	a	97
Start of text	2	"	34	B	66	b	98
End of text	3	#	35	C	67	c	99
End of transmission	4	$	36	D	68	d	100
Enquiry	5	%	37	E	69	e	101
Acknowledge	6	&	38	F	70	f	102
Bell (beep)	7	'	39	G	71	g	103
Backspace	8	(	40	H	72	h	104
Horizontal tab	9	)	41	I	73	i	105
Line feed, new line	10	*	42	J	74	j	106
Vertical tab	11	+	43	K	75	k	107
Form feed, new page	12	,	44	L	76	l	108
Carriage return	13	-	45	M	77	m	109
Shift out	14	.	46	N	78	n	110
Shift in	15	/	47	O	79	o	111
Data link escape	16	0	48	P	80	p	112
Device control 1	17	1	49	Q	81	q	113
Device control 2	18	2	50	R	82	r	114
Device control 3	19	3	51	S	83	s	115
Device control 4	20	4	52	T	84	t	116
Negative acknowledge	21	5	53	U	85	u	117
Synchronous idle	22	6	54	V	86	v	118
End of transmission block	23	7	55	W	87	w	119
Cancel	24	8	56	X	88	x	120
End of medium	25	9	57	Y	89	y	121
Substitute	26	:	58	Z	90	z	122
Escape	27	;	59	[	91	{	123
File separator	28	<	60	\	92	|	124
Group separator	29	=	61	]	93	}	125
Record separator	30	>	62	^	94	~	126
Unit separator	31	?	63	_	95	Delete	127

parity bit, as described in Chapter 1. For example, the character code for letter A is 65 or 1000001. If the parity bit is 0, the complete code for letter A is 01000001.

ASCII includes 128 characters. Unicode and Universal Character Set covers more characters. For example, Unicode uses 16 bits and covers mathematical symbols, characters in languages other than English, and so on.

2.3 Error Detection and Correction

As described in Section 2.2, the code for a character includes a parity bit for error detection. When data are stored on computers or transmitted over networks, errors may occur.

Detecting and correcting errors are required for reliable data storage and transmission. The parity bit method can detect an odd number of errors but cannot detect an even number of errors or identify where errors occur to correct errors. This section describes the Hamming algorithm of detecting and correcting errors.

An ASCII code uses 7 bits representing a character and 1 parity bit. If k check bits are used to detect errors in m data bits, there are, totally, $n = m + k$ code bits. The Hamming distance of two codes is the number of bits in which two codes are different. For example, 1000001 (65), along with its parity bit 0, produces the ASCII code 01000001 for letter A, and 1000011 (67), along with its parity bit 1, produces the ASCII code 11000011 for letter C. The two codes, 01000001 and 11000011 differ in two bit positions and have a Hamming distance of 2. For a code system such as ASCII, the minimum Hamming distance, D_{min}, is the smallest distance among all pairs of codes in the code system. If the number of bit errors that occur to a code is no more than D_{min}, the errors can be detected because D_{min} is the smallest Hamming distance among all pairs of valid codes, and thus, the code with no more than D_{min} bits of errors is different from any valid codes in the system. We detect the presence of error(s) in a given code if the given code is different from any valid code in the system.

If the presence of error(s) in a given code is detected, we correct error(s) in a given code to the valid code that is closest to the given code with error(s) in the Hamming distance. Considering that two valid codes, x and y, have the smallest Hamming distance D_{min} in the code system and p bits of errors occur to x, producing x', correcting x' to its original valid code x requires that x' has a smaller distance to x than y. Because the Hamming distance of x' and x is p, the Hamming distance of x' and y must be greater than p, with the smallest value being $p + 1$. Hence, the Hamming distance between x and y, $D_{min} \geq p + p + 1$ or $D_{min} \geq 2p + 1$, is required to correct p bits of errors such that the correction to the original valid code is guaranteed. That is, D_{min} of a code system must be $2p + 1$ to correct p bits of errors.

Given m data bits, we need to determine how many check bits are required to detect and correct errors. Let us consider the simple case of detecting and correcting 1 bit of error. Let k denote the number of check bits and n denote the total numbers of bits in a code: $n = m + k$. Using m data bits, we can create 2^m valid codes with $D_{min} = 1$. For each valid code including m data bits and k check bits, there are n possible positions where 1 bit of error can occur, and thus, there are n possible invalid codes with 1 bit of error. Hence, for each code, there are $n + 1$ bit patterns (including 1 valid code and n invalid codes). For 2^m valid codes, there are $(n + 1)2^m$ bit patterns. For n bits, there are totally 2^n possible bit patterns. The following inequality should hold:

$$(n + 1)2^m \leq 2^n$$

$$(m + k + 1)2^m \leq 2^{m+k}$$

$$m + k + 1 \leq 2^k$$

This inequality can be used to determine k, the number of check bits to detect and correct 1 bit of error.

Example 2.7

Determine the number of check bits needed for a code with 4 data bits to detect and correct 1 bit of error.

For $m = 4$, we have:

$$4 + k + 1 \leq 2^k$$
$$5 + k \leq 2^k$$

Comparing the values of $5 + k$ and 2^k, we obtain $k \geq 3$. Hence, we can let $k = 3$.

After determining the number of check bits for detecting and correcting 1 bit of error, the Hamming algorithm is used to construct Hamming codes. The Hamming algorithm has the following steps:

1. Number n bit positions from right to left, starting with 1.
2. Let each bit position, which is a power of 2, be a check bit; let other bits be data bits, and put in the values of data bits.
3. Write each bit position as the sum of the numbers that are powers of 2, using the highest power of 2 first, and then determine which bit positions each check bit contributes to the sum of the numbers for these bit positions.
4. Use the parity bit to determine the value of each check bit by considering the values at bit positions where the check bit contributes to the sum of the numbers.

Example 2.8 illustrates these steps.

Example 2.8

Construct the Hamming code with data bits 0100, using the Hamming algorithm and the even parity bit such as one in Table 1.6.

Using the result from Example 2.7, we know that we need 3 check bits for 4 data bits, a total of 7 bits, for the code. In Step 1 of the Hamming algorithm, we number 7 bit positions from right to left, starting from 1, as follows:

Bit position	7	6	5	4	3	2	1
Value							

In Step 2, we let bit positions 1, 2, and 4 be the check bits because these bit positions are powers of 2, and put in the values of data bits 0100 as follows:

Bit position	7	6	5	4	3	2	1
Value	0	1	0		0		

In Step 3, we write each bit position as the sum of the numbers that are powers of 2 as follows:

$$1 = 1$$
$$2 = 2$$
$$3 = 1 + 2$$
$$4 = 4$$
$$5 = 1 + 4$$
$$6 = 2 + 4$$
$$7 = 1 + 2 + 4$$

Hence, the check bit at position 1 contributes to the sum of the numbers for bit positions 1, 3, 5, and 7. The check bit at position 2 contributes to the sum of the numbers for bit positions 2, 3, 6, and 7. The check bit at position 4 contributes to bit positions 4, 5, 6, and 7. In Step 4, we compute the value of each check bit using the even parity. Because the check bit at position 1 contributes to the sum of numbers for bit positions 1, 3, 5, and 7:

Bit position	7	5	3	1
Value	0	0	0	?

the even parity bit at bit position 1 is 0. Because the check bit at position 2 contributes to the sum of numbers for bit positions 2, 3, 6, and 7:

Bit position	7	6	3	2
Value	0	1	0	?

the even parity bit at bit position 2 is 1. Because the check bit at position 4 contributes to the sum of numbers for bit positions 4, 5, 6, and 7:

Bit position	7	6	5	4
Value	0	1	0	?

the even parity bit at bit position 4 is 1. Putting all the check bits and data bits together, we have the Hamming code 0101010 as follows:

Bit position	7	6	5	4	3	2	1
Value	0	1	0	1	0	1	0

After constructing the code, we can detect a 1-bit error using the check bits, as shown in Example 2.9.

Example 2.9

Detect and correct a 1-bit error in a code 0101110. Note that the code 0101110 is produced by introducing a 1-bit error to the Hamming code 0101010 from Example 2.8 at bit position 3.

The parity bit at bit position 1 checks bit positions 1, 3, 5, and 7, which have values of 0, 1, 0, and 0, respectively, and violate the even parity. Hence, a 1-bit error occurs at bit position 1, 3, 5, or 7. The parity bit at bit position 2 checks bit positions 2, 3, 6, and 7, which have values of 1, 1, 1, and 0, respectively, and violate the even parity. Hence, a 1-bit error occurs at bit position 2, 3, 6, or 7. The parity bit at bit position 4 checks bit positions 4, 5, 6, and 7, which have values of 1, 0, 1, and 0, respectively, and conform to the even parity. Hence, there is no 1-bit error at bit positions 4, 5, 6, and 7. Taking all the results of the parity checks:

- A 1-bit error at bit position 1, 3, 5, or 7, using the check bit at position 1
- A 1-bit error at bit position 2, 3, 6, or 7, using the check bit at position 2
- No 1-bit error at bit positions 4, 5, 6, and 7, using the check bit at position 4

A 1-bit error at position 3 is detected because there is no 1-bit error at positions 4, 5, 6, and 7, and both check bits 2 and 3 indicate the possibility of the 1-bit error at position 3. A straightforward method of detecting a 1-bit error is using the sum of the bit positions of the check bits that indicate an error. Because check bits 1 and 2 indicates an error, $1 + 2 = 3$. Hence, the 1-bit error occurs at bit position 3. To correct the 1-bit error, we simply flip the bit at position 3 from 1 to 0.

2.4 Summary

This chapter illustrates how objects (e.g., numbers and characters) used by end users are supported on computer hardware using binary data. Because the data representation of objects uses check bits to detect and correct errors that may occur to stored or transmitted data, error detection and correction methods such as the Hamming algorithm are also introduced in this chapter. Based on the understanding of materials in this chapter, the next chapter gives an overview of how computer system software, especially the operating system and network software, manages computer and network hardware resources and allows application software to provide services to end users without directly handling the implementation details of computer and network hardware.

EXERCISES

1. What decimal integers can be represented by 3-bit binary integers? Show the representation of these decimal integers by using 3-bit binary integers.
2. Convert 111011.01111 into a decimal number.
3. Use the repeated subtraction method to convert 12.34 into a binary number, with two digits after the decimal point.
4. Use the division remainder method to convert 12.34 into a binary number, with two digits after the decimal point.
5. Use the signed magnitude method and an 8-bit binary number to represent –12 and 12.
6. Use one's complement method and an 8-bit binary number to represent –12 and 12.
7. Use two's complement method and an 8-bit binary number to represent –12 and 12.
8. Perform the addition of one 8-bit positive number and one 8-bit negative number, 00010111 – 00001001, using one's complement method.
9. Perform the addition of one 8-bit positive number and one 8-bit negative number, 00001001 – 00010111, using one's complement method.
10. Perform the addition of one 8-bit positive number and one 8-bit negative number, 00010111 – 00001001, using two's complement method.
11. Perform the addition of one 8-bit positive number and one 8-bit negative number, 00001001 – 00010111, using two's complement method.
12. Determine the number of check bits that are needed for a code system with 8 data bits to detect and correct 1 bit of error.
13. Construct the Hamming code with 8 data bits 01001011, using the Hamming algorithm.
14. Detect and correct a 1-bit error in the Hamming code 010111010110.

3

Computer and Network System Software

System software manages computer hardware resources and provides libraries of system functions that application software can use to create a variety of services to end users without handling computer hardware directly. This chapter introduces two important kinds of system software: the operating system that manages data processing and storage on a computer and network software that manages data communication between computers.

3.1 The Operating System

Figure 3.1 gives the von Neumann model of computer architecture (Null and Lobur, 2006), which puts computer hardware devices into three categories: CPU (which consists of the ALU, registers, and control unit), memory, and input/output (I/O) devices. The hardware implementation of the ALU and memory is described in Chapter 1. I/O devices include the hard drive, network interface card (NIC), keyboard, mouse, USB, printer, etc. The CPU gets program instructions and data from memory, executes program instructions, places the results in registers or memory, and responds to events such as requests from I/O devices. Interactions of the CPU, memory, and I/O devices are managed by the operating system. This section describes three main functions of the operating system:

- Process management
- Storage management
- I/O management

The operating system also carries out other functions for managing networking and communication, security, quality of service, etc. Networking and communication management is described in Section 3.2.

3.1.1 Process Management

A process is a program in execution. A process has its execution environment to keep track of the process status and data, which include the running state (e.g., ready or waiting), the memory address of the next program instruction, the values of variables, the contents of CPU registers, page tables, resources being used, etc. A process may create multiple threads to perform multiple tasks simultaneously. A thread is the smallest unit that can be scheduled for the CPU to execute. Threads share the same execution environment as their parent process.

Since the CPU executes only one process/thread at a given time, the operating system schedules processes and threads to determine which process/thread uses the CPU at a given time. Examples of process/thread scheduling methods are round robin; first come,

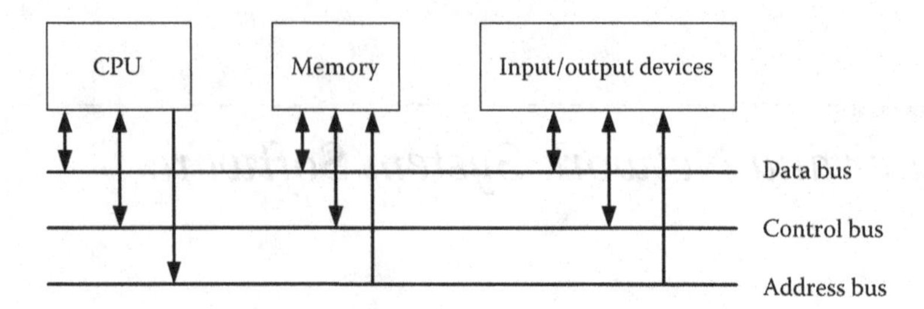

FIGURE 3.1
The von Neumann model of computer architecture.

first serve; shortest job first; shortest remaining time first; and priority-based scheduling. The round robin method is the most popular method, which lets all processes/threads share the CPU time by taking turns in using the CPU for a given amount of time. In the round robin method, more processes/threads will result in more waiting time of each process/thread before its turn of using the CPU. In the first come, first serve method, a process/thread that comes to request for the CPU first will be executed and completed by the CPU first, and thus, processes/threads use the CPU one by one instead of sharing CPU time. In the shortest job first method, a process/thread that requires the shortest CPU processing time is scheduled to use the CPU first. In the shortest remaining time first method, a process/thread that requires the least amount of CPU time to complete the process is scheduled to use the CPU first. If it is difficult to estimate the CPU processing time or the remaining CPU processing time of a process/thread, the shortest job first method or the shortest remaining time first method should not be employed. The priority-based scheduling method schedules processes/threads based on the priority of each process/thread that may be assigned to reflect the importance or another aspect of the process/thread.

The round robin method is a preemptive scheduling method because a process/thread is interrupted, pushed out of running on the CPU, and placed into the waiting state from the running state when the process/thread uses all its allocated amount of CPU time. The first come, first serve method; the shortest job first method; the shortest remaining time first method; and the priority-based method are non-preemptive scheduling methods because a process/thread is not interrupted before it completes its execution on the CPU. When a process/thread is interrupted before its completion of execution, the operating system performs the context switch to store the execution environment of the interrupted process/thread and bring in the execution environment of the next process/thread. When it is the turn for a process/thread to use the CPU again, the operating system restores the execution environment of the process/thread to resume the execution of the process/thread. When a process completes its execution, the operating system deletes the process and cleans up after the process termination.

3.1.2 Storage Management

The operating system manages the storage of data in memory, the hard drive, the USB drive, and other data storage devices. The operating system breaks a data file into pages of standard size for data storage efficiency. If data file is stored entirely on a single block of memory or storage space, the block of memory or storage space left after a data file is deleted may not be usable for another data file of a different size, resulting in unused memory or storage space. Breaking a data file into pages of standard size allows a page of any data file to fill in a page space left by a deleted page, maximizing the use of memory or storage space. Hence, pages of a data file may be stored at different locations of memory or permanent storage. Pages, which are currently used by a process, are stored in memory. If the pages needed are not in memory, they need to be brought into memory from a permanent storage such as the hard drive, the USB drive, etc.

The operating system creates a virtual memory system to manage pages and performs the translation of a virtual memory address and a physical memory address. The memory/storage management performed by the operating system includes the following:

- Breaking files into pages
- Translating virtual memory addresses to physical memory addresses
- Transferring pages between memory and a permanent storage device
- Maintaining page tables to locate pages
- Allocating memory/storage space to pages
- Releasing memory/storage space of deleted pages
- Keeping track of free memory/storage space

The operating system also performs file management, including file creation, file deletion, directory creation, and directory deletion.

3.1.3 I/O Management

I/O devices usually have device drivers, which are software packages implementing functions for particular I/O devices to allow interactions with I/O devices. For example, a particular printer has its special driver that may come with the operating system to allow application software (e.g., Microsoft Word) to send data to the printer for printing. The operating system provides system calls that application software can use to interact with I/O devices.

3.2 Networking and Communication Software

Figure 3.2 shows the open systems interconnect (OSI) reference model, which defines the functions required to implement networking and data communication between computers. The OSI reference model organizes networking functions in seven layers: application, presentation, session, transport, network, data link, and physical. Transmission control protocol/internet protocol (TCP/IP) is the most common software suite that implements networking functions in the OSI reference model and has four layers: application, transport, network, and link. Figure 3.2 shows the mapping of OSI and TCP/IP layers. The OSI reference model and TCP/IP are described in the following sections.

3.2.1 OSI Reference Model

At the application layer, a network application (e.g., Web application) on a computer at the source of data communication creates data packets containing application data (e.g., data containing a uniform resource locator (URL) entered by a user of the Web application to search for a website) and sends data packets to the same network application on a

OSI:	TCP/IP:
Application	Application, e.g., HTTP, SMTP, etc.
Presentation	
Session	Transport
Transport	
Network	Network
Data link	Link
Physical	

FIGURE 3.2
The OSI reference model and the TCP/IP.

computer at the destination of the data communication. The client–server architecture is commonly used by a network application to have the client software and the server software of the application work together. For example, Internet Explorer is the client software of the Web application in the Windows operating system and works together with Internet Information Server, which is the server software of the Web application.

The presentation layer manages the presentation of communicated data by defining the data presentation format (e.g., the format of floating point numbers) that is understood by computers of different kinds on the network. The session layer is responsible for establishing synchronized dialogs through data communication sessions between two computers on the network. Many types of data communication on the network involve dialogs between a client and its server, and there are many data packets being transmitted between the client and the server. Because all client–server dialogs share the same network, data packets for different client–server dialogs are mixed in the network bandwidth. A data communication session is established to identify all data packets that belong to a synchronized dialog between the client and the server.

The transport layer takes care of the reliability of data communication by providing acknowledgment, error checking, and retransmission. The network layer is responsible for finding a routing path to relay data packets from the source computer to the destination computer. The network layer also breaks down a data packet into smaller datagrams that are suitable to be transmitted over a particular physical transmission medium. The data link layer detects and corrects errors of physical data transmission that occur in the physical layer. The physical layer transmits binary bits across the physical transmission medium.

3.2.2 Transmission Control Protocol/Internet Protocol

TCP/IP covers networking functions in the OSI reference model in four layers: application, transport, network, and link. The application layer of the TCP/IP contains network application program suites (e.g., hypertext transfer protocol (HTTP) for the Web application, simple mail transfer protocol (SMTP) for the email application, and a domain name server for translating a network domain name to an IP address), each of which implements networking functions at the application and presentation layers of the OSI reference model. The transport layer of TCP/IP contains program suites such as TCP, user datagram protocol (UDP) and Internet Control Message Protocol (ICMP), which implement the networking functions at the session and transport layers of the OSI reference model. The term "protocol" is used to refer to a program suite. TCP is used for connection-oriented reliable data delivery through acknowledgment, error checking, and retransmission. Network applications such as HTTP and SMTP need to use TCP for connection-oriented reliable data delivery. UDP is used to provide connectionless, efficient data transmission without guarantee for reliable data delivery. Network applications such as domain name server (DNS) do not require reliable data delivery and, thus, use UDP for data transmission. ICMP is used to exchange command and control information between two computers on a network. The network layer of TCP/IP contains mainly IP, which implements the networking functions at the network layer of the OSI reference model to handle the routing of network data and break down network data to fit into the physical transmission medium. The link layer of TCP/IP contains network drivers that implement the networking functions at the data link and physical layers of the OSI reference model.

Figure 3.3 illustrates how data generated by the client program of the HTTP Web application on a computer at the data source are transmitted over the Ethernet to the server program of the HTTP Web application on a computer at the data destination through the TCP/IP. On the source computer, the HTTP client at the application layer of the TCP/IP generates the HTTP data packet and passes it down to the TCP at the transport layer of the TCP/IP. The TCP generates the TCP segment by adding the TCP header to the HTTP data packet. The following are the important fields in the TCP header to assist connection-oriented reliable data delivery:

- Source port: indicates which application sends the data
- Destination port: indicates which application should receive the data
- Sequence number: works with the acknowledgement number to identify the data stream in a network session between the application programs on the source computer and the destination computer

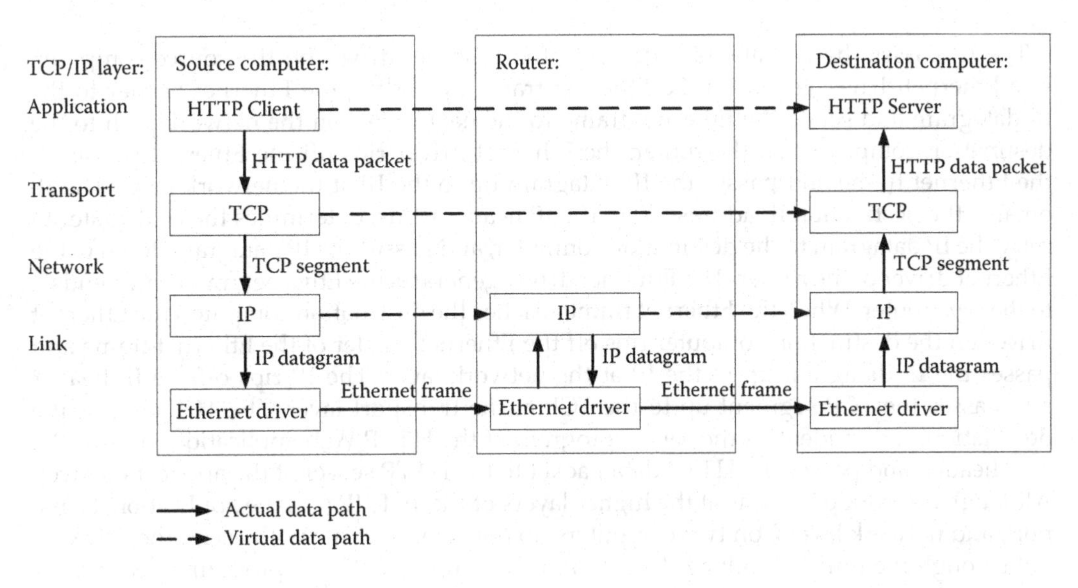

FIGURE 3.3
Data transmission through TCP/IP layers.

- Acknowledgement number: works with the sequence number to identify the data stream in a network session
- Data offset: indicates the header length
- Flags: contain six control bits—URG, ACK, PSH, RST, SYN, and FIN
- Window: indicates the number of bytes allowed in a segment
- Header checksum: applies to the TCP header fields to ensure the correct transmission of the TCP header fields

TCP passes the TCP segment down to the IP at the network layer of TCP/IP. The IP generates the IP datagram by adding the IP header to the TCP segment. The following are the important fields in the IP header:

- Packet ID: a serial number assigned to each IP datagram
- Header length: indicates the length of the IP header
- Total length of the datagram: indicates the length of the IP datagram
- Flags: indicate whether the datagram may be fragmented by intermediate network nodes between the source computer and the destination computer
- Fragment offset: indicates the location of a fragment within a datagram
- Time to live: specifies the maximum number of hops over the network path to reach the destination computer
- Protocol number: indicates the protocol (TCP, UDP, etc.) at the transport layer that calls the IP
- Source IP address: contains the IP address of the source computer
- Destination IP address: contains the IP address of the destination computer
- Header checksum: applies to the IP header fields to ensure the correct transmission of the IP header fields

The IP passes the IP datagram down to the Ethernet driver on the source computer. The Ethernet driver generates the Ethernet frame by adding the Ethernet header to the IP datagram and sends the Ethernet frame to the next router on the network path to the destination computer. On the router, the Ethernet driver rips off the Ethernet header of the Ethernet frame and passes the IP datagram up to the IP at the network layer. The IP obtains the destination IP address from the IP header fields, determines the next router to relay the IP datagram to the destination computer, and passes the IP datagram down to the Ethernet driver of the router. The Ethernet driver generates the Ethernet frame and sends it to the next router. When the Ethernet frame reaches the destination computer, the Ethernet driver on the destination computer rips off the Ethernet header of the Ethernet frame and passes the IP datagram up to the IP at the network layer. The IP rips off the IP header and passes the TCP segment up to the TCP at the transport layer. The TCP obtains the destination port to identify the server program of the HTTP Web application, rips off the TCP header, and passes the HTTP data packet to the HTTP server at the application layer. Although network programs at the higher layers of the TCP/IP (i.e., the application, transport, and network layers) on two computers do not communicate with each other directly but through the link layer along the actual data path, the network program at each layer responds to data only at that layer after data are passed up from the link layer, making the

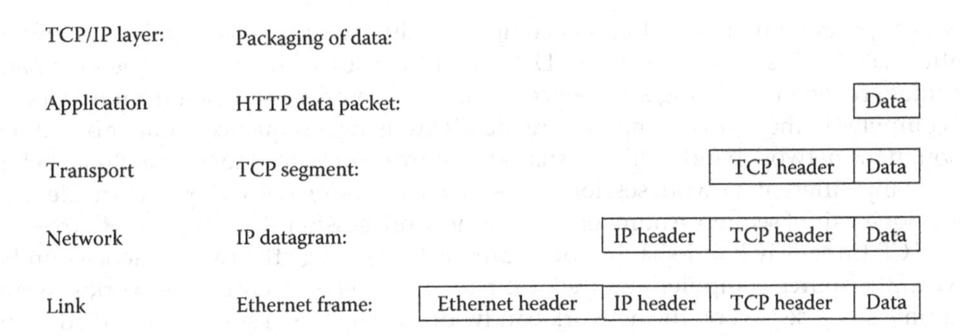

FIGURE 3.4
Data packaging through TCP/IP layers.

virtual path at each layer between two computers. Figure 3.4 shows the packaging of data by TCP/IP layers, each of which adds its header to data.

When a user requests a website to start a session of interactive data stream with the website, a network session needs to be established for identifying the data stream in this network session between the client program of the Web application on the source computer and the server program of the Web application on the destination computer. The network session is established through a three-way hand-shaking procedure, which includes the exchange of three TCP segments between the source computer and the destination computer, as shown in Figure 3.5. At first, the source computer sends to the destination computer a TCP segment with the control bit SYN being set to the TRUE value and a given sequence number m to request opening a network session. The destination computer responds to this request by sending a TCP segment with SYN being set to TRUE, a sequence number n, ACK being set to TRUE, and the acknowledgement number generated using the sequence number m (e.g., $m + 1$). Because the purpose of ACK along with the acknowledgement number is to acknowledge receiving the sequence number m from the source computer, the acknowledgement number must be related to the sequence number m. In Figure 3.5, the method of adding 1 to the sequence number m is used to create the acknowledgement number. A more complex method of creating an acknowledgement number related to a sequence number is used in TCP/IP software. When the source computer receives the TCP segment from the destination computer, the source computer knows that its request for opening a network session is received and sends a TCP segment to the destination computer to acknowledge receiving the sequence number n. When the destination computer receives this TCP segment for acknowledging the receipt of the sequence number n, a network session is established between the client program on the source computer and

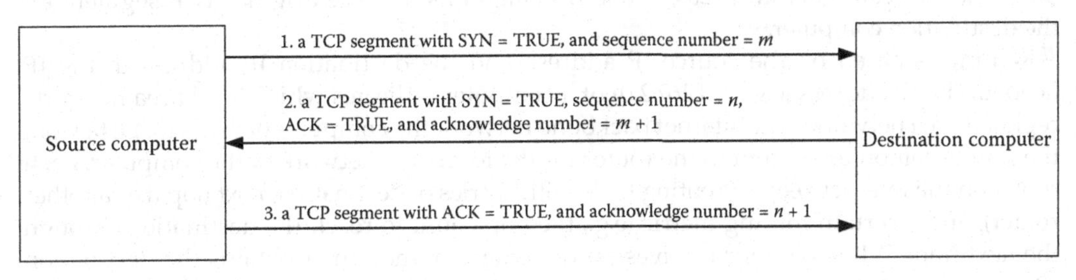

FIGURE 3.5
TCP three-way hand-shaking procedure.

the server program on the destination computer, with two sequence numbers *m* and *n* to identify data in this network session. Data from the source computer to the destination computer are identified using a sequence number based on *m*, and data from the destination computer to the source computer are identified using a sequence number based on *n*. Although the network bandwidth is a shared resource and is used to transmit data belonging to many different network sessions, data in a particular network session are identified by a unique pair of sequence numbers for this network session. The purpose of three steps in the TCP three-way hand-shaking procedure is to exchange the two sequence numbers between the source computer and the destination computer. Hence, a network session is not a time slot reserved on the network bandwidth but a virtual session identified by two sequence numbers.

After a network session is established, the sequence number and the acknowledgement number in a TCP header also take part in reliable data delivery. Suppose that application data are sent from the source computer to the destination computer through a sequence of TCP segments that have the related sequence numbers, m, $m + 100$, $m + 200$, $m + 300$, and so on. When the destination computer receives the first TCP segment from the source computer with the sequence number *m*, the destination computer sends to the source computer a TCP segment with the acknowledgement number $m + 100$ to acknowledge receiving the TCP segment with the sequence number *m* and expecting to receive the next TCP segment with the sequence number $m + 100$. If the TCP segment with the sequence number *m* is lost on the network and does not reach the destination computer, the destination computer will not send the TCP segment for acknowledgement. If the source computer sends out the TCP segment with the sequence number *m* but does not receive the acknowledgement within a time period, the source computer knows that the TCP segment is lost and will retransmit the TCP segment with the sequence number *m* again. Hence, the sequence number and the acknowledgement number work together for reliable data delivery through acknowledgement and retransmission.

UDP provides connectionless data transmissions and supports network applications that do not require reliable data delivery. For example, DNS lets domain name servers periodically send updates of domain name mapping information to each other. DNS does not require reliable data delivery of those updates. DNS uses UDP to send updates. The UDP header includes fields such as source and destination ports, header length, and checksum but does not include the sequence number and the acknowledgement number.

The IP header fields support routing and breaking a TCP segment into IP datagrams whose size fits to the transmission size allowed by the physical transmission medium. The flags and fragment offset fields of the IP header are used to break up a TCP segment into smaller IP datagrams that fit to the transmission size of the physical transmission medium on the source computer and reassemble IP datagrams into the original TCP segment on the destination computer.

Routing is based on the source IP address and the destination IP address in the IP header. The IP address (e.g., 191.167.2.0) of a computer indicates which local area network, regional area network, and Internet backbone network the computer belongs to. Data from the source computer are sent to the router for the local area network of the computer. Each router on the Internet keeps a routing table with entries of destination, next hop (i.e., another router), and a certain routing metric (e.g., the hop count to reach the destination through the next hop). When a router receives an IP datagram, the router obtains the destination IP address from the IP header, looks up the routing table using the destination IP address, selects the next hop to the destination, and sends data to the next hop. In general, data are routed from the source computer to the router of the local area network for the source computer, to the router of the regional area network for the local area network, to the Internet backbone router for the regional area network, to the regional area network for the local area network of the destination computer, to the local area network for the destination computer, and finally, to the destination computer. However, some routers on the Internet may contain incorrect routing entries, which may cause a data packet to fall into an endless routing loop on the Internet instead of reaching the destination computer. To prevent a data packet from staying on the Internet forever in an endless routing loop, the time to live (TTL) field of the IP header sets the maximum number of hops that a data packet can travel before reaching the destination computer. When a data packet reaches a router, the router decreases the TTL value by 1. When the TTL value of a data packet becomes 0 and the data packet has not reached the destination, the data packet is dropped by a router.

The link layer of the TCP/IP contains the software driver for the NIC on a computer to handle the physical transmission of signals representing binary data across a physical transmission medium. The link layer of the TCP/IP includes the address resolution protocol to map an IP address to a medium address control address, which is built into each NIC and uniquely identifies each individual computer on the network.

3.3 Summary

This chapter describes how system software, especially the operating system and network software, manages computer and network resources and enables application software to provide end-user services of data processing, data storage, and data communication without directly dealing with the implementation details of the computer and network hardware. The operating system manages the scheduling of multiple processes/threads on the CPU, the storage and retrieval of data in memory and other storage devices, and interactions with I/O devices. The network software, especially the TCP/IP, allows heterogeneous computers on the Internet to locate each other and transmit data over the shared network bandwidth. The next chapter introduces a variety of computer and network applications that are built on system software, including information system applications.

EXERCISES

1. What scheduling method is commonly used on a computer to schedule the execution of processes/threads on the CPU? Explain the pros and cons of using this CPU scheduling method in comparison with other CPU scheduling methods described in this chapter.
2. Discuss the advantages of breaking a data file into pages and storing pages in memory or other storages such as the hard drive.
3. Describe how the sequence number and the acknowledgement number in the TCP header are used in the three-way hand-shaking procedure to establish a network session.
4. Describe how the sequence number and the acknowledgement number in the TCP header are used for reliable data delivery through acknowledgement and retransmission.
5. Describe why the time to live field in the IP header is needed for the proper routing of data on the Internet.

4

Overview of Information Systems

4.1 Information System Concepts

An information system is an organized combination of people, hardware, software, communication networks, and data resources that collects and transforms data into meaningful information and knowledge to assist the decision-making process in an organization. Figure 4.1 is a schematic view of an information system structure that shows that people, technology, and organization are interconnected within a societal and business environment.

In an organization, information is one of the most valuable resources for decision making at different levels (e.g., operational decisions, tactical decisions, and strategic decisions). However, people often get confused between information and data. Taking a hospital as an example, raw data may include the patient's name; the patient's ID; the patient's address; and the patient's medical history, medical exam results, and treatments, just to name a few. When the raw facts are organized in a meaningful manner, they become information that adds additional value beyond the aggregated individual facts. For instance, the medical physicians may find the information about the patient's history and medical exam results to be more critical for treatment planning. The financial executives may find the information about the types of medical exams serve more purposes for billing and reimbursement. Transforming the raw data into meaningful information for different purposes thus becomes a critical procedure that requires specific knowledge related to the task or decision making (Ravishankar et al., 2011). As shown in Figure 4.2, from data to information to knowledge is an iterative process, where the focus lies in understanding the relationships among the data items to gather the information and discovering the underlying pattern in the information to obtain the knowledge.

During the transformation process, people sometimes organize or process data mentally or manually, while in most cases, a computer is utilized to automate such processes. No matter how the data are handled, the key is to ensure the resulting information is useful and valuable. Stair and Reynolds (2013) comprehensively review the value of information and summarize the basic characteristics that valuable information requires to support decisions (see Table 4.1).

The valuable information has been used in a massive number of applications across all levels of activities in a business setting. These will be reviewed in the next section.

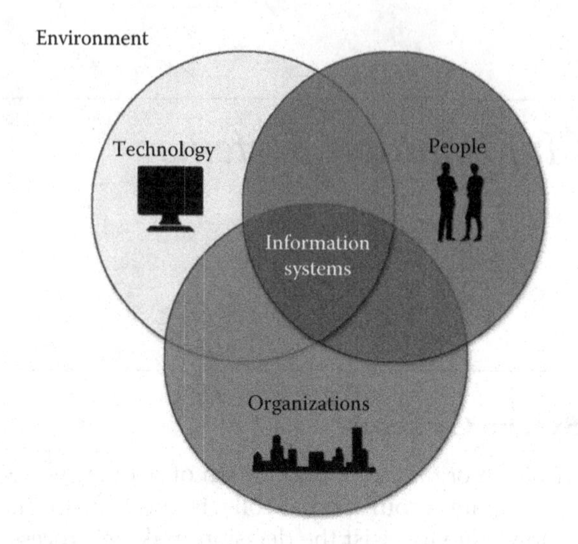

FIGURE 4.1
Information system components.

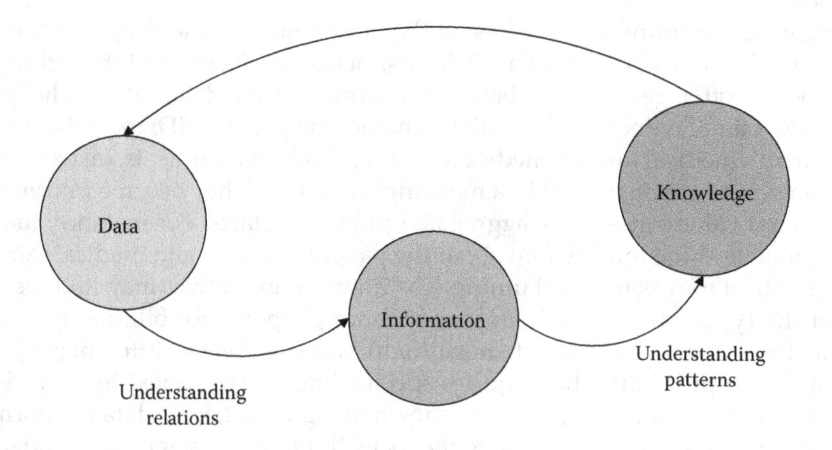

FIGURE 4.2
Data, information, and knowledge.

4.2 The Role of Information System in Business

In general, information systems used in an organization can be classified into three levels (Figure 4.3):

- Operational information system: It delivers information to the point of business where information is used as part of an operational process. For example, most retail stores now use a computer-based system for point-of-sale transactions. This operational information system can be used to help record customer purchases and keep track of product inventory levels.

TABLE 4.1
Characteristics of Valuable Information

Characteristics	Description
Accessible	Information should be easily accessible to authorized users so that they can obtain it in the right format and at the right time to meet their needs.
Accurate	Accurate information is error free. Errors could occur due to different reasons. If there are some problems in the knowledge required for the process, the information rendered may have errors. In other cases, inaccurate information is generated because the data contain errors. This is commonly known as garbage in, garbage out.
Complete	Complete information contains all the important facts for decision making. For example, a patient report that does not include patient demographic information (e.g., age, gender) is not complete.
Economical	Information should be relatively economical to produce. There is a trade-off between the value of information and the costs of producing it. For example, if collecting the data takes lots of resources and time, it is not economical.
Flexible	Information should be flexible to be used for different purposes. For example, patient information can be used by physicians for treatment planning. Such information can also be used by the quality assurance committee for workflow improvements and the patient safety committee to manage patient safety events.
Relevant	Relevant information is important to the decision maker. It is determined based on the usefulness of information with respect to the decision making process.
Reliable	Reliable information refers to the correctness of the information. If the reliability of data collection is poor, it will directly impact the information produced. Thus, it is questionable to trust such information.
Secure	The value of information could be lost due to issues such as unauthorized access or intentionally damaging its existence. Information should be secured from access by unauthorized users.
Simple	Information should be simple, not complex. Sophisticated and detailed information might not be needed. In fact, too much information can cause information overload, whereby a decision maker has too much information and is unable to determine what is really important.
Timeliness	Decisions should be made at the right time to achieve effectiveness.
Verifiable	Information should be verifiable. This means that the user can check it to make sure that it is correct, perhaps by checking many sources for the same information.

Source: Stair, R., Reynolds, G., *Fundamentals of Information Systems, Course Technology*, 7th edn., 2013.

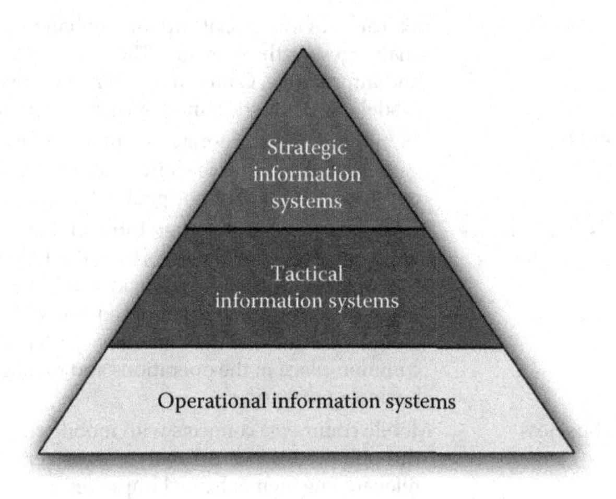

FIGURE 4.3
Information systems in business.

- Tactical information system: It is the application to analyze business trends, frequently compare a specific metric (such as sales) to the same metric from a previous month or year, and generate a report to support the decision making of employees and managers. For example, a retail store manager uses the tactical information system to make decisions such as which merchandise needs to be added or discontinued.
- Strategic (executive) information system: It supports strategic decisions for competitive advantages. For example, some retail stores (e.g., Kohl's) decide to install touch-screen kiosks to help customers locate the merchandise. In a similar manner, some healthcare providers decide to install kiosks to increase efficiency and patient satisfaction in emergency departments, waiting rooms, and ambulatory settings. Patients can easily check in, perform payment transactions, and get help in locating the service unit through complicated facility layouts.

In a typical organization, these three systems often interact with each other. The information system that supports operational decisions may also provide data to, or may be accepting data from, tactical and strategic information systems (SIS), and vice versa.

Under these three categories, the business applications of information systems have expanded significantly over the years, from traditional transactional processing systems (TPS), management information systems (MIS), and decision support systems (DSS), to executive information system/strategic information system (EIS/SIS), to the more recently emerged mobile business systems (M-business). The general description on each milestone of the information evolution is provided in Table 4.2.

TABLE 4.2

Evolution of Information System Applications

Milestone	Application System	Description
1950s–1960s	TPS	TPS functions for information storage and retrieval. The system is used for transaction processing, record keeping, accounting, etc.
1960s–1970s	MIS	MIS can process the data into useful information reports to help the manager make decisions.
1970s–1980s	DSS	DSS can provide special supports to handle different management challenges for the manager. The report in DSS must be interactive and informative. Compared to MIS, DSS usually has a decision model (e.g., analytical models) component to realize the support.
1980s–1990s	EIS/SIS	EIS/SIS provides corporate executives with simple information gathered from multiple sources across the organization, which are opportune to the strategic goal of the company.
1990s–2000s	E-business/E-commerce	The rapid growth of Internet, Intranet, Extranet and other interconnected global networks in the 1990s dramatically changed the capabilities of information systems in business at the beginning of the 21st century. Internet-based and Web-enabled enterprise and global E-business and E-commerce systems are becoming common-place in the operations and management of today's business enterprises.
2000s–present	M-business	Mobile commerce connects with mobile customers, builds insights through more powerful analytics, delivers more convenient and relevant engagements, and improves management and customer service with a seamless integration of front-end functionality and back-end data.

4.2.1 Transaction Processing System

Since the 1950s, computers have been more commonly used for business applications. Many of these early systems were designed to reduce costs by automating routine, labor-intensive business transactions. A business transaction initiates the accounting cycle of the business activities. They are recorded to maintain accurate records to ensure accountability, to establish historical business activity data, and to provide information to decision makers for determining business strategies. TPS is an information system that involves the collection, modification, and retrieval of all these transaction data.

The major functions of a TPS usually include paying employees and suppliers, controlling inventory, sending invoices, and ordering supplies. Today, a majority of business organizations use enterprise resource planning (ERP) systems for these tasks that are, indeed, a set of integrated TPS. An ERP system facilitates the flow of information between the functional areas of a business and manages the important business operations for an entire multi-site, global organization. Using smartphones and mobile devices, a timely report can be generated from an ERP system these days. According to Forbes 2013 data (http://www.forbes.com), the worldwide ERP software market size is about $24.5 billion, and the top two vendors are SAP and Oracle, with the market shares being 25% and 13%, respectively.

4.2.2 Management Information System

MIS is a system that is used to design and implement procedures, processes, and routines, providing detailed reports in an accurate, consistent, and timely manner. In an MIS, modern, computerized systems continuously collect relevant data which, in turn, are processed, integrated, and stored. Valuable information transformed from the data is made available to users who have the authority to access it, in the form that fits their needs. An MIS typically provides standard reports generated with data from the ERP/TPS and is used to improve operational efficiency (Munoz et al., 2011). For example, Dell used manufacturing MIS software to develop a variety of reports in its manufacturing processes and to account for costs (http://www.dell.com). As a result, Dell was able to double its product variety while saving about $1 million annually in manufacturing costs.

4.2.3 Decision Support System

DSS is a collection of integrated software applications that form the backbone of an organization's decision-making process. It is a data-driven application that focuses on the collection and availability of data to analyze. Whereas an MIS helps an organization to "do things right," a DSS helps a manager to "do the right things." In general, a DSS consists of a collection of models used to support decision makers or users (models), a collection of facts and information to assist in decision making (database), and systems and procedures (user interface) that help decision makers and other users interact with the system. In this book, some basic DSS models will be discussed in Chapter 25.

4.2.4 Executive Information System and Strategic Information System

To some extent, the EIS and the SIS serve the same purpose for an organization. An EIS/SIS is an infrastructure (a combination of software and hardware) that supplies an enterprise's executives up-to-the-minute operational data gathered and sifted from various data sources. The typical information mix presented to the executives may include financial information, work in process, inventory figures, sales figures, market trends, industry statistics and market price of the firm's shares, etc. It differs from a DSS as it targets top-level executives and not managers.

4.2.5 Electronic Business and Electronic Commerce

Electronic business (E-business) and electronic commerce (E-commerce) involve any business transaction executed electronically between companies (business-to-business), companies and customers (business-to-customer), customers and other customers (customer-to-customer), a business and the public sector, and customers and the public sector. This term was first introduced by IBM in 1997, and now, the business model built around it has been widely adopted. E-business and E-commerce have offered tremendous opportunities for businesses of all sizes to enter the global market with competitive advantages. With the security built into today's browsers and with digital certificates now available for individuals and companies, much of the early concern about the security of business transaction on the Web has been addressed, and E-business/E-commerce is accelerating.

4.2.6 Mobile Commerce

Mobile commerce (M-commerce) is a newly emerging business model that uses mobile devices to browse, buy, and sell products and services. With the boom of smartphone and tablet ownership, M-commerce has followed the wave. In 2012, M-commerce sales in the United States hit $24.66 billion, which was an 81% increase from $13.63 billion in 2011. More impressively, looking at the 2012 holiday season alone, mobile visitors accounted for one-third of the holiday E-commerce traffic. BMW, the German car company, has invested about $100 million in developing mobile applications for its cars and other products (Murphy, 2011). Today, M-commerce has exploded in popularity with advances in smartphones and tablets.

As various information applications used in business are reviewed, an emerging trend is observed: the development of advanced applications from PC-based applications to Web applications to mobile applications has fundamentally shifted the types of information technologies being used. This unveils the new information age: post-PC. Apparently, understanding the basic characteristic of this new information age offers the opportunity to develop and use information systems for the future.

4.3 Post-PC Information Age: Five Trends in the Future Information System Applications

There are five intertwined trends that have been observed in the new information age (Hinchcliffe, 2011): mobile, social media, big data, cloud computing, and consumerization of IT (Figure 4.4). Each is explained in the following section.

4.3.1 Mobile

In the post-PC era, one of the biggest trends being seen today is the move toward mobile devices. In most developed countries, the vast majority of adults have a mobile phone, and

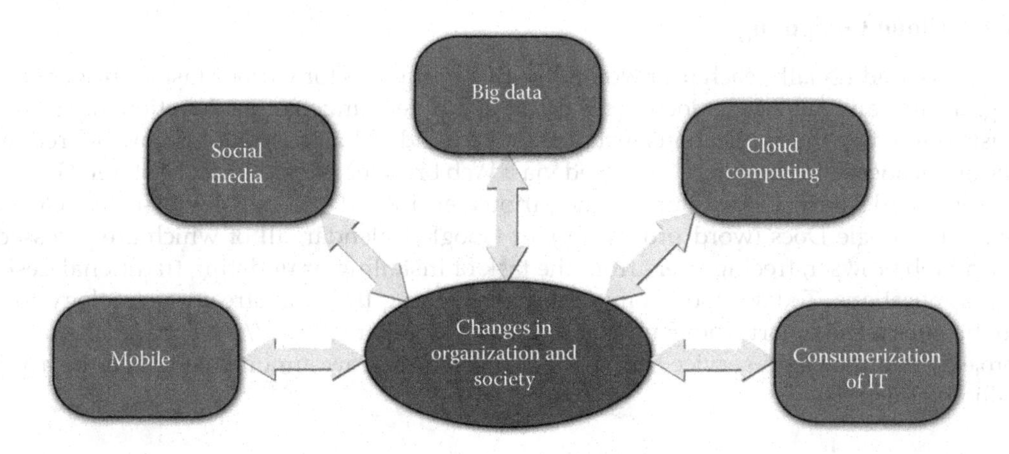

FIGURE 4.4
Five trends of the new information age.

typically, people have their mobile phones within their reach 24/7, as compared to having an access to the laptop or PC. Another mobile device gaining popularity is the tablet. For organizations, this increase in mobility has a wide range of implications, from increased collaboration to the ability to manage a business in real-time to a change in the way that new (or existing) customers can be reached.

4.3.2 Social Media

Another notable trend is social media. We may be one of the over 1.15 billion (and still growing) Facebook users who share status updates or pictures with friends and family or we may be using a social network such as Google+ to stay tuned with the activities of our social "circles." A recent survey of over 8000 faculty members done by Pearson Learning Solutions found that 41% of college professors use social media as a teaching tool, up from around 34% in 2012. Similarly, companies are using social media to get people (employees, customers) to participate in innovation and other activities.

4.3.3 Big Data

The transformation of social and work interactions enabled by this 24/7 connectivity has given rise to a third trend—big data. A study by the research firm International Data Corporation (IDC) estimates that, in 2011, 1.8 zettabytes of data were generated and consumed. The number is forecast to be 50 times more by the end of the decade. For many organizations in this new information age, value and business opportunities are created from the big data. In the "old economy," the largest/most valuable organizations (such as GE or Ford) have 100,000 to 200,000 employees and the largest organizations in the "new economy" (such as Microsoft, HP, or Oracle) have 50,000 to 100,000 employees, whereas in the "new information age economy," thriving companies such as Facebook and Twitter have a mere 5000 to 20,000 employees (Hofmann, 2011). However, it should be noted that the big data poses tremendous challenges in terms of data storage, data analysis, and transforming the data to useful information and knowledge, which leads to the fourth trend—cloud computing.

4.3.4 Cloud Computing

Whereas, traditionally, each user would install applications for various tasks—from creating, storing, and managing documents on the computer—much of the functionality previously offered by the applications installed on each individual computer is now offered by the applications "in the cloud," accessed via a Web browser (e.g., Internet Explorer, Google Chrome, and Safari). One example is the various services offered by Google, such as Gmail (e-mail), Google Docs (word processing), or Google Calendar, all of which are accessed via a Web browser, freeing users from the task of installing or updating traditional desktop applications. Today, cloud computing is becoming the mainstream technology tool for business. The report from Parallels, a software company (http://www.parallels.com/), projects that the cloud service business is expected to grow annually at 28% to be a $95 billion industry.

4.3.5 Consumerization of IT

Lastly, the most significant trend according to the research firm Gartner is the consumerization of IT. Driven by the advances in consumer-oriented mobile devices and the ability to access data and applications in the cloud, today's employees are increasingly using their own devices for work-related purposes and/or are using software that they are used to (e.g., social media) in the workplace. Although this trend may raise concerns related to security or compliance or may increase the need to support the employees' own devices, it may also provide opportunities, such as increased productivity, higher retention rates of talented employees, or higher customer satisfaction (TrendMicro, 2011).

4.4 Summary

This chapter gives an overview of the information system concept, including data, information, and knowledge. The basic characteristics of valuable information are reviewed. Various types of information systems using valuable information are discussed. Observing the emergence of the new information age, five trends for future information system applications are introduced. We note that the core of any information system is data. Throughout the remainder of this book, the focus is on the design and implementation of information system applications to store, retrieve and manage data (database), and use the data for different end users in various forms (VB.NET and VBA for Windows applications, ASP.NET for Web applications) to support decisions.

EXERCISES

1. An information system is an organized combination of the following except:
   a. Environment
   b. Data
   c. Technology
   d. Organizations
   e. All of the above are valid components of information systems

2. Information, data, and knowledge mean the same thing. True or false?
3. Collecting data and analyzing its patterns produces knowledge. True or false?
4. An information system should be flexible enough and economical during implementation. True or false?
5. Which of the following statements are true?
   a. A tactical information system delivers information from point-of-sale transactions to provide tactical reports for middle management.
   b. An SIS supports top management, which affects a wide range of business departments, such as finance, inventory management, and warehousing.
   c. An operational information system compares a performance metric trend for a couple of months to develop operational advantages over competitors.
   d. All of the above are true.
   e. None of the above is true.
6. M-business is preceded by E-business in the evolution of information systems. True or false?
7. A set of integrated transaction processing systems is called "material requirements planning" or MRP. True or false?
8. What are the two typical sources of information for MIS?
9. What is the main difference between a SIS and a DSS?
10. E-business is a business model that offers business to a market and sells their products online through a mobile device.
11. The following trend examples are observed during the post-PC information age except:
    a. Facebook
    b. Smartphones
    c. Big data algorithms
    d. Gmail
    e. All of the above are valid examples
12. The purpose of big data is to analyze huge amounts of data for the purpose of improving business value proposition to customers. True or false?
13. An information system hosted entirely online is an example of an application of cloud computing. True or false?
14. An application that lets you navigate the streets of New York on a smartphone is an example of consumerization of IT. True or false?

Section II

Database Design and Development

A database is a major component of information systems. A database contains an organized collection of related data. The relational data model is widely used to organize data in a database. There are many database software packages, such as Microsoft Access, Microsoft Structured Query Language (SQL) Server, MySQL, and Oracle, which implement database management systems (DBMSs) to support the definition, creation, access, and management of a relational database by end users. This book introduces the use of Microsoft Access and MySQL for database development. The development of a relational database includes four phases:

- Enterprise modeling: Specifies data needed for a database and data queries for using data in an enterprise environment
- Conceptual data modeling: Uses the entity–relationship (E-R) modeling to define the conceptual organization of related data, which can easily be understood by end users in the enterprise environment without reference to the specific relational data structure
- Logical database design: Uses the relational modeling to transform the ER model of a database into the relational database design, which can be directly implemented using a relational DBMS
- Physical database implementation: Implements the relational data model to create a relational database with data storage and query capacities

Section II of this book covers the entire process of developing a database. Enterprise modeling is illustrated using a case study for a health care information system. Chapter 5 describes the E-R modeling for conceptual data modeling. Chapter 6 presents relational modeling along with database normalization for logical database design. Chapter 7 shows the use of Microsoft Access to create a relational database and data queries. Chapter 8 introduces the SQL, which is a standard database language supported by many different DBMS software packages. Because SQL is interoperable among different DBMS, it allows us to define, create, access, and manage a new database, or transport an existing database, using even a DBMS that we are not familiar with. Chapter 9 shows the use of MySQL to create a relational database and data queries. Chapter 10 introduces the concepts of object-oriented database.

5

Conceptual Data Modeling: Entity-Relationship Modeling

An E-R model organizes data using entities, relationships, and their attributes. Such an organization of data is natural and easy to understand but is not the way in which data are stored on computers. An E-R model will need to be transformed into a relational model, which organizes data using the data structure called "relation," which is similar to a table with columns and rows. Because concepts for a relational model involve the implementation of data storage on computers and may not be natural to our understanding, we first use an E-R model to describe a conceptual organization of data and then transform the E-R model into a relational model.

This chapter describes the elements and concepts of E-R modeling, including types, attributes, and instances of entities; types, attributes, instances, and degrees of relationships between entities; maximum and minimum cardinalities of relationships; associative entities and weak entities; and superclass and subclass entities. A case study of a database for a healthcare information system is used to illustrate the development of an E-R model. Chapter 26 gives the requirements for the healthcare database in Section 26.1.

5.1 Types, Attributes, and Instances of Entities

An entity can represent a person, a place, an object, an event, a concept, or others. The following are some examples of entities that may be included in an E-R model:

- Person: student, employee, client
- Object: restaurant, movie theater, machine
- Place: city, park, warehouse, room
- Event: marriage, lease, war
- Concept: project, account, course

An entity is represented by a rectangular box in an E-R diagram, which is used to define an E-R model. An attribute of an entity describes a property or characteristic of the entity and holds data about the entity. An attribute is represented by an ellipse in an E-R model. An entity may have one or more attributes.

Example 5.1

Use an E-R diagram to represent the Patient and the InsurancePlan entities in the healthcare database. The Patient entity has the attributes PatientID, SSN, PatientName, Address, Phone, Gender, DateOfBirth, Employer, Position, E-Name, and E-Phone. The InsurancePlan entity has the attributes PlanID, PlanName, Description, MemberPremium, and EmployerPremium.

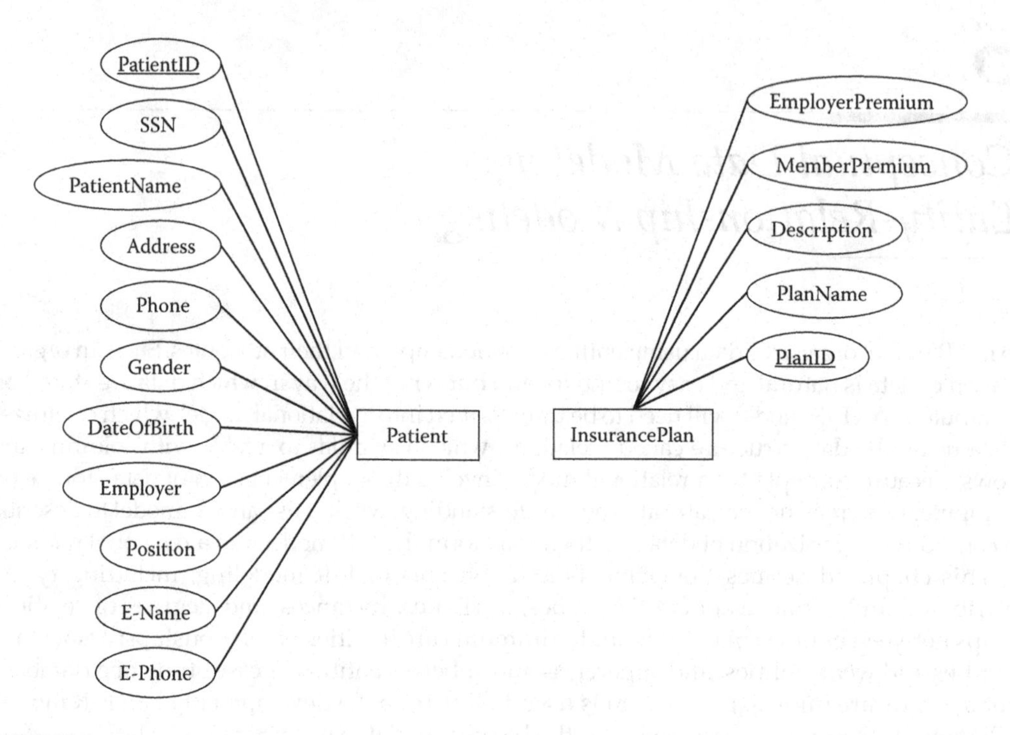

FIGURE 5.1
An E-R model containing the Patient entity, the InsurancePlan entity, and their attributes in the healthcare database.

Figure 5.1 gives the E-R diagram containing the Patient and the InsurancePlan entities, and their attributes.

The set of attributes for an entity can also be represented in the text format. For example, the InsurancePlan entity and its attributes in Example 5.1 can be expressed as follows:

InsurancePlan = {PlanID, PlanName, Description, MemberPremium, and EmployerPremium}

In an E-R model, an entity box represents a type or collection of entities that have the same attributes. For example, in the E-R model of the healthcare database, the Patient entity represents a collection of all patients in the database. Each individual patient is an instance of the entity type Patient. All individual patients have the same attributes. For example, the InsurancePlan entity in the E-R model of the healthcare database represents a collection of all insurance plans in the healthcare database. A particular insurance plan is an instance of the entity type InsurancePlan.

Example 5.2

Give two instances of the Patient and the InsurancePlan entities described in Example 5.1.

Table 5.1 gives two instances of the Patient entity. Table 5.2 gives two instances of the InsurancePlan entity.

Hence, an entity and its attributes in an E-R model of a database describe the data organization for the entity type (i.e., what type of attributes for what entity type are kept in the database), while instances of the entity type and attribute values for instances are the data

TABLE 5.1

Two Instances of the Patient Entity

PatientID	SSN	Patient Name	Address	Phone	Gender	DateOfBirth	Employer	Position	E-Name	E-Phone
PA1	111-11-1111	John Smith	321 Mill Avenue	480-965-1111	M	1/1/1983	ASU	Staff	Helen Smith	480-965-1112
PA2	222-22-2222	Mary Davis	432 Mill Avenue	480-965-2222	F	2/2/1995	NULL	Student	Jack Davis	480-965-2223

TABLE 5.2

Two Instances of the InsurancePlan Entity

PlanID	PlanName	Description	MemberPremium	EmployerPremium
PL1	Care HMO	A HMO plan	$100	$800
PL2	Care PPO	A PPO plan	$300	$600

stored in the database. When the value of an attribute for an instance is not known or not available, a NULL value can be used for the attribute of the instance. For example, a particular patient (e.g., PA2), which is an instance of the Patient entity in Example 5.2, may not have an employer. This instance of the Patient entity can have NULL for the Employer attribute.

Each entity type should have the key attribute(s) whose value is unique for each instance of the entity and, thus, can be used to uniquely identify each instance. A key attribute is underlined in an E-R model. For example, PatientID is the key attribute of the Patient entity and is underlined in the E-R model shown in Figure 5.1. Each patient has a unique value of PatientID so that each patient can be uniquely identified by the value of PatientID. Consider the City entity in Figure 5.2 that has the three attributes: CityName, StateName, and Population. Because two different cities may have the same city name (e.g., Miami in Arizona and Miami in Florida), using CityName alone cannot uniquely identity each of the two cities that have the same city name. Hence, both CityName and StateName can be used together as the key attributes to uniquely identify each instance of the City entity. The key attribute of an entity cannot have the NULL value because the NULL value of the key attribute cannot identify a particular instance of the entity.

There are various types of attributes: simple versus composite, single-valued versus multi-valued, and stored versus derived. Because different types of attributes are implemented in a database on computers differently, each attribute type must be identified in an E-R model.

A simple or atomic attribute does not have any other attributes as components. A composite attribute has some other attributes as components. For example, we can have EmergencyContact, which contains E-Name and E-Phone, as a composite attribute of the Patient entity in the E-R model of the healthcare database, as shown in Figure 5.3.

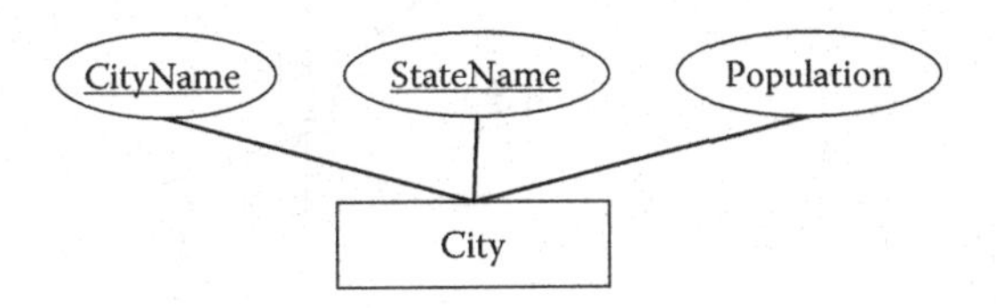

FIGURE 5.2
An E-R model containing the City entity and its attributes.

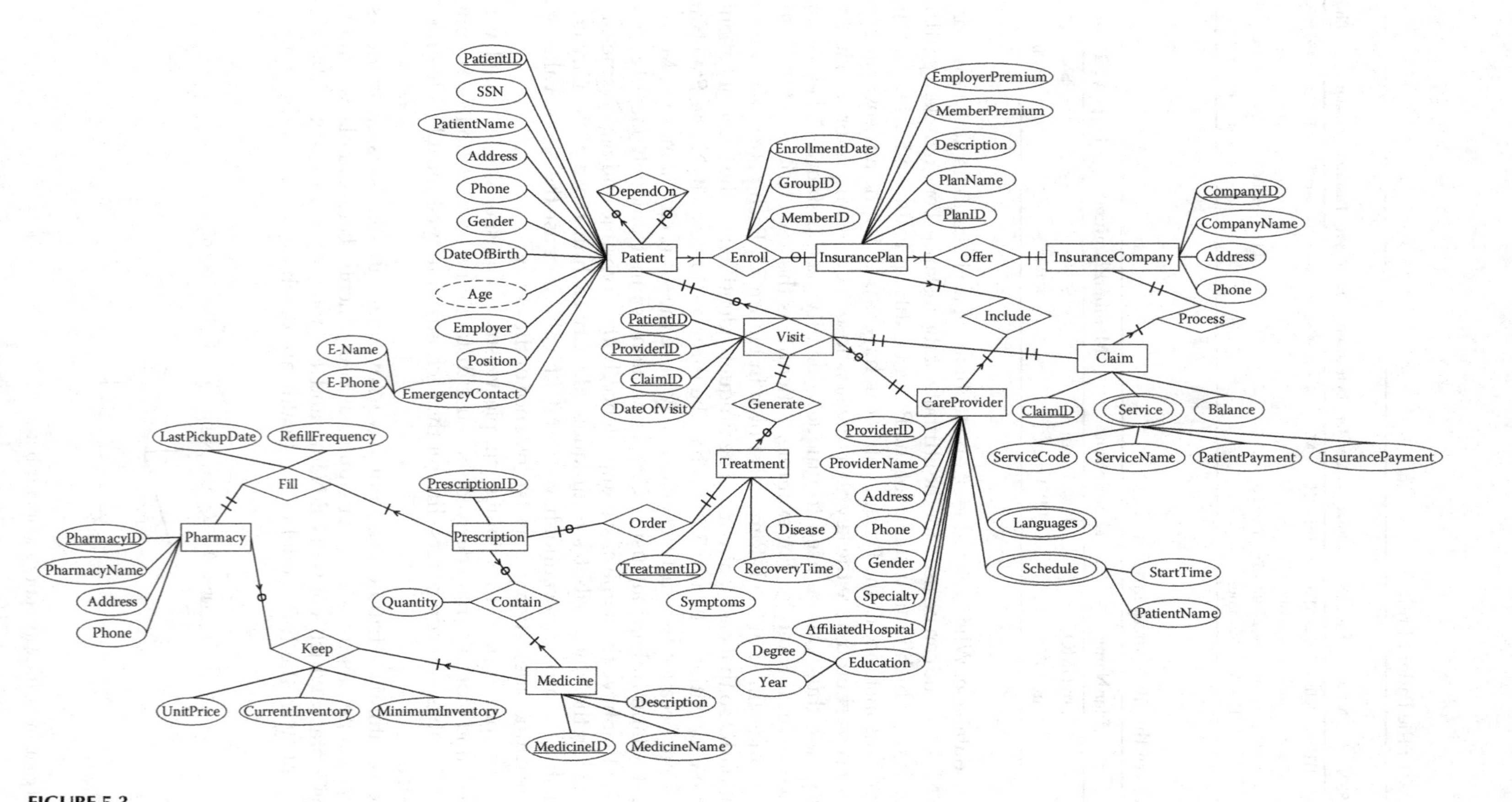

FIGURE 5.3
The E-R model of the healthcare database.

That is, the EmergencyContact attribute consists of the E-Name and E-Contact attributes. In an E-R model, a line is used to connect a composite attribute to its component attribute. All the other attributes of the Patient entity are the simple attributes. In the E-R model of the healthcare database (see Figure 5.3), the following attributes are the composite attributes: the EmergencyContact attribute of the Patient entity, the Education and Schedule attributes of the CareProvider entity, and the Service attributes of the Claim entity. All the other attributes in the E-R model of the healthcare database are the simple attributes.

A single-valued attribute of an entity has only one value for an instance of the entity. A multi-valued attribute of an entity may have more than one value for an instance of the entity. For example, in the E-R model of the healthcare database, the Language attribute of the CareProvider entity is a multi-valued attribute because an instance of the CareProvider entity, for example, a physician, may speak both English and Spanish and, thus, have two values for the Language attribute. A multi-valued attribute is denoted by a double-lined ellipse (see Figure 5.3). In the E-R model of the healthcare database, the Language and Schedule attributes of the CareProvider entity and the Service attribute of the Claim entity are the multi-valued attributes, and all the other attributes in the E-R model are the single-valued attributes.

The value of a derived attribute can be derived from values of some stored attribute(s) and, thus, does not need to be stored in a database. For example, in the E-R model of the healthcare database, we may add and consider the Age attribute of the Patient entity as a derived attribute because the value of Age can be computed from the DateOfBirth attribute of the Patient entity. A derived attribute is denoted by a dash-lined ellipse (see Figure 5.3). The value of a stored attribute cannot be derived from the values of some other attributes and must be stored in a database.

In summary, there are three pairs of different attribute types: simple versus composite, single-valued versus multi-valued, and stored versus derived. An entity attribute falls into one of the two types in each pair. For example, in the E-R model of the healthcare database in Figure 5.3, the Schedule attribute of the CareProvider entity is a composite, multi-valued, stored attribute, and the Gender attribute of the Patient entity is a simple, single-valued, stored attribute.

5.2 Types, Attributes, Instances, and Degrees of Relationships between Entities

A relationship represents an association between entities. For example, in the E-R model of the healthcare database (see Figure 5.3), there is the Enroll relationship between the Patient and the InsurancePlan entities because a patient is enrolled in an insurance plan and the Enroll relationship links a patient to an insurance plan. A relationship is represented by a diamond-shaped box with lines connecting entities involved in the relationship (see Figure 5.3). A relationship may have no attributes or may have a set of attributes. Unlike an entity type, a relationship type does not need a key attribute. An attribute of a relationship stores information about the relationship. For example, the Enroll relationship between the Patient and the InsurancePlan entities has the attributes EnrollmentDate, GroupID, and MemberID. Each of the three attributes is not an attribute of either the Patient or the InsurancePlan entity because the value of EnrollmentDate, GroupID, and MemberID

TABLE 5.3
Two Instances of the Enroll Relationship

PatientID	PlanID	EnrollmentDate	GroupID	MemberID
PA1	PL1	1/1/2013	10000	1234567
PA2	PL1	1/1/2013	10000	2345678

depends on both a specific instance of the Patient entity and a specific instance of the InsurancePlan entity involved in the Enroll relationship. For example, MemberID cannot be an attribute of the Patient entity because the MemberID of a given patient depends on which insurance plan this patient is enrolled in. MemberID cannot be an attribute of the InsurancePlan entity because an insurance plan has many members and a particular MemberID is linked to a particular patient.

A relationship in an E-R model represents a relationship type or a collection of relationship instances. An instance of a relationship type involves an instance of each entity type involved in the relationship. To identify an instance of a relationship, we should identify the instance of each entity type involved in the instance of the relationship, which can be done by using the value of the key attribute(s) of each entity type. For example, when a patient is enrolled in an insurance plan, an instance of the Enroll relationship is created. This instance of the Enroll relationship can be identified by the value of PatientID and the value of PlanID involved in the instance of the relationship.

Example 5.3

Give two instances of the Enroll relationship.

Table 5.3 gives two instances of the Enroll relationship.

The number of entity types involved in a relationship is called the "degree" of a relationship. In the E-R model of the healthcare database (see Figure 5.3), the DependOn relationship represents the relationship between a patient and his/her dependents. Because the DependOn relationship involves only the Patient entity, the DependOn relationship is a degree-1 or unary relationship. The Enroll relationship between the Patient and the InsurancePlan entities is a degree-2 or binary relationship. The Visit relationship associates the Patient, the CareProvider, and the Claim entities, and is a degree-3 or ternary relationship, which is converted into an associative entity, as shown in Figure 5.3 (see the description of associative entity in Section 5.4).

5.3 Maximum and Minimum Cardinalities of a Relationship

Given entity A, entity B, and their relationship R in Figure 5.4, the maximum cardinality of relationship R describes the maximum number of entity B's instances that can be associated with any instance of entity A (see Figure 5.4a and b) and the maximum number of entity A's instances that can be associated with any instance of entity B (see Figure 5.4c and d). There are only two values for the maximum cardinality: one or many (more than one).

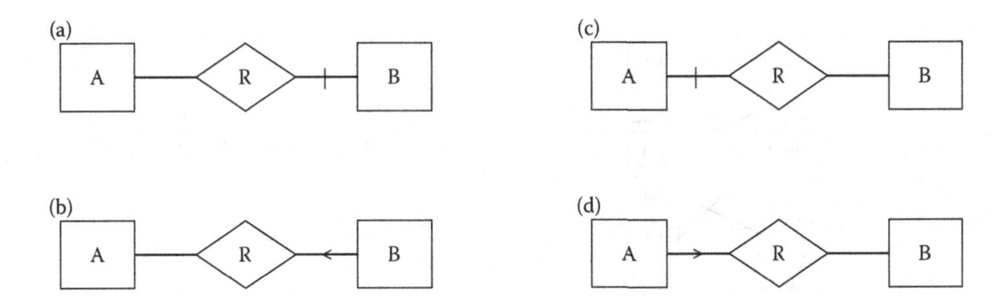

FIGURE 5.4
Representation of the maximum cardinality of a relationship. (a) Given any instance of entity A, the maximum number of entity B's instances that can be associated with the instance of entity A is one (denoted by |). (b) Given any instance of entity A, the maximum number of entity B's instances that can be associated with the instance of entity A is many (denoted by <). (c) Given any instance of entity B, the maximum number of entity A's instances that can be associated with the instance of entity B is one (denoted by |). (d) Given any instance of entity B, the maximum number of entity A's instances that can be associated with the instance of entity B is many (denoted by <).

Figure 5.5 gives examples of one-to-one, one-to-many, many-to-one, and many-to-many relationships for the maximum cardinality. Suppose that all instances of entity A, entity B, and relationship R are shown in each example in Figure 5.5. In Figure 5.5a, for the one-to-one relationship between entities A and B, any instance of entity A is associated with, at maximum, one instance of entity B (a1 with b1, a2 with b3, a3 with b5, a5 with b4, and no association of a4 and a6 with any instance of entity B), and any instance of entity B is associated with, at maximum, one instance of entity A (b1 with a1, b3 with a2, b4 with a5, b5 with a3, and no association of b2 with any instance of entity A). In Figure 5.5b, for the one-to-many relationship, there is an instance of entity A that is associated with more than one instance of entity B (e.g., a2 with b2 and b3), and any instance of entity B is associated with, at maximum, one instance of entity A (b1 with a1, b2 with a2, b3 with a2, b4 with a3, and b5 with a6). In Figure 5.5c, for the many-to-one relationship, any instance of entity A is associated with, at maximum, one instance of entity B (a1 with b1, a2 with b3, a3 with b3, a4 with b1, a5 with b5, and a6 with b1), and there is an instance of entity B that is associated with more than one instance of entity A (e.g., b1 with a1, a4 and a6). In Figure 5.5d, for the many-to-many relationship, there is an instance of entity A that is associated with more than one instance of entity B (e.g., a1 with b1 and b2), and there is an instance of entity B that is associated with more than one instance of entity A (e.g., b3 with a2 and a3).

The minimum cardinality of relationship R between entities A and B describes the minimum number of entity B's instances that can be associated with any instance of entity A (see Figure 5.6a and b) and the minimum number of entity A's instances that can be associated with any instance of entity B (see Figure 5.6c and d). There are only two values for the minimum cardinality: mandatory or optional. The mandatory value means that the minimum number of instances is at least 1. The optional value means that the minimum number of instances is 0. Figure 5.6a shows that the association of any instance of entity A with an instance of entity B is optional, that is, there is an instance of entity A that is not associated with any instance of entity B. Figure 5.6b shows that the association of any instance of entity A with an instance of entity B is mandatory, that is, any instance of entity A is associated with at least one instance of entity B. Figure 5.6c shows that the

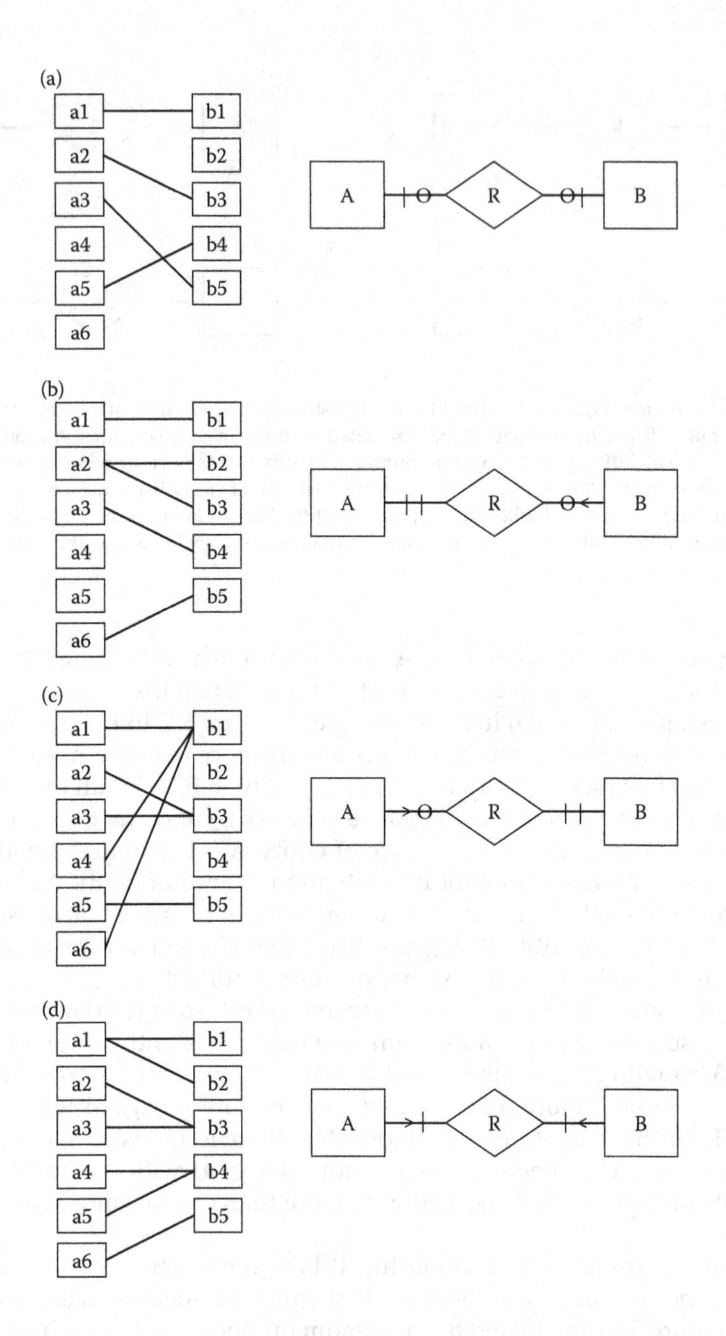

FIGURE 5.5
Examples of one-to-one, one-to-many, many-to-one, and many-to-many relationships for the maximum cardinality, and optional-to-optional, mandatory-to-optional, optional-to-mandatory, and mandatory-to-mandatory relationships for the minimum cardinality. (a) One-to-one and optional-to-optional. (b) One-to-many and mandatory-to-optional. (c) Many-to-one and optional-to-mandatory. (d) Many-to-many and mandatory-to-mandatory.

association of any instance of entity B with an instance of entity A is optional, that is, there is an instance of entity B that is not associated with any instance of entity A. Figure 5.6d shows that the association of any instance of entity B with an instance of entity A is mandatory, that is, any instance of entity B is associated with at least one instance of entity A.

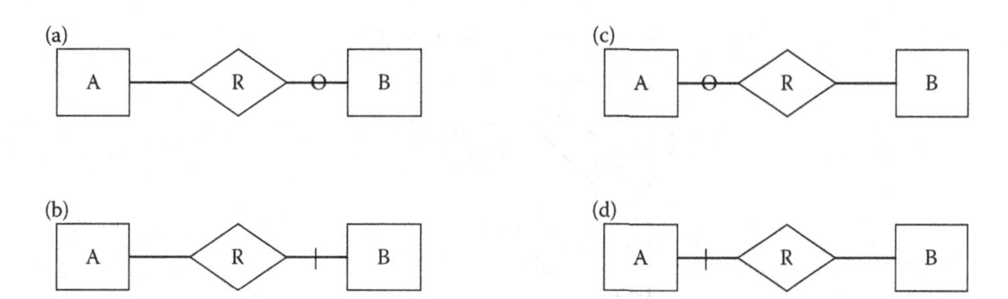

FIGURE 5.6
Representation of the minimum cardinality of a relationship. (a) Given any instance of entity A, the minimum number of entity B's instances that can be associated with the instance of entity A is 0 (denoted by a circle). (b) Given any instance of entity A, the minimum number of entity B's instances that can be associated with the instance of entity A is at least one (denoted by |). (c) Given any instance of entity B, the minimum number of entity A's instances that can be associated with the instance of entity B is 0 (denoted by a circle). (d) Given any instance of entity B, the minimum number of entity A's instances that can be associated with the instance of entity B is at least one (denoted by |).

Figure 5.5 shows examples of optional-to-optional, mandatory-to-optional, optional-to-mandatory, and mandatory-to-mandatory relationships for the minimum cardinality. As shown in Figure 5.5, the pair of symbols for the minimum cardinality are placed close to the diamond-shaped relationship, and the pair of symbols for the maximum cardinality are placed close to the entity boxes. In Figure 5.5a, for the optional-to-optional relationship, there is an instance of entity A (e.g., a4) that is not associated with any instance of entity B, and there is an instance of entity B (e.g., b2) that is not associated with any instance of entity A. In Figure 5.5b, for the mandatory-to-optional relationship, every instance of entity B is associated with an instance of entity A, indicating that the association of entity B with entity A is mandatory, and there is an instance of entity A (e.g., a4) that is not associated with any instance of entity B. In Figure 5.5c, for the optional-to-mandatory relationship, every instance of entity A is associated with an instance of entity B, indicating that the association of entity A with entity B is mandatory, and there is an instance of entity B (e.g., b4) that is not associated with any instance of entity A. In Figure 5.5d, for the mandatory-to-mandatory relationship, every instance of entity A is associated with an instance of entity B, and every instance of entity B is associated with an instance of entity A, indicating that the association between entities A and B is mandatory.

The maximum and minimum cardinalities of a relationship determine how instances of a relationship are stored in a database on computers (see Chapter 6 for detailed discussions). Hence, the maximum and minimum cardinalities of a relationship must be identified in an E-R model.

Example 5.4

Figure 5.7 gives five instances of the Enroll relationship between the Patient and the InsurancePlan entities. Based on these instances, determine the maximum and minimum cardinalities of the Enroll relationship.

For the maximum cardinality, an insurance plan may have many patients enrolled, for example, IP1 has PA1, PA2, and PA3 enrolled, and a patient is enrolled in one insurance at maximum. Hence, the maximum cardinality of the Enroll relationship between the Patient and the InsurancePlan entities is many-to-one, as shown in Figure 5.3. For

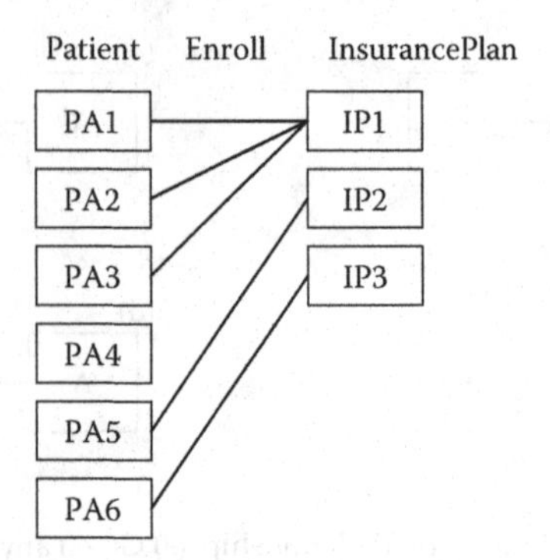

FIGURE 5.7
Examples of instances for the Enroll relationship between the Patient and the InsurancePlan entities.

the minimum cardinality, each insurance plan has at least one patient enrolled, and a patient (e.g., PA4) may not enroll in any insurance plan. Hence, the minimum cardinality of the Enroll relationship between the Patient and the InsurancePlan entities is mandatory to optional, as shown in Figure 5.3.

Example 5.5

The DependOn relationship for the Patient entity in the E-R model of the healthcare database is a unary relationship representing which patient is a dependent of another patient. Figure 5.8 gives some instances of the DependOn relationship for the Patient entity. In Figure 5.8, we use a copy of the Patient entity and its instances (PA1, PA2, PA3, PA4, PA5, and PA6) to represent the unary relationship as a binary relationship so that we can clearly see which patient depends on which patient. Figure 5.8 shows that PA1 and PA2 depend on PA3, PA4 depends on PA5, and PA6 has no dependent and does not depend on anyone. Based on these instances, determine the maximum and minimum cardinalities of the DependOn relationship.

For the maximum cardinality, one patient may have many dependents (e.g., PA3 has PA1 and PA2 as dependents), and one patient depends on only one other patient. Hence, the DependOn relationship is a many-to-one relationship, as shown in Figure 5.9. For the minimum cardinality, there is a patient (e.g., PA6) who has no dependent and does not depend on another patient. Hence, the DependOn relationship is an optional-to-optional

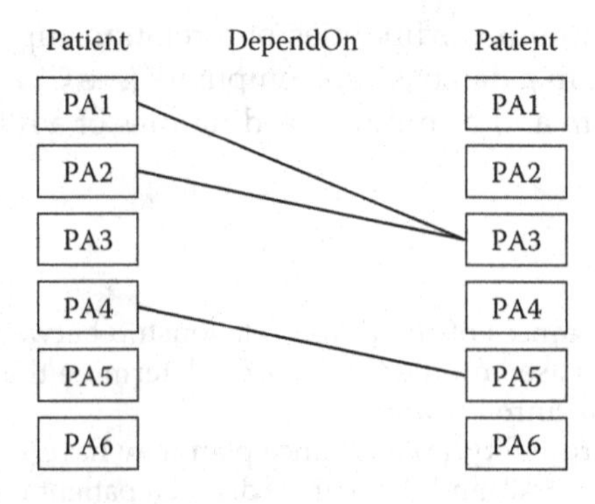

FIGURE 5.8
Examples of instances for the DependOn relationship for the Patient entity.

FIGURE 5.9
Maximum and minimum cardinalities of the DependOn relationship for the Patient entity.

relationship, as shown in Figure 5.9. Flipping the copy of the Patient entity back to the Patient entity produces the representation of the maximum and minimum cardinalities for the DependOn relationship in Figure 5.3.

We cannot show the maximum and minimum cardinalities on a relationship of degree-3 or higher, such as the Visit relationship in Figure 5.3 and the degree-3 relationship shown in Figure 5.10. For example, if we place the Many value (<) for the maximum cardinality next to entity A in Figure 5.10, it is not clear whether the Many value is for the relationship with entity B or with entity C. Furthermore, the degree-3 relationship of entities A, B, and C is not equivalent to the binary relationship of entities A and B, the binary relationship of entities A and C, and the binary relationship of entities B and C all together. Before we can mark the maximum and minimum cardinalities of a relationship of degree-3 or more, we need to first convert a relationship of degree-3 or higher to an associative entity, which is described in Section 5.4.

Example 5.6

Determine the maximum and minimum cardinalities for each relationship in the E-R model of the healthcare database.

Figure 5.3 shows the maximum and minimum cardinalities of each relationship in the E-R model of the healthcare database. The DependOn relationship of the Patient entity is a many-to-one and optional-to-optional relationship, as described in Example 5.5. As described in Example 5.4, the Enroll relationship of the Patient and the InsurancePlan entities is a many-to-one and a mandatory-to-optional relationship because an insurance plan must have one patient as a member and may have many patients as members. We assume that a patient has, at maximum, one insurance plan and may not have any insurance plan. The Offer relationship of the InsurancePlan and the InsuranceCompany entities is a many-to-one and a mandatory-to-mandatory relationship because an insurance plan must be offered by only one insurance company, and an insurance company must offer an insurance plan and may offer many insurance plans. The Include relationship of the InsurancePlan and the CareProvider entities is a many-to-many and a mandatory-to-mandatory relationship because an insurance plan includes at least one healthcare provider and may include many healthcare providers, and a healthcare

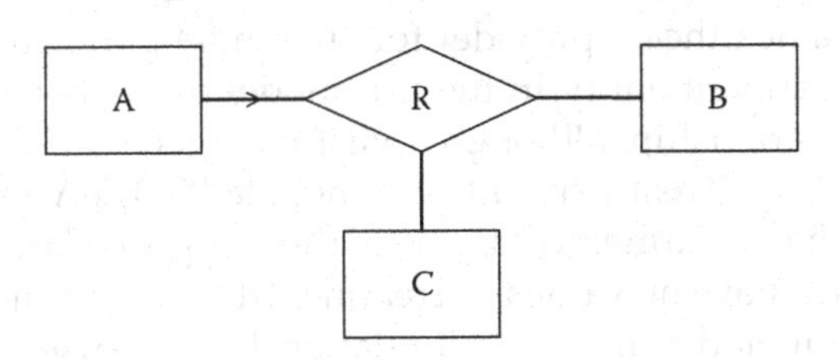

FIGURE 5.10
Example of a degree-3 relationship.

provider is included in at least one insurance plan and may be included in many insurance plans. The Visit relationship of the Patient, the Claim, and the CareProvider entities is converted into an associative entity, which is described in Section 5.4. The Process relationship between the InsuranceCompany and the Claim entities is a one-to-many and a mandatory-to-mandatory relationship because an insurance company must process at least one claim and may process many claims, and a claim must be processed by one insurance company. The Generate relationship of the Visit associative entity and the Treatment entity is a one-to-many and a mandatory-to-optional relationship because a visit to a healthcare provider may generate many treatments due to many diseases diagnosed or may generate no treatment if the visit is for a physical exam, and a treatment must be generated by one visit. The Order relationship of the Treatment and the Prescription entities is a one-to-one and a mandatory-to-optional relationship because a treatment may or may not produce a prescription order, and a prescription must be ordered for a treatment. The Contain relationship of the Prescription and the Medicine entities is a many-to-many and an optional-to-mandatory relationship because a prescription must contain at least one medicine and may contain many medicines, and a medicine may be contained in many prescriptions or may not be contained in any prescription. The Fill relationship between the Prescription and the Pharmacy entities is a many-to-one and a mandatory-to-mandatory relationship because a pharmacy store fills at least one prescription and may fill many prescriptions, and a prescription must be filled by only one pharmacy store. The Keep relationship between the Pharmacy and the Medicine entities is a many-to-many and an optional-to-mandatory relationship because the pharmacy store keeps at least one medicine and may keep many medicines, and a medicine may be kept by many pharmacy stores but may not be kept by any pharmacy store.

5.4 Associative Entities

An associative entity is created by converting a many-to-many binary relationship or a relationship of degree-3 or higher into an entity. No other type of relationship than a many-to-many binary relationship and a relationship of degree-3 or higher can be converted into an associative entity. An associative entity is denoted by adding a rectangular box to the diamond representation of the relationship.

The associative entity keeps the attributes of the relationship as the attributes of the associative entity. The key attributes of entities involved in the relationship are used as the composite key attributes for the associative entity. Hence, any instance of an associative entity must have a value for each of the composite key attribute to uniquely identify the instance of the associative entity. This also means that any instance of an associative entity must involve an instance of each entity involved in the relationship. Considering that a patient's visit to a healthcare provider for an annual physical exam does not generate a treatment, the Treatment entity in the E-R model of the healthcare database is not included in the Visit relationship, although a visit may generate a treatment (e.g., a visit for treating a disease). If the Treatment entity is included in the Visit relationship and the Visit associative entity has TreatmentID as one of the composite key attributes, a visit generating no treatment will have no value for TreatmentID, which is not allowed. Hence, the Treatment entity is not included in the Visit relationship because a visit may not involve a treatment.

In general, it is not necessary to convert a many-to-many binary relationship or a relationship of degree-3 or higher to an associative entity because an E-R model with or without the conversion is transformed into the same relational model. However, we have to convert the Visit relationship in the E-R model of the healthcare database into the Visit associative entity so that we can establish a relationship between the Visit associative entity and the Treatment entity.

Example 5.7

Figure 5.11 shows the Visit relationship of the Patient, the CareProvider, and the Claim entities in the E-R model of the healthcare database. Convert the Visit relationship into an associative entity.

Figure 5.3 shows the associative entity Visit. This Visit associative entity keeps the attribute of the Visit Relationship: DateOfVisit. The key attributes of the entities involved in the Visit relationship, PatientID, ProviderID, and ClaimID, are also added as the composite key attributes of the Visit associative entity.

A new key attribute can be created and used for an associative entity without using the key attributes of the entities involved in the relationship as the composite key attributes. For example, instead of using PatientID, ProviderID, and ClaimID as the composite key attributes of the Visit associative entity in Example 5.7, we can create a new attribute, VisitID, and use it as the key attribute of the Visit associative entity, as shown in Figure 5.12.

After converting a relationship into an associative entity, we can describe the maximum and minimum cardinalities between the associative entity and each of the other entities.

Example 5.8

For the Visit associative entity with the composite key attributes, specify the maximum and minimum cardinalities between the Visit associative entity and the Patient entity, between the Visit associative entity and the CareProvider entity, and between the Visit associative entity and the Claim entity.

For the maximum and minimum cardinalities of the Visit associative entity with the Patient entity, each instance of the Visit associative entity must involve one instance of the Patient entity, producing the One value for the maximum cardinality and the Mandatory value for the minimum cardinality on the side of the Patient entity. Because a patient may have many visits or no visit, we have the Many value for the maximum cardinality and the Optional value for the minimum cardinality on the side of the Visit associative entity. For the maximum and minimum cardinalities of the Visit associative entity with the CareProvider entity, each instance of the Visit associative entity must involve one instance of the CareProvider entity, producing the One value for the maximum cardinality and the Mandatory value for the minimum cardinality on the side of the CareProvider entity. Because a healthcare provider may be involved in many visits or no visit, we have the Many value for the maximum cardinality and the Optional value for the minimum cardinality on the side of the Visit associative entity. For the maximum and minimum cardinalities of the Visit associative entity with the Claim entity, each instance of the Visit associative entity must involve one instance of the Claim entity, producing the One value for the maximum cardinality and the Mandatory value for the minimum cardinality on the side of the Claim entity. Because a claim must be for one visit, we have the One value for the maximum cardinality and the mandatory value for the minimum cardinality on the side of the Visit associative entity. Figure 5.13 shows all the values of the maximum and minimum cardinalities.

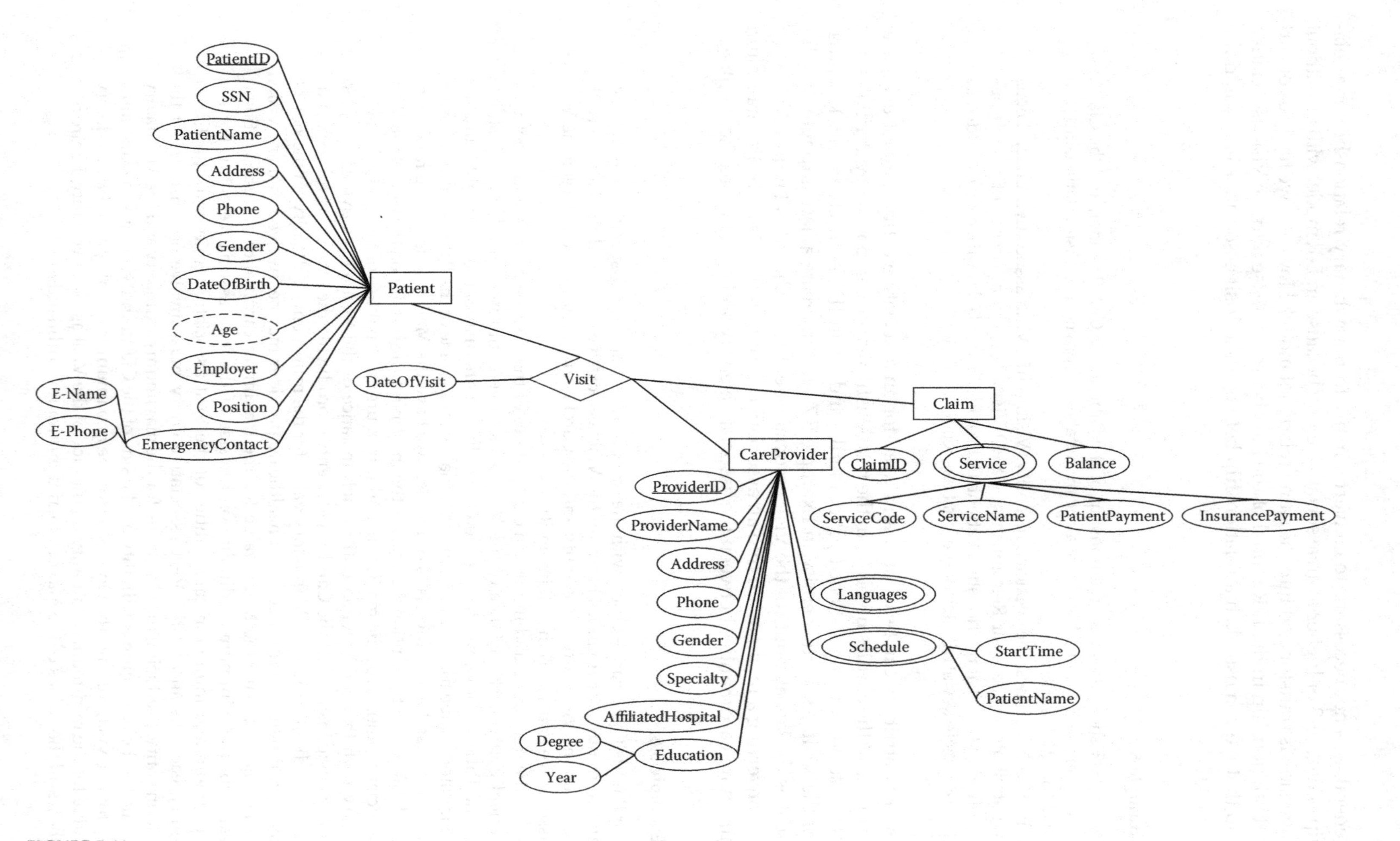

FIGURE 5.11
E-R model containing the Visit relationship of the Patient, the CareProvider, and the Claim entities in the healthcare database.

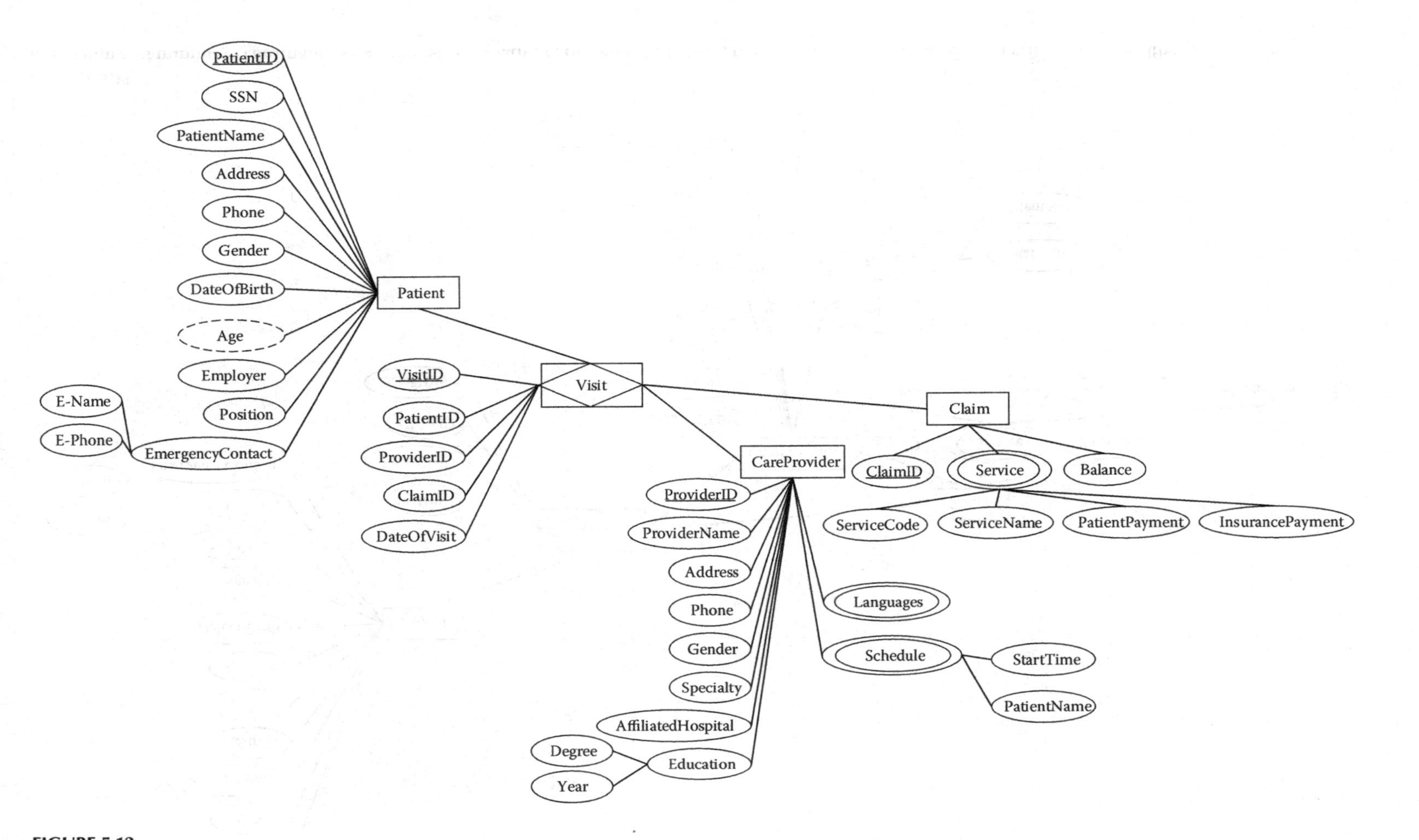

FIGURE 5.12
Conversion of the Visit relationship to an associative entity in the healthcare database.

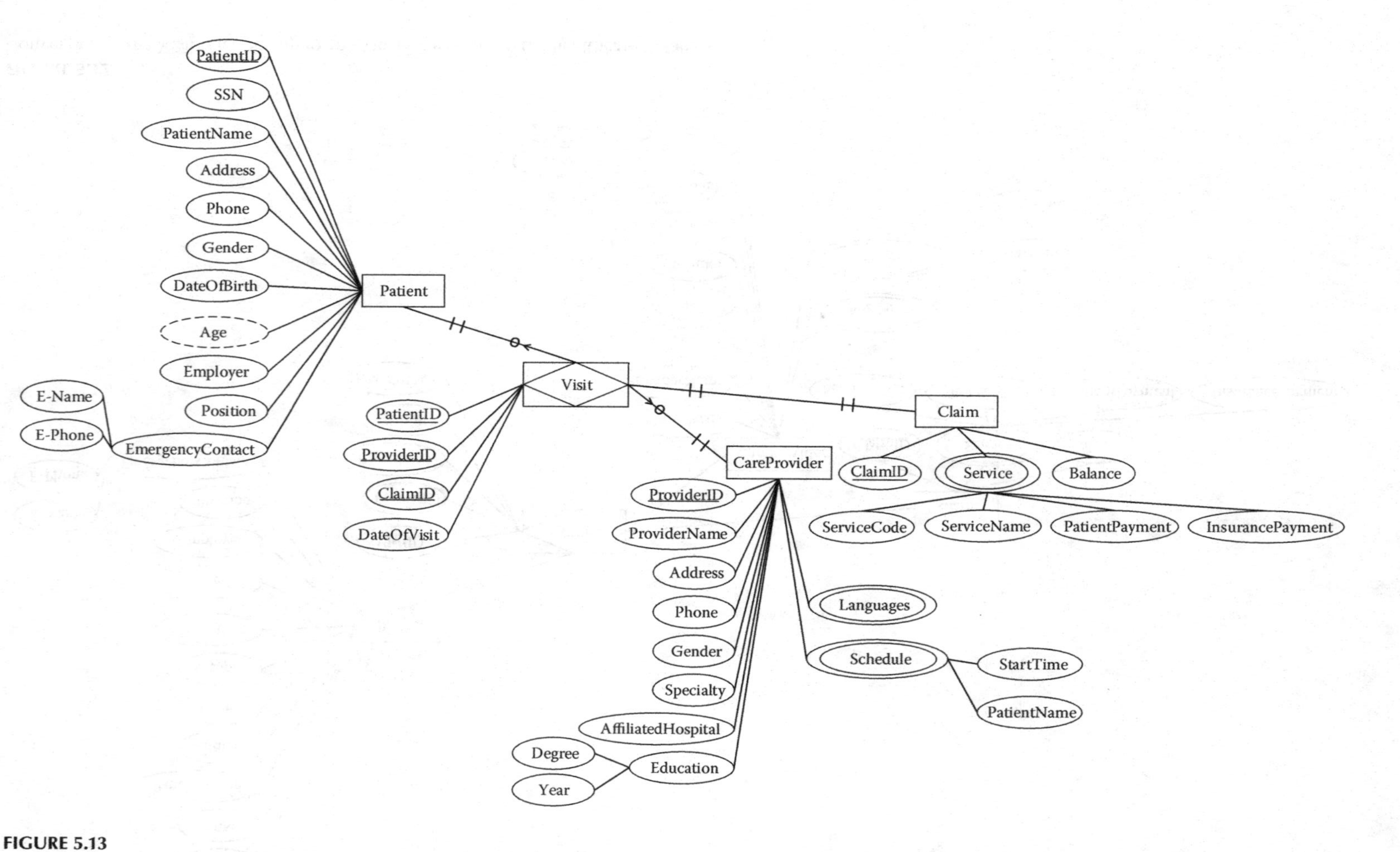

FIGURE 5.13
Maximum and minimum cardinalities for the Visit associative entity, and the Patient, the CareProvider, and the Claim entities in the healthcare database.

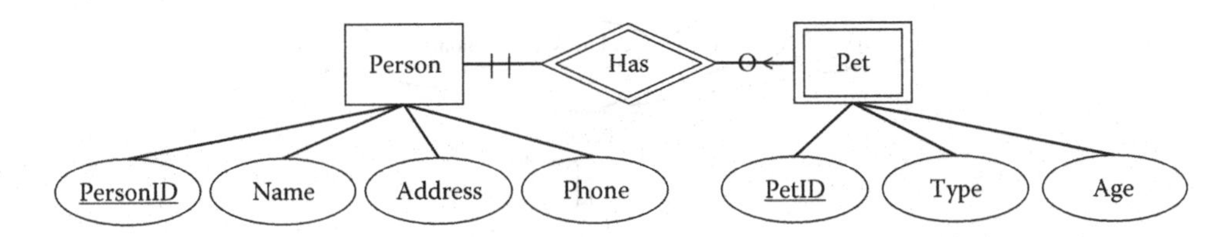

FIGURE 5.14
Example of a weak entity type.

5.5 Weak Entities

A weak entity type depends on a strong entity type to exist in a database. A strong entity type exists independently of other entity types. The examples of entity types introduced in Sections 5.1 to 5.4 are all strong entity types. A weak entity type is denoted by a box with double lines (see Figure 5.14). The diamond representing the relationship between a weak entity type and its associated strong entity type also has double lines. Suppose that a database for a neighborhood community has an entity type Person. If we want to add the pet(s) of a person to the database, a weak entity type, Pet, is added because no pet(s) of a person can exist in the database if the person does not exist in the database. Figure 5.14 gives the E-R model containing the Person entity, the Pet entity, and their relationship. Hence, the existence of an instance of a weak entity type depends on the existence of an instance of a strong entity type in a database.

5.6 Superclass and Subclass of Entities

In the E-R model of the healthcare database shown in Figure 5.3, the Patient and the CareProvider entities have a common set of attributes (ID, Name, Address, Phone, and Gender) and their own special attributes (SSN, DateOfBirth, Age, Employer, Position, and EmergencyContact for the Patient entity, and Specialty, AffiliatedHospital, Education, Languages, and Schedule for the CareProvider entity). As shown in Figure 5.15, a superclass, the Person entity, which has the attributes common to the Patient and the CareProvider entities, can be created and added to the E-R model of the healthcare database while making the Patient and the CareProvider entities as the subclass entities of the superclass entity Person. Figure 5.15 shows how the connection between a superclass entity and its subclass entities is represented in an E-R model.

Although only the attributes special to each subclass entity are placed as the attributes of the subclass entities, all the attributes of the superclass entity are inherited by each subclass entity of the superclass entity. This property between a superclass entity and each of its subclass entities is called "attribute inheritance." Because of attribute inheritance, the key attribute(s) of the superclass entity is also the key attribute(s) of each subclass entity. Hence, there is no key attribute(s) placed under each subclass entity in an E-R model. An E-R model using superclass and subclass entities is called an "enhanced E-R model." In this book, we refer to both an E-R model with superclass and subclass entities and an E-R model without superclass and subclass entities as an E-R model.

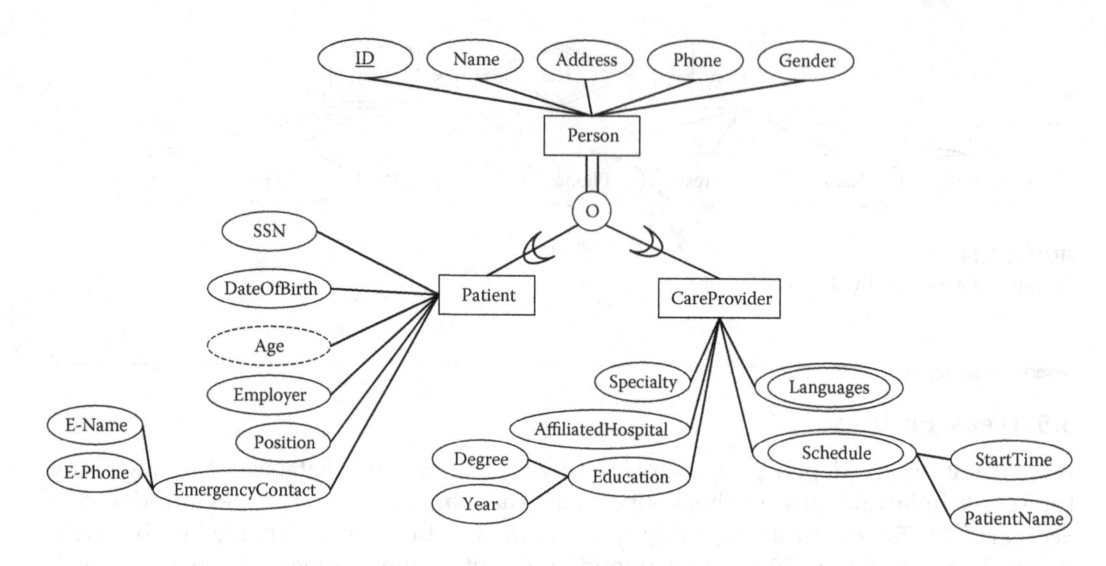

FIGURE 5.15
E-R model of the superclass entity, Person, and its subclass entities, Patient and CareProvider, with the total participation and overlap properties.

The participation property of a superclass entity with its subclass entities describes whether an instance of a superclass entity must participate as an instance of a subclass entity. There are two types of participation: total participation and partial participation. In the total participation, each instance of a superclass entity must be an instance of at least one subclass entity. The total participation is denoted by double lines between the superclass entity and the circle connecting the superclass entity to its subclass entities, as shown in Figure 5.15. The total participation in Figure 5.15 indicates that each instance of the Person entity must be an instance of the Patient entity, the CareProvider entity, or both the Patient and the CareProvider entities. In the partial participation, an instance of a superclass entity does not have to be an instance of any subclass entity. The partial participation is denoted by changing the double lines for the total participation into a single line, as shown in Figure 5.16. Figure 5.16 shows an example of the partial participation between the superclass entity Customer and the subclass entity Member in a database for an online store. The Customer entity has only one subclass entity, Member, which has two attributes for the login name and the password used to log in to an online account. Because a customer of the store may not sign up as a member, an instance of the Customer entity may not be an instance of the Member entity.

If a superclass entity has at least two subclass entities, the disjoint property of subclass entities for a superclass entity defines whether an instance of the superclass entity can be an instance of only one subclass entity or multiple subclass entities. There are two types of disjoint: disjoint and overlap. In the disjoint type (denoted by "D," as shown in Figure 5.17, with the assumption that a patient cannot be a healthcare provider and a healthcare provider cannot be a patient), an instance of the superclass entity can be an instance of only one subclass entity. In the overlap type (denoted by "O," as shown in Figure 5.15, with the assumption that a healthcare provider can be a patient), an instance of the superclass entity can be an instance of multiple subclass entities.

For the overlap property in Figure 5.15, if there are 100 patients (10 of 100 patients are also healthcare providers) and 20 healthcare providers, the Person superclass entity has 110

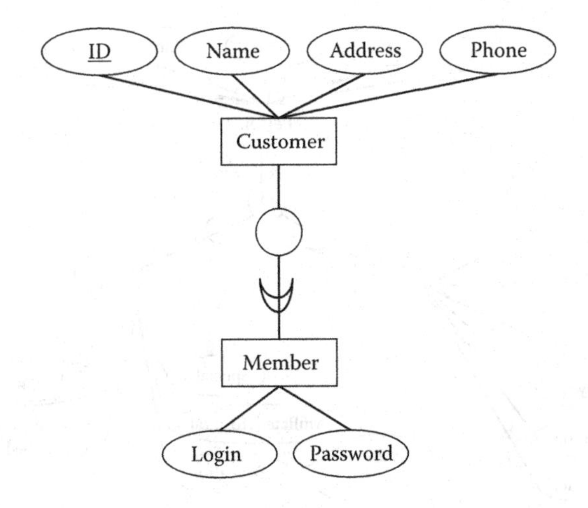

FIGURE 5.16
E-R model of the superclass entity, Customer, and its subclass entity, Member, with the partial participation.

instances, including 90 instances who are patients only, 10 instances who are healthcare providers only, and 10 instances who are both patients and healthcare providers. For each instance of the Person entity who is both a patient and a healthcare provider, we have an instance of the Patient entity and an instance of the CareProvider entity for that person.

However, if we assume that a patient cannot be a healthcare provider and a healthcare provider cannot be a patient in the healthcare database, it is now the disjoint property for the Person entity and its subclass entities, as shown in Figure 5.17. Suppose that there are 100 instances of the Patient entity and 20 instances of the CareProvider entity, the Person superclass entity has 120 instances for 100 patients and 20 healthcare providers. For each of the 120 instances for the Person entity, we can find an instance of either the Patient or the CareProvider entity for the same person. If an instance of the Person entity is a patient,

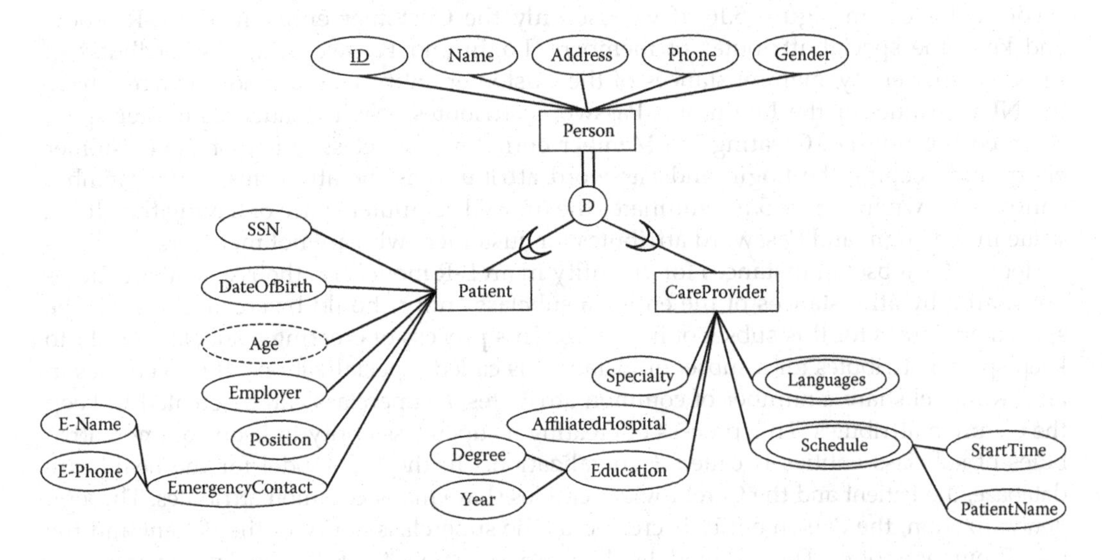

FIGURE 5.17
E-R model of the superclass entity, Person, and its subclass entities, Patient and CareProvider, with the total participation and disjoint properties.

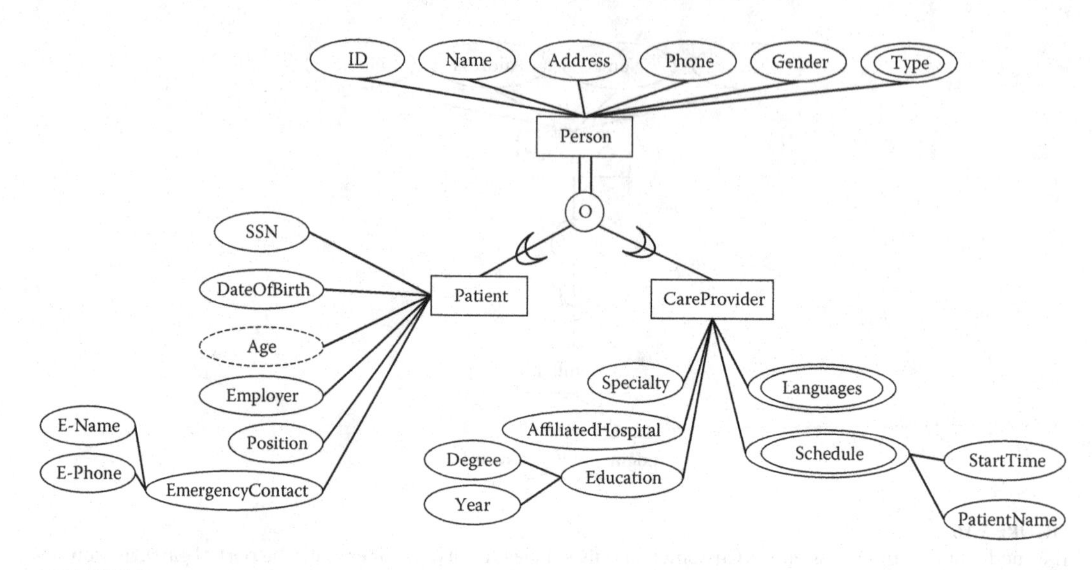

FIGURE 5.18
E-R model of the superclass entity, Person, and its subclass entities, Patient and CareProvider, with the subclass discriminator.

we have an instance of the Patient entity for that patient. If an instance of the Person entity is a healthcare provider, we have an instance of the CareProvider entity for that person.

We may add a new attribute of a superclass entity to indicate which subclass an instance of the superclass entity falls into. This new attribute plays the role of the subclass discriminator. If we have the disjoint property for the superclass entity and its subclass entities, the new attribute is a single-valued attribute. If we have the overlap property for the superclass entity and its subclass entities, the new attribute is a multi-valued attribute, as shown by the Type attribute in Figure 5.18.

For customers in Figure 5.16, if we use only the Customer entity in the E-R model and keep the special attributes of members, Login and Password, as the attributes of the Customer entity, many instances of the Customer entity who are not members have the NULL value for the Login and Password attributes, leaving much computer space reserved but unused. Creating the Member entity as a subclass entity of the Customer entity and keeping the Login and Password attributes as the attributes of the Member entity, as shown in Figure 5.16, eliminates the unused computer space of having the NULL value in the Login and Password attributes for customers who are not members.

Hence, if a subset of instances for an entity in an E-R model has their special attributes not shared by all instances of the entity, a subclass entity should be created to keep the special attributes for this subset of instances. This process of creating a subclass entity to keep special attributes for a subset of instances is called "specialization." If two entities in an E-R model share a number of common attributes, a superclass can be created to keep the common attributes. This process of creating a superclass entity to keep common attributes of subclass entities is called "generalization." In the E-R model for the healthcare database, the Patient and the CareProvider entities share many common attributes. Through generalization, the Person entity is created as the superclass entity of the Patient and the CareProvider entities. The E-R model in Figure 5.3 becomes the E-R model with superclass and subclass entities in Figure 5.19.

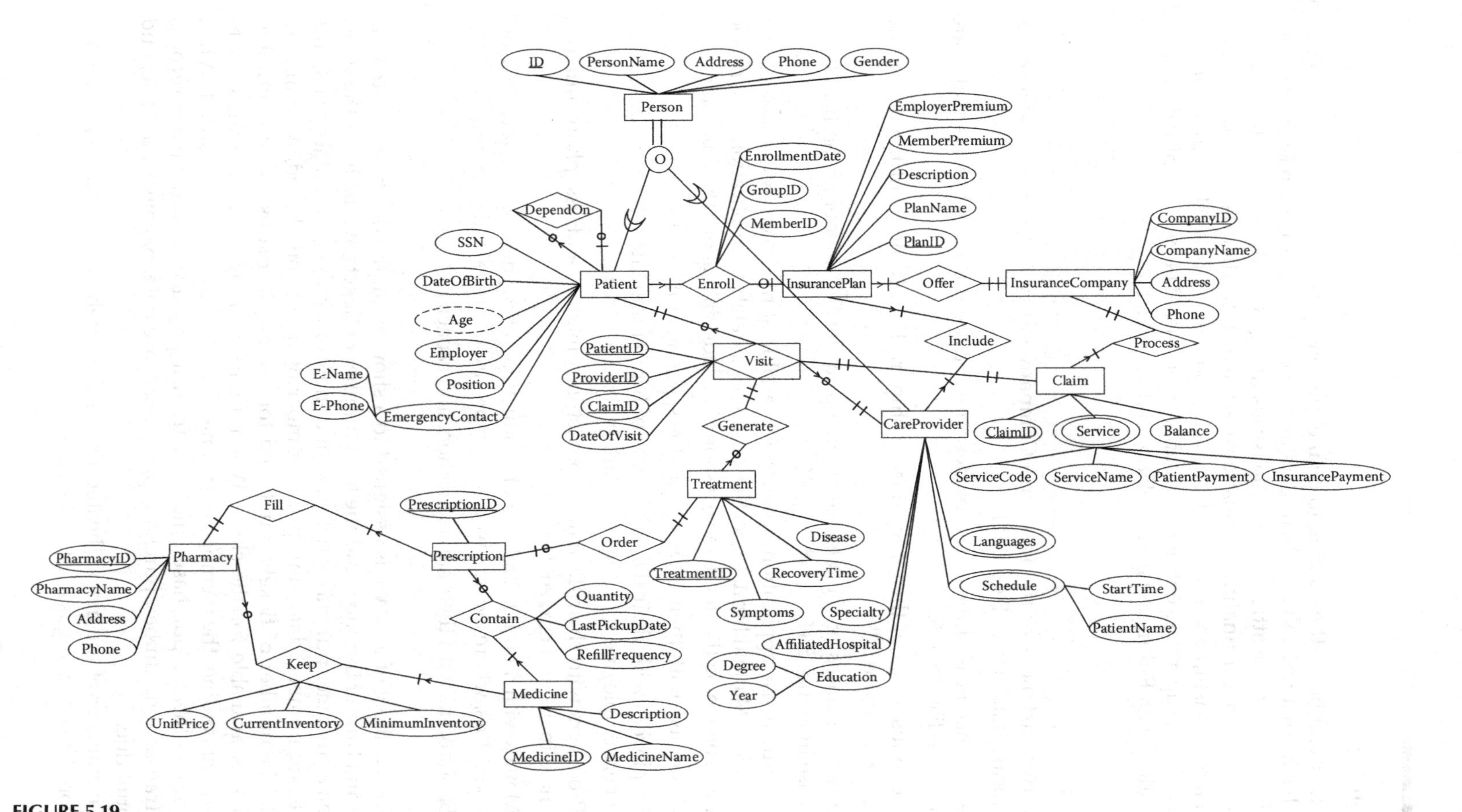

FIGURE 5.19
E-R model of the healthcare database with superclass and subclass entities.

5.7 Summary

This chapter describes E-R modeling to define the conceptual organization of data using concepts that are natural and easy to understand. The data organization is specified using entities along with their attributes, relationships along with their attributes, and maximum and minimum cardinalities of relationships, which further define how instances of entities in a relationship are associated. An E-R model may also include associative entities, weak entities, superclass entities, and subclass entities. The next chapter describes the transformation of an E-R model into a relational data model.

EXERCISES

1. Give two instances of the InsuranceCompany entity in the E-R model of the healthcare database in Figure 5.19.
2. Give two instances of the CareProvider entity in the E-R model of the healthcare database in Figure 5.19.
3. Give two instances of the Order relationship in the E-R model of the healthcare database in Figure 5.19.
4. Give two instances of the Visit relationship in the E-R model of the healthcare database in Figure 5.11.
5. A list of database application projects is available at http://www.dssbooks.com/web/ProjectsManual.html. Project 48 describes a database application for a movie theater. The movie theater database should include data about the following:
   - Customers with their ID, name, address, phone number, and e-mail
   - Employees with their ID, name, date of hire, employment history, and salary, where employment history is a multi-valued attribute
   - Producers with their ID, name, contact information (address, phone number, and e-mail), and current balance
   - Movies with their ID, title, year of production, rating, description, awards won, and actors, where awards won and actors are multi-valued attributes
   - Showrooms with their name, location, and capacity

   When a show of a movie is assigned to a showroom, the database records the date and time of the show and the total number of tickets available for the show. When a customer purchases ticket(s) for a show of a movie, the database records the ticket number, date, unit price, number of tickets purchased, and amount paid. The number of tickets purchased for a show decreases the total number of tickets available for the show. When a customer signs up as a member to buy tickets online, the member's login name and password are recorded. When the movie theater purchases movie(s) from a producer, the database records the transaction number, purchase price, purchase date, payment due date, and amount due.

   The manager of the movie theater queries the database to obtain certain information. The following is a list of data queries:

1. Financial information
   a. Create a query that presents monthly revenues from ticket sales, monthly expenses from salaries, monthly expenses from movie purchases, and monthly earnings during the current year.
   b. Create a query that presents yearly revenues, salary expenses, and movie expenses during the current year.
2. Create a query that lists the five best movies of the current year based on the number of awards won.
3. Create a query that lists the five most expensive movies of the current year.
4. Create a query that displays the total number of tickets sold per movie in decreasing order of tickets sold.
5. Create a query that presents the average room usage per showroom in the current year.
6. Create a query that lists the 100 most preferred customers based on the amounts that the customers spent in the movie theater.
7. Create a query to present information about the producer that the movie theater did the most business during the current year.
8. Create a query that prompts for a date and a movie title and then presents the number of tickets available per show for the movie on that date.
9. Create a query that prompts for a movie title and then presents the weekly show schedule of the movie and the total number of tickets available per show.
10. Create a query that prompts for a customer name and presents information about the tickets purchased by the customer during the current month.

Given the above requirements for the movie theater database, use an E-R diagram to describe the Customer, the Employee, the Producer, the Movie, and the Showroom entities, along with their attributes. Give two instances of each entity.

6. Given the requirements of the movie theater database in Exercise 5, use an E-R diagram to describe all entities and relationships in this database without the maximum and minimum cardinalities of each relationship.
7. Add the maximum and minimum cardinalities to each relationship in the E-R model produced in Exercise 6.
8. Create superclass and subclass entities, if needed, in the E-R model produced in Exercise 7.

6

Logical Database Design: Relational Modeling and Normalization

The relational model was created by Codd (1971, 1990) for database management and has been used predominantly to design databases. In this chapter, we first define a "relation" as the data structure to store data in the relational database model and introduce several data integrity constraints. We then describe the methodology of transforming an entity-relationship (E-R) model into a relational model of database. We also describe normalization, which is used to identify and remove data anomalies in a relational model.

6.1 Relational Model of Database and Data Integrity Constraints

In a relational database model, relations are used to store data in a database. A relation is a data structure (called a "data schema") that consists of columns and rows. A column specifies a data attribute, and a row specifies a data record. Figure 6.1 shows the InsurancePlan relation to represent the InsurancePlan entity in the E-R model of the healthcare database in Figure 5.19. In Figure 6.1, the InsurancePlan relation has five columns to describe the five attributes of the InsurancePlan entity and has two rows or two data records to store two instances of the InsurancePlan entity. The relational schema for InsurancePlan can also be defined in the following text form:

InsurancePlan (<u>PlanID</u>, PlanName, Description, MemberPremium, and InsurancePremium)

A relation must have the primary key consisting of one or more attributes as the key attribute(s). The primary key should uniquely identify each record in the relation. That is, each record in the relation has a unique value of the primary key. The primary key is underlined in a relation. The InsurancePlan relation has PlanID as the primary key.

Example 6.1

Give the InsuranceCompany relation that represents the InsuranceCompany entity and its attributes in the E-R model of the healthcare database in Figure 6.2a (see also Figure 5.19).

Figure 6.2b gives the InsuranceCompany relation with CompanyID as the primary key and two records for two instances of the InsuranceCompany entity.

When an entity in an E-R model is represented by a relation in a relational model, the key attribute(s) of the entity is typically taken as the primary key of the relation. The following procedure can be followed to select the primary key for a relation or determine whether the selected primary key is appropriate:

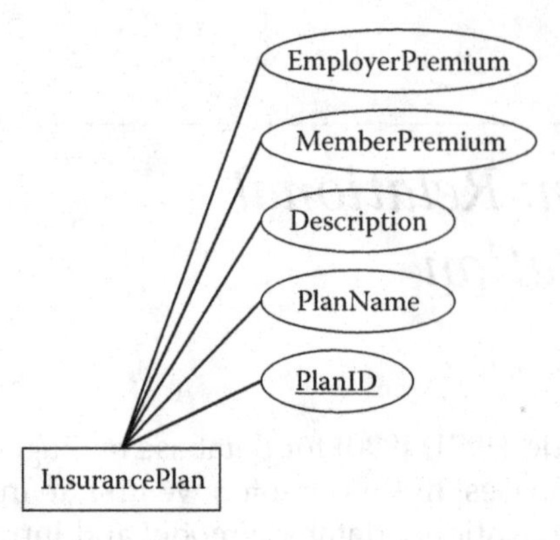

InsurancePlan

PlanID	PlanName	Description	MemberPremium	EmployerPremium
PL1	Care HMO	An HMO plan	$100	$500
PL	Care PPO	A PPO plan	$300	$600

FIGURE 6.1
Representation of the InsurancePlan entity by the InsurancePlan relation in the relational model of the healthcare database.

(a)

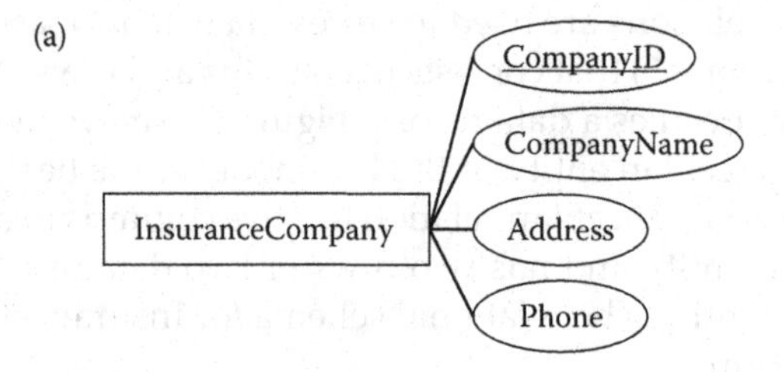

(b) InsuranceCompany

CompanyID	CompanyName	Address	Phone
C1	ABC	111 Mill Avenue	480-965-1111
C2	XYZ	222 Mill Avenue	480-965-2222

FIGURE 6.2
Representation of the InsuranceCompany entity by the InsuranceCompany relation in the relational model of the healthcare database. (a) The InsuranceCompany entity in the E-R model of the healthcare database. (b) The InsuranceCompany relation representing the InsuranceCompany entity.

1. Identify all super keys, where a super key is a set of one or more attributes that uniquely identify each record in the relation.
2. Determine the candidate key(s) from all super keys, where a candidate key is a super key with the smallest number of attribute(s).
3. Select the primary key from the candidate key(s) to meet the following criteria:
   a. The primary key should be unique at all times.
   b. The primary key should have a value that never changes for each record.
   c. The primary key should never have the NULL value.

Example 6.2

Determine the super key(s), the candidate key(s), and the primary key for the InsuranceCompany relation in Figure 6.2.

There are 13 super keys, each of which can uniquely identify a record of the InsuranceCompany relation:

1. CompanyID
2. Address
3. Phone
4. CompanyID, Address
5. CompanyID, Phone
6. CompanyID, CompanyName
7. Address, Phone
8. Address, CompanyName
9. Phone, CompanyName
10. CompanyID, Address, Phone
11. CompanyID, Address, CompanyName
12. CompanyID, Phone, CompanyName
13. CompanyID, CompanyName, Address, Phone

Because there may exist two companies that have the same name, the CompanyName attribute cannot uniquely identify a record of the InsuranceCompany relation. Among the 13 super keys, there are 3 candidate keys that have only one attribute:

1. CompanyID
2. Address
3. Phone

Considering that a company's address and phone number may change, CompanyID is selected as the primary key of the InsuranceCompany relation. CompanyID is created to be unique for each company at all times, does not change, and does not have the NULL value.

A relation has the following properties:

- The relation is uniquely identified by its name. Hence, different relations must use different names.
- Each cell, which is defined by a given column and a given row, contains exactly one atomic value.
- Each record is unique. Hence, we cannot have two identical records.
- Different attributes of the relation have different attribute names. However, an attribute of one relation and an attribute of another relation may use the same attribute name, for example, ID.
- Values of an attribute are from the same domain, for example, integer.
- The order of attributes is irrelevant.
- The order of records is irrelevant.

An E-R model has entities and relationships of entities. As shown in Figure 6.1, an entity in an E-R model can be represented by a relation in a relational model. Because a relational model consists only of relations, a relationship of entities in an E-R model must also be represented using the data structure of a relation. A relational model uses the pairing of the primary key and the foreign key in a relation to represent a relationship of entities in an E-R model. As shown in Figure 6.3, the Offer relationship of the InsurancePlan and the InsuranceCompany entities is represented in a relational model by adding CompanyID (the primary key of the InsuranceCompany relation) to the InsurancePlan relation as the foreign key that is paired with PlanID (the primary key of the InsuranceCompany relation) for representing the Offer relationship of the InsurancePlan and the InsuranceCompany entities. With the pairing of PlanID and CompanyID in the InsurancePlan relation, we can see that the company, C1, offers two insurance plans, PL1 and PL2. CompanyID in the InsurancePlan relation is called the "foreign key" because it comes from the primary key of another relation, the InsuranceCompany relation. In a relational model, a line is drawn, starting from the foreign key and pointing back to its primary key, as shown in Figure 6.3, to indicate where the foreign key comes from.

Data in a relational database model are subject to three types of data integrity constraints: domain constraint, entity constraint, and referential constraint. The domain constraint requires that all the values of an attribute must be from the same domain. For example, in the InsurancePlan relation, all the values of MemberPremium must be currency values. The entity constraint requires that every relation in a relational database model must have the primary key and that a value of the primary key cannot be NULL. For example, if a Customer relation has Email as an attribute, we cannot use Email as the primary key of the Customer relation because some customers may not have an e-mail address, and we cannot let the primary key take the NULL value to represent an unknown or unavailable e-mail address. The referential constraint requires that a foreign key must take either one of the values that its primary key has or the NULL value. Figure 6.4 shows the violation of the referential constraint because CompanyID as a foreign key in the InsurancePlan relation has a value of C3, whereas CompanyID as the primary key of the InsuranceCompany relation does not have the value of C3.

InsurancePlan

PlanID	PlanName	Description	MemberPremium	EmployerPremium	CompanyID
PL1	Care HMO	An HMO plan	$100	$800	C1
PL2	Care PPO	A PPO plan	$300	$600	C1

InsuranceCompany

CompanyID	CompanyName	Address	Phone
C1	ABC	111 Mill Avenue	480-965-1111
C2	XYZ	222 Mill Avenue	480-965-1111

FIGURE 6.3
Pairing of a primary key and a foreign key and the referential integrity constraint to represent the Offer relationship of the InsuranceCompany and the InsurancePlan entities.

InsurancePlan

PlanID	PlanName	Description	MemberPremium	EmployerPremium	CompanyID
PL1	Care HMO	A HMO plan	$100	$800	C1
PL2	Care PPO	A PPO plan	$300	$600	C1
PL3	HealthPlan	A PPO plan	$350	$650	C3

InsuranceCompany

CompanyID	CompanyName	Address	Phone
C1	ABC	111 Mill Ave	480-965-1111
C2	XYZ	222 Mill Ave	480-965-2222

FIGURE 6.4
Illustration of a violation of the referential integrity constraint.

6.2 Transformation of an E-R Model to a Relational Model

This section describes the transformation of entities and their attributes, the transformation of superclass and subclass entities and their attributes, the transformation of relationships and their attributes, the transformation of associative entities and their attributes, and the transformation of weak entities and their attributes.

6.2.1 Transformation of Entities and Their Attributes

A regular entity in an E-R model is represented by a relation in a relational model. The name of the entity can be used as the name of the relation. If the key attribute(s) of the entity meets the requirements of the primary key, as described in Section 6.1, the key attribute(s) of the entity becomes the primary key of the relation. Each simple, stored, and single-valued attribute of the entity becomes an attribute of the relation with the same attribute name. Figure 6.2 shows the transformation of the InsuranceCompany entity to the InsuranceCompany relation.

A derived attribute of an entity will not be included as an attribute of the relation representing the entity because the value of a derived attribute can be obtained from values of stored attributes through data queries, as described in Chapter 7. If an entity has a composite attribute, the relation representing the entity will include only the simple, stored, and single-valued attributes that make up the composite attribute.

Example 6.3

Transform the Patient entity in the E-R model of the healthcare database in Figure 6.5a (see also Figure 5.19) to a relation.

Figure 6.5b shows the Patient relation representing the Patient entity. Because the Patient entity is a subclass of the Person entity and inherits the ID attribute from the Person entity as the key attribute, the ID attribute is included in the Patient relation

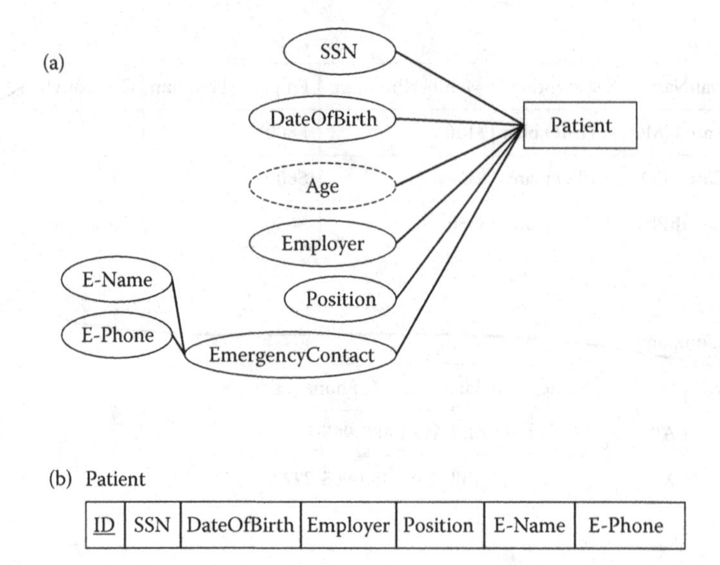

FIGURE 6.5
Representation of a derived attribute and a composite attribute in a relational model. (a) The Patient entity in the E-R model of the healthcare database. (b) The Patient relation representing the Patient entity.

as the primary key. The Age attribute is a derived attribute that is not included in the Patient relation. The EmergencyContact attribute is a composite attribute that is not included in the Patient relation. Instead, the E-Name attribute and the E-Phone attribute, which make up the EmergencyContact attribute, are included in the Patient relation.

To represent a multi-valued attribute of an entity in an E-R model, we create a new relation that includes the multi-valued attribute and the key attribute(s) of the entity as the composite key attributes in the primary key of the new relation.

Example 6.4

Transform the Claim entity in the E-R model of the healthcare database in Figure 6.6a (see also Figure 5.19) to a relational model.

Figure 6.6b shows the Claim and the ClaimService relations created for the multi-valued attribute of Service. The claim relation represents the Claim entity. Because the Service attribute is a multi-valued attribute, we create a new relation, the ClaimService relation, which includes ClaimID (the primary key of the Claim relation) and the ServiceCode, ServiceName, PatientPayment, and InsurancePayment attributes that make up the composite attribute of Service.

Because the ServiceCode attribute can uniquely identify each service, only ServiceCode (instead of all the attributes making up the composite Service attribute) is used together with ClaimID as the composite primary key for the ClaimService relation.

Figure 6.6c shows three records of the Claim relation to represent a claim containing three services if a new relation is not created for the multi-valued Service attribute, but instead, the Service attribute is kept in the Claim relation. The Balance attribute of this claim, which has the value of $600, repeats three times in the three records for the three services in this claim. If the Claim entity has other simple, single-valued attributes, these attributes also have to repeat the same value in the three records for the three services of the claim. Having the repetitive storage of the same data (the Balance of a

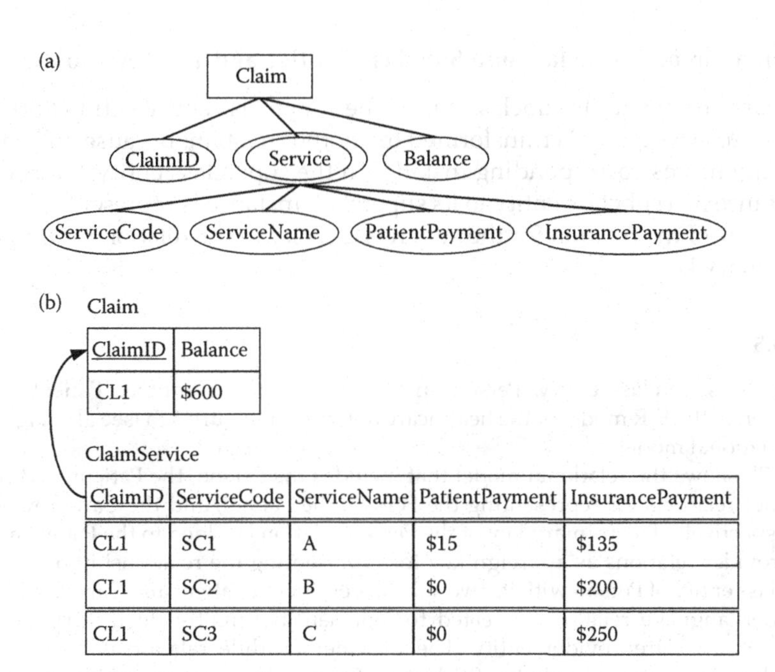

Claim

ClaimID	Balance
CL1	$600

ClaimService

ClaimID	ServiceCode	ServiceName	PatientPayment	InsurancePayment
CL1	SC1	A	$15	$135
CL1	SC2	B	$0	$200
CL1	SC3	C	$0	$250

(c) ClaimService

ClaimID	Balance	ServiceCode	ServiceName	PatientPayment	InsurancePayment
CL1	$600	SC1	A	$15	$135
CL1	$600	SC2	B	$0	$200
CL1	$600	SC3	C	$0	$250

FIGURE 6.6
Representation of a multi-valued attribute in a relational model. (a) The Claim entity in the E-R model of the healthcare database. (b) The Claim relation and the Claim service relation representing the Claim entity. (c) The same value of the Balance attribute is repeated three times in three records for representing multiple values of the multi-valued Service attribute if a new relation is not created for the multi-valued Service attribute.

claim) creates data redundancy in the database, which is undesirable because of the following:

- Redundant data waste precious storage space on the computer.
- It takes more computing time to update data (e.g., updating of a claim's balance in multiple records).
- There is a possibility of data inconsistency (e.g., the balance of a claim updated in one record but not in other records containing the balance of the claim) and other types of data anomalies described in Section 6.3.1.

Note that, because the value of ClaimID is the same for the three records in the ClaimService relation, ServiceCode must join ClaimID as the composite primary key of the ClaimService relation in order to uniquely identify each record of the ClaimService relation.

Hence, without creating a new relation for a multi-valued attribute, all the single-valued attributes of an entity have to repeat the same value in multiple records created for storing multiple values of the multi-valued attribute. Figure 6.6b shows the representation of the same claim and its three services when a new relation is created for the multi-valued attribute. The representation in Figure 6.6b has no data redundancy.

6.2.2 Transformation of Superclass and Subclass Entities and Their Attributes

Given a superclass entity and its subclass entity, the superclass entity is transformed to a relation, and the subclass entity is transformed to another relation. Because an instance of the subclass entity has its corresponding instance in the superclass entity, the correspondence of an instance in a subclass entity to its superclass instance is represented by including the primary key of the relation for the superclass entity in the relation for the subclass entity as the primary key.

Example 6.5

Transform the superclass entity, Person, and its two subclass entities, Patient and CareProvider, in the E-R model of the healthcare database in Figure 6.7a (see also Figure 5.19) to a relational model.

Figure 6.7b shows the relational model that includes the Person, the Patient, and the CareProvider relations for representing the Person, the Patient, and the CareProvider entities, respectively. The primary key of the Person relation is added to the Patient and the CareProvider relations as a foreign key for representing the relationship between the superclass entity of Person with its two subclass entities of Patient and CareProvider. The ProviderLanguage relation is created to represent the multi-valued attribute of Language for the CareProvider entity. The ProviderSchedule relation is created to represent the multi-valued attribute of Schedule for the CareProvider entity. Because the Schedule attribute is made up of the StartTime attribute and the PatientName attribute and the StartTime attribute alone can uniquely identify each value of the Schedule attribute, the StartTime attribute is used together with the ID attribute in the ProviderSchedule relation as the composize primary key.

6.2.3 Transformation of Relationships and Their Attributes

In this section, we first look at transforming binary relationships and then extend the methods of transforming binary relationships to transforming unary relationships and relationships of degree-3 or higher.

As described in Section 6.1, a relationship is represented in a relational model through the pairing of a primary key and a foreign key in a relation. For a binary many-to-one or one-to-many relationship, we use the "one migrates to many" rule to add the primary key of the relation on the "one" side of the many-to-one or the one-to-many maximum cardinality to the relation on the "many" side as a foreign key so that the pairing of the primary key and the foreign key is included in the relation on the "many" side of the many-to-one or the one-to-many maximum cardinality. Any attribute of the relationship goes where the foreign key is. Example 6.6 further explains the "one migrates to many" rule.

Example 6.6

Transform the binary, many-to-one Offer relationship of the InsurancePlan and the InsuranceCompany entities in the E-R mode of the healthcare database (see Figures 5.19 and 6.8a) to a relational model.

At first, we transform the InsurancePlan and the InsuranceCompany entities to the InsurancePlan and the InsuranceCompany relations, respectively. To represent the Offer relationship, there are two relations to add a foreign key and, thus, create the pairing of PlanID and CompanyID: adding CompanyID as a foreign key to the CompanyPlan relation (see Figure 6.8b) or adding PlanID as a foreign key to the InsuranceCompany relation (see Figure 6.8c).

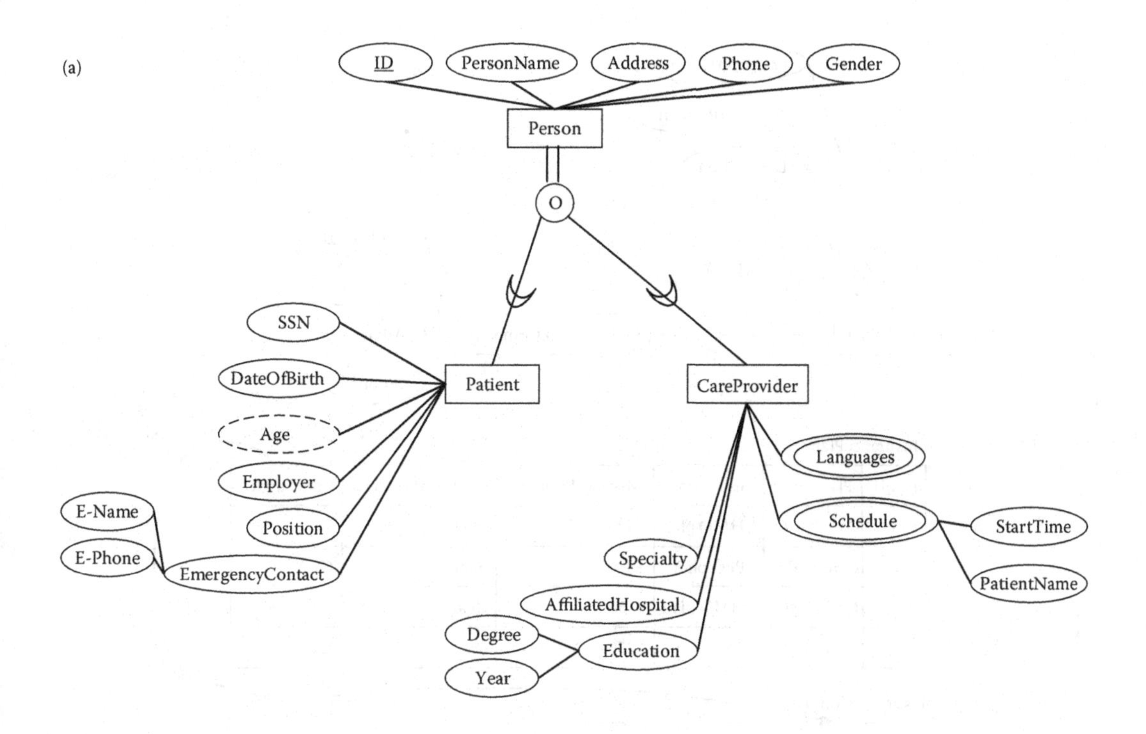

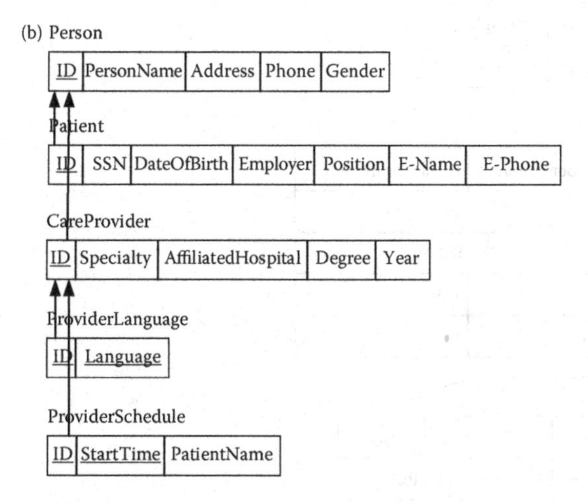

FIGURE 6.7
Representation of superclass and subclass entities in a relational model. (a) The Person, Patient, and CareProvider entities in the E-R model of the healthcare database. (b) The relational model representing the E-R model in (a).

The Offer relationship of the InsurancePlan and the InsuranceCompany entities has the many-to-one maximum cardinality. That is, an insurance company may offer many insurance plans, and one insurance plan is offered by only one insurance company. If we add PlanID as a foreign key to the InsuranceCompany relation, as shown in Figure 6.8c, this many-to-one maximum cardinality causes the record of an insurance company to repeat in the InsuranceCompany relation as many times as the number of the insurance plans that the insurance company offer, for example,

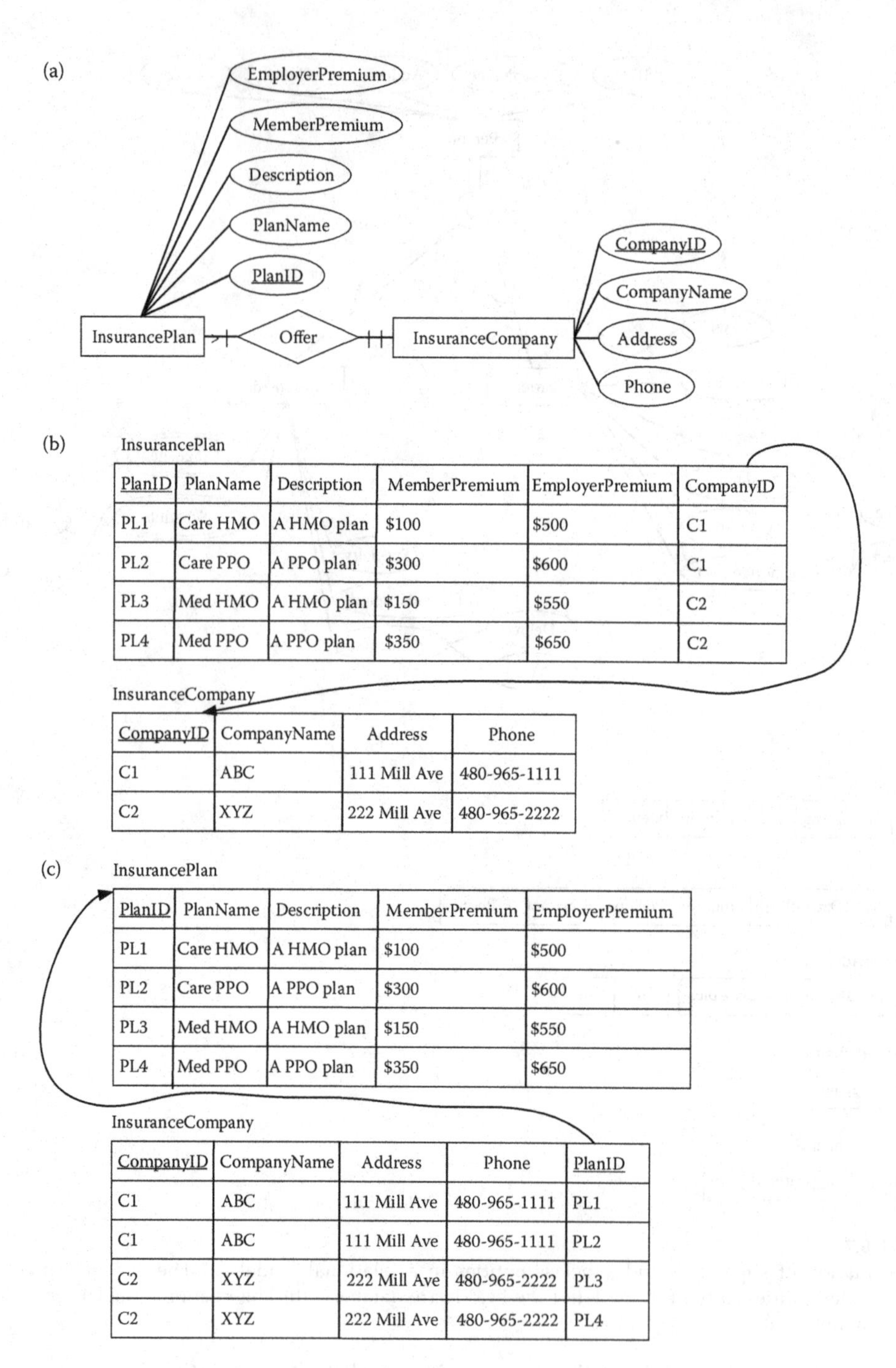

(b)

InsurancePlan

PlanID	PlanName	Description	MemberPremium	EmployerPremium	CompanyID
PL1	Care HMO	A HMO plan	$100	$500	C1
PL2	Care PPO	A PPO plan	$300	$600	C1
PL3	Med HMO	A HMO plan	$150	$550	C2
PL4	Med PPO	A PPO plan	$350	$650	C2

InsuranceCompany

CompanyID	CompanyName	Address	Phone
C1	ABC	111 Mill Ave	480-965-1111
C2	XYZ	222 Mill Ave	480-965-2222

(c)

InsurancePlan

PlanID	PlanName	Description	MemberPremium	EmployerPremium
PL1	Care HMO	A HMO plan	$100	$500
PL2	Care PPO	A PPO plan	$300	$600
PL3	Med HMO	A HMO plan	$150	$550
PL4	Med PPO	A PPO plan	$350	$650

InsuranceCompany

CompanyID	CompanyName	Address	Phone	PlanID
C1	ABC	111 Mill Ave	480-965-1111	PL1
C1	ABC	111 Mill Ave	480-965-1111	PL2
C2	XYZ	222 Mill Ave	480-965-2222	PL3
C2	XYZ	222 Mill Ave	480-965-2222	PL4

FIGURE 6.8
Representation of a one-to-many or many-to-one binary relationship in a relational model. (a) The Offer relationship of the InsurancePlan entity and the InsuranceCompany entity in the E-R model of the healthcare database. (b) Adding CompanyID as a foreign key to the InsurancePlan relation. (c) Adding PlanID as a foreign key to the InsuranceCompany relation.

two times of CI and two times of C2 in Figure 6.8c. The repetitive storage of the same data (CompanyID, CompanyName, Address, and Phone of an insurance company) in the database creates data redundancy, which is undesirable, as discussed in Section 6.2. In Figure 6.8c, because there are multiple records in the InsuranceCompany relation containing the same data of an insurance company, CompanyID alone cannot uniquely identify each of these multiple records for the same insurance company. PlanID needs to be included in the primary key of the InsuranceCompany relation so that CompanyID and PlanID together can uniquely identify each record of the InsuranceCompany relation.

Various problems of adding PlanID as a foreign key to the InsuranceCompany relation in Figure 6.8c do not exist if we add CompanyID as a foreign key to the InsurancePlan relation according to the "one migrates to many" rule (see Figure 6.8b) because an insurance plan is offered by only one insurance company and CompanyID of the insurance company can be simply added to the existing record of the insurance plan without adding any new records to the InsurancePlan relation.

Hence, to transform a many-to-one or one-to-many relationship, we add the primary key of the relation on the "one" side of the maximum cardinality (e.g., the InsuranceCompany relation in the Offer relationship) to the relation on the "many" side of the maximum cardinality (e.g., the InsurancePlan relation in the Offer relationship). That is, with the goal of eliminating data redundancy and associated problems, we use the "one migrates to many" rule to transform a many-to-one or one-to-many relationship and determine where to add a foreign key for creating the pairing of the primary key and the foreign key to represent the relationship. If a relationship has attribute(s), the attribute(s) of the relationship goes where the foreign key is and, thus, is included in the relation where the pairing of the primary key and the foreign key is because this is where the relationship is represented. The Offer relationship in Figure 6.8 does not have any attribute.

When transforming a many-to-many binary relationship, we first represent two entities involved in the binary relationship by two relations. To represent the many-to-many relationship, adding the primary key of one relation to another relation as a foreign key creates data redundancy and associated problems. Hence, a new relation is created to include the primary keys of both relations in the new relation as the foreign keys whose pairing represents the relationship. The two foreign keys in the new relation are used as the composite primary key for the new relation. Any attribute of the relationship goes where the foreign keys are.

Example 6.7

Transform the binary many-to-many Keep relationship of the Pharmacy and the Medicine entities in the E-R model of the healthcare database (see Figures 5.19 and 6.9a) to a relational model.

Figure 6.9b gives the relational model with the Pharmacy, the Medicine, and the Keep relations. The Keep relationship has the attributes of UnitPrice, CurrentInventory, and MinimumInventory. These three attributes go along with the foreign keys and are included in the Keep relation.

For a binary one-to-one relationship, adding a foreign key to either relation does not make any difference with regard to the maximum cardinality. If the minimum cardinality of the relationship is mandatory-to-mandatory or optional-to-optional, adding a foreign key to either relation does not make any difference with regard to the minimum cardinality. However, if the minimum cardinality of the relationship is

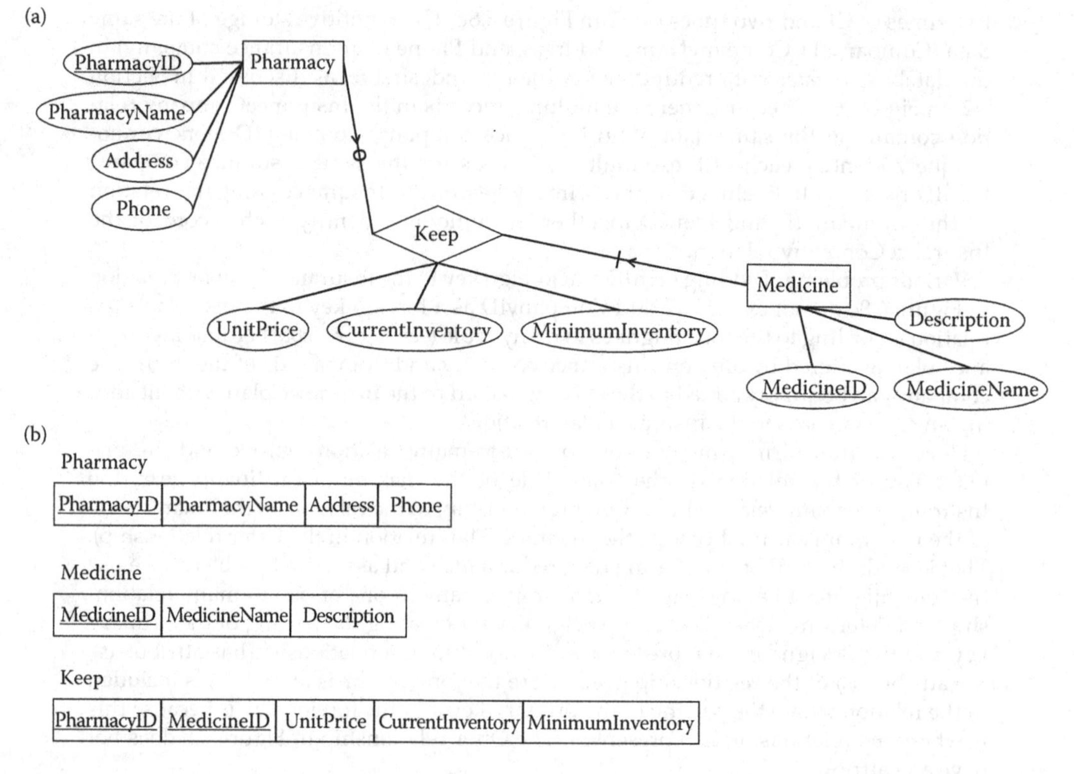

FIGURE 6.9
Representation of a many-to-many binary relationship in a relational model. (a) The Keep relationship of the InsurancePlan entity and the InsurancePlan entity in the E-R model of the healthcare database. (b) The transformation of the Pharmacy entity, the Medicine entity, and the Keep relationship to a relational model.

mandatory-to-optional or optional-to-mandatory, we use the "mandatory migrates to optional" rule to add the primary key of the relation on the "mandatory" side of the minimum cardinality to the relation on the "optional" side as the foreign key. Any attribute of the relationship goes into the relation containing the foreign key. This rule is further explained in Example 6.8.

Example 6.8

Transform the Order relationship of the Treatment and the Prescription entities in the E-R model of the healthcare database (see Figures 5.19 and 6.10a) to a relational model.

We first transform the Prescription and the Treatment entities to the Prescription and the Treatment relations, respectively. Figure 6.10c shows one way of representing the Order relationship in the relational model by adding PrescriptionID (the primary key of the Prescription relation) to the Treatment relation as the foreign key. This representation of the relationship causes the unused computer storage space for the PrescriptionID attribute in the records of TR3 and TR4 because a prescription is optional for a treatment, and the two treatments, TR3 and TR4, do not produce an order of a prescription. Computer space still needs to be reserved for the PrescriptionID attribute in the Treatment relation even for treatments that do not produce a prescription.

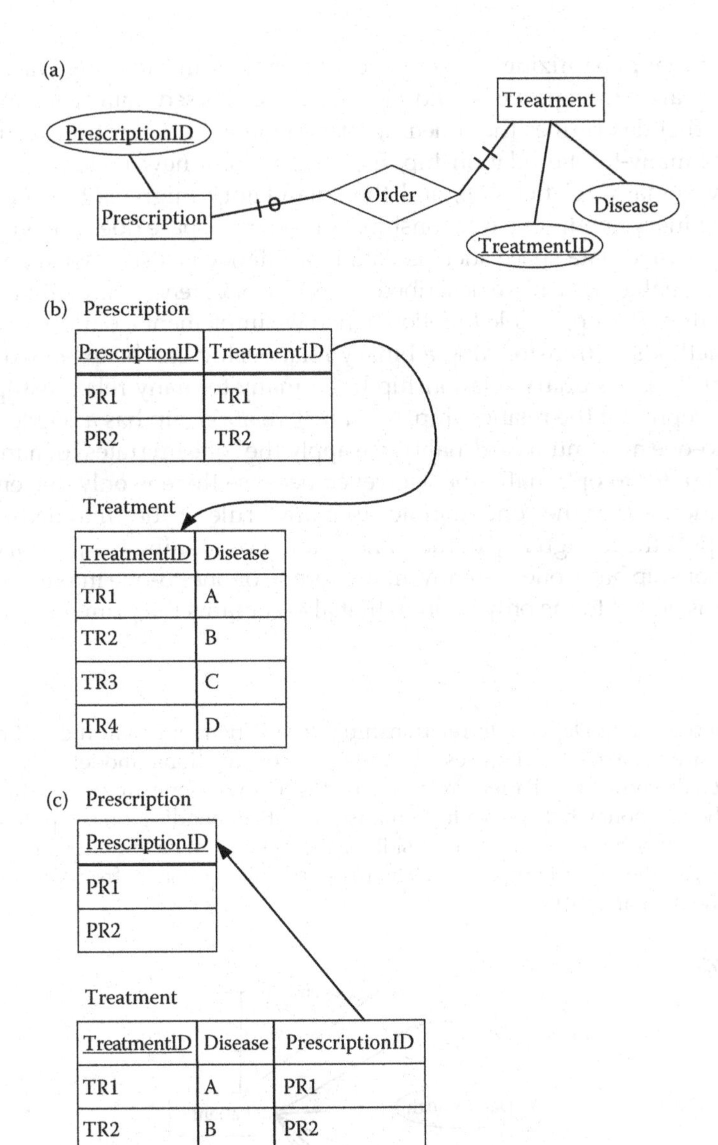

(b) Prescription

PrescriptionID	TreatmentID
PR1	TR1
PR2	TR2

Treatment

TreatmentID	Disease
TR1	A
TR2	B
TR3	C
TR4	D

(c) Prescription

PrescriptionID
PR1
PR2

Treatment

TreatmentID	Disease	PrescriptionID
TR1	A	PR1
TR2	B	PR2
TR3	C	
TR4	D	

FIGURE 6.10
Representation of a one-to-one, mandatory-to-optional/optional-to-mandatory relationship in a relational model. (a) The Order relationship of the Prescription entity and the Treatment entity in the E-R model of the healthcare database. (b) Adding a foreign key to the Prescription relation according to the "mandatory migrates to optional" rule. (c) Adding a foreign key to the Treatment relation.

Figure 6.10b shows another way of representing the Order relationship by following the "mandatory migrates to optional" rule and adding TreatmentID (the primary key of the Treatment relation on the "mandatory" side of the minimum cardinality) to the Prescription relation on the "optional" side of the minimum cardinality as the foreign key. This representation of the relationship in Figure 6.10b does not produce any unused computer space as seen in Figure 6.10c.

It can be seen that maximizing data storage efficiency or minimizing wasted data space (e.g., data redundancy in Figure 6.6c and Figure 6.8c or unused data space in Figure 6.10c) is the principle that determines the "one migrates to many" rule for transforming a binary one-to-many or many-to-one relationship, the creation of a new relation for representing a binary many-to-many relationship, and the "mandatory migrates to optional" rule for transforming a binary one-to-one relationship. Wasted data space does not only increase the database cost—wasted data space such as data redundancy is also associated with various kinds of data anomalies, which are described in Section 6.3. Hence, maximizing data storage efficiency is an important principle to follow when we implement a database on computers.

The above methods of transforming a binary relationship can be extended to transform a unary relationship. If a unary relationship has a many-to-many relationship, we create a new relation to represent the relationship. If a unary relationship has a one-to-many/many-to-one or one-to-one maximum cardinality, we apply the "one migrates to many" rule or the "mandatory migrates to optional" rule. However, because there is only one entity involved in a unary relation, either the "one migrates to many" rule or the "mandatory migrates to optional" rule puts the foreign key in the same relation containing the primary key. That is, if a unary relationship has a one-to-many/many-to-one or one-to-one maximum cardinality, the foreign key is added to the only relation that also contains the primary key.

Example 6.9

Transform the unary DependOn relationship of the Patient entity in the E-R model of the healthcare database (see Figures 5.19 and 6.11a) to a relational model.

Figure 6.11b shows the Patient relation. In the many-to-one unary relationship, a patient who is a policy holder can have many patients depending on the policy holder. Hence, the policy holder on the "one" side of the maximum cardinality migrates as a foreign key, as shown in Figure 6.11b. The primary key, ID, comes from the superclass entity of the patient entity.

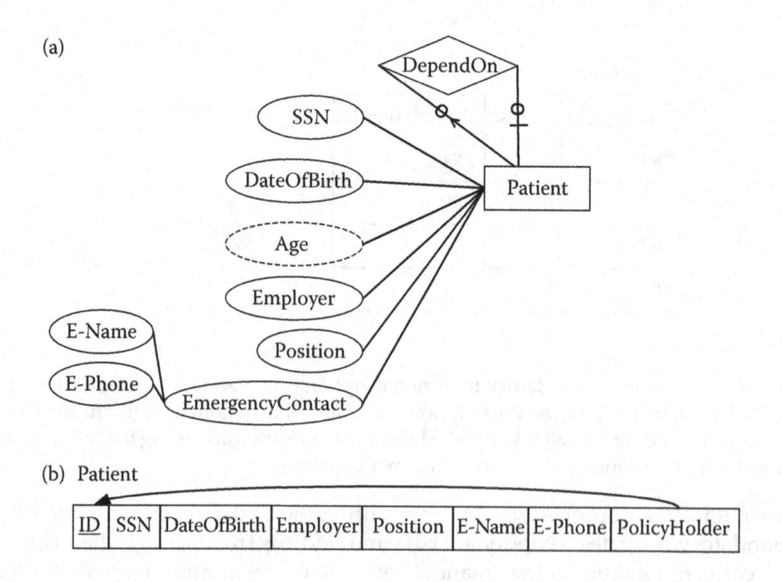

FIGURE 6.11
Representation of a one-to-many or many-to-one unary relationship in a relational model. (a) The Patient entity and the DependOn relationship in the E-R model of the healthcare database. (b) The transformation of the Patient entity and the DependOn relationship to a relational model.

To transform a relationship of degree-3 or higher, each entity involved in the relationship is represented by a relation. In addition, a new relation is created for representing the relationship. The new relation contains and uses the primary key of each relation representing each entity as the composite primary key of the new relation. Example 6.10 illustrates the transformation of a degree-3 relationship. A new relation is created to represent a relationship of degree-3 or higher for the same reason why we create a new relation to represent a binary many-to-many relationship for preventing data redundancy.

Example 6.10

Transform the degree-3 Visit relationship of the Patient, the CareProvider, and the Claim entities in the E-R model of the healthcare database (see Figure 5.11, which is copied in Figure 6.12a) to a relational model.

Figure 6.12b shows the relational model for the Visit relationship of the Patient, the CareProvider, and the Claim entities. The Visit relation is created to represent the Visit relationship. The Visit relation includes PatientID, ProviderID, and ClaimID as the foreign keys and use them together as the composite primary key.

6.2.4 Transformation of Association Entities

As described in Chapter 5, a binary many-to-many relationship or a relationship of degree-3 or higher can be converted to an associate entity. As a binary many-to-many relationship or a relationship of degree-3 or higher is represented by a new relation in a relational model, an associative entity is represented by a new relation in a relational model. Example 6.11 illustrates the transformation of an associative entity to a relational model.

Example 6.11

Transform the associative entity, Visit, in the E-R model of the healthcare database (see Figures 5.19 and 6.13a) to a relational model.

Figure 6.13b shows the Visit relation that represents the associative entity and contains three foreign keys, PatientID, ProviderID, and ClaimID, as the composite primary key. Hence, the Visit relationship in Example 6.10 and the Visit associative entity in Example 6.11, which is converted from the Visit relationship, have the same representation in the relational model.

6.2.5 Transformation of Weak Entities

Like a regular entity, a weak entity is represented by a relation in a relational model. Because a weak entity depends on a strong entity, the primary key of the strong entity needs to be included in the relation for the weak entity as a foreign key, which is included in the primary key of the relation for the weak entity, so that no instance of the weak entity can be stored in the database without association with an instance of the strong entity. Example 6.12 illustrates the transformation of a weak entity to the relational model.

Example 6.12

Transform the E-R model containing the weak entity, Pet, in Figure 5.14, which is copied in Figure 6.14a to a relational model.

The E-R model is transformed to a relational model shown in Figure 6.14b. In the Pet relation, PersonID (the primary key of the strong entity) is included as a foreign key and used together with PetID as the primary key. If PersonID is not included in the primary key of the Pet relation, a record of the Pet relation can exist, with PersonID having

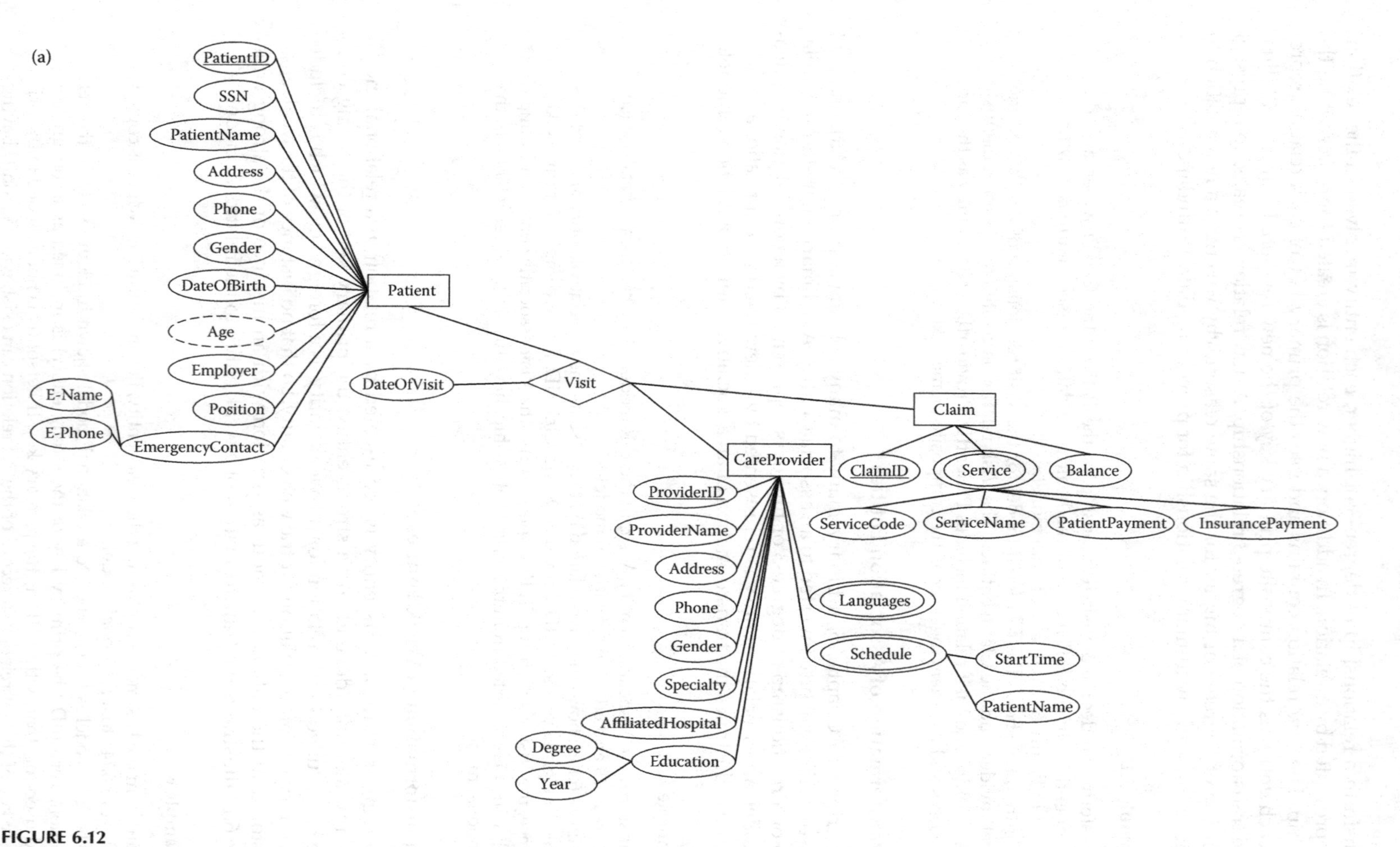

FIGURE 6.12
Representation of a degree-3 relationship in a relational model. (a) The Visit relationship of the Patient entity, the CareProvider entity, and the Claim entity in the E-R model of the healthcare database. (b) The transformation of the Visit relationship of the Patient entity, the CareProvider entity, and the Claim entity to a relational model.

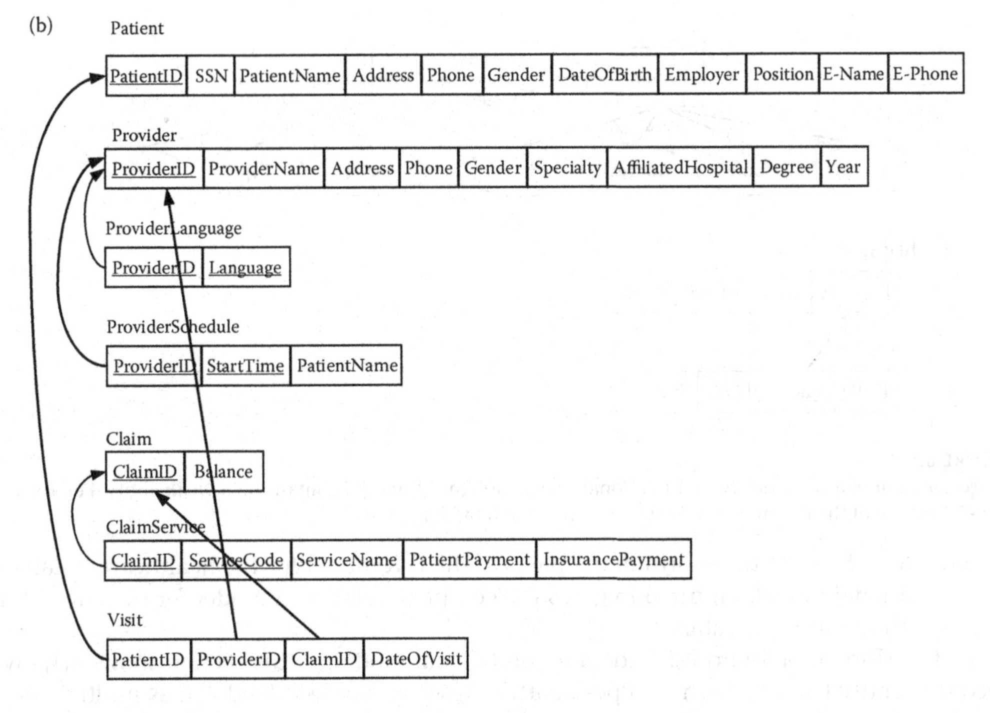

FIGURE 6.12 (Continued)
Representation of a degree-3 relationship in a relational model. (a) The Visit relationship of the Patient entity, the CareProvider entity, and the Claim entity in the E-R model of the healthcare database. (b) The transformation of the Visit relationship of the Patient entity, the CareProvider entity, and the Claim entity to a relational model.

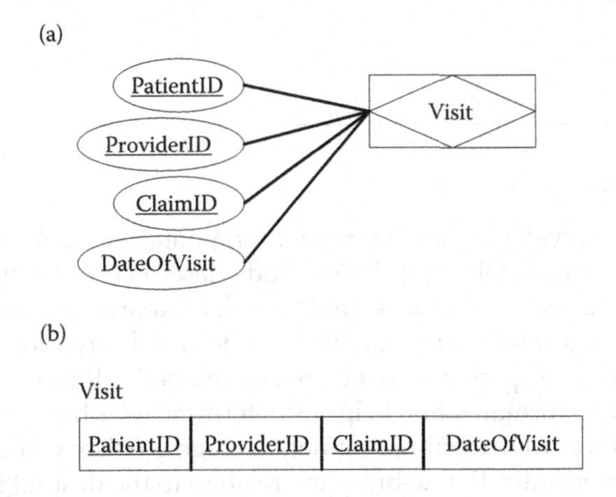

FIGURE 6.13
Representation of an associate entity in a relational model. (a) The associative entity, Visit, in the E-R model of the healthcare database. (b) The transformation of the associative entity, Visit, to a relational model.

the NULL value. This means that a pet exists without a person whom the pet depends on, which is not allowed because a pet is a weak entity and has to depend on a person. Having PersonID join PetID as the composite primary key of the Pet relation enforces that PersonID in the Pet relation cannot have the NULL value since a primary key attribute cannot have the NULL value. Hence, a record of the Pet relation must have a specific value of PetID and a specific value of PersonID, enforcing the dependence of a pet on a person.

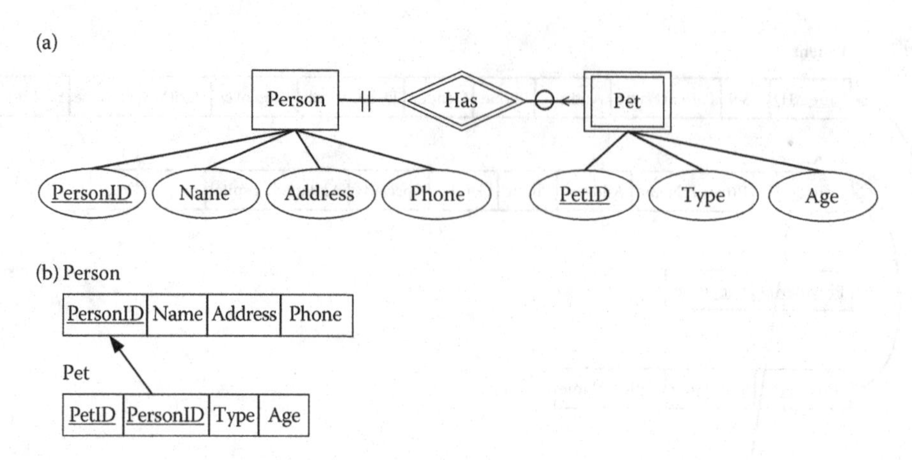

FIGURE 6.14
Representation of a weak entity in a relational model. (a) An E-R model containing a weak entity, Pet. (b) The transformation of the E-R model in Part (a) to a relational model.

Examples 6.1 to 6.11 transform many parts of the E-R model for the healthcare database to the relational model. Figure 6.15 gives the complete relational model for the entire E-R model of the healthcare database.

The transformation of an E-R model to a relational model in this section shows why we need to identify features such as types of attributes (e.g., single-valued versus multi-valued, simple versus composite, and stored versus derived), the degree of a relationship, the maximum and minimum cardinalities of a relationship, superclass and subclass entities, and weak entities in an E-R model. These features identified in an E-R model determine how data are organized into relations in a relational model to maximize data storage efficiency.

6.3 Normalization

If we do a good job of developing an E-R model for a database and then transforming the E-R model to a relational model using the methods described in Section 6.2, the resulting relational model is expected to be in a desired form for maximizing data storage efficiency. However, we may have a relational model that is not in a desired form and contains data redundancy or wasted data space due to a poorly developed E-R model or an existing database that is poorly built. Normalization helps us determine whether a relation in a relational model is in a desired form for maximizing data storage efficiency. If a relation is not in a desired form, we use normalization to bring the relation to the desired form.

In this section, we first look into data redundancy and associated data anomalies to explain why data redundancy should be eliminated. We then introduce the concept of functional dependency and use this concept to define normal forms. We finally describe methods of normalization to bring a relation to a desired normal form.

6.3.1 Data Redundancy and Data Anomalies

Figure 6.16 gives the PrescriptionFilling relation that contains data about pharmacies filling medicine prescriptions. Four records of the relation are also included in Figure 6.16.

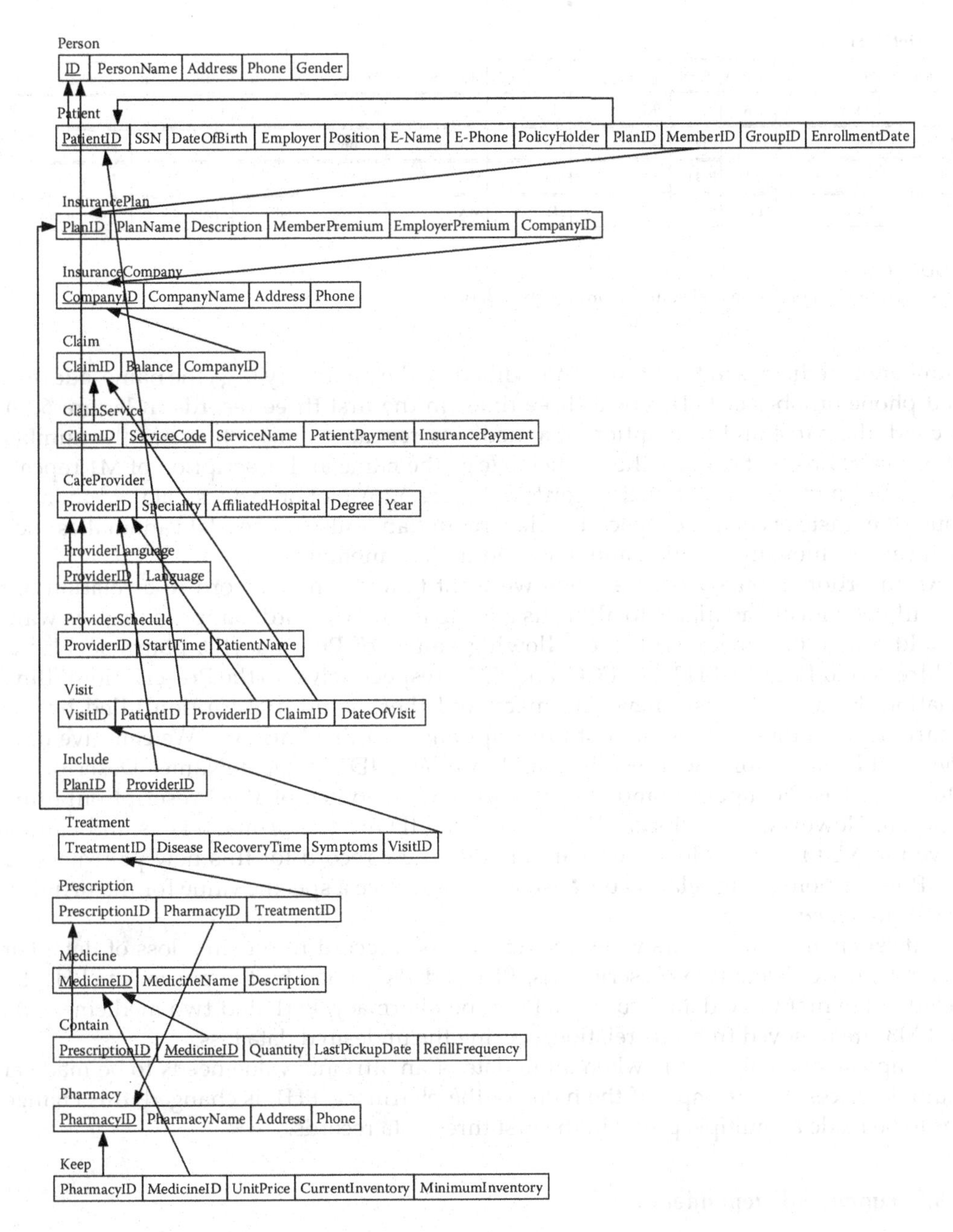

FIGURE 6.15
Relational model of the healthcare database.

As indicated by the records of the relation, a pharmacy can fill multiple prescriptions (e.g., PH1 fills PR1 and PR2), a prescription can contain multiple medicines (e.g., PR1 contains M1 and M2), a medicine can be in multiple prescriptions (e.g., M1 is in PR1 and PR3), and a prescription can be filled by only one pharmacy.

Data redundancy exists in several places in the PrescriptionFilling relation. First of all, the name, address, and phone number of a pharmacy repeat as many times as the

PrescriptionFilling

PharmacyID	PharmacyName	Address	Phone	PrescriptionID	MedicineID	MedicineName	Description	Quantity	LastPickupDate	RefillFrequency
PH1	PA	PAA	P111	PR1	M1	MA	MAAA	30	1/1/2013	Monthly
PH1	PA	PAA	P111	PR1	M2	MB	MBBB	20	1/3/2013	Quarterly
PH1	PA	PAA	P111	PR2	M3	MC	MCCC	50	1/1/2013	Monthly
PH2	PB	PBB	P222	PR3	M1	MA	MAAA	60	1/30/2013	Semi-annually

FIGURE 6.16
PrescriptionFilling relation and some records of the relation.

number of medicines in the prescriptions filled by the pharmacy (e.g., the name, address, and phone number of PH1 repeat three times in the first three records in Figure 6.16). Second, the name and description of a medicine repeats as many times as the number of prescriptions that contain the medicine (e.g., the name and description of M1 repeats two times in the first and fourth records in Figure 6.16). Not only does data redundancy cause the waste of computer space, but data redundancy also causes data anomalies such as insertion anomaly, deletion anomaly, and update anomaly.

An insertion anomaly occurs when we want to add a new record to a relation but not all the data are available to allow us adding the new record. Suppose that we want to add a new pharmacy with the following values of PharmacyID, PharmacyName, Address, and Phone—PH3, PC, PCC, and P333, respectively—to the PrescriptionFilling relation. Because this is a new Pharmacy and there is no prescription filled by the Pharmacy yet, there are no data of prescriptions for this pharmacy. We can give only the NULL value for the PrescriptionID, MedicineID, MedicineName, Description, Quantity, LastPickupDate, and RefillFrequency attributes of the PrescriptionFilling relation. However, PrescriptionID and MedicineID are the primary keys and cannot have the NULL value. Hence, we cannot add a new record for this new pharmacy to the PrescriptionFilling relation because we do not have a specific value for the primary key of this new record.

A deletion anomaly occurs when the deletion of a record results in a loss of data. For example, if we delete two prescriptions, PR1 and PR2, filled by the pharmacy, PH1, by deleting the first three data records, data of the pharmacy, PH1, and two medicines, M2 and M3, are removed from the relation, causing the undesired data loss.

An update anomaly occurs when an update of an attribute value needs to be made at multiple places. For example, if the name of the pharmacy, PH1, is changed, this change has to be made at multiple places in the first three data records.

6.3.2 Functional Dependency

Normalization is performed based on the analysis of functional dependencies that are present in a relation. A functional dependency is defined in this section. The next section describes normalization, which includes a series of normal forms based on the analysis of functional dependencies.

Given that A and B are two sets of attributes, B functionally depends on A, or in other terms, A determines B, if a given value of A uniquely determines the value of B. The functional dependency of B on A is denoted as $A \rightarrow B$ or represented using a dependency diagram as shown in Figure 6.17. A is called "determinant(s)," and B is called "dependent(s)."

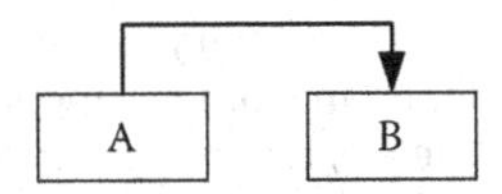

FIGURE 6.17
Dependency diagram showing a functional dependency of B depending on A.

Example 6.13

Identify the functional dependencies that exist in the PrescriptionFilling relation in Figure 6.16.

Because PrescriptionID and MedicineID are the primary keys of the PrescriptionFilling relation, a given value of the primary key uniquely identifies one record of the relation, and each of the non-key attribute(s) has only one value in the record. That is, the primary key attributes, PrescriptionID and MedicineID, uniquely determine all the non-key attributes. Hence, we have the following functional dependency:

PrescriptionID, MedicineID → PharmacyID, PharmacyName, Address, Phone, MedicineName, Description, Quantity, LastPickupDate, and RefillFrequency

For any relation, these always is a functional dependency that has the primary key attribute(s) as the determinant(s) and the non-key attribute(s) as the dependents.

In addition to the functional dependency with the primary key as the determinant(s), we identify the functional dependencies that have a smaller number of determinants than that in the functional dependency with the primary key as the determinant(s). The following two functional dependencies are identified:

PharmacyID → PharmacyName, Address, Phone

MedicineID → MedicineName, Description

Figure 6.18 gives a dependency diagram that shows the above three functional dependencies in the PrescriptionFilling relation.

Each of the three functional dependencies in Figure 6.18 can be changed to another functional dependency by making a dependent to a determinant. For example, the following functional dependency is created from FD2 in Figure 6.18 by changing PharmacyName from a dependent to a determinant:

PharmacyID, PharmacyName → Address, Phone

For normalization, however, we are not interested in functional dependencies that have the same or larger number of determinants than the number of determinants in the primary key of the relation. For the purpose of normalization, we are interested

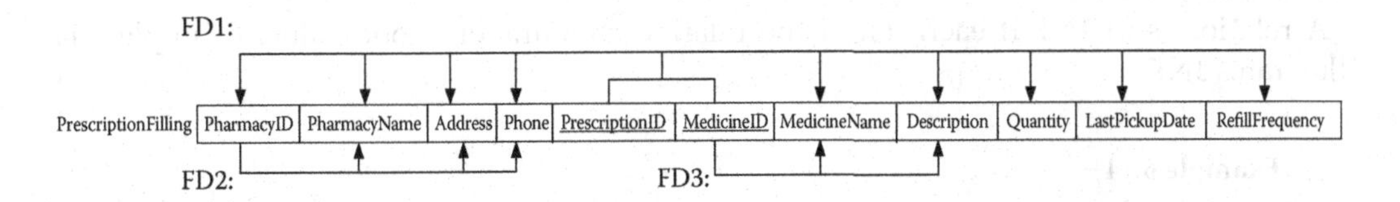

FIGURE 6.18
Functional dependencies in the PrescriptionFilling relation.

in the functional dependency with the primary key as the determinant(s) and all the non-key attributes as the dependents and other functional dependencies that have a smaller number of determinants. Hence, for the PrescriptionFilling relation, only three functional dependencies in Figure 6.18 are identified.

Among the three functional dependencies that exist in the PrescriptionFilling relation as shown in Figure 6.18, FD3 is a partial functional dependency. In a partial functional dependency, a non-key attribute depends on some, but not all, of the primary key attributes. In FD3, two non-key attributes are determined by MedicineID, which is one of the two primary key attributes. FD2 is a transitive functional dependency. A transitive functional dependency involves no primary key attribute(s). In FD2, none of the PharmacyID, PharmacyName, Address, and Phone attributes are a primary key attribute.

The only functional dependency that should be present in a relation is the functional dependency with the primary key as the determinant and all the non-key attributes as the dependents. Additional functional dependencies in the relation, such as partial functional dependency, transitive functional dependency, and any other type of functional dependency, cause data redundancy and data anomalies and should be removed from the relation through normalization. For example, a functional dependency with a non-key attribute as the determinant and a primary key attribute as the dependent is neither a partial functional dependency nor a transitive functional dependency but should be removed because it causes data redundancy and data anomalies.

6.3.3 Normalization and Normal Forms

In general, if no functional dependencies other than the functional dependency with the primary key as the determinant(s) and all the non-key attributes as the dependents can be identified in a relation, the relation is in a normal form without data redundancy and data anomalies. If multiple functional dependencies are identified in a relation, normalization needs to be performed on the relation to remove the functional dependencies other than the one with the primary key as the determinant(s) and all the non-key attributes as the dependents and thus bring the relation to a normal form without data redundancy and data anomalies. That is, if there is only one functional dependency that has the primary key as the determinant(s) and all the non-key attributes as the dependents in a relation, there are no data redundancy and data anomalies in the relation, and no normalization needs to be performed. If there are multiple functional dependencies in a relation, normalization needs to be performed to remove those functional dependencies that do not have the primary key as the determinant(s) because those functional dependencies cause data redundancy and data anomalies. Normalization defines the first normal form (1NF), the second normal form (2NF), the third normal form (3NF), the Boyce–Codd normal form (BCNF), and so on.

A relation is in 1NF if each cell of the relation contains only one value. Example 6.14 illustrates 1NF.

Example 6.14

Figure 6.19 shows the PrescriptionFilling2 relation, which is the same as the PrescriptionFilling relation in Figure 6.16, except that the PrescriptionFilling2 relation

PrescriptionFilling2

PharmacyID	PharmacyName	Address	Phone	PrescriptionID	MedicineID	MedicineName	Description	Quantity	LastPickupDate	RefillFrequency
PH1	PA	PAA	P111	PR1	M1, M2	MA, MB	MAAA, MBBB	30, 20	1/1/2013, 1/3/2013	Monthly, Quarterly
PH1	PA	PAA	P111	PR2	M3	MC	MCCC	50	1/1/2013	Monthly
PH2	PB	PBB	P222	PR3	M1	MA	MAAA	60	1/30/2013	Semi-annually

FIGURE 6.19
PrescriptionFilling2 relation that is not in the 1NF.

has only one primary key attribute, PrescriptionID. Is the PrescriptionFilling2 relation in 1NF? If not, perform the normalization to bring the relation to 1NF.

Because the PrescriptionFilling2 relation uses only PrescriptionID as the primary key and the primary key should uniquely identify each record of the relation, there is only one record for PR1. A prescription can have more than one medicine. For example, the prescription, PR1, has two medicines, M1 and M2. Two medicines, M1 and M2, in PR1 have to be put in the same cell for the MedicineID attribute in the record for PR1. Hence, this cell contains two values. Similarly, the cells for the MedicineName, Description, Quantity, LastPickupDate, and RefillFrequency attributes in the record for PR1 also contain two values. Hence, the PrescriptionFilling2 relation is not in 1NF.

To bring the PrescriptionFilling2 relation to 1NF, we need to add an attribute that contains multiple values (e.g., MedicineID) to the primary key to separate multiple values into different cells. By adding MedicineID to the primary key and, thus, using both PrescriptionID and MedicineID as the primary keys, as shown in Figure 6.16, each cell of the relation in Figure 6.16 contains only one value.

A relation is in 2NF if the relation is in 1NF and has no partial functional dependencies. Example 6.15 illustrates 2NF.

Example 6.15

Is the PrescriptionFilling relation in Figure 6.18 in 2NF? If not, perform the normalization to bring the relation to 2NF.

The PrescriptionFilling relation shown in Figure 6.18 has a partial functional dependency, FD3. The normalization of removing FD3 and bringing the relation to 2NF is performed by decomposing the PrescriptionFilling relation, specifically by taking out the attributes in FD3 and putting them into a new relation, Medicine, while keeping MedicineID in the PrescriptionFilling relation as the foreign key. Figure 6.20a shows the PrescriptionFilling and the Medicine relations after performing the normalization to take out the attributes in FD3 and put these attributes in the Medicine relation. Figure 6.20b shows two functional dependencies in the PrescriptionFilling relation and one functional dependency in the Medicine relation. The Medicine relation has only one functional dependency with the primary key (MedicineID) as the determinant and the non-key attributes (MedicineName and Description) as the dependents. There is no partial functional dependency in the Medicine relation. Hence, the Medicine relation is in 2NF. The PrescriptionFilling relation has two functional dependencies but no partial functional dependency. Hence, the PrescriptionFilling relation is also in 2NF.

A relation is in 3NF if the relation is in 2NF and has no transitive functional dependency. Example 6.16 illustrates 3NF.

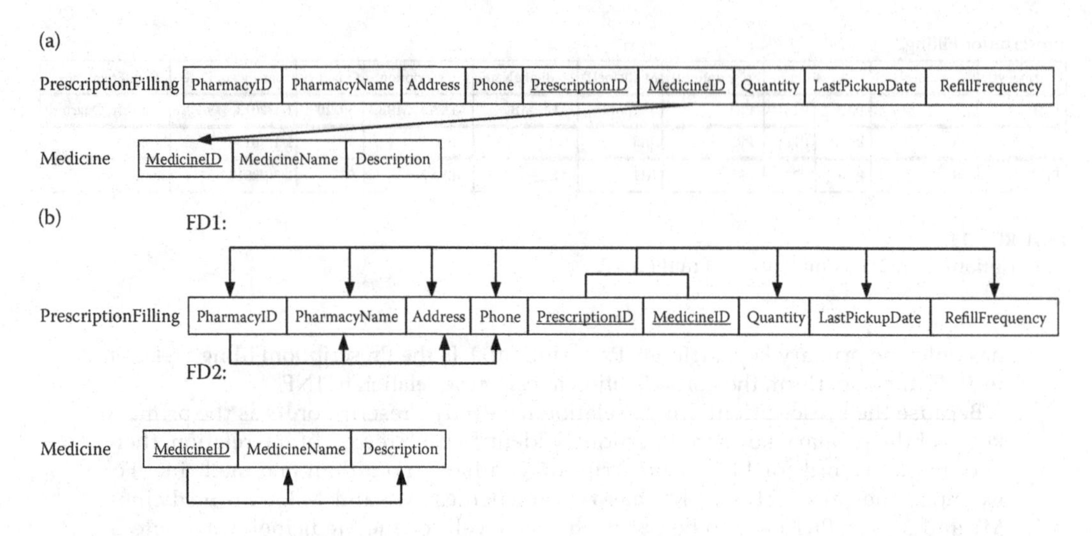

FIGURE 6.20
Normalization to bring the PrescriptionFilling relation to the 2NF. (a) The PrescriptionFilling relation and the Medicine relation. (b) The dependency diagrams for the PrescriptionFilling relation and the Medicine relation.

Example 6.16

Are the PrescriptionFilling and the Medicine relations in Figure 6.20b in 3NF? If not, perform the normalization to bring the relations to 3NF.

The Medicine relation has one functional dependency with the primary key as the determinant and all the non-key attributes as the dependents. As shown in Example 6.15, the Medicine relation is in 2NF. The Medicine relation does not have any transitive functional dependency. Hence, the Medicine relation is in 3NF.

The PrescriptionFilling relation has two functional dependencies. As shown in Example 6.15, the PrescriptionFilling relation is in 2NF. Because the PrescriptionFilling relation has a transitive functional dependency, FD2, the PrescriptionFilling relation is not in 3NF. We perform the normalization to take out all the attributes in FD2 and put them in a new relation, Pharmacy, while keeping PharmacyID in the PrescriptionFilling relation as a foreign key. Figure 6.21a shows the PrescriptionFilling and the Pharmacy relations along with the Medicine relation after the normalization. Figure 6.21b shows the dependency diagrams for these relations. Each relation is in 3NF and has only one functional dependency with the primary key as the determinant and all the non-key attributes as the dependents.

A relation is in BCNF if the relation is in 3NF and does not have any functional dependency whose determinant(s) is not the primary key. Example 6.17 illustrates BCNF.

Example 6.17

Figure 6.22a shows the Kid relation containing data about the programs that each kid participates in and the mentors assigned to each program. As shown by records in the Kid relation (see Figure 6.22a), a child can participate in multiple programs, multiple mentors can be assigned to a program, and a mentor can be assigned to only one program. Figure 6.22b shows the functional dependencies in the Kid relation, FD1 and FD2. FD2 has the determinant that is not the primary key. FD2 is neither a partial functional dependency nor a transitive functional dependency. Hence,

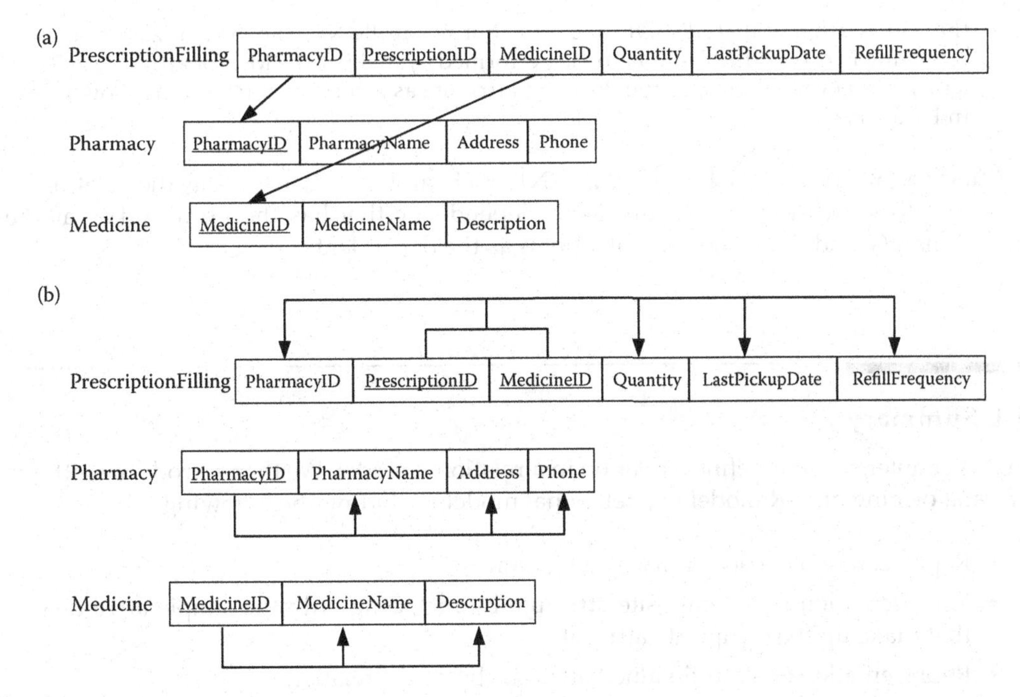

FIGURE 6.21
Normalization to bring the PrescriptionFilling relation to the 3NF. (a) The PrescriptionFilling relation, the Pharmacy relation, and the Medicine relation after normalization to 3NF. (b) The dependency diagrams for the PrescriptionFilling relation, the Pharmacy relation, and the Medicine relation.

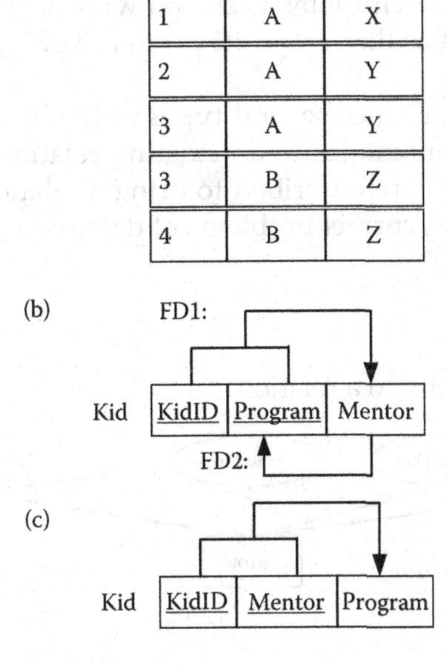

FIGURE 6.22
Normalization to bring the Kid relation to the BCNF. (a) The Kid relation with some records. (b) The functional dependencies in the Kid relation. (c) The Kid relation after normalization to BCNF.

the Kid relation is in 1NF, 2NF, and 3NF, but not in BCNF. The normalization of removing FD2 from the Kid relation is performed by making the Mentor attribute as a primary key attribute and the Program attribute as a non-key attribute, as shown in Figure 6.22c.

After normalizing a relation to 1NF, 2NF, 3NF, and BCNF, we bring the relation in a normal form, with only one functional dependency that has the primary key as the determinant(s) and all the non-key attributes as the dependents.

6.4 Summary

In this chapter, we first define a relation in a relational model. We then introduce methods of transforming an E-R model to a relational model, including the following:

- Representation of each entity by a relation
- Representation of a composite attribute by including only the simple attributes that make up the composite attributes
- Representation of a multi-valued attribute by a new relation
- Rule of "mandatory migrates to optional" to place a foreign key for a one-to-one, optional-to-mandatory/mandatory-to-optional relationship
- Rule of "one migrates to many" to place a foreign key for a one-to-many/many-to-one relationship
- Representation of a many-to-many binary relationship, a relationship of degree-3 or higher or an associative entity by a new relation
- Representation of a weak entity by a relation with the key attributes of both the weak the strong entities as the composite primary key

In this chapter, we also introduce several types of functional dependencies to detect data redundancy and data anomalies in an existing relation that is constructed incorrectly. Steps of normalization are described to bring a relation to the 1NF, the 2NF, the 3NF, and the BCNF and, thus, correct problems of data redundancy and data anomalies in a relation.

EXERCISES

1. Transform the Show entity to a relation.

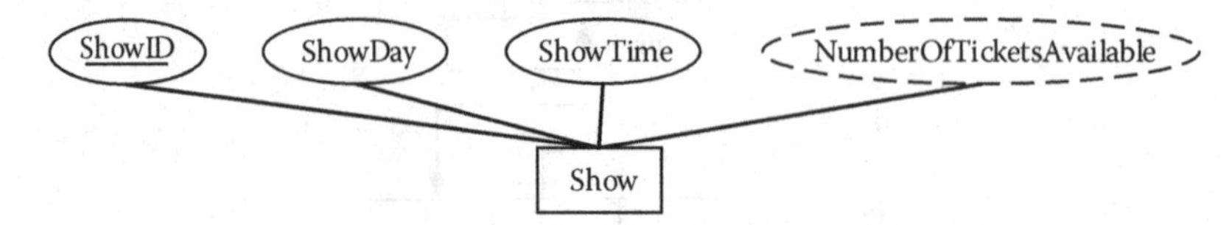

2. Transform the Movie entity to a relation.

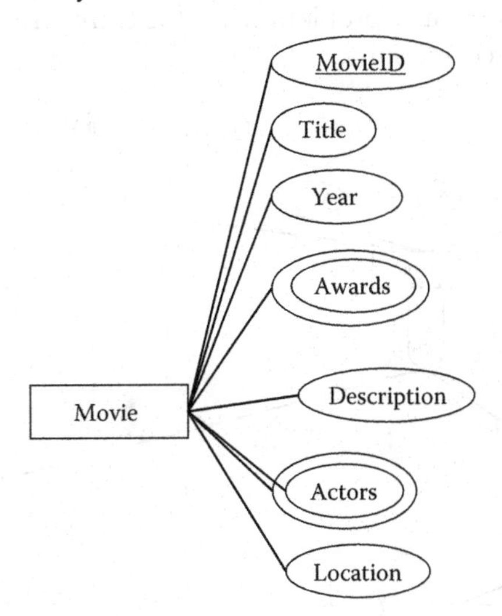

3. Transform the following Person entity to a relation.

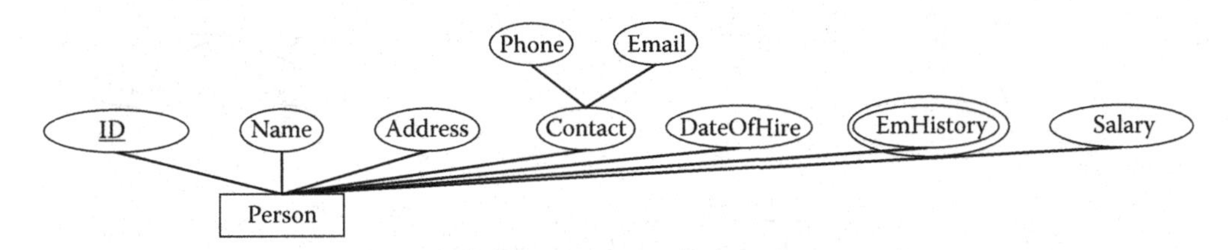

4. Transform the following E-R model to a relational model.

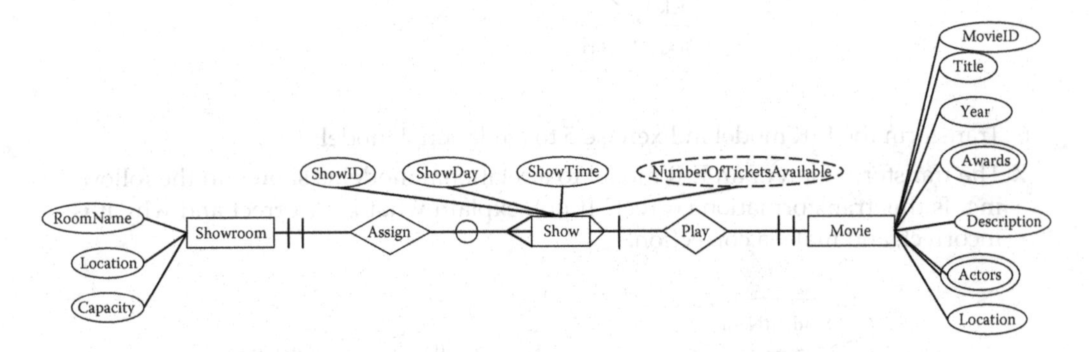

5. The Buy relationship in the following E-R model is transformed into a relational model as follows. Is there any problem with this transformation? If so, discuss the problem(s) and make corrections.

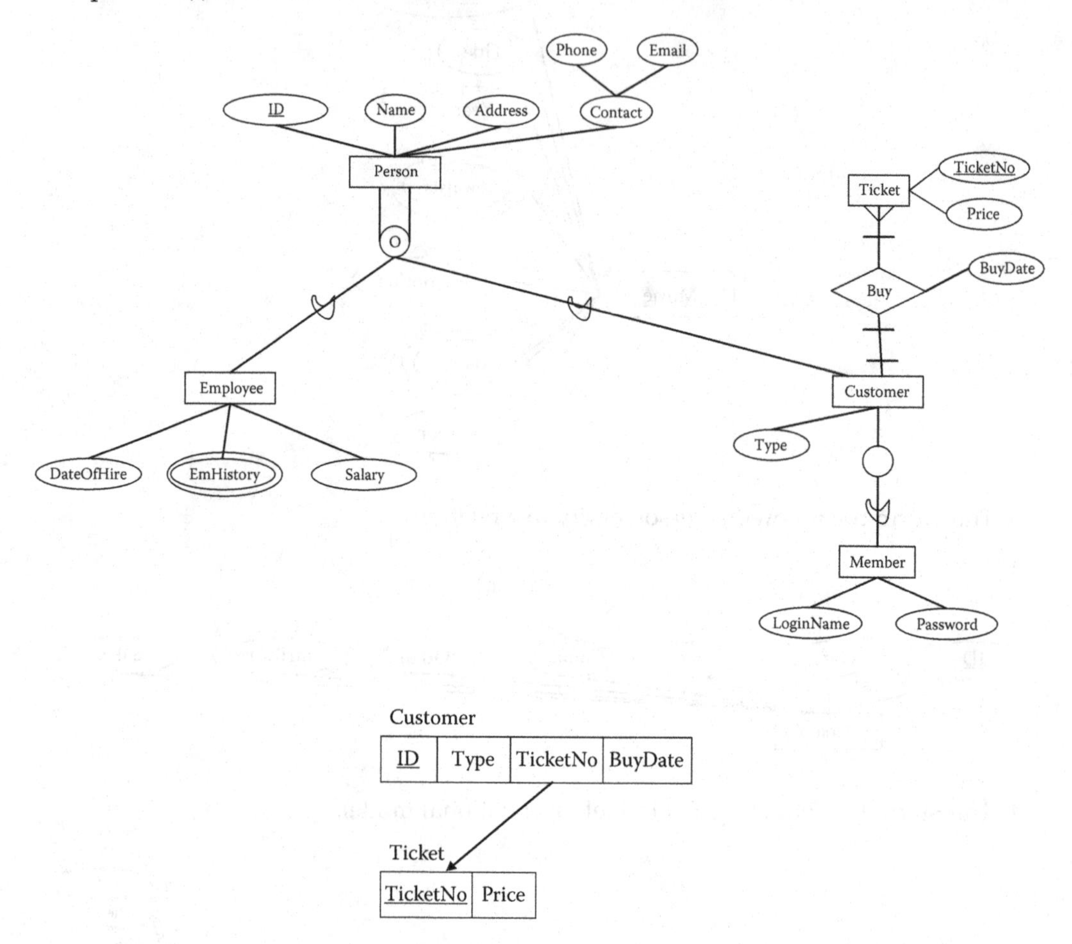

6. Transform the E-R model in Exercise 5 to a relational model.
7. The transformation of an E-R model to a relational model is shown in the following. Is this transformation correct? If not, explain what is incorrect and why it is incorrect, and make a correction.

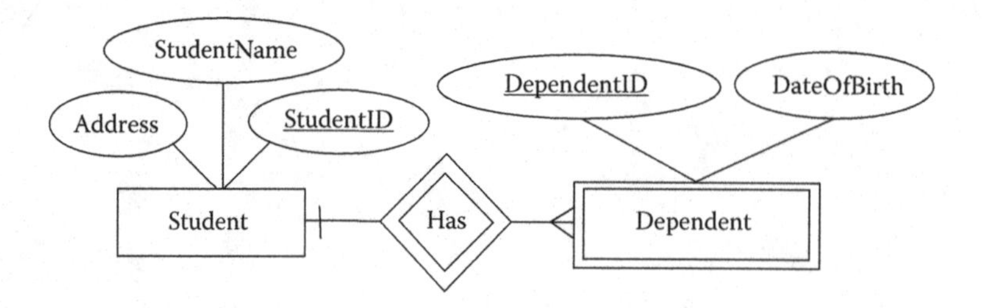

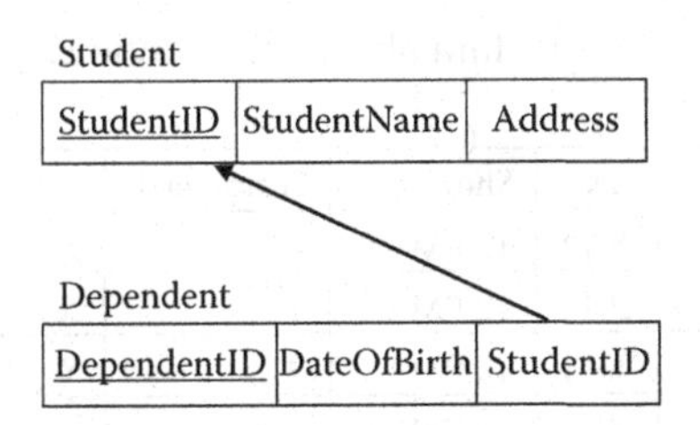

8. Transform the E-R model of the movie theater database from Exercise 8 in Chapter 5 to the relational model of the movie theater database.
9. Transform the following E-R model to a relational model.

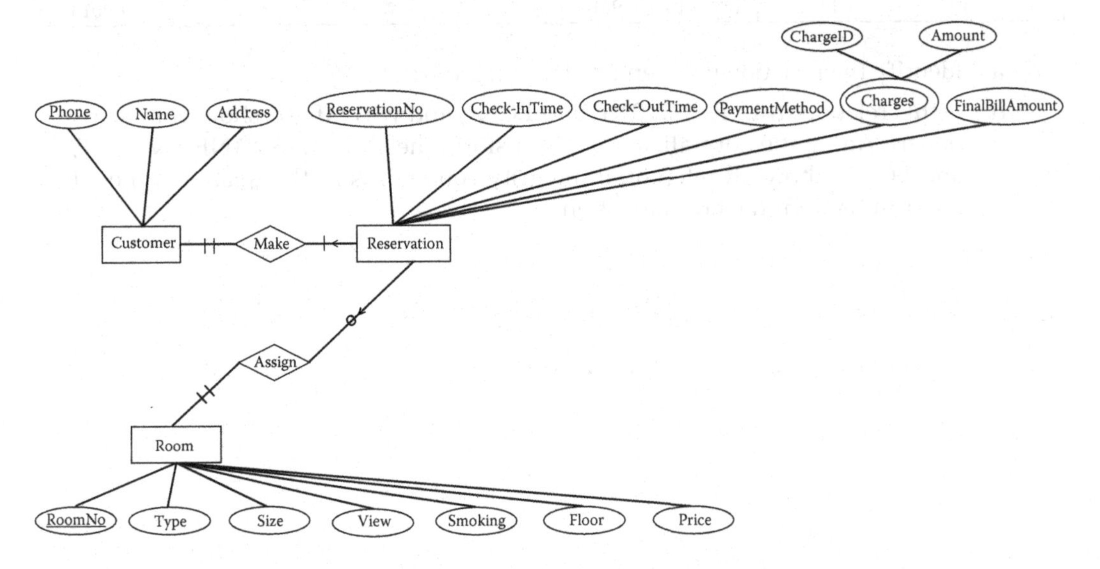

10. The PatientVisit relation keeps information about the patients' visits to doctors.
    a. Identify the functional dependencies in the relation.
    b. Is this relation in 2NF, 3NF, and BCNF, and why? If not, show your steps of performing the normalization to transform the relation into 2NF, 3NF, and BCNF. Show all referential integrity constraints in the relational model that result from the normalization.
    c. Identify the functional dependency in each relation in the final relational model produced in b.

PatientVisit

PatientID	Name	Address	VisitDate	DoctorID	DoctorName
P1209	J. Johns	134 West	3/3/02	D1256	A. Frank
P1209	J. Johns	134 West	9/10/02	D1256	A. Frank
P1215	J. Johns	200 West	3/3/02	D4523	D. Gomez
P1221	F. Brown	223 South	6/7/02	D6712	G. Kelly
P8912	F. Dong	45 East	9/12/02	D8917	M. Julius

11. The MovieTicket relation keeps data about the movie tickets bought by customers.

MovieTicket

MovieID	MovieTitle	ShowID	ShowDay	ShowTime	CustomerID	CustomerName	TicketNo	BuyDate
1	AAA	1	1/1/2013	12 PM	1	X	1	12/31/2012
1	AAA	1	1/1/2013	12 PM	1	X	2	12/31/2012
1	AAA	1	1/1/2013	12 PM	2	Y	3	1/1/2013
1	AAA	1	1/1/2013	12 PM	2	Y	4	1/1/2013
1	AAA	2	1/2/2013	2 PM	3	Z	1	1/1/2013
1	AAA	2	1/2/2013	2 PM	3	Z	2	1/2/2013
2	BBB	1	1/3/2013	6 PM	4	W	1	1/2/2013
2	BBB	1	1/3/2013	6 PM	4	W	2	1/3/2013

a. Identify the functional dependencies in the relation.
b. Is this relation in 1NF, 2NF, 3NF, and BCNF, and why? If not, show your steps of performing the normalization to transform the relation into 1NF, 2NF, 3NF, and BCNF. Show all referential integrity constraints in the relational model that results from the normalization.

7

Database Implementation in Microsoft Access

Microsoft Access 2013 (called "Access" in the following text) is a database management system (DBMS) that is included in Microsoft Office to support the implementation of a database on computers based on a relational model of database. An Access database contains a collection of data objects: tables to store records of relations, queries to retrieve data selectively in various ways, and forms and reports to present retrieved data in user-preferred formats. This chapter describes the use of Access to implement data storage and data queries in a database. Access forms and reports are not included in this chapter because the graphical user interface for the data presentation in user-preferred formats is created using Microsoft Visual Studio and is described in Chapters 11, 12, 13, and 16. Visual Studio is used to develop a Windows-based or Web-enabled database application with an Access database embedded.

7.1 Tables for Data Storage

Each relation in a relational model is implemented by a table in Access. To create a table in Access, we first open Access and then click on "Blank Desktop Database," which brings up a pop-up window, as shown in Figure 7.1.

For the prompt of the file name in the pop-up window, we can select a directory and a file name for a database and then click on the Create button, which brings up a Datasheet View of a default table, which is named "Table1." As shown in Figure 7.2, the screen is divided into three areas: the menus and icons in the top area, the list of all Access objects in the left area, and the work space in the right area now showing the Datasheet View of Table1. We need to change to the Design View of the table for creating and defining the table. To change to the Design View of the table, we go to the View icon at the top-left corner of the screen (see Figure 7.2). Clicking on the down arrow in the View icon, we see two choices: Datasheet View and Design View. Selecting the Design View brings up a pop-up window with a prompt for the table name (see Figure 7.2). We enter "Patient" to create the table for the Patient relation in the relational model of the healthcare database in Figure 6.15.

Each attribute of the Patient relation becomes a data field of the Patient table. The Design View of the Patient table shows a default data field named "ID," with AutoNumber as the data type of the ID field. We select the ID field and change "ID" to "PatientID," as shown in Figure 7.3. We add other data fields of the Patient table.

7.1.1 Setting the Primary Key for a Table

There is a key symbol next to the PatientID field to indicate that the PatientID field is the primary key of the Patient table. In the Design menu, there is the Primary Key icon next to the View icon. To set the primary key of a table, we select a data field and then click on

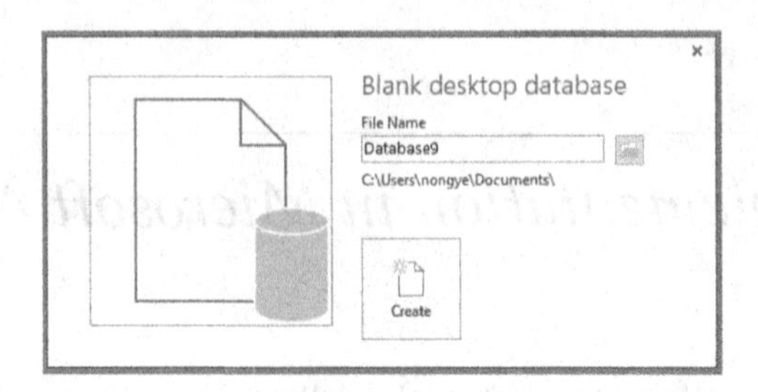

FIGURE 7.1
Pop-up window in Access to select a directory and name a database file.

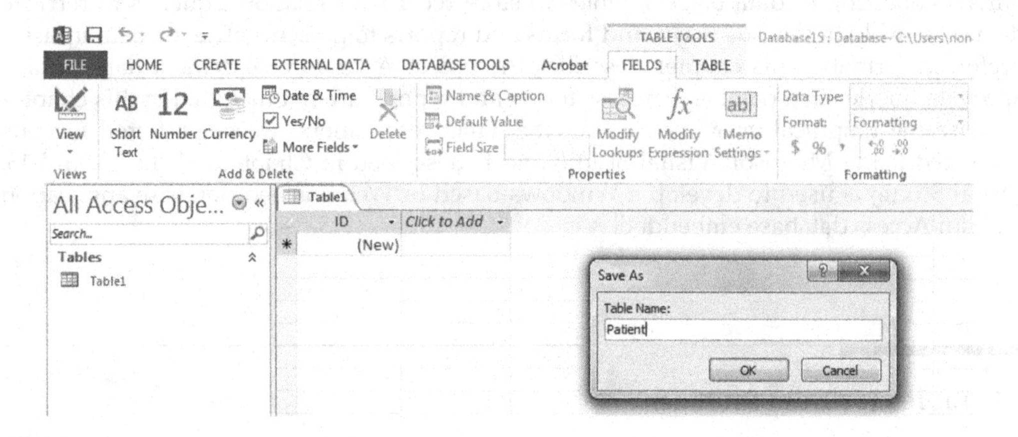

FIGURE 7.2
Screen in Access to save and name a table.

the Primary Key icon. We may deselect the data field as the primary key by clicking on the Primary Key icon again. If a table has two or more data fields as the primary key, we can select all of these data fields at the same time and click on the Primary Key icon to make these data fields together as the primary key.

7.1.2 Data Types of a Data Field

In Access, we can select one of the following data types for a data field:

- Short Text
- Long Text
- Number
- Date/Time
- Currency
- AutoNumber
- Yes/No
- Object linking and embedding (OLE) object (Note that the data type of OLE object can be used for object such as Microsoft Word document, Excel spreadsheet, etc.)
- Hyperlink

Patient

Field Name	Data Type
PatientID	Short Text
SSN	Short Text
DateOfBirth	Date/Time
Employer	Short Text
Position	Short Text
E-Name	Short Text
E-Phone	Short Text
PolicyHolder	Short Text
PlanID	Short Text
MemberID	Short Text
GroupID	Short Text
EnrollmentDate	Date/Time

General | Lookup

Field Size	255
Format	
Input Mask	
Caption	
Default Value	
Validation Rule	
Validation Text	
Required	Yes
Allow Zero Length	Yes
Indexed	Yes (No Duplicates)
Unicode Compression	No
IME Mode	No Control
IME Sentence Mode	None
Text Align	General

FIGURE 7.3
Design View of the Patient table in the healthcare database.

- Attachment
- Calculated
- Lookup

The data type of each data field in the Patient table is shown in Figure 7.3. Because the PolicyHolder field is a foreign key linked to the primary key, PatientID, the data type of the PolicyHolder field is set to Lookup. Selecting Lookup as the data type of a data field brings up a pop-up window for the Lookup Wizard. We then use the Lookup Wizard to select the primary key that the data field looks up. Figure 7.4 shows the steps of using the Lookup Wizard for the PolicyHolder field of the Patient table to look up to the PatientID field of the Patient table. Using the Lookup data type for a data field that is a foreign key allows to select one of the values for the primary key as a value of the foreign key. This makes it easy to comply with the referential integrity constraint between a foreign key and its primary key.

7.1.3 Field Size Property of a Data Field

When we select the PatientID field, the Field Properties area at the bottom of the workspace shows the other properties of the PatientID field (see Figure 7.3). For the Field Size property, the default value is 255 characters for Short Text. We may change 255 characters for the PatientID field to a shorter length, e.g., 9 characters. For the data type of Number, if we click on the down arrow button in the value of the Field Size property, we see several

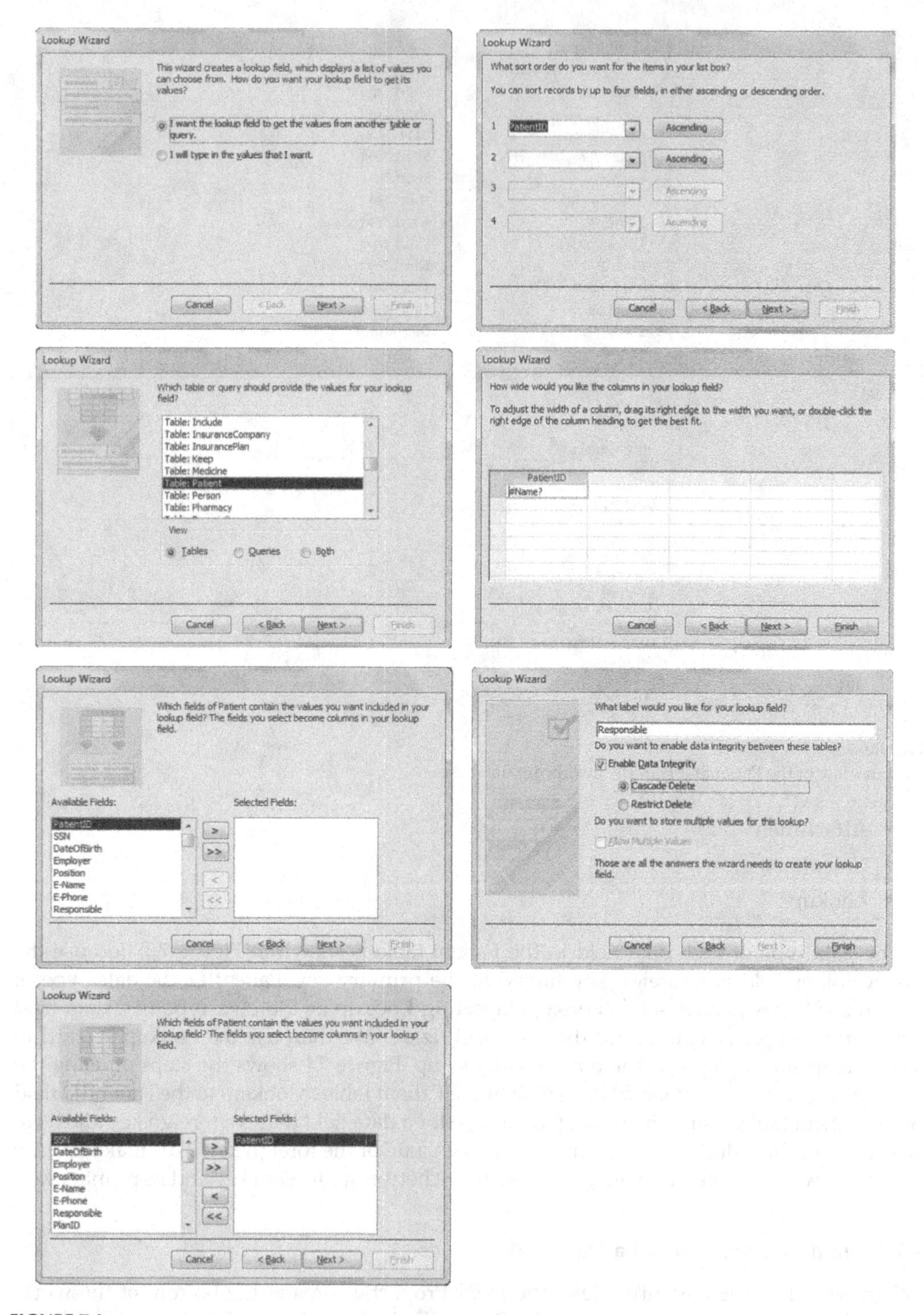

FIGURE 7.4
Illustration of using the Lookup Wizard in Access to define the Lookup data type.

choices: Byte, Integer, Long Integer, Single, Double, Replication ID, and Decimal. For a data field with a data type of AutoNumber, there are two choices of values for the Field Size property: Long Integer and Replication ID.

7.1.4 Format Property of a Data Field

For certain data types, the Format property of a data field is available for us to choose among several different formats to store data. If the data type is Time/Date, clicking on the down arrow button in the value of the Format property shows us several format choices, as shown in Figure 7.5: General Date, Long Date, Medium Date, Short Date, Long Time, Medium Time, and Short Time. For the DateOfBirth field of the Patient table, we set the Format property to Short Date since we do not need the time but only the date for this data field. If the data type is Number, clicking on the down arrow button in the value of the Format property shows us several format choices: General Number, Currency, Euro, Fixed, Standard, Percent, and Scientific.

7.1.5 Input Mask Property of a Data Field

The Input Mask property of a data field masks a data field in a form to assist us in entering a value for the data field but does not change the way in which the value is stored in the data field. If the data type is Short Text, clicking on the down arrow button in the value of the Input Mask property brings up several pre-defined input masks for Social Security

Patient

Field Name	Data Type	Description (Optional)
PatientID	Short Text	
SSN	Short Text	
DateOfBirth	Date/Time	
Employer	Short Text	
Position	Short Text	
E-Name	Short Text	
E-Phone	Short Text	
Responsible	Short Text	
PlanID	Short Text	
MemberID	Short Text	
GroupID	Short Text	
EnrollmentDate	Date/Time	

Field Properties

General | Lookup

Format	Short Date	
Input Mask	General Date	11/12/2015 5:34:23 PM
Caption	Long Date	Thursday, November 12, 2015
Default Value	Medium Date	12-Nov-15
Validation Rule	Short Date	11/12/2015
Validation Text	Long Time	5:34:23 PM
Required	Medium Time	5:34 PM
Indexed	Short Time	17:34
IME Mode	No Control	
IME Sentence Mode	None	
Text Align	General	
Show Date Picker	For dates	

FIGURE 7.5
Format property of the Date/Time data type.

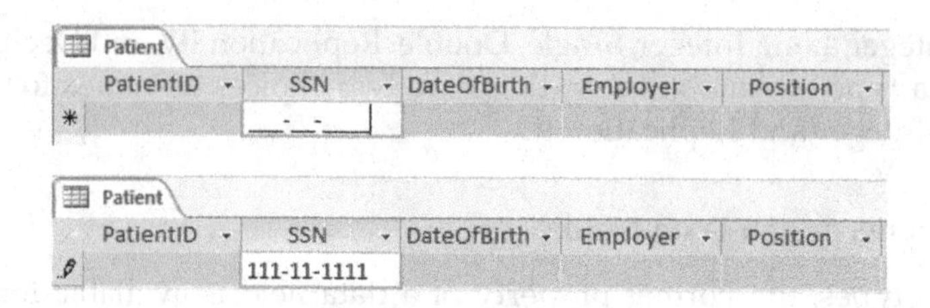

FIGURE 7.6
Illustration of the Input Mask property for the Short Text data type.

Number (SSN), Phone, Zip Code, Password, Data, and Time. For example, for the SSN field of the Patient table in Figure 7.3, we set the Input Mask for the SSN field of the Patient table to Social Security Number. With this setting, a form of SSN with "-" (see Figure 7.6) is displayed when we enter a value of SSN. Although an SSN is displayed like 111-11-1111, the stored value of the SSN is 111111111 without "-." When performing a query (see discussions in Section 7.2) to look for a specific value of SSN, the specific value of SSN should be expressed in the query without "-"; otherwise, no match will be found because all values of SSN are stored without "-."

If the data type is Date/Time, clicking on the down arrow button in the value of the Input Mask property shows us several pre-defined input masks: Long Time, Short Date, Short Time, Medium Time, and Medium Date.

7.1.6 Default Value Property of a Data Field

The Default Value property of a data field automatically sets the value of a data field to the default value if we do not enter another value to the data field. When we click on the button with a mark of "…" in the value of the Default Value property, the Expression Builder window is popped up to allow us to enter an expression for the default value (see Figure 7.7).

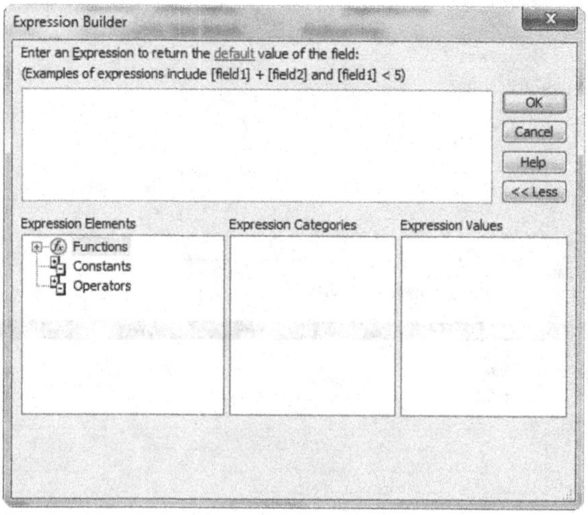

FIGURE 7.7
Expression Builder.

7.1.7 Validation Rule and Validation Text Properties of a Data Field

The Validation Rule property of a data field allows us to specify the range or type of values that the data field can take. For example, we may specify the following validation rule:

```
Like "PA*"
```

for the PatientID field of the Patient table to require that a PatientID value starts with "PA" followed by 0 or more characters. Note that "*" in "PA*" matches any string of 0 or more characters. This validation rule does not allow us to enter any value that does not comply with the pattern of "PA*". The Validation Text property allows us to enter the text describing the validation rule.

7.1.8 Required Property of a Data Field

The Required Property of a data field can have one of two possible values: Yes and No. If the Yes value is selected, we must enter a value for the data field. If the No value is selected, we do not need to enter a value for the data field. The data field as the primary key is automatically set to have the Yes value. A non–key data field has the No value as the default for the Required property.

7.1.9 Indexed Property of a Data Field

The Indexed property of a data field allows us to set an index on the data field. If an index is set on a data field of a table, a sorted order of all the records in the table by values of the indexed field is kept in the database. Hence, indexing a data field takes more memory space but allows finding a data record with a specific value of the indexed field efficiently without checking every record of the table. Hence, there is a trade-off between less search time and more memory space with regard to indexing a data field. Indexing a data field is justified if the data field is queried frequently. Clicking on the down arrow button in the value of the Indexed property shows three possible values of the Indexed property: No, Yes (Duplicates OK), and Yes (No Duplicates).

The data field(s) as the primary key of every table is automatically indexed with the value of Yes (No Duplicates) because the primary key should uniquely identify each record of the table, and thus, no duplicate values are allowed in the primary key. The value of Yes (Duplicates OK) can be used to set the Indexed property of a non–key data field if the data field is queried frequently. A non–key attribute has the default value of No for the Indexed property.

Example 7.1

Complete the design of the Patient table in the healthcare database.

Figure 7.3 shows the design of the Patient table. Table 7.1 gives some properties of some data fields in the Patient table.

7.1.10 Adding and Deleting Records of a Table

To add a record to a table, we need to use the Datasheet View of the table by clicking on the View icon in the Home or the Design menu. When we are in a Datasheet View of a table, we enter a record in a row of the table by entering a value in each cell.

TABLE 7.1
Data Fields and Field Properties of the Patient Table in the Healthcare Database

Data Field	Property	Value for the Property
SSN	Input Mask	SSN
DateOfBirth	Format	Short Date
DateOfBirth	Input Mask	Short Date
E-Phone	Input Mask	Phone Number
PlanID	Data Type	Lookup (to PlanID in the InsurancePlan table)
EnrollmentDate	Format	Short Date
EnrollmentDate	Input Mask	Short Date

Example 7.2

Add two data records to the Patient table in the healthcare database.

Figure 7.8 shows two data records that we add to the Patient table of the healthcare database.

When adding data records to tables, we should pay attention to the values of a foreign key and its primary key. As discussed in Section 7.1.2, using the Lookup data type for a data field that is a foreign key makes it easy to maintain the referential integrity constraint.

To delete a record of a table, we select the row for the record and right-click in the row to bring up the pop-up menu that contains a menu item, Delete Record, as shown in Figure 7.9. Clicking on this menu item deletes the record.

The name of a table can be changed by right-clicking the table in the list of All Access Objects at the left of the screen and selecting the Rename item in the pop-up menu. The name of an Access object in other types (e.g., query) can be changed similarly.

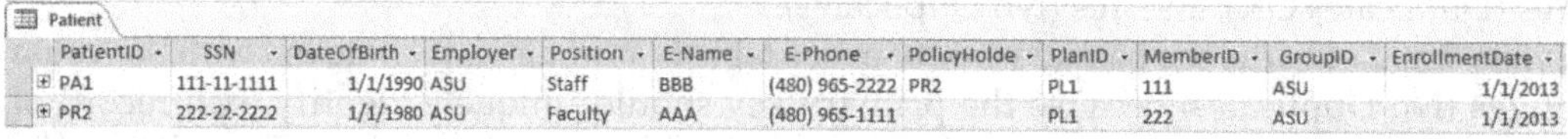

Patient

PatientID	SSN	DateOfBirth	Employer	Position	E-Name	E-Phone	PolicyHolde	PlanID	MemberID	GroupID	EnrollmentDate
PA1	111-11-1111	1/1/1990	ASU	Staff	BBB	(480) 965-2222	PR2	PL1	111	ASU	1/1/2013
PR2	222-22-2222	1/1/1980	ASU	Faculty	AAA	(480) 965-1111		PL1	222	ASU	1/1/2013

FIGURE 7.8
Datasheet View of the Patient table with two records.

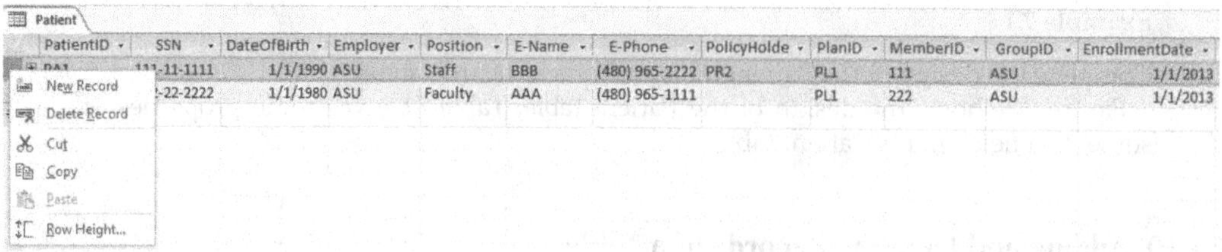

Patient

PatientID	SSN	DateOfBirth	Employer	Position	E-Name	E-Phone	PolicyHolde	PlanID	MemberID	GroupID	EnrollmentDate
PA1	111-11-1111	1/1/1990	ASU	Staff	BBB	(480) 965-2222	PR2	PL1	111	ASU	1/1/2013
	-22-2222	1/1/1980	ASU	Faculty	AAA	(480) 965-1111		PL1	222	ASU	1/1/2013

New Record
Delete Record
Cut
Copy
Paste
Row Height...

FIGURE 7.9
Deleting a record in a table.

7.2 Relationships of Tables

As relations are linked through primary key and foreign key pairings and referential integrity constraints in a relational model, tables in an Access database are linked through relationships of tables, which are based on primary key and foreign key pairings. To establish a relationship between two tables for the pairing of a foreign key and a primary key, we select the Database Tools menu and click on the Relationships icon in this menu, which brings up the workspace for the layout of relationships and a pop-up window, as shown in Figure 7.10. The pop-up window allows us to add selected tables to the relationships layout. For example, we select the Patient and the InsurancePlan tables, click on the Add button to add these tables to the relationships layout, and then close the pop-up window (see Figure 7.10). We can bring back the pop-up window and add more tables to the relationships layout by clicking on the Show Table icon in the Design menu, or right clicking in the area of the relationship layout.

The relationships layout now has the Patient and the InsurancePlan tables, as shown in Figure 7.11. The PlanID field in the Patient table is a foreign key, with its primary key being the PlanID field in the InsurancePlan table. We call the table with the primary key as the "parent table" and the table with the foreign key as the "child table." Hence, there is a relationship between the Patient and the InsurancePlan tables through the pairing of PlanID in the Patient table and PlanID in the InsurancePlan table. To establish this relationship, we drag PlanID in the Patient table over PlanID in the InsurancePlan table, or drag PlanID in the InsurancePlan table over to PlanID in the Patient table. This brings up the Edit Relationships window, as shown in Figure 7.11. This window shows two related data fields that are linked together and a list of checks for Enforce Referential Integrity, Cascade Update Related Fields, and Cascade Delete Related Records. We want to check Enforce Referential Integrity for enforcing the referential integrity constraint between the two related data fields. If we do not check Cascade Update Related Records and Cascade Delete Related Records, we are not allowed to delete a record or update a value of PlanID

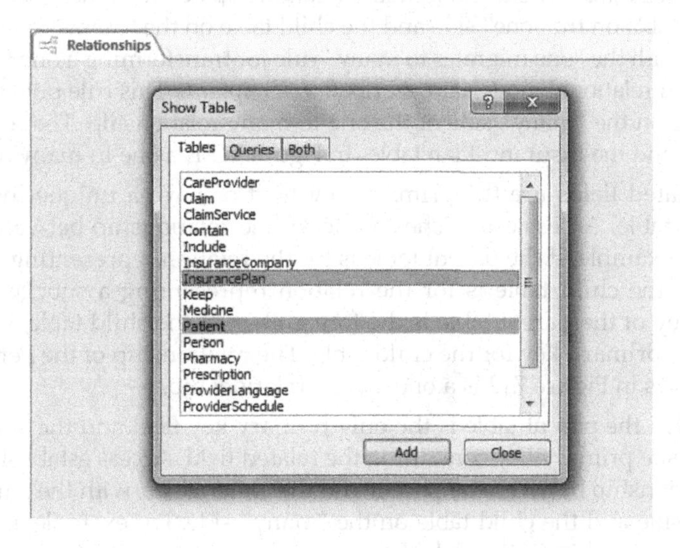

FIGURE 7.10
Show Table window.

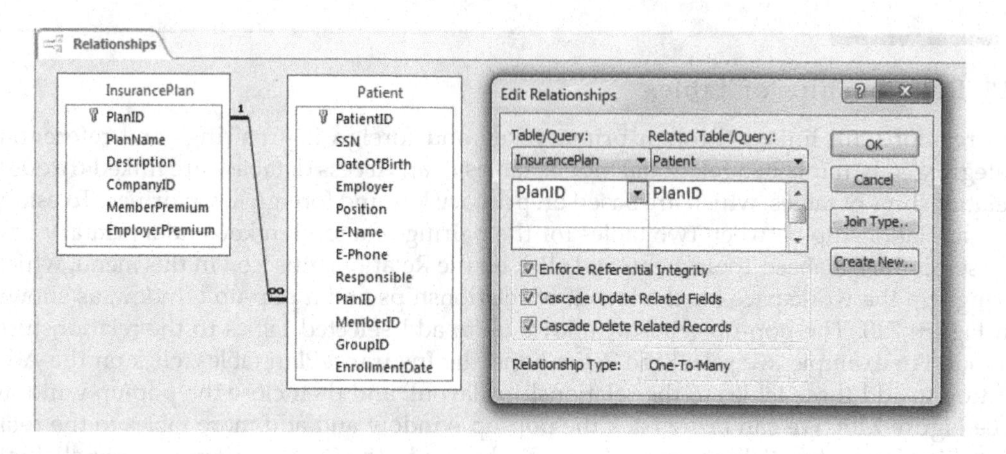

FIGURE 7.11
Edit Relationships window.

in the parent table (the InsurancePlan table) if the value of PlanID exists in the PlanID field of the child table (the Patient table). If we check Cascade Update Related Fields and Cascade Delete Related Records, as shown in Figure 7.11, Access automatically deletes the related data record or updates the related value of PlanID in the child table when we delete a record in the parent table or update the value of PlanID in the parent table.

At the bottom of the Edit Relationships window in Figure 7.11, it is shown that the relationship type is one-to-many. This one-to-many relationship is identified automatically by Access. Access uses the following rules to identify the type of a relationship.

- If the data field in the parent table is the primary key or has a unique index (i.e., having the Yes [No Duplicates] value for the Indexed property), and the related data field in the child table is a foreign key (not being the primary key or having no unique index), Access establishes a one-to-many relationship between these two tables, with the parent table on the "one" side and the child table on the "many" side. This rule is consistent with the "one migrates to many" rule for transforming a one-to-many relationship to a relational model, as described in Chapter 6. This rule puts a foreign key in the table on the "many" side of the one-to-many relationship. The relationship of the Patient and the InsurancePlan tables in Figure 7.11 is a one-to-many relationship.
- If both related fields are the primary key field or have a unique index in their respective table, Access establishes a one-to-one relationship between these two tables. For example, if the parent table is for the relation representing a superclass entity and the child table is for the relation representing a subclass entity, the primary key of the parent table is the foreign key in the child table, which is also used as the primary key for the child table. The relationship of the Person and the Patient tables in Figure 7.12 is a one-to-one relationship.
- If the field in the parent table is the only primary key field and the child table has the composite primary key containing the related field, Access establishes a one-to-many relationship between the parent and the child tables, with the parent table on the "one" side and the child table on the "many" side. For example, if an entity in an entity-relationship (E-R) model has a multi-valued attribute, there are the parent table for the entity and the child table having the primary key of the parent table and the multi-valued attribute as the composite primary key. For one record in the parent

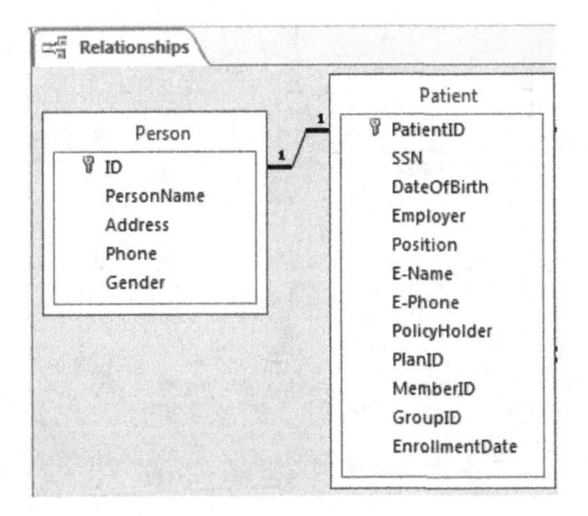

FIGURE 7.12
A one-to-one relationship in the Relationship layout.

table, there may exist multiple records with the same value of the related field in the child table due to the multi-valued attribute. The relationship of the CareProvider and the ProviderSchedule tables is a one-to-many relationship that falls in this rule.

- If none of the related fields is a primary key or has a unique index, Access establishes an indeterminate relationship between these two tables, and we cannot enforce referential integrity constraints for such a relationship.

In the Edit Relationships shown in Figure 7.11, there is the Join Type button. The join type of a relationship is described in Section 7.3.

To bring up the Edit Relationships window after we close it, we can right-click on the relationship line linking the related fields. This brings up a menu with two items: Edit Relationship and Delete. Selecting the Edit Relationship item brings up the Edit Relationships window. Selecting the Delete item deletes the relationship. We can also delete a relationship in the Relationships layout by selecting the relationship and pressing the Delete button on the keyboard. To remove a table in the Relationships layout, we select the table and press the Delete button on the keyboard. To add more tables to the Relationships layout for establishing more relationships, we go to the Database Tools menu, click on the Relationships icon, and then click on the Show Table icon to add more tables to the Relationship layout.

To save the Relationships layout, we click on the Save icon at the top-left corner of the Access screen or select the Save item in the File menu. We can also click on the Clear Layout icon in the Design menu to clear the entire relationships layout.

If we add two tables with the related data fields to the Relationships layout, and one of the related fields uses the Lookup data type to look up the values of another related field, a line linking the related fields is shown in the Relationships layout without us dragging one field over another field. However, the relationship type is not shown on the line until we bring the Edit Relationships window and check Enforce Referential Integrity.

Example 7.3

Establish the relationships of all the tables in the healthcare database.

Figure 7.13 shows the relationships layout that includes all the tables and their relationship to implement the relational model of the healthcare database in Figure 6.15.

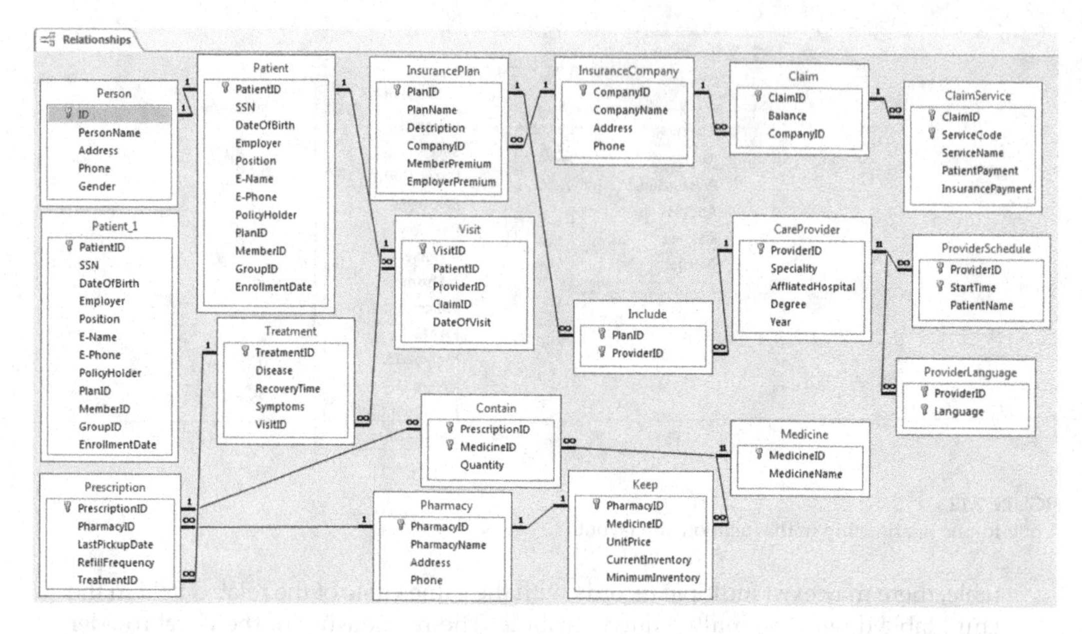

FIGURE 7.13
Relationships of all the tables in the healthcare database.

7.3 Queries for Data Retrieval

A query allows to retrieve selected data from a database. This section introduces the graphical user interface (GUI) of Access to construct a query. There are several types of queries in Access: Select query, Crosstab query, and Action query (e.g., Delete, Update, Append, or Make Table queries). We describe how to construct these queries in the following sections. Select queries include regular Selection queries, Select queries with parameters, and Select queries with Group-By.

7.3.1 Select Queries Using One Table

We can retrieve data from one table or multiple tables in a database. We show first how to construct a Select query to retrieve data from one table in Example 7.4.

Example 7.4

Construct a Select query to present the address and phone number of a pharmacy named "Super Pharmacy" in the healthcare database.

The Pharmacy table in the healthcare database has the name, address, and phone number of all pharmacies. Figure 7.14a shows records that we store in the Pharmacy table. Two of the three pharmacies have the name "Super Pharmacy." We need to create a Select query to present the address and phone number of these two pharmacies.

Figure 7.15 shows the steps of creating this Select query. We first select the Create menu and then select the Query Design icon, which brings up the Design View of a query and

(a)

Pharmacy

PharmacyID	PharmacyName	Address	Phone
PH1	Super Pharmacy	1 Mill Avenue	(480) 111-1111
PH2	Super Pharmacy	1 University Avenue	(480) 222-2222
PH3	AZ Pharmacy	1 Apach Road	(480) 123-4567

(b)

QueryPharmacy

Address	Phone
1 Mill Avenue	(480) 111-1111
1 University Avenue	(480) 222-2222

FIGURE 7.14
Select query to present selected data from one table. (a) Records in the Pharmacy table. (b) The result of executing QueryPharmacy to present the selected data from the Pharmacy table.

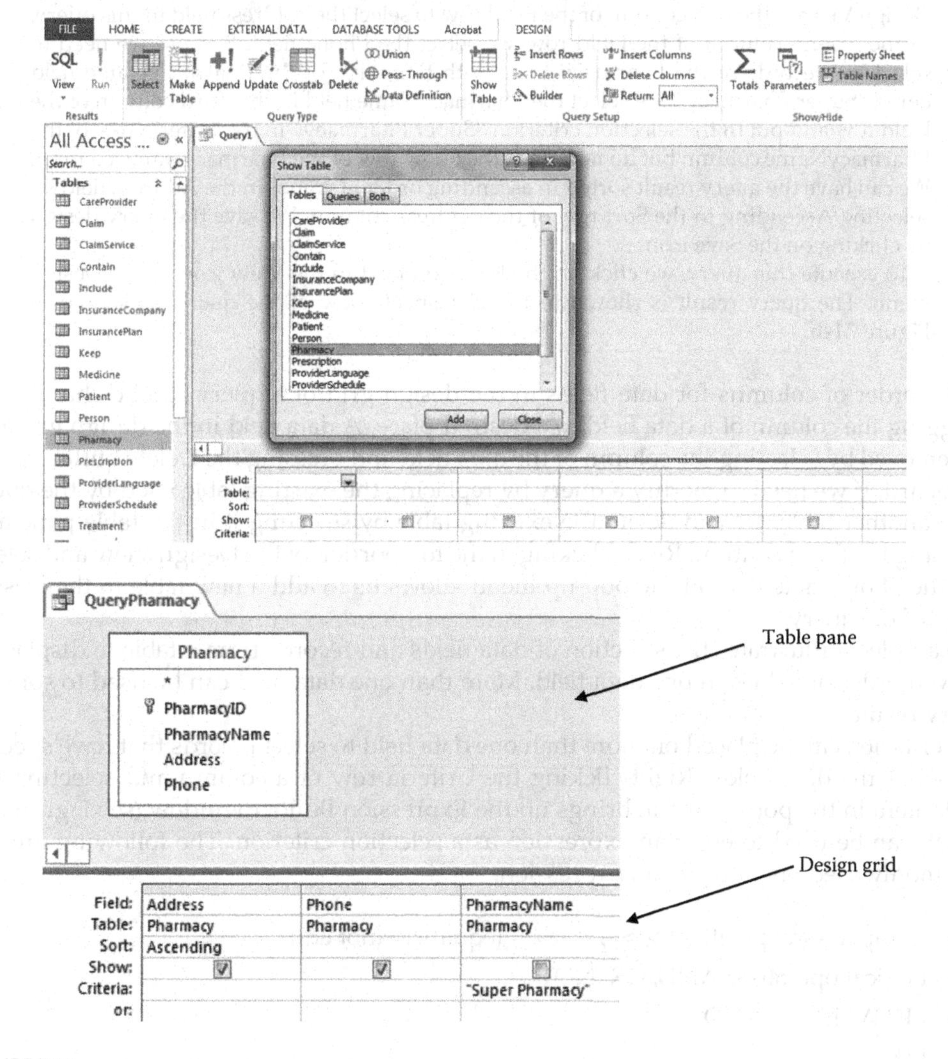

FIGURE 7.15
Design View of a Select query using one table.

the Show Table window. Using the Show Table window, we select the Pharmacy table and then close the Show Table window. We may change the name of this query from Query1 to another name, QueryPharmacy, by clicking on the Save icon at the top-left corner of the screen and then using the pop-up window to change the name of the query. We can also change the name of the query by right-clicking the query in the list of All Access Objects at the left side of the screen and selecting the Rename item in the pop-up menu.

Now, the Pharmacy table appears in the table pane in the top portion of the Design View for QueryPharmacy. The bottom portion of the Design View for QueryPharmacy is the design grid that has the rows for Field, Table, Sort, Show, and Criteria. Clicking on the down arrow in a column of the Field row, we see a list of the data fields in the Pharmacy table along with another field named "Pharmacy.*." The asterisk (*) is a wildcard character that matches any string of characters. In Pharmacy.*, the asterisk matches every data field in the Pharmacy table and, thus, represents all the data fields in the Pharmacy table. In the first column of the Field row, we select the Address field. We can also drag the Address field in the Pharmacy table in the top portion of the Design View to the first column of the Field row to select the Address field in this query. In the second column of the Field row, we select the Phone field. Because we need to select the records of the Pharmacy table with the value of the PharmacyName field being "Super Pharmacy," we select the PharmacyName field in the third column of the Field row and put in the selection criterion "Super Pharmacy" in the Criteria row of the PharmacyName column but do not check the Show row of the PharmacyName column. We can have the query result sorted in ascending order of values in the Address field by selecting Ascending in the Sort row of the Address column. We save the query design by clicking on the Save icon.

To execute this query, we click on the Run icon next to the View icon in the Design menu. The query result is shown in the Datasheet View of the query, as shown in Figure 7.14b.

The order of columns for data fields in the design grid of a query can be changed by dragging the column of a data field to a desired place. A data field in the design grid can be removed by selecting the column of the data field and pressing the Delete button on the keyboard. If we need to modify a query by replacing the existing table used by the query with another table, we can delete the existing table by selecting it in the table pane and pressing the Delete button. Right-clicking in the top portion of the Design View and selecting the Show Table item in the pop-up menu allows us to add a new table to the Design View of the query.

Example 7.4 illustrates the selection of data fields and records from a table to display by a given order of values in one data field. More than one data field can be used to sort the query result.

A criterion can be placed on more than one data field to select records that have specific values of the data fields. Right-clicking the Criteria row of a column and selecting the Build item in the pop-up menu brings up the Expression Builder window (see Figure 7.7), which can be used to enter an expression as a selection criterion. The following are the commonly used operators in an expression:

- Comparison operators: >, >=, <, <=, = (equal), != (not equal)
- Logical operators: AND, OR, NOT
- BETWEEN … AND …
- LIKE

Some examples of expressions are given as follows:

- Year = 2010
- GPA >= 3.0
- Class = "Junior" OR Class = "Senior"
- Salary > 40,000 AND Type = "Full Time"
- BETWEEN #1/1/2010# AND #12/31/2010#
- BETWEEN #09:00:00 AM# AND #12:00:00 PM#
- BETWEEN 1 AND 10
- LIKE "*Database*"

Note that a date or time is enclosed by the pound sign (#). "*Database*" matches any string of characters that contains "Database." There are also build-in functions in the Expression Builder. Some commonly used date/time functions are as follows:

- Now(): returns the current date and time
- Date(): returns the current date
- Year(): returns the year in a date/time
- Month(): returns the month in a date/time
- Day(): returns the day in a date/time

For example, we put the expression = Year(Now()) in the Criteria row of a data field with a data type of Date/Time to select the records with the value of that data field falling in the current year.

7.3.2 Select Queries with Joins of Multiple Tables

This section shows how to construct a Select query to retrieve data from multiple tables.

Example 7.5

Design Query4 for patients in the healthcare database (see Chapter 26) to present the name, address, and phone number of all the healthcare providers who specialize in heart diseases and whose address is in Phoenix, Arizona.

This query needs the CareProvider table to select the healthcare providers who specialize in heart diseases and needs the Person table to retrieve the name, address, and phone number of these healthcare providers. Figure 7.16 shows the query design. Figure 7.17 shows the query result in comparison with the data records in the Person and the CareProvider tables.

This query also shows how to change a column heading in the query result (see the Field row of the PersonName field in Figure 7.16). The column heading for the data field of PersonName is changed to Name, as seen in the query result in Figure 7.17.

When we put two tables in the table pane of a query design, we join the data fields of the two tables. There are three types of join: inner join, left outer join, and right outer join. In the table pane of the Design View of the query shown in Figure 7.16, right-clicking the line linking the ID field of the Person table and the ProviderID field of the CareProvider table brings up a

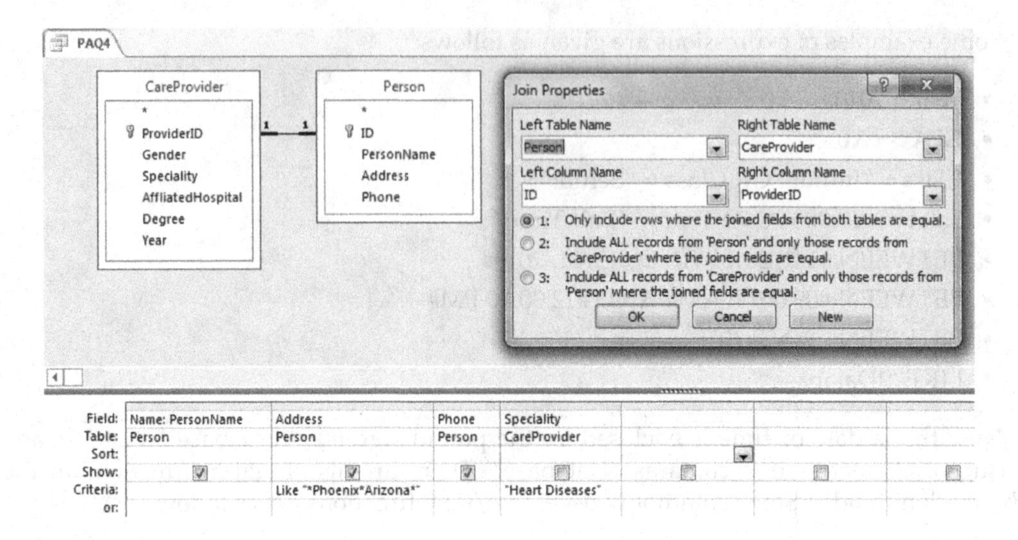

FIGURE 7.16
Design of a Select query, PAQ4, using multiple tables and the join properties.

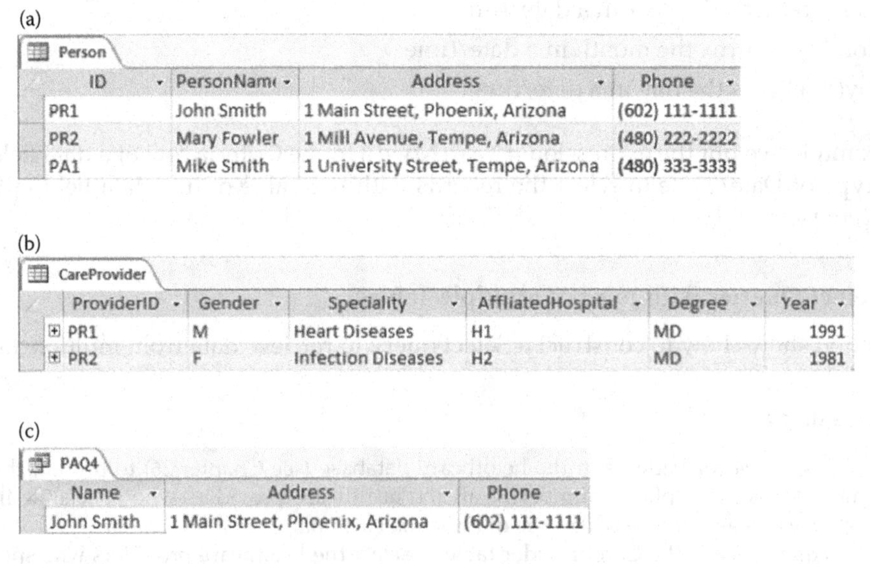

(a)

Person

ID	PersonName	Address	Phone
PR1	John Smith	1 Main Street, Phoenix, Arizona	(602) 111-1111
PR2	Mary Fowler	1 Mill Avenue, Tempe, Arizona	(480) 222-2222
PA1	Mike Smith	1 University Street, Tempe, Arizona	(480) 333-3333

(b)

CareProvider

ProviderID	Gender	Speciality	AffliatedHospital	Degree	Year
PR1	M	Heart Diseases	H1	MD	1991
PR2	F	Infection Diseases	H2	MD	1981

(c)

PAQ4

Name	Address	Phone
John Smith	1 Main Street, Phoenix, Arizona	(602) 111-1111

FIGURE 7.17
Query result of a Select query, PAQ4, using multiple tables. (a) The records in the Person table. (b) The records in the CareProvider table. (c) The query result for PAQ4.

pop-up menu with two items: Join Properties and Delete. Selecting the Join Properties brings up the Join Properties window, as shown in Figure 7.16. The window has three checks:

1. Only include rows where the joined fields from both tables are equal
2. Include ALL records from "Person" and only those records from "CareProvider" where the joined fields are equal
3. Include ALL records from "CareProvider" and only those records from "Person" where the joined fields are equal

The first check is for the inner join, which is the default join type. In Figure 7.16, the inner join of the Person and the CareProvider tables is checked. Because the inner join includes in the query result rows or records where the joined fields from both tables are equal, the inner join includes only the first two records of the Person table whose ID field has a value equal to a value of the ProviderID field in a record in the CareProvider table. The third record of the Person table is not included in the query result because there is no record in the CareProvider table that has a value of ProviderID equal to the value of ID in the third record of the Person table. For a record of the Person and the CareProvider tables with the same value of the joined fields, the inner join creates a record including all the data fields of the Person and the CareProvider tables in the query result. Figure 7.18a shows the resulting two records of the inner join in the query result. After the join operation, the other query operations (e.g., selection of columns and rows, and sorting) specified in the design grid are executed on the join results to produce the query result. That is, when we put two or more tables in the table pane of a query design, the join operation is first performed, which is followed by other operations specified by the design grid.

The second and third checks in the Join Properties window in Figure 7.16 are for the outer joins. In the Join Properties window, the Person table is taken as the left table, and the CareProvider table is taken as the right table. The second check is for the left outer join, and the third check is for the right outer join. Because the left outer join includes all the records from the left table (the "Person" table in Figure 7.16) and only those records from the right table (the "CareProvider" table) where the joined fields are equal, the left outer join produces the join result, as shown Figure 7.18b. In Figure 7.18b, there are three records instead of two in Figure 7.18a because all the three records of the Person table are included in the join result. The record with the ID value of PA1, which does not have the equal value of ProviderID in the CareProvider table, is included in the join result of the left outer join but without the values of the data fields in the CareProvider table.

(a)

ID	PersonName	Address	Phone	ProviderID	Gender	Speciality	AffliatedHospital	Degree	Year
PR1	John Smith	1 Main Street, Phoenix, Arizona	(602) 111-1111	PR1	M	Heart Diseases	H1	MD	1991
PR2	Mary Fowler	1 Mill Avenue, Tempe, Arizona	(480) 222-2222	PR2	F	Infection Diseases	H2	MD	1981

(b)

ID	PersonName	Address	Phone	ProviderID	Gender	Speciality	AffliatedHospital	Degree	Year
PA1	Mike Smith	1 University Street, Tempe, Arizona	(480) 333-3333						
PR1	John Smith	1 Main Street, Phoenix, Arizona	(602) 111-1111	PR1	M	Heart Diseases	H1	MD	1991
PR2	Mary Fowler	1 Mill Avenue, Tempe, Arizona	(480) 222-2222	PR2	F	Infection Diseases	H2	MD	1981

(c)

ID	PersonName	Address	Phone	ProviderID	Gender	Speciality	AffliatedHospital	Degree	Year
PR1	John Smith	1 Main Street, Phoenix, Arizona	(602) 111-1111	PR1	M	Heart Diseases	H1	MD	1991
PR2	Mary Fowler	1 Mill Avenue, Tempe, Arizona	(480) 222-2222	PR2	F	Infection Diseases	H2	MD	1981

(d)

ID	PersonName	Address	Phone	ProviderID	Gender	Speciality	AffliatedHospital	Degree	Year
PR1	John Smith	1 Main Street, Phoenix, Arizona	(602) 111-1111	PR1	M	Heart Diseases	H1	MD	1991
PR1	John Smith	1 Main Street, Phoenix, Arizona	(602) 111-1111	PR2	F	Infection Diseases	H2	MD	1981
PR2	Mary Fowler	1 Mill Avenue, Tempe, Arizona	(480) 222-2222	PR1	M	Heart Diseases	H1	MD	1991
PR2	Mary Fowler	1 Mill Avenue, Tempe, Arizona	(480) 222-2222	PR2	F	Infection Diseases	H2	MD	1981
PA1	Mike Smith	1 University Street, Tempe, Arizona	(480) 333-3333	PR1	M	Heart Diseases	H1	MD	1991
PA1	Mike Smith	1 University Street, Tempe, Arizona	(480) 333-3333	PR2	F	Infection Diseases	H2	MD	1981

FIGURE 7.18

Effects of inner join, left outer join, and right outer join. (a) The join result of the inner join between the Person table and the CareProvider table. (b) The join result of the left outer join between the Person table and the CareProvider table. (c) The join result of the right outer join between the Person table and the CareProvider table. (d) The join result involving two tables that are not linked.

Figure 7.18c shows the join result of the right outer join. This join result is the same as the join result of the inner join in Figure 7.18a because the ProviderID value in every record of the CareProvider table has the equal value in a record of the Person table. In the following text, the inner join is used as the default in a query design.

If two tables in the table pane of a query design do not have joined fields or are not linked, the join of the two tables produces all the combinations of records in the two tables. For example, if we remove the relationship between the Person and the CareProvider tables and put these two tables in the table pane of a query design, the join result includes 3 × 2 records, as shown in Figure 7.18d, because the Person table has 3 records and the CareProvider table has 2 records. If three or more tables are put in the table pane of a query design, the join result includes the data fields from all the tables.

7.3.3 Select Queries with Parameters and Calculated Fields

Many queries require the user of a database to enter a parameter. For example, a user enters the title of a book to search for the book at http://www.amazon.com. We give several examples of designing Select queries with parameters.

Example 7.6

Design Query1 for an employer in the healthcare database (see Chapter 26), which is to list the name of insurance companies and their insurance plans whose employer premium is smaller than or equal to a given premium level.

For this query, the employer who wants to retrieve the data needs to enter a given premium level as a parameter. That is, this query takes a parameter input from the user of the database. Figure 7.19 shows the design of this query. It also shows the prompt for the input parameter when the query is executed.

Example 7.7

Design Query4 for healthcare providers in the healthcare database (see Chapter 26) to present appointment times and patient names for a given healthcare provider on a given date.

Figure 7.20 shows the query design. Note that the StartTime field appears in two columns—one for displaying StartTime and another for selecting the records with

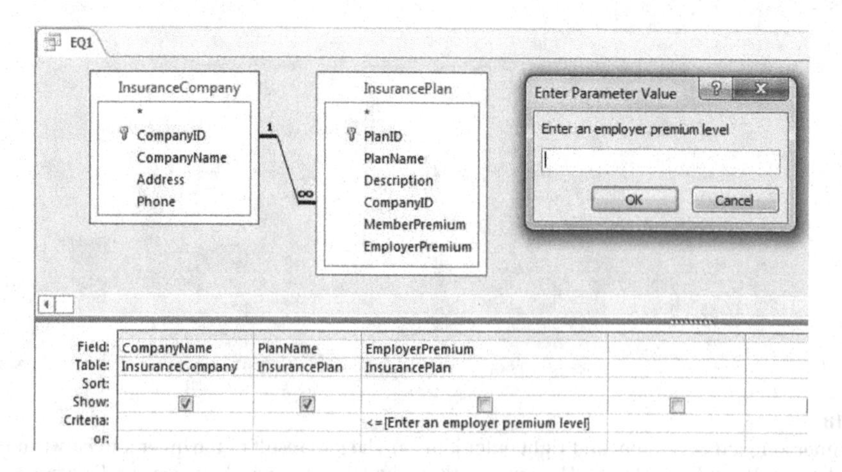

FIGURE 7.19
Design View of a Select query, EQ1, with a parameter.

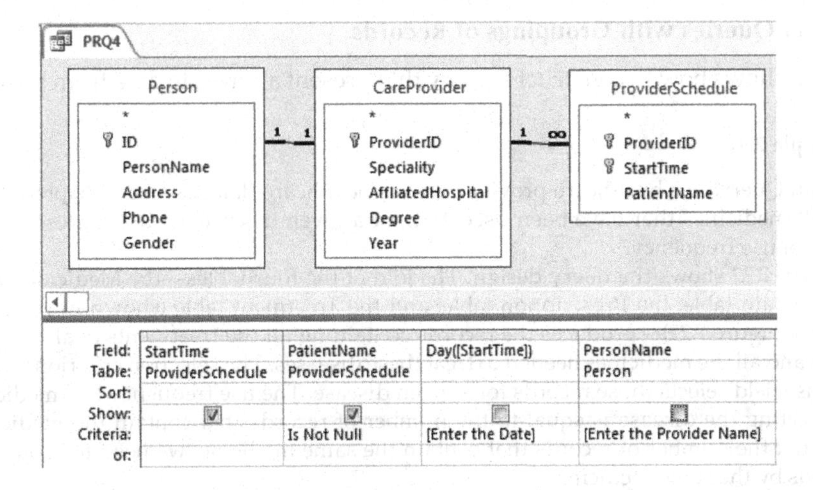

FIGURE 7.20
Design View of a Select query, PRQ4, with two parameters.

Day[StartTime] equal to the date embed by the user. The selection criterion for the PersonName field selects the records with a specific value of the CareProvider name entered by the user. The selection criterion for the PatientName field, Is Not Null, ensures that a specific patient has been scheduled at an appointment time.

Example 7.8 shows a query that uses a calculated field.

Example 7.8

Design Query2 for pharmacies in the healthcare database (see Chapter 26) to give the name and ID of all medicines whose inventory level falls below the minimum level at a given pharmacy.

Figure 7.21 shows the query design. In the design grid, the column with the heading of Difference is a calculated field that computes the difference between the values of two data fields, CurrentInventory and MinimumInventory, for a medicine. If the difference is less than 0, the query presents the name and ID of the medicine.

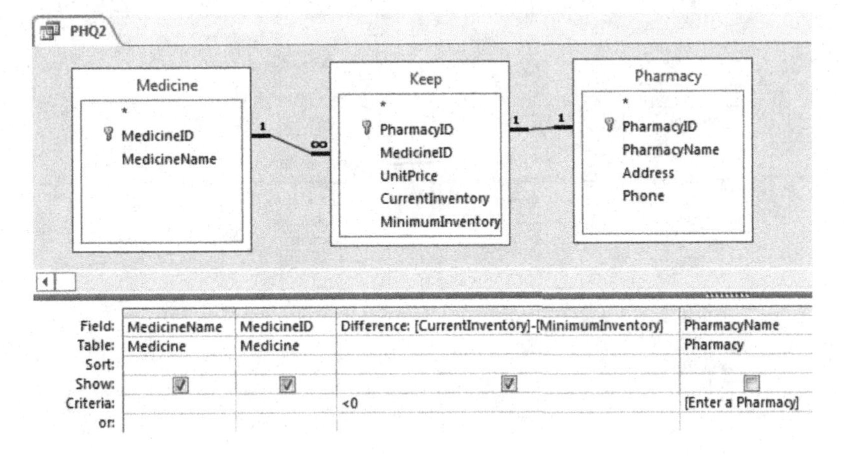

FIGURE 7.21
Design View of a Select query, PHQ2, with a calculated field.

7.3.4 Select Queries with Groupings of Records

This section shows how to construct a query that present aggregate data from the database.

Example 7.9

Design Query3 for healthcare providers in the healthcare database (see Chapter 26) to list all medicines that have been used to treat a given disease, sorted in descending order of use frequency.

Figure 7.22 shows the query design. The join of the four tables—the Medicine table, the Contain table, the Prescription table, and the Treatment table (shown in the table pane in Figure 7.22)—produces the records containing all the treatments of all the diseases and all the medicines needed to treat these diseases. The selection criterion for the Disease field selects those records for a given disease. The use frequency of a medicine for treating the disease is equal to the number of records that contain the medicine. To count the number of records that contain the same medicine, we need to group the records by the same medicine.

To enable the grouping of records by the medicine, we click on the Totals icon in the Design menu, as shown in Figure 7.22. This adds the Total row in the design grid of the query (see Figure 7.22). Clicking on the down arrow in the Total row of a column for a data field in the design grid shows the following choices for the Total row:

- Group By: use it if we want to group records by values of this data field
- Sum: use it if we want to get the sum of values from this data field in each group of records
- Avg: use it if we want to get the average of values from this data field in each group of records
- Min: use it if we want to get the minimum of values from this data field in each group of records
- Max: use it if we want to get the maximum of values from this data field in each group of records
- Count: use it if we want to get the count of values from this data field in each group of records

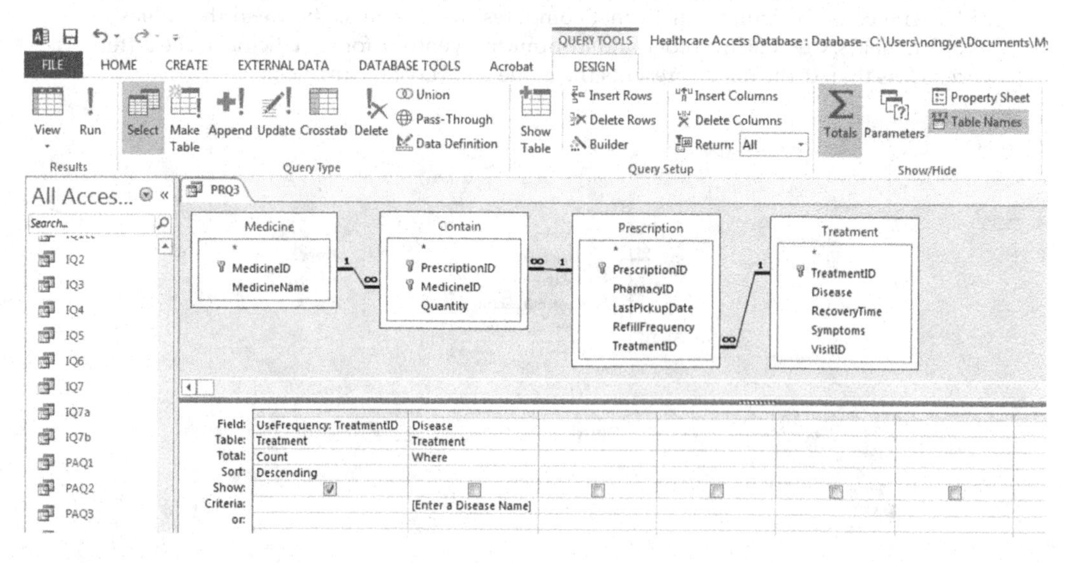

FIGURE 7.22
Design View of a Select query, PRQ3, with grouping and aggregate data.

- StDev: use it if we want to get the standard deviation of values from this data field in each group of records
- Var: use it if we want to get the variance of values from this data field in each group of records
- First: use it if we want to get the first value from this data field in each group of records
- Last: use it if we want to get the last value from this data field in each group of records
- Expression: use it if an expression is put in the data field as a calculated field
- Where: use it if the data field is used for a selection criterion

We use Group By in the Total row of the MedicineName field in the design grid to group the records in the join result by MedicineName. We use Count in the Total row of the TreatmentID field to count how many treatment records each MedicineName group contains, give the heading of UseFrequency for this count, and sort the query result in descending order of UseFrequency. We use Where in the Total row of the Disease field, which is used to specify the selection criterion. Figure 7.23 shows the records in the

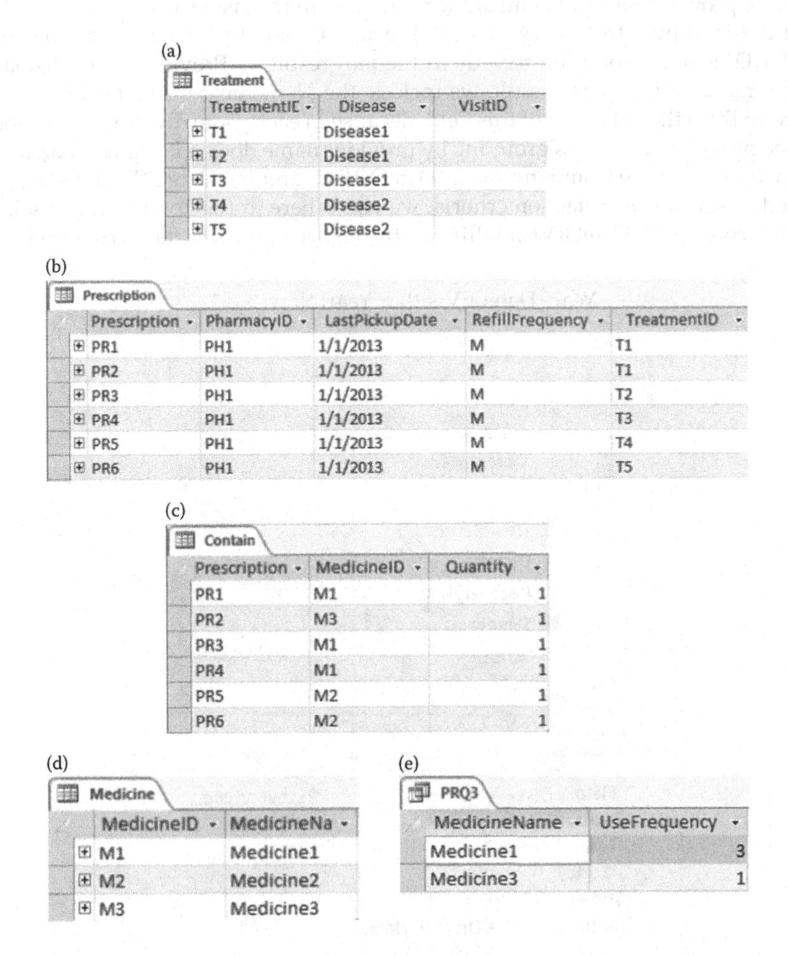

(a)

Treatment

TreatmentIE	Disease	VisitID
T1	Disease1	
T2	Disease1	
T3	Disease1	
T4	Disease2	
T5	Disease2	

(b)

Prescription

Prescription	PharmacyID	LastPickupDate	RefillFrequency	TreatmentID
PR1	PH1	1/1/2013	M	T1
PR2	PH1	1/1/2013	M	T1
PR3	PH1	1/1/2013	M	T2
PR4	PH1	1/1/2013	M	T3
PR5	PH1	1/1/2013	M	T4
PR6	PH1	1/1/2013	M	T5

(c)

Contain

Prescription	MedicineID	Quantity
PR1	M1	1
PR2	M3	1
PR3	M1	1
PR4	M1	1
PR5	M2	1
PR6	M2	1

(d)

Medicine

MedicineID	MedicineNa
M1	Medicine1
M2	Medicine2
M3	Medicine3

(e)

PRQ3

MedicineName	UseFrequency
Medicine1	3
Medicine3	1

FIGURE 7.23
Result of a Select query, PRQ3, with grouping and aggregate data. (a) Records in the Treatment table. (b) Records in the Prescription table. (c) Records in the Contain table. (d) Records in the Medicine table. (e) The query result for Example 7.9.

Treatment table, the Prescription, the Contain, and the Medicine tables, and the query result when we enter Disease1, as the input parameter.

A selection criterion can be applied to aggregate data. For example, we can put the selection criterion, >2, in the Criterion row of the UseFrequency column to display only the medicines whose use frequency is greater than 2.

Example 7.10

Design Query6 for insurance companies in the healthcare database (see Chapter 26) to give the average recovery time of a given disease.

Figure 7.24 shows the query design.

Example 7.11

Design Query2 for insurance companies in the healthcare database (see Chapter 26) to give the total number of patients who are 65 years or older and are treated by each healthcare provider on a given insurance company in the last year.

Figure 7.25 shows the query design. We use Group By in the Total row of the ProviderID field to group the records in the join result by ProviderID. To display the provider name in the query result, we include the Name field of the Person table and use Group By in the Total row of this field. Because in each group by ProviderID, there is only one provider name, this grouping by provider name does not change the grouping by ProviderID. For the CompanyName, DateOfVisit, and the DateOfBirth fields, which are used to specify the selection criteria, we use Where in the Total row. For selecting the join records with DateOfVisit falling in the last year, the selection criterion is

Year([DateOfVisit]) = Year(Now()) – 1.

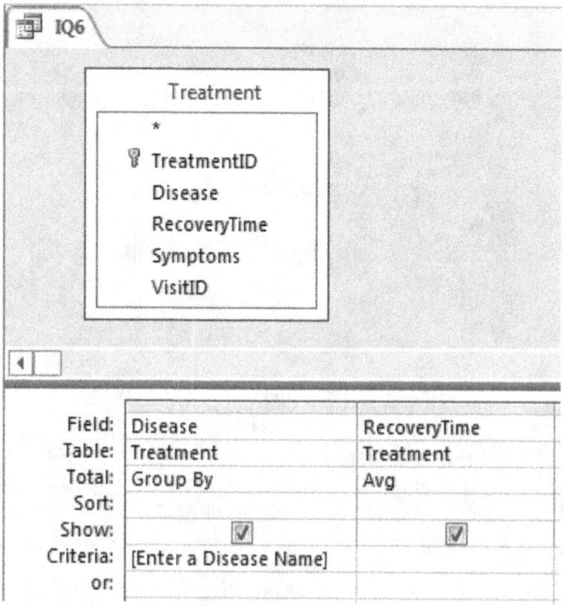

IQ6

Treatment

*
TreatmentID
Disease
RecoveryTime
Symptoms
VisitID

Field:	Disease	RecoveryTime
Table:	Treatment	Treatment
Total:	Group By	Avg
Sort:		
Show:	☑	☑
Criteria:	[Enter a Disease Name]	
or:		

FIGURE 7.24
Design View of a Select query, IQ6, with grouping and aggregate data.

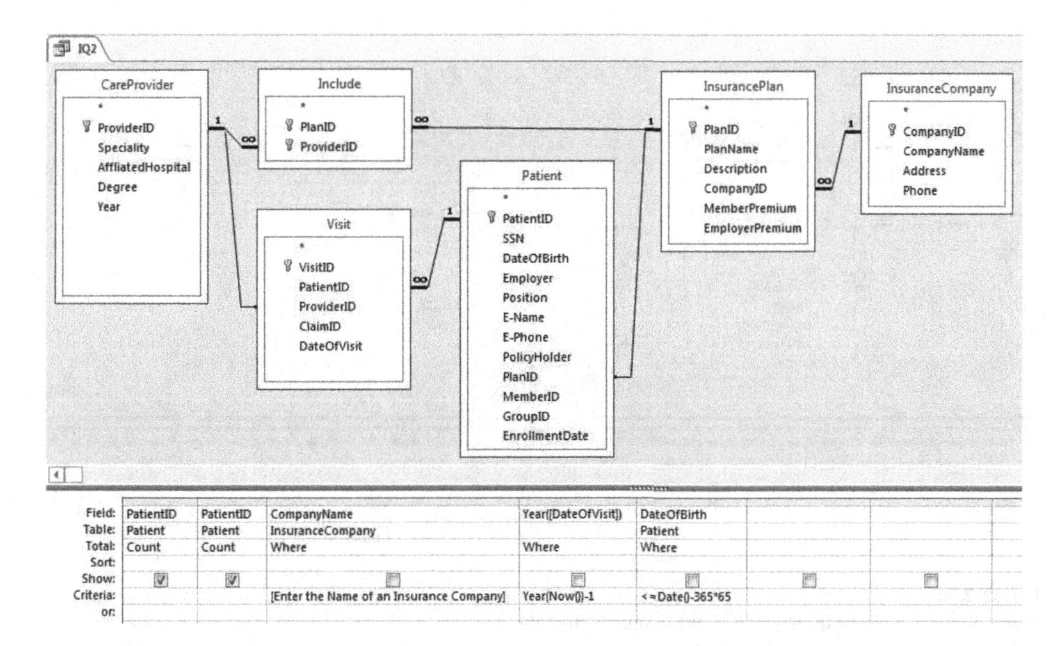

FIGURE 7.25
Design View of a Select query, IQ2, with grouping and aggregate data.

The existing field name is enclosed in square brackets. For selecting the join records for patients who are 65 years or older, the selection criterion is

DateOfBirth <= Date() – 365*65, or equivalently Date() – DateOfBirth >= 365*65

which means that the difference between the current date and the date of birth is equal to or more than 65 years.

Example 7.12

Design Query1 for pharmacies in the healthcare database (see Chapter 26) to give the total number of patients who are treated for a given disease in a given month and the total amount of a given medicine used to treat the disease in that month.

Figure 7.26 shows the query design. Note that this query does not use Group By in the Total row of any field. This means that all the records in the join result are treated as one group and that aggregate data (i.e., the count of PatientID and the Sum of Quantity for medicine) are obtained from this group.

Example 7.13

Design Query5 for healthcare providers in the healthcare database (see Chapter 26) to give the total amount of payments that a given healthcare provider received from patient and insurance payments in the current year.

Figure 7.27 shows the query design. The first field in the design grid is a calculated field to compute the total amount of payments that the selected provider received from patient and insurance payments. For this field, we use Expression in the Total row.

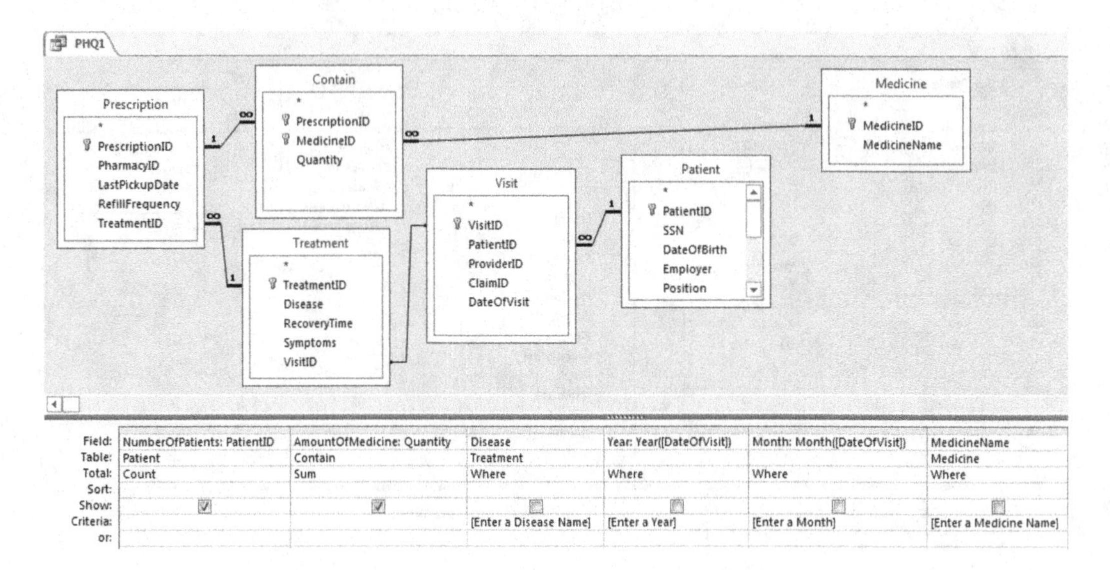

FIGURE 7.26
Design View of a Select query, PHQ1, with aggregate data.

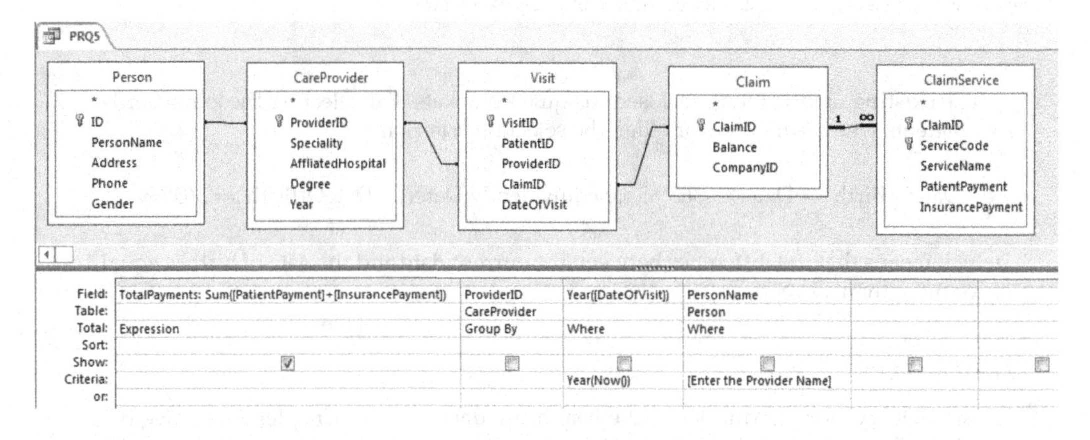

FIGURE 7.27
Design View of a Select query, PRQ5, with grouping and use of an aggregate data function in a calculated field.

7.3.5 Crosstab Queries

A Crosstab query is similar to a Select query with Group By in two data fields but presents the query result in a table format. Example 7.14 illustrates a Crosstab query.

Example 7.14

Design Query3 for employers in the healthcare database (see Chapter 26) to present a table showing the total number of healthcare providers in the insurance plan of each insurance company.

Figure 7.28a shows the design of a Select query with Group By in the Total row for the CompanyName and the PlanName fields. Figure 7.28b shows the query result.

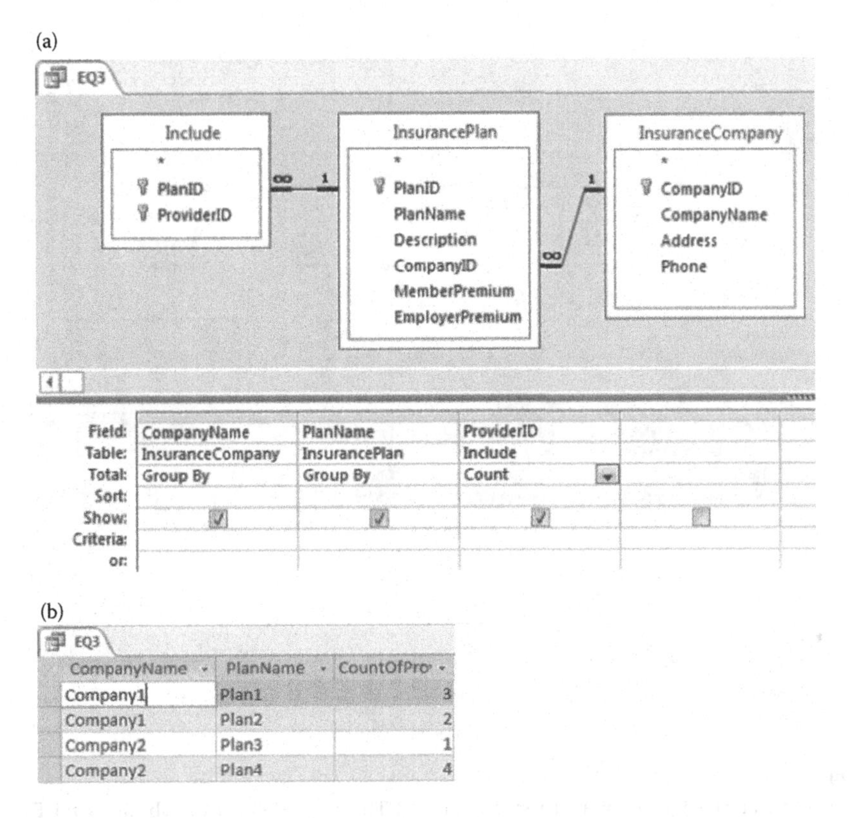

CompanyName	PlanName	CountOfPro
Company1	Plan1	3
Company1	Plan2	2
Company2	Plan3	1
Company2	Plan4	4

FIGURE 7.28
Design View and Datasheet View of a Select query, EQ3, with two groupings and aggregate data. (a) The design of a Select query for EQ3. (b) The query result of the Select query in Part a.

To create a Crosstab query, we first bring up the Design View of a Select query and then click on the Crosstab icon in the Design menu. This adds the Total and the Crosstab rows in the design grid, as shown in Figure 7.29a. Clicking on the down arrow button in the Crosstab row of a field brings up the following three choices:

- Row Heading
- Column Heading
- Value

Figure 7.29a shows the design of a Crosstab query, with CompanyName as the row heading, PlanName as the column heading, and the count of ProviderID as the value. In the Total row, we use Group By for the CompanyName field, Group By for the PlanName field, and Count for the ProviderID field. Figure 7.29b shows the query result. By comparing the query results in Figures 7.28b and 7.29b, we can see that the Crosstab query presents the same data as the Select query with two groupings in Figure 7.28 but uses a different format to display the query result. The Crosstab query puts each value of the data field as the row heading in a row, each value of the data field as the column heading in a column, and the value of the data field as the value in the cell of a row and a column. In the Crosstab query result in Figure 7.29b, the cells for Company1 and Plan3, Company1 and Plan4, Company2 and Plan1, and Company2 and Plan2 do not have a value because Company1 does not offer Plan3 or Plan4 and Company2 does not offer Plan1 or Plan2.

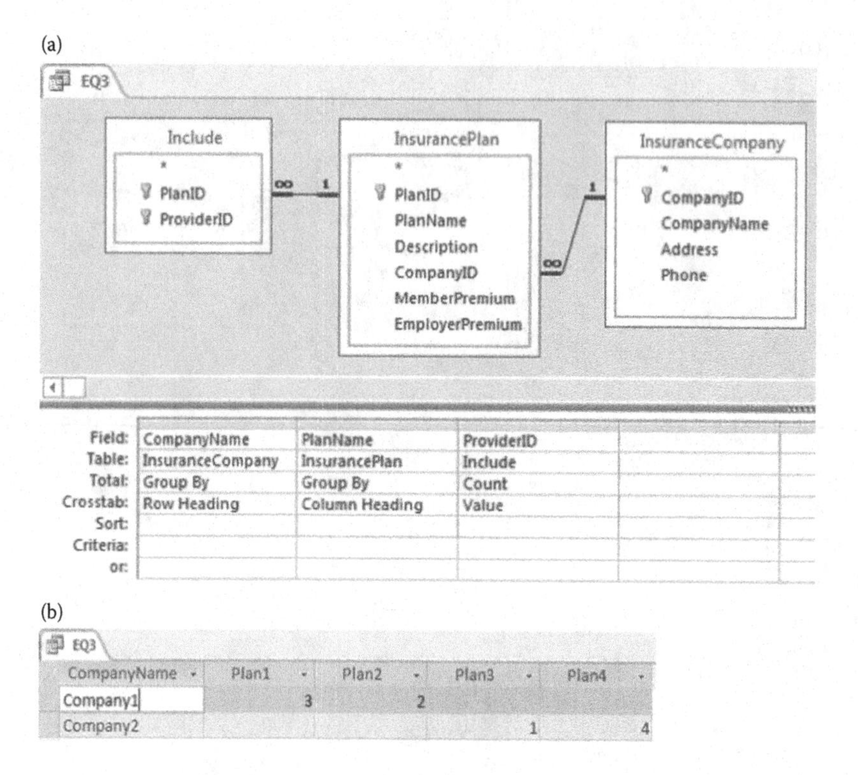

FIGURE 7.29
Design View and Datasheet View of a Crosstab query. (a) The design of a Crosstab query for Example 7.14. (b) The query result of the Crosstab query in Part a.

7.3.6 Embedded Select Queries

The Select queries in the previous sections retrieve data from tables in a database. A Select query can be designed to retrieve data from the result of another Select query, creating embedded queries. Examples 7.15 and 7.16 show how to design embedded queries.

Example 7.15

Design Query1 for insurance companies in the healthcare database (see Chapter 26) to give the total number of all patients, female patients, and male patients treated by each healthcare provider, sorted in alphabetical order of provider name.

The group of all patients, the group of female patients, and the group of male patients are three different groups. We cannot create three different groups in one query design. Instead, we create sub-queries (see Figure 7.30a–e) to create three different groups and count for each group. Then, we create a query (see Figure 7.30f), which adds the sub-queries in the table pane of the query design and puts together the results of the sub-queries. Because the Gender field is in the Person table, which is linked to both the Patient and the CareProvider tables, using a selection criterion of letting the Gender field of the Person table equal to "F" will produce records in the join result containing female patients treated by female healthcare providers instead of female patients treated by all healthcare providers. To select female patients treated by all healthcare providers, we first create a sub-query to retrieve PatientID of female patients using the Patient and the Person tables (see Figure 7.30b). We then create a

(a)

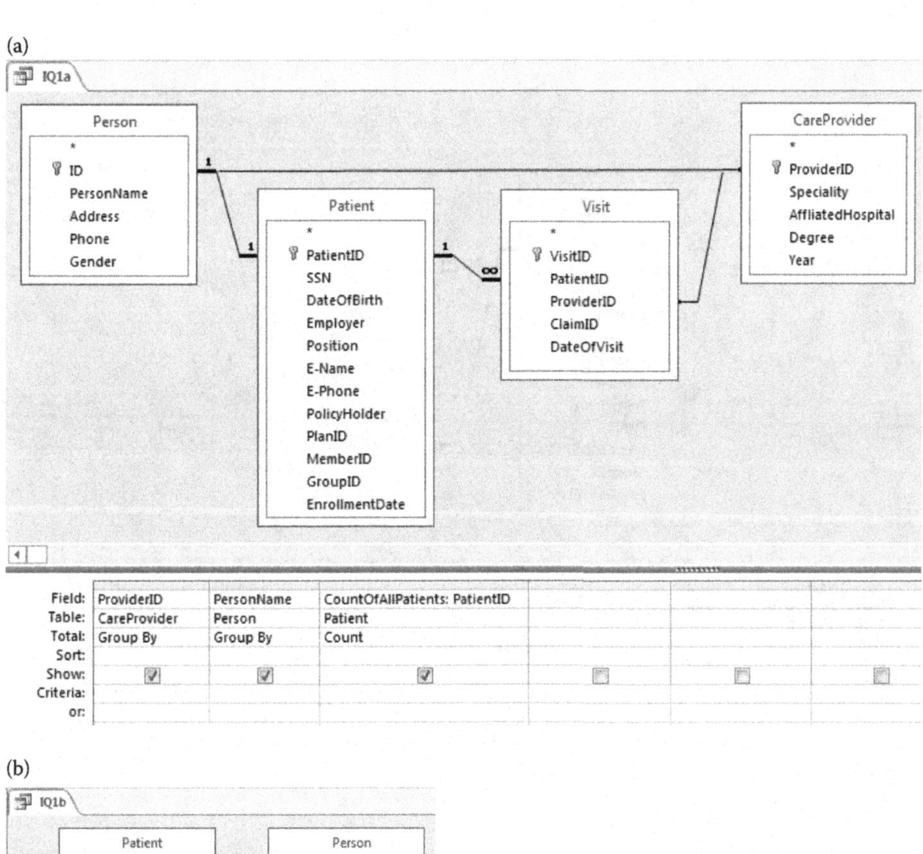

(b)

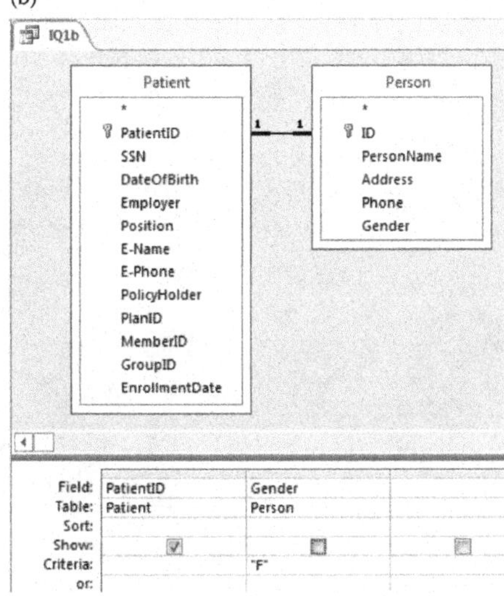

FIGURE 7.30
Design View of embedded queries for IQ1. (a) The design of a subquery to count the number of all patients treated by each healthcare provider. (b) The design of a subquery to retrieve the PatientID of female patients.

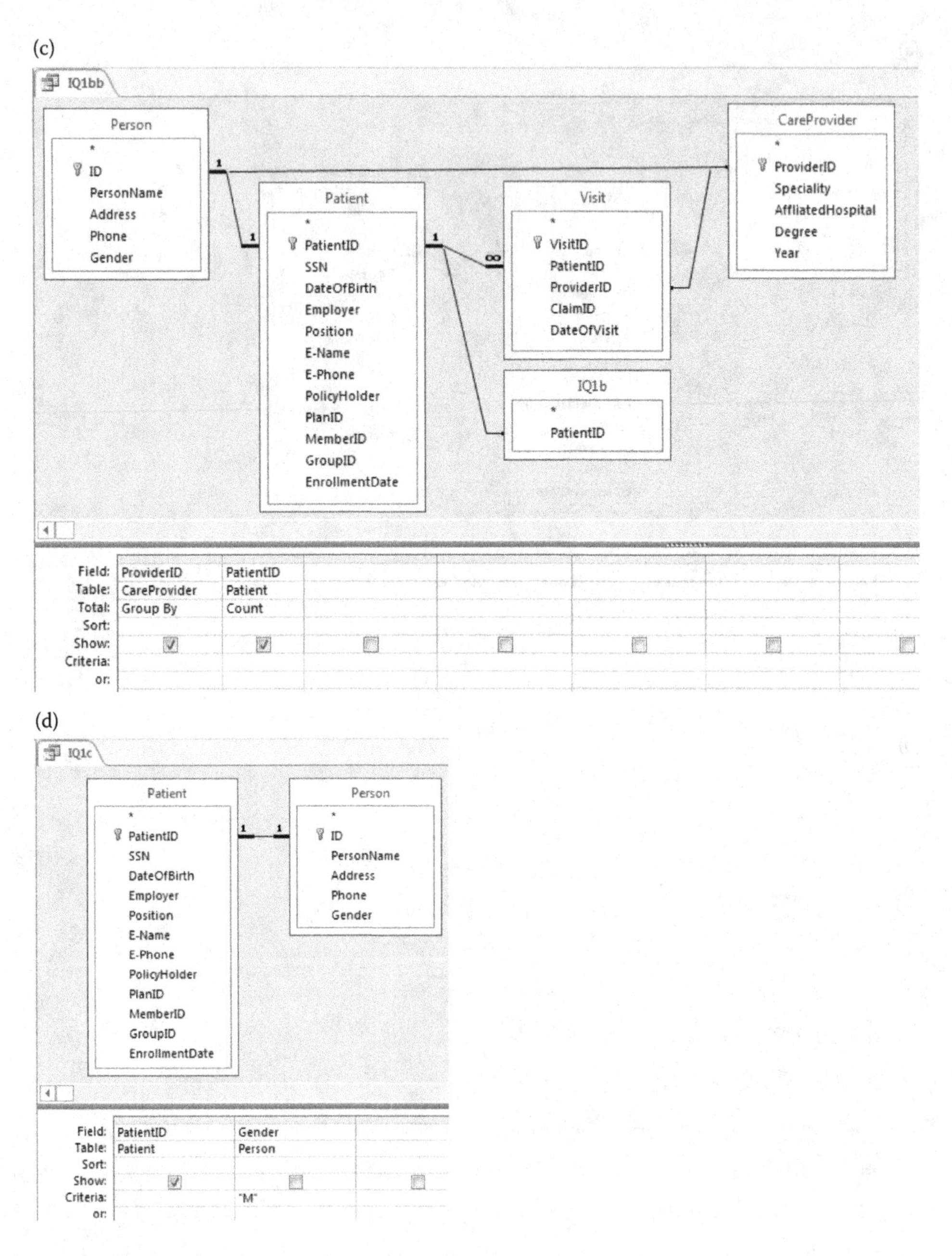

FIGURE 7.30 (Continued)
Design View of embedded queries for IQ1. (c) The design of a subquery to produce the group and count of female patients treated by each healthcare provider. (d) The design of a subquery to retrieve the PatientID of male patients.

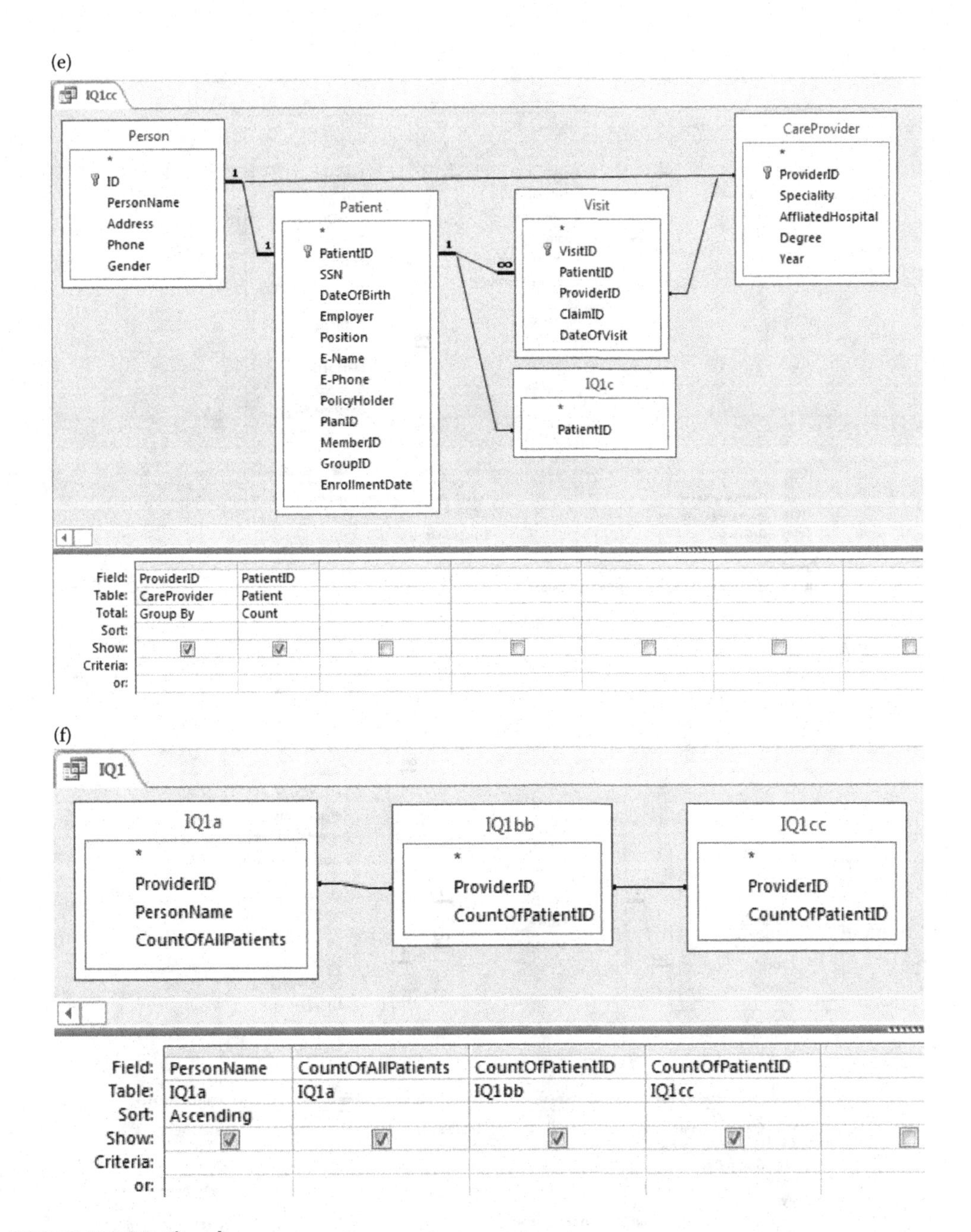

FIGURE 7.30 (Continued)
Design View of embedded queries for IQ1. (e) The design of a subquery to produce the group and count of male patients treated by each healthcare provider. (f) The design of the query to put together the counts of all patients, female patients, and male patients treated by each healthcare provider using the three subqueries in Figure 7.30a, c, and e.

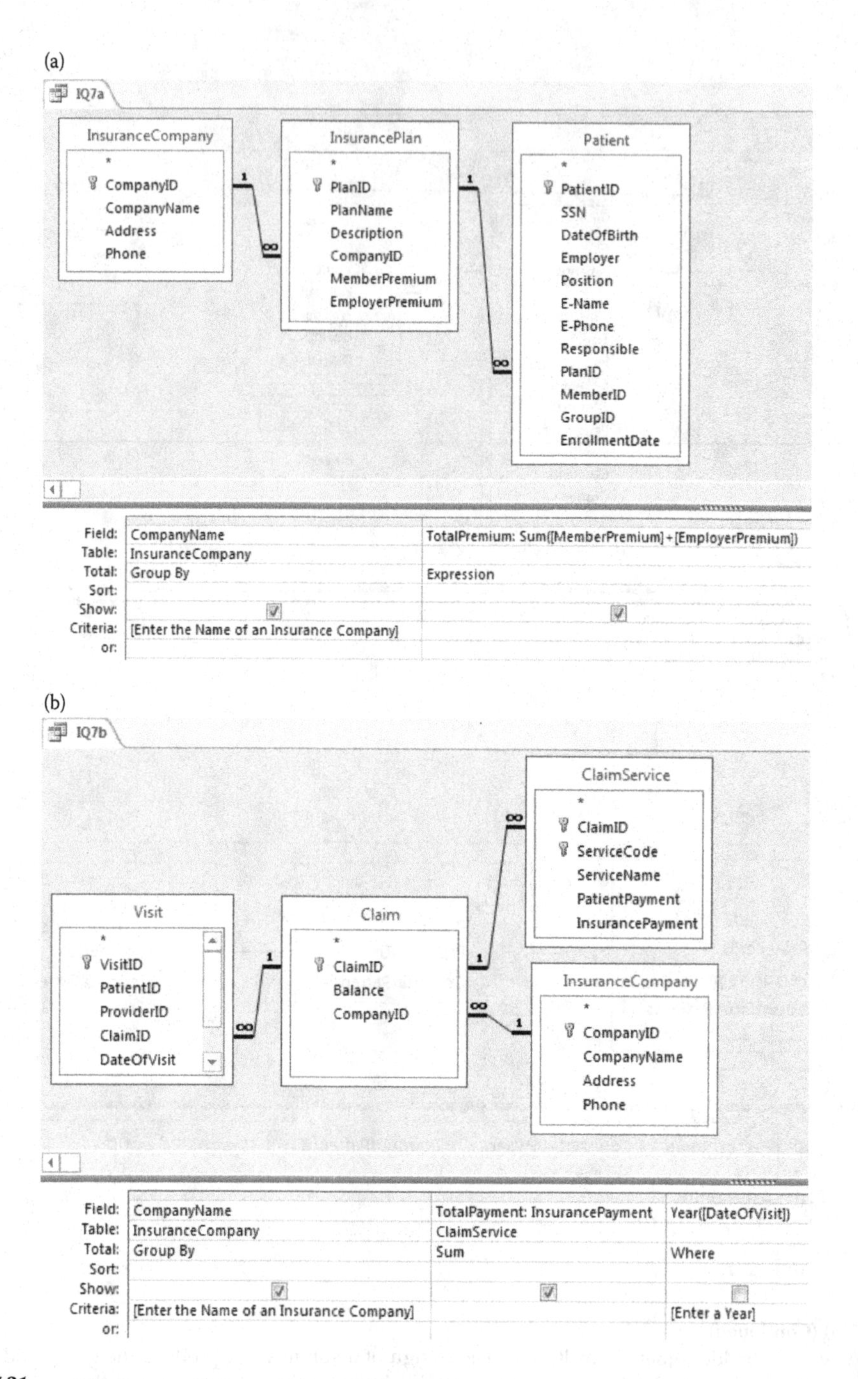

FIGURE 7.31
Design View of embedded queries for IQ7. (a) The design of a subquery to compute the total amount of premiums for IQ7. (b) The design of a subquery to compute the total insurance payments for IQ7.

sub-query to produce the group and count of female patients by including the sub-query in the table pane along with the Person, Patient, CareProvider, and Visit tables (see Figure 7.30c). To add the sub-query in Figure 7.30b to the table pane, we right-click anywhere in the table pane area, select the Show Table item, click on the Queries tab in the Show Table window, and select the sub-query. Including PatientID of only female patients from the sub-query in the table pane of the sub-query in Figure 7.30c produces the join result containing the records for female patients only due to the inner join of the tables and the sub-query in Figure 7.30b. Figure 7.30c and d shows two sub-queries working together to provide the group and count of male patients treated by each healthcare provider. Finally, the query in Figure 7.30f puts together the counts of all patients, female patients, and male patients treated by each healthcare provider using the three sub-queries in Figure 7.30a, c, and e. We establish the relationships of the three sub-queries in the table pane of the query in Figure 7.30f through the common fields of PatientID in the three sub-queries to have the inner join of the query results from the three sub-queries.

Example 7.16

Design Query7 for insurance companies in the healthcare database (see Chapter 26) to give the total amount of premiums received, the total amount of insurance payments, and the total amount of profit for a given insurance company in a given year.

Data from different tables are needed to compute the total amount of premiums and the total amount of insurance payments. Hence, we create two sub-queries to compute the total amount of premiums and the total amount of insurance payments. Then, we design a query that puts together the results of the two sub-queries, link the two sub-queries to have the inner join through the CompanyName field, and compute the total amount of profit. Note that, in the sub-query in Figure 7.31a, to compute the total amount of premiums, we use a calculated field including the sum function and Expression in the Total row for this calculated field. Because the premiums are annual premiums, we do not need to specify a particular year when computing the total amount of premiums for the year.

(c)

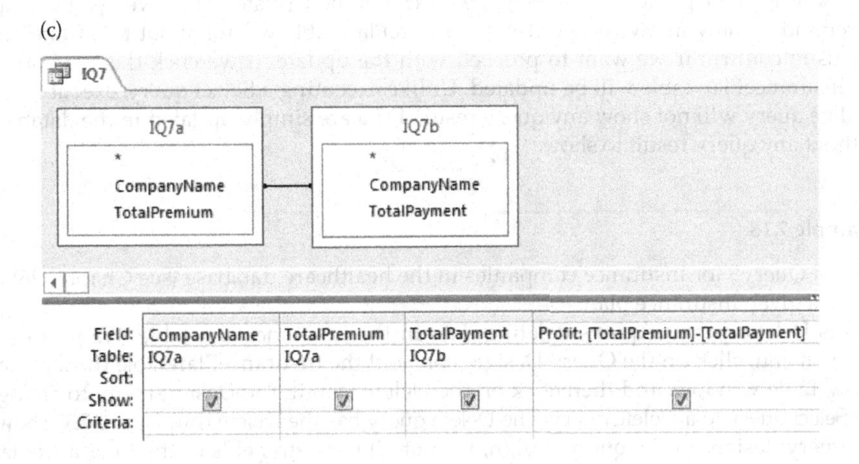

FIGURE 7.31 (Continued)
Design View of embedded queries for IQ7. (c) The design of the query to put together the results of the subqueries in Parts a and b.

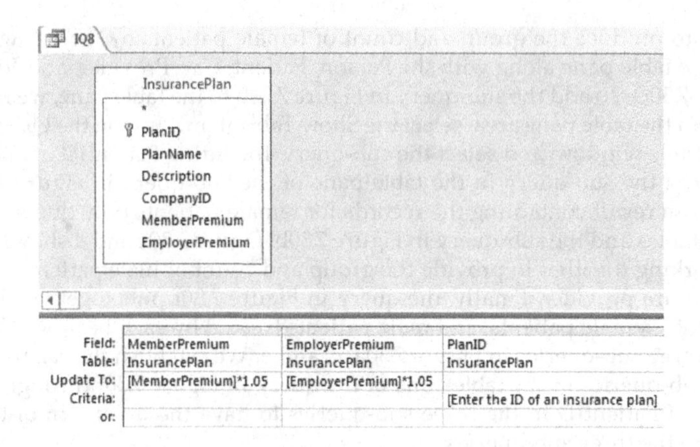

FIGURE 7.32
Design View of an Update query, IQ8.

7.4 Action Queries

An action query changes data stored in a database. We give examples of an Update query and a Delete query.

Example 7.17

Design Query8 for insurance companies in the healthcare database (see Chapter 26) to increase the member premium and the employer premium by 5%.

This Update query requires updating data in the InsurancePlan table. We go to the Create menu, click on the Query Design icon, add the InsurancePlan table through the Show Table window, and then click on the Update icon in the Design menu to change the Select query to an Update query. The Update query has the Update To row. Figure 7.32 shows the query design. In the query design, we create two data fields, MemberPremium and EmployerPremium, and put the expression of updating each data field in the Update To row. When we execute the Update query, we get a prompt to remind us how many rows of the InsurancePlan table we are about to update and ask us to confirm if we want to proceed with the update. If we click the Yes button, the InsurancePlan table will be updated. Unlike executing a Select query, executing an Update query will not show any query result. Data are simply updated in the database without any query result to show.

Example 7.18

Design Query9 for insurance companies in the healthcare database (see Chapter 26) to delete a given insurance plan.

This Delete query requires deleting data in the InsurancePlan table. We go to the Create menu, click on the Query Design icon, add the InsurancePlan table through the Show Table window, and then click on the Delete icon in the Design menu to change the Select query to a Delete query. The Delete query has the Delete row. Figure 7.33 shows the query design. In the query design, we put all the data fields of the InsurancePlan table in one column of the design field by selecting InsurancePlan.* and use From in the Delete row of this column. When we execute the Delete query, we get a prompt to remind us how many rows of the specified table we are about to delete and ask us to

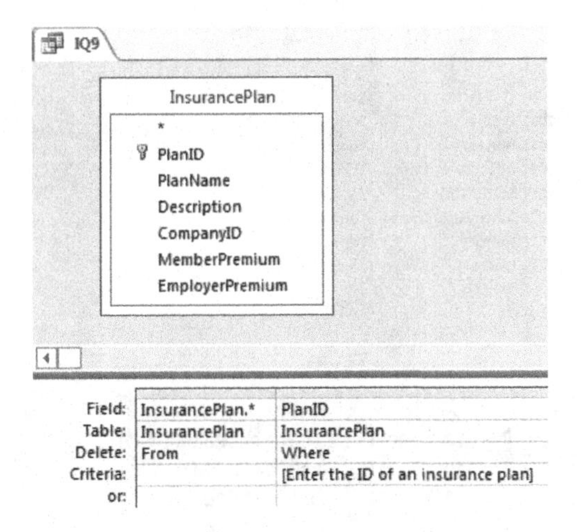

FIGURE 7.33
Design view of a Delete query, IQ9.

confirm if we want to proceed with the deletion. If we click the Yes button, the affected rows of the InsurancePlan table will be deleted. Unlike executing a Select query, executing a Delete query will not show any query result. Data are simply deleted in the database without any query result to show.

7.5 Summary

In this chapter, we introduce the use of tables in Access for data storage. Each relation in a relational model is implemented by a table in Access to store data. The properties of each data field in a table are described to define the primary key, the data type, the format, the input mask, the validation rule, and the required and indexed properties.

Each referential integrity constraint through primary key and foreign key pairings in a relational model is implemented by a relationship between two tables. The relationship type and the rules that Access uses to determine the relationship type are described. The use of the enforcement of referential integrity for a relationship is introduced.

This chapter also describes how to create various types of Select queries to retrieve data from a database, including Select queries using only one table, Select queries joining multiple tables, Select queries with parameters and calculated fields, Select queries to produce groupings of records and compute aggregate data from a group of records, and Crosstab queries. Action queries, specifically Update and Delete queries, are described to enable us to update and delete data in a database.

EXERCISES

1. Create tables and relationships of tables in Access to implement the relational model of the movie theater database from Exercise 8 in Chapter 6.
2. Create queries for the movie theater database described in Exercise 5 in Chapter 5.
3. Suppose that tblMovie, tblShow, and tblShowroom in the movie theater database from Exercise 1 have the following records.

tblMovie:

MovieID	Title	Year	Description	Location	ProducerID	TransactionNo	PurchasePrice	PurchaseDate	Pay
1	Moneyball	2011	Drama	Tempe	1	1	$603.00	10/1/2011	
2	The Thing	2011	Horror	Tempe	2	2	$500.00	9/30/2011	
3	The Three Musketeers	2011	Adventure/ Drama	Tempe	2	3	$600.00	9/10/2011	
4	In Time	2011	Sci-Fi/ Thriller	Tempe	4	4	$800.00	8/25/2011	
5	The Rum Diary	2011	Adventure/ Drama	Tempe	1	5	$750.00	9/1/2011	
6	Paranormal Activity 3	2011	Horror	Tempe	1	6	$820.00	8/30/2011	
7	The Descendants	2010	Drama	Tempe	3	7	$560.00	11/23/2011	

tblShow:

ShowID	ShowDay	ShowTime	RoomName	MovieID
1	10/31/2011	12:00:00 PM	Room 1	2
2	10/31/2011	4:30:00 PM	Room 1	2
3	10/31/2011	1:30:00 PM	Room 3	1
4	10/31/2011	2:20:00 PM	Room 4	3
5	10/31/2011	5:30:00 PM	Room 4	4
6	11/1/2011	10:00:00 AM	Room 1	5
7	11/1/2011	10:00:00 AM	Room 2	2
8	11/1/2011	12:30:00 PM	Room 1	5
9	11/2/2011	5:30:00 PM	Room 3	3
10	11/2/2011	7:30:00 PM	Room 3	3
11	11/2/2011	4:30:00 PM	Room 1	6
12	11/2/2011	4:00:00 PM	Room 4	7
13	11/2/2011	7:30:00 PM	Room 1	6
14	11/2/2011	10:00:00 PM	Room 1	6
15	11/2/2011	6:30:00 PM	Room 4	7

tblShowroom:

RoomName	Location	Capacity
Room 1	Tempe	130
Room 2	Tempe	150
Room 3	Tempe	120
Room 4	Tempe	100

a. How many records will be produced by the following query, given that the inner join is used for each relationship? Show these data records.

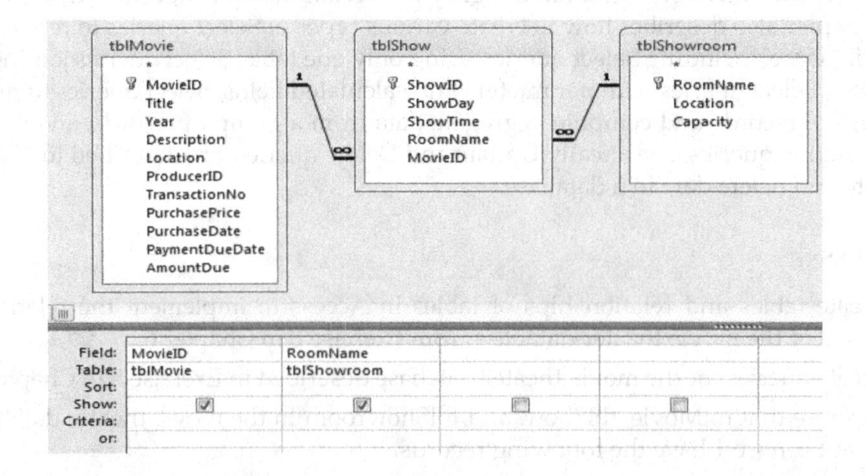

b. How many records will be produced by the following query, given that the inner join is used for each relationship? Show these data records.

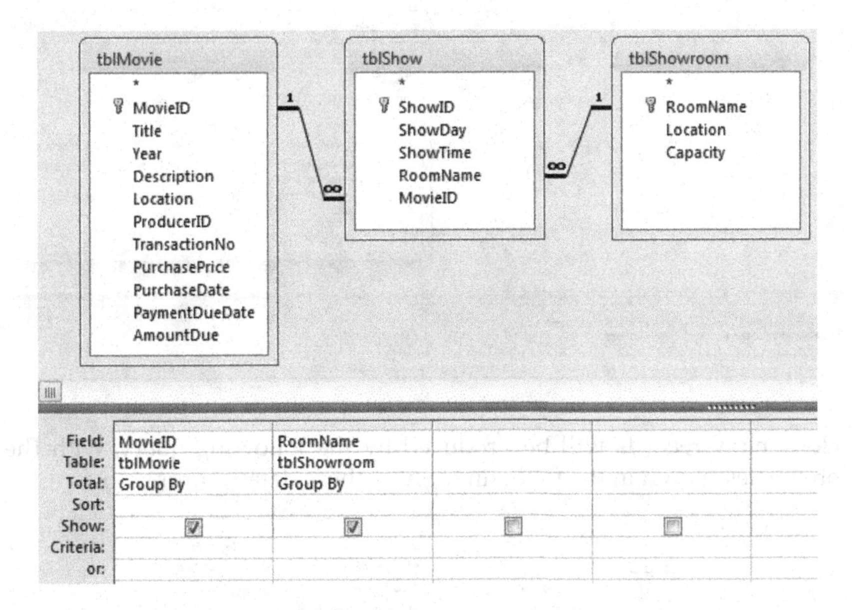

4. For the movie theater database from Exercise 1, create a query that prompts for a date and return the movie show schedule for the selected date and the total number of tickets per show as follows.

ShowDay	MovieID	Title	ShowTime	TicketsSold	Capacity	AvailableTickets
11/2/2011	3	The Three Musketeers	5:30:00 PM	1	120	119
11/2/2011	3	The Three Musketeers	7:30:00 PM	3	120	117
11/2/2011	6	Paranormal Activity 3	7:30:00 PM	1	130	129
11/2/2011	6	Paranormal Activity 3	10:00:00 PM	1	130	129
11/2/2011	7	The Descendants	4:00:00 PM	1	100	99

5. Suppose that tblCustomer and tblMember in the movie theater database from Exercise 1 have the following records.

tblCustomer:

CustomerID	CuName	CuAddress	CuPhone	CuEmail
1	James	Tempe AZ 8528	(480) 516-3542	jjames@gmail.com
2	Susan	Mesa AZ 85201	(408) 657-9842	susanlee@gmail.com
3	Eric	Chandler AZ 85	(510) 236-4789	ericyoung20@gmail.com
4	Jing	Tempe AZ 8528	(408) 675-1245	jing20@gmail.com
5	Josh	CA 85221	(510) 364-7812	josh@hotmail.com
6	Apple	AZ 85281	(480) 129-3112	apple@gmail.com
7	Yang	AZ 85281	(480) 194-3310	Yang100@hotmail.com
8	Hong	AZ 85201	(408) 194-9321	Hong25@gmail.com

tblMember:

CustomerID	LoginName	Password
1	Jsmith	jsh2130
2	Slee29	sle9!34
3	EricYoung	ecyng12094

a. How many records will be produced by the following query, with the join properties shown in the following? Show these data records.

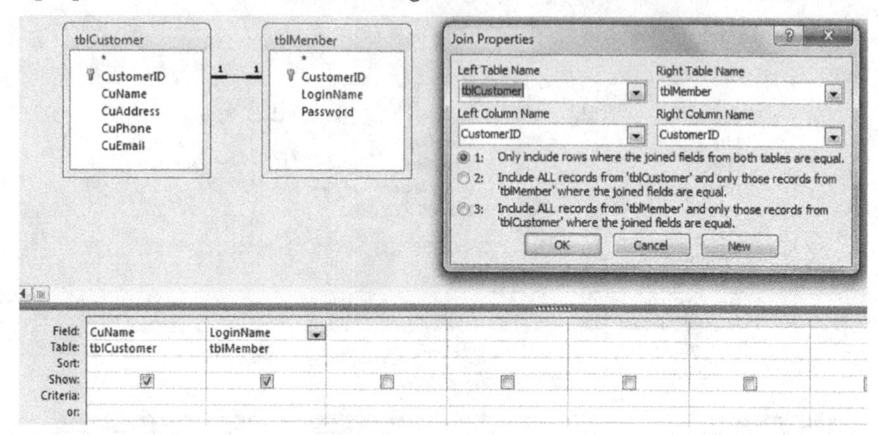

b. How many records will be produced by the following query, with the join properties shown in the following? Show these data records.

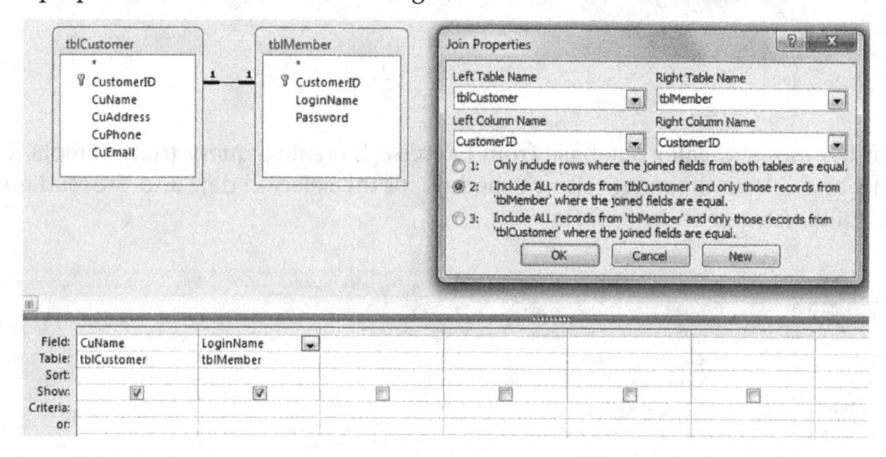

c. How many records will be produced by the following query, with the join properties shown in the following? Show these data records.

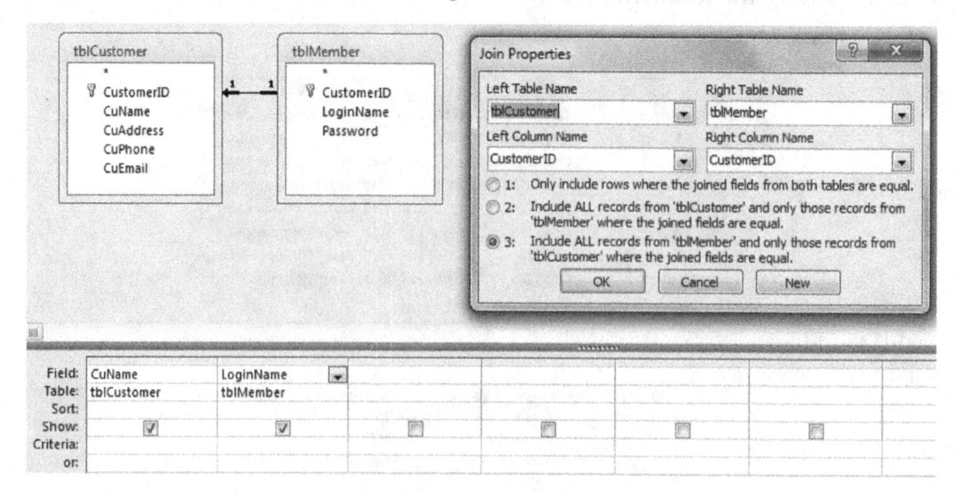

d. Records in tblTicket are shown in the following. How many records will be produced by the following query?

tblTicket:

TicketNo	Price	ShowID	CustomerID	BuyDate
1	$9.50	1	1	10/31/2011
2	$9.50	2	2	10/31/2011
3	$12.50	3	3	10/31/2011
4	$9.50	5	3	10/31/2011
5	$9.50	7	5	11/1/2011
6	$12.50	6	8	11/1/2011
7	$12.50	6	7	11/1/2011
8	$9.50	7	1	11/1/2011
9	$12.50	8	2	11/1/2011
10	$9.50	9	1	11/2/2011
11	$9.50	10	2	11/2/2011
12	$9.50	10	3	11/2/2011
13	$9.50	10	4	11/2/2011
14	$12.50	12	5	11/2/2011
15	$12.50	13	6	11/2/2011
16	$12.50	14	7	11/2/2011

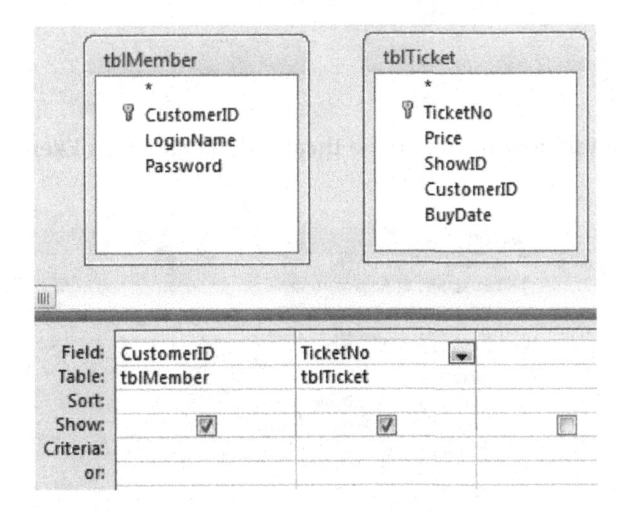

6. The design and the data records of tblCustomer in the movie theater database from Exercise 1 are shown in the following. Use the design template to design a query that gives the name and the e-mail address of the customers whose address has the zip code of 85281.

tblCustomer

Field Name	Data Type
CustomerID	AutoNumber
CuName	Text
CuAddress	Text
CuPhone	Text
CuEmail	Text

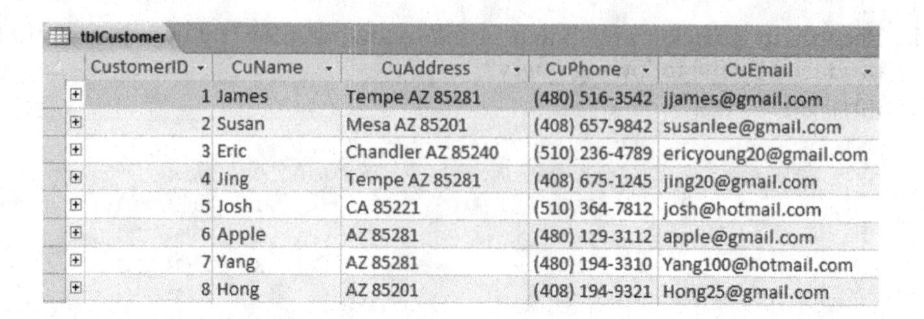

tblCustomer

CustomerID	CuName	CuAddress	CuPhone	CuEmail
1	James	Tempe AZ 85281	(480) 516-3542	jjames@gmail.com
2	Susan	Mesa AZ 85201	(408) 657-9842	susanlee@gmail.com
3	Eric	Chandler AZ 85240	(510) 236-4789	ericyoung20@gmail.com
4	Jing	Tempe AZ 85281	(408) 675-1245	jing20@gmail.com
5	Josh	CA 85221	(510) 364-7812	josh@hotmail.com
6	Apple	AZ 85281	(480) 129-3112	apple@gmail.com
7	Yang	AZ 85281	(480) 194-3310	Yang100@hotmail.com
8	Hong	AZ 85201	(408) 194-9321	Hong25@gmail.com

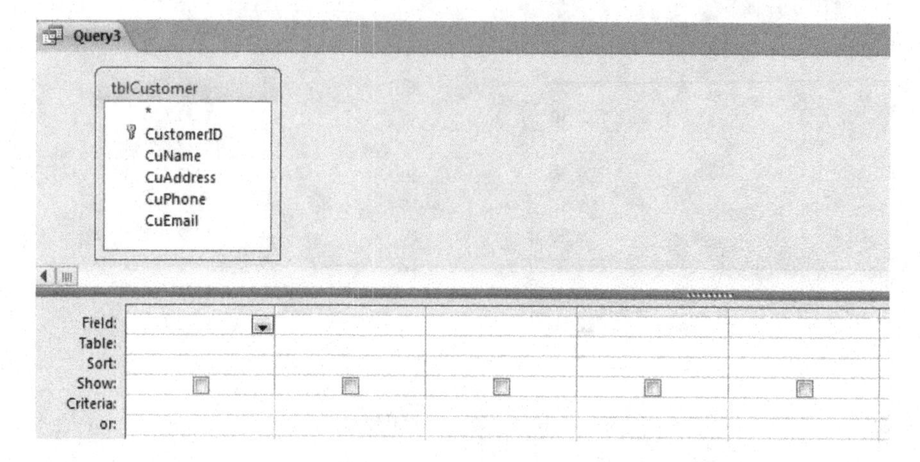

7. The records of tblShow in the movie theater database from Exercise 1 are shown as follows.

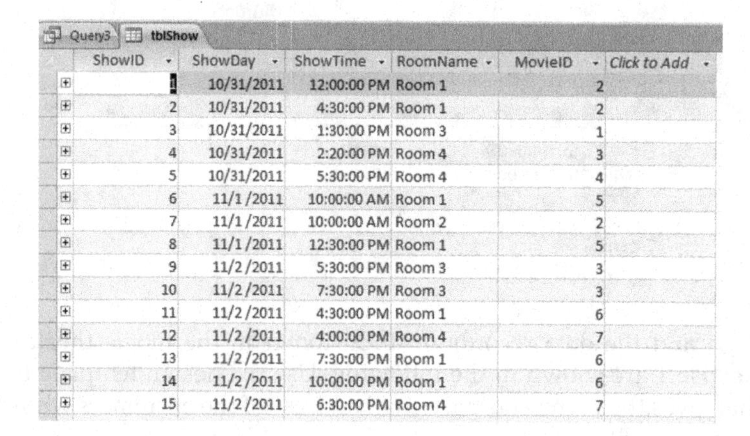

Query3 | tblShow

ShowID	ShowDay	ShowTime	RoomName	MovieID	Click to Add
1	10/31/2011	12:00:00 PM	Room 1	2	
2	10/31/2011	4:30:00 PM	Room 1	2	
3	10/31/2011	1:30:00 PM	Room 3	1	
4	10/31/2011	2:20:00 PM	Room 4	3	
5	10/31/2011	5:30:00 PM	Room 4	4	
6	11/1/2011	10:00:00 AM	Room 1	5	
7	11/1/2011	10:00:00 AM	Room 2	2	
8	11/1/2011	12:30:00 PM	Room 1	5	
9	11/2/2011	5:30:00 PM	Room 3	3	
10	11/2/2011	7:30:00 PM	Room 3	3	
11	11/2/2011	4:30:00 PM	Room 1	6	
12	11/2/2011	4:00:00 PM	Room 4	7	
13	11/2/2011	7:30:00 PM	Room 1	6	
14	11/2/2011	10:00:00 PM	Room 1	6	
15	11/2/2011	6:30:00 PM	Room 4	7	

a. How many records are produced by the following query? List all the records in the query result.

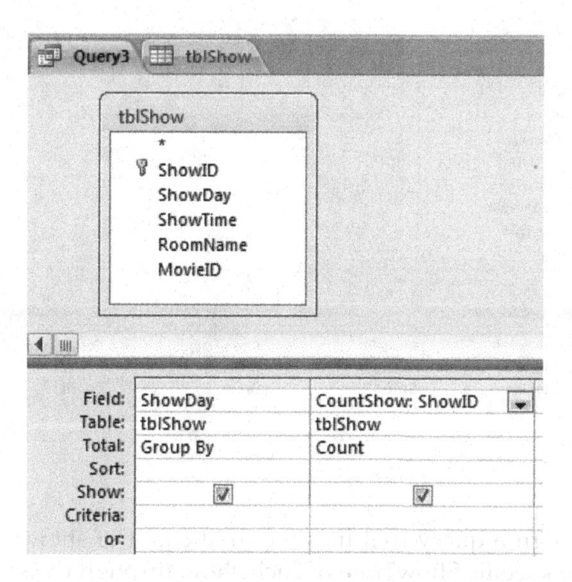

b. How many records are produced by the following query? List all the records in the query result.

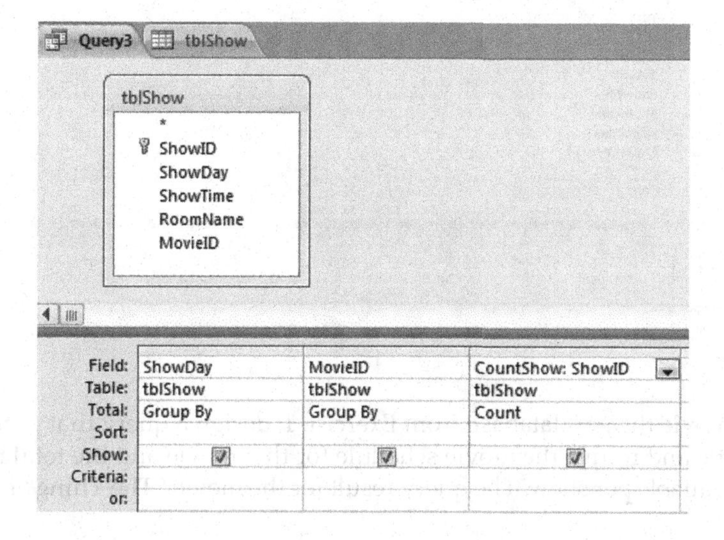

c. How many records are produced by the following query? List all the records in the query result.

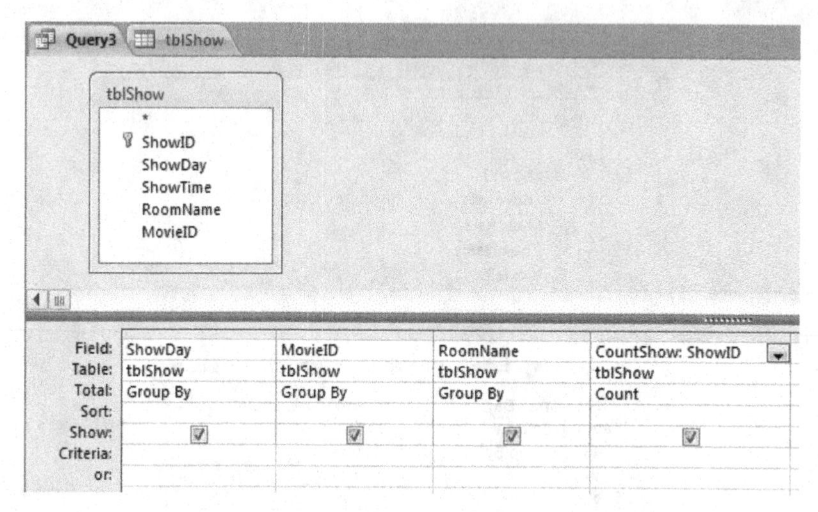

d. Can you design a query that lists the total count of shows by ShowDay and displays the specific ShowTime of each show through the same query? If yes, show your query design in the following; if no, explain why.

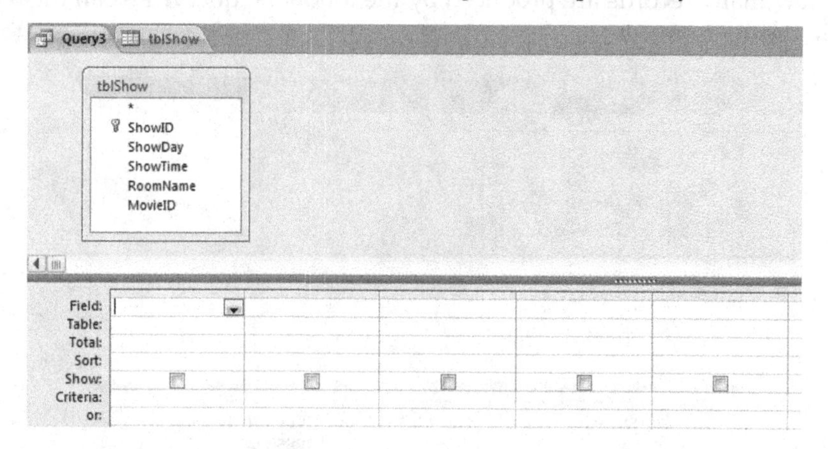

8. For the movie theater database from Exercise 1, design a query that prompts for a movie title and return the movie schedule for that movie and the total number of tickets available per show. The query result for the movie "The Thing" is shown as follows.

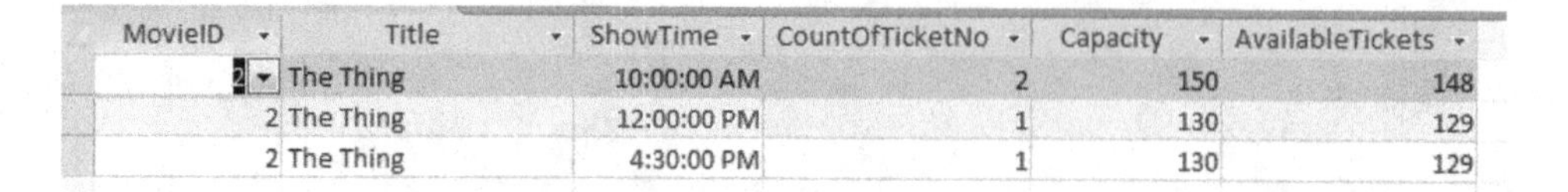

MovieID	Title	ShowTime	CountOfTicketNo	Capacity	AvailableTickets
2	The Thing	10:00:00 AM	2	150	148
2	The Thing	12:00:00 PM	1	130	129
2	The Thing	4:30:00 PM	1	130	129

9. For the movie theater database from Exercise 1, create a query that prompts for an actor in a movie and a show period with a beginning date and an ending date and returns the schedule of the movie with the selected actor for the selected period.

The schedule is sorted in ascending order of the show day and the show time. Note that the user input for the prompt of an actor may contain only the first name or the last name of the actor, whereas an actor stored in the database has the actor's full name. The format of the query result is shown as follows.

Actor	ShowDay	ShowTime
Jonah Hill	10/31/2011	12:00:00 PM
Jonah Hill	10/31/2011	1:30:00 PM
Jonah Hill	10/31/2011	4:30:00 PM
Jonah Hill	11/1/2011	10:00:00 AM

10. For the movie theater database from Exercise 1, create a Totals query that prompts for an actor in a movie and a show period with a beginning date and an ending date and returns the title of each movie including the selected actor and the total number of shows for each movie in the selected period. Note that the user input for the prompt of an actor may contain only the first name or the last name of the actor, whereas an actor stored in the database has the actor's full name. The format of the query result is shown as follows.

Title	TotalShows
Moneyball	1
The Thing	3

8

Structured Query Language

Structured query language (SQL) is the most influential query language of relational databases. This chapter first reviews the background of SQL followed by a brief discussion on Backus–Naur form (BNF) and extended BNF (EBNF), a main notation technique used in illustrating SQL syntax. The basic SQL statements, including CREATE, UPDATE, DELETE, and QUERY data, are then explained with examples.

8.1 Introduction to SQL

SQL was initially developed by IBM as a part of the System R project in the early 1970s and was named structured English query language (SEQUEL). Later, it evolved to today's acronym: SQL. In 1986, the American National Standards Institute first published the SQL standard, called "SQL-86," followed by the International Organization for Standardization (ISO) in 1987. Since then, SQL has been enhanced several times. The first major revision was released in 1992, known as "SQL2" or "SQL-92," which greatly enhanced the standard specification. The SQL syntax discussed in this chapter follows the SQL2 standard. SQL3 was released with support for object-oriented data management in 1999. In 2003, extensible markup language (XML)-related features were first introduced into SQL, resulted into version SQL-2003. As of today, there are seven revisions, and the most recent one is SQL-2011 (ISO 9075:2011), which was formally adopted in December 2011.

According to the SQL standard, SQL is composed of five main languages:

- *Data definition language (DDL).* DDL is used to create and modify the relation schemas within a database. Example commands are the CREATE, ALTER, and DROP statements.
- *Data manipulation language (DML).* DML is used to retrieve, store, modify, delete, insert, and update data in the database. Examples of this are the SELECT, UPDATE, and INSERT statements.
- *Data control language (DCL).* DCL is used to create roles, permissions, and referential integrity constraints for a database. Updates that violate integrity constraints are disallowed. It is also used to control secured access to the database. Examples of DCL are the GRANT and REVOKE statements.

TABLE 8.1
BNF and EBNF Symbols

Notation	Description	Example
{ }	Required element	{a}
[ ]	Term enclosed is optional	[a]: a may or may not be used
(|)	Enclose groups of alternative terms	(a | b | c): either a or b or c
...	Repetition: term enclosed is used zero or more times	

- *Transaction control language (TCL).* TCL is used to manage different transactions occurring within a database. It can specify the beginning and the ending of a transaction, and undo transactions. The COMMIT and ROLLBACK statements are examples.
- *Embedded and dynamic SQL.* Embedded and dynamic SQL defines how SQL statements can be embedded within general-purpose programming languages, such as C, C++, Java, C#, VB.NET, etc.

This chapter will mainly discuss the first two languages: DDL and DML. We will visit DCL and TCL when we discuss MySQL in Chapter 9.

8.2 Backus–Naur Form

Before introducing the SQL syntax, it is necessary to understand BNF and EBNF, one main notation technique for context-free grammars, often used to describe the syntax of languages in computing. In Table 8.1, the general symbols in BNF and EBNF used to describe SQL syntax are summarized.

For example, the EBNF statement: {mandatory_1 | mandatory_2} (optional,...) indicates that it is required to have either mandatory_1 or mandatory_2; in addition, there may be multiple optionals, separated by commas. Please remember the connotation of these symbols to interpret the SQL commands in the following sections.

8.3 SQL Syntax

We will adopt part of the healthcare system schema discussed in Chapter 5, the entity-relationship (E-R) and enhanced entity-relationship (EER) design, to illustrate the SQL syntax throughout the chapter. Specifically, the three relations that will be used here are as follows:

Drugs

Drug_ID	Name	Price	Description
T_001	Tylenol	$17.99	100 325-mg tablets bottle
T_003	Tylenol	$8.99	
T_002	Tri_Vitamin	$14.99	1500-400-35 solution 50-mL bottle
C_001	Claritin	$35.99	10 tablets

Stores

Store_ID	Name	Phone_Number
S_001	Walgreens	1-800-746-7287
S_002	CVS Pharmacy	1-800-925-4733
S_003	Fry's Pharmacy	1-866-221-4141

Sales

Transaction_ID	Drug_ID	Quantity	Store_ID
ST_001	T_001	25	S_001
ST_002	T_001	300	S_002
ST_003	C_002	150	S_001

Note: SQL uses the terms "table," "row," and "column" for the formal relational model termed "relation," "tuple," and "attribute," respectively. The corresponding terms will be used interchangeably. In addition, uppercase is used for the SQL statements. However, most databases now accept both uppercase and lowercase.

8.3.1 SQL Data Definition Language and Data Types

The main SQL command for data definition is the **CREATE** statement, which can be used to create schemas, tables (relations), and domains.

8.3.1.1 CREATE SCHEMA

An SQL schema includes an authorization identifier to indicate the user or account who owns the schema, as well as descriptors for each element in the schema. It is identified by a schema name. These elements include tables, views, and domains that describe the schema. The earliest version of SQL did not include the concept of a relational database schema. All tables are considered to be part of a single schema. In SQL2, a schema was first introduced intending to group tables that belong to the same database application together.

To create a schema, the syntax is

```
CREATE SCHEMA schemaname [AUTHORIZATION username] [schema _ element [...]]
```

The *schemaname* is the name for the schema. The **AUTHORIZATION** command is optional and is used to specify the user, *username*, who owns the schema. The *schema_ element* statement declares an optional list of elements (e.g., table, view).

Example 8.1

Create a schema named "myschema":

```
CREATE SCHEMA myschema;
```

Example 8.2

Create a schema for user Joe:

```
CREATE SCHEMA myschema AUTHORIZATION joe
```

Note: These two examples illustrate creating simple schemas. We will discuss the syntax of adding element descriptors for the schema when we learn how to create views, domains, and tables.

8.3.1.2 CREATE VIEW

The CREATE VIEW statement creates a new database view, which is an SQL query stored in the database.

The syntax for creating a view is

```
CREATE VIEW viewname [(column-list)] AS query
       [WITH [CASCADED|LOCAL|] CHECK OPTION]
```

The *viewname* statement is the name for the new view. Additionally, the *column-list* declaration is an optional list of names for the columns of the view, separated by commas. The *query* statement is any **SELECT** statement without an **ORDER BY** clause. Additionally, the *column-list* code must have the same number of columns as the select list in the query. The optional **WITH CHECK OPTION** clause is a constraint on updatable views. It affects SQL **INSERT** and **UPDATE** statements. The **CASCADED** and **LOCAL** specifiers apply when the underlying table for query is another view. The **CASCADED** statement requests that the **WITH CHECK OPTION** clause applies to all underlying views (to any level). On the other hand, the **LOCAL** statement requests that the **WITH CHECK OPTION** applies only to the current view. The **LOCAL** statement is the default.

Example 8.3

Create a view element that lists which drugs cost more than $20:

```
CREATE VIEW myviews AS
   SELECT Name FROM drugs WHERE price > 20
```

Example 8.4

Create a schema authorized to Joe, and create the above view element within the schema:

```
CREATE SCHEMA myschema AUTHORIZATION joe
   CREATE VIEW myviews AS
   SELECT Name FROM drugs WHERE price > 20
```

8.3.1.3 CREATE DOMAIN

A domain essentially defines the data type of an attribute. These optional constraints apply the attribute to be restricted to only a specific set of values. In SQL, a variety of built-in domain data types are supported (see Table 8.2).

The syntax for creating a domain is

```
CREATE DOMAIN [AS] domainname datatype [[NOT] NULL]
       [DEFAULT default-value] [CHECK (condition)]
```

The *domainname* command is the identifier of the domain. The *datatype* is the built-in SQL data type chosen from Table 8.2. The **NULL** clause allows one to specify the nullability

TABLE 8.2

Domain Data Types

Data Types	Description
char (*n*)	A fixed-length character string with user-specified length *n*
varchar (*n*)	A variable-length character string with user-specified maximum length *n*.
Int	An integer
smallint	A small integer
numeric (*p*, *d*)	A fixed-point number with user-specified precision. The number consists of *p* digits, and *d* of the *p* digits are to the right of the decimal point. For example, numeric (3, 1) allows 23.9 to be stored, but neither 239.0 nor 0.239
Real	Floating-point number
double precision	Double-precision floating-point numbers
float (*n*)	A floating-point number, with precision of at least *n* digits
Date	A calendar date containing a year, month, and day in the form of YYYY/MM/DD
Time	The time of a day, in hours, minutes, and seconds. It is in the form of HH:MM:SS

of the domain. If it is not explicitly specified, the allow_nulls_by_default option is used. Furthermore, the **CHECK** clause specifies integrity constraints, that is, which attributes of the domain must satisfy a user-defined *condition*. The following are two examples of creating a domain and making it an element of the schema.

Example 8.5

Create a domain named "store," which holds a 10-character string and may be NULL:

```
CREATE DOMAIN store CHAR(10) NULL;
```

Example 8.6

Create a schema authorized to Joe, and create the above domain element within the schema:

```
CREATE SCHEMA myschema AUTHORIZATION joe
   CREATE DOMAIN store CHAR(10) NULL
```

8.3.1.4 CREATE TABLE

Tables are the basic units of organization and the storage of data in a database. Each table represents an entity such as a list of stores, drug sales, or purchase orders. The **CREATE TABLE** statement is the SQL command that adds a new table to a relational database.

The basic syntax is

```
CREATE TABLE tableName
{(columnname datatype) [NOT NULL][UNIQUE]
[DEFAULT defaultoption] [CHECK (searchcondition)] [,...]}
[PRIMARY KEY (listofcolumns),]
[FOREIGN KEY (listofforeignkeycolumns)
REFERENCES parenttablename [(listofcandidatekeycolumns)]
```

Here, the *tablename* statement is the name for the new table. The *columnname* clause is required to list the names of the columns within a table. Commas are used to separate multiple columns. The *datatype* clause is similar to the declaration of the **DOMAIN** statement and is obtained from standard SQL data types (as shown in Table 8.2). The optional constraints **NULL** and **UNIQUE** may also apply. The optional **DEFAULT** constraint is used to set a default value of the column. The optional **CHECK** constraint is again used to limit the range of permissible values that can be added into the column. The optional **PRIMARY KEY** constraint specifies the list of columns in the table that forms the primary key, separated by commas. The optional **FOREIGN KEY** constraint is used to list the columns within the table as the foreign key, followed by the **REFERENCES** constraint, which specifies the other table where the foreign key columns are linked.

Example 8.7

Use **CREATE TABLE** statement to create three relations: Drugs, Stores, and Sales:

```
CREATE TABLE Drugs (
   Drug_ID          char (20) NOT NULL,
   Name     varchar (255),
   Price            numeric (10, 2),
   Description      varchar (255),
   CHECK (Price > 0),
   PRIMARY KEY (Drug_ID))

CREATE TABLE Stores (
   Store_ID         char (20) NOT NULL,
   Name             varchar (255),
   Phone_Number     varchar (10),
   PRIMARY KEY (Store_ID))

CREATE TABLE Sales (
   Transaction_ID char (20) NOT NULL,
   Drug_ID          char (20),
   Quantity         int      DEFAULT 0,
   Store_ID         char (20),
   PRIMARY KEY (ST_ID)
   FOREIGN KEY (Store_ID) REFERENCE Stores (Store_ID)
   FOREIGN KEY (Drug_ID) REFERENCE Drugs (Drug_ID))
```

Example 8.8

Add the Drugs table element to the *myschema* created in Example 8.2:

```
CREATE SCHEMA myschema AUTHORIZATION joe
   CREATE TABLE Drugs (
            Drug_ID          char (20) NOT NULL,
            Name     varchar (255),
            Price            numeric (10, 2),
            Description      varchar (255),
            CHECK (Price > 0),
            PRIMARY KEY (Drug_ID))
```

Now, we have learned to create schema, table. We will learn how to interact (retrieve, modify) the data in the next section.

8.3.2 SQL Data Manipulation Language: Data Queries

8.3.2.1 Basic Structure of the SELECT Command

A relational database consists of a collection of relations (tables), each of which is assigned with a unique name. Each relation has a structure similar to that presented in Chapter 5. To create a query or a report that spans tables and to take in a subset of attributes, the SQL **SELECT-FROM** clause is used.

The basic syntax when creating an SQL **SELECT-FROM** query is

```
SELECT [DISTINCT | ALL]
        {* | [columnexpression [AS newname]] [,...]}
FROM          tablename [alias] [, ...]
[WHERE        condition]
[GROUP BY     columnlist] [HAVING condition]
[ORDER BY     columnlist]
```

The **SELECT** clause specifies the list of columns desired in the result of a query. The **FROM** clause specifies the table(s) to be used in the query. The optional **WHERE** clause filters the rows based on the defined conditions. Additionally, the optional **GROUP BY** clause forms groups of rows with the same column value, while the optional **HAVING** clause filters the group subject to some conditions. The optional **ORDER BY** clause specifies the order of the output, with the default being descending order.

Example 8.9

Query all information about drugs in the Drugs relation:

```
SELECT * FROM Drugs;
```

The result is a relation consisting of all four columns with four rows as follows:

Drug_ID	Name	Price	Description
T_001	Tylenol	$17.99	100 325-mg tablets bottle
T_003	Tylenol	$8.99	
T_002	Tri_Vitamin	$14.99	1500-400-35 solution 50-mL bottle
C_001	Claritin	$35.99	10 tablets

Example 8.10

Query all the names of all drugs stored in the Drugs relation:

```
SELECT Name FROM Drugs;
```

The result of the query is a relation consisting of a single column with the heading *Name* as follows:

Name
Tylenol
Tylenol
Tri_Vitamin
Claritin

As observed, of the four queried rows, the drug "Tylenol" shows up twice. This result is the same as running the query:

```
SELECT ALL Name FROM Drugs;
```

To eliminate duplicates, we insert the **DISTINCT** statement after the **SELECT** clause.

Example 8.11

Query all the names of all drugs in the Drugs relation with no duplicate records:

```
SELECT DISTINCT Name FROM Drugs;
```

The result is a relation consisting of a single column with the heading Name. Since DISTINCT eliminates duplicate results, three records will be returned as follows:

Name
Tylenol
Tri_Vitamin
Claritin

Example 8.12

Query which drugs would cost less than $20:

```
SELECT Name FROM Drugs
WHERE Price < 20;
```

A relation consisting of one column with heading Name, two records will be returned when executing the query:

Name
Tylenol
Tri_Vitamin

Example 8.13

Query information about a specific set of drugs that costs between $10 and $20:

```
SELECT * FROM Drugs
WHERE Price <= 20 and Price >= 10;
```

or

```
SELECT * FROM Drugs
WHERE Price BETWEEN 20 AND 10;
```

The result of the query is a relation consisting of four columns, and only three records will be returned. Note that the **BETWEEN** and **AND** clauses are inclusive conditions.

Drug_ID	Name	Price	Description
T_001	Tylenol	$17.99	100 325-mg tablets bottle
T_003	Tylenol	$8.99	
T_002	Tri_Vitamin	$14.99	1500-400-35 solution 50-mL bottle

Example 8.14

Find the price as well as the name of the drug with the characters "Tyle" in their name:

```
SELECT Price FROM Drugs
WHERE Name LIKE "%Tyle%";
```

After executing the query, the result is a relation consisting of two columns with the headings Name and Price, and two records:

Name	Price
Tylenol	$17.99
Tylenol	$8.99

SQL has two special pattern matching symbols: specifically, the "%" character is used for a sequence of zero or more characters. As an example, the statement *Name* **LIKE** "%*Tyle*%" means to return any rows in which the *Name* column contains the string "*Tyle*" of any length. On the other hand, in the case that it is known, the *Name* string starts with "*Tyle*" or ends with "*Tyle*," the commands *Name* **LIKE** "*Tyle*%" or *Name* **LIKE** "%*Tyle*" can be used, respectively.

Note: The pattern matching symbol may vary from different databases. For example, Microsoft Access database uses "*" as the wildcard character, while MySQL and Oracle databases both use the "%" character as the pattern matching symbol.

Example 8.15

Query information of all drugs where a description has not been filled:

```
SELECT * FROM Drugs
WHERE Description IS NULL;
```

The result is a relation consisting of four columns, and one record will be returned as follows:

Drug_ID	Name	Price	Description
T_003	Tylenol	$8.99	

Example 8.16

Obtain a list of the details of all drugs, arranged in ascending order of price:

```
SELECT * FROM Drugs
ORDER BY Price ASC;
```

The query result is a relation consisting of four columns and four records, which are listed in the order of price from low to high:

Drug_ID	Name	Price	Description
T_003	Tylenol	$8.99	
T_002	Tri_Vitamin	$14.99	1500-400-35 solution 50-mL bottle
T_001	Tylenol	$17.99	100 325-mg tablets bottle
C_001	Claritin	$35.99	10 tablets

Example 8.17

Assume that the drug "Tylenol" is having a 20% discount. Query the information of "Tylenol" after the promotion:

```
SELECT Drug_ID, Name, Price*0.8, Description FROM Drugs
WHERE Name = "Tylenol"
```

A relation that has the same four columns as the Drugs relation, except for the column Price multiplied by 0.8, is queried from the Drugs relation.

Since the new price is the promotion price, we can use the **AS** clause to rename the column as follows:

```
SELECT Drug_ID, Name, Price*0.8 AS NEW_Price, Description FROM Drugs
WHERE "Name" = "Tylenol"
```

Drug_ID	Name	New_Price	Description
T_001	Tylenol	$14.39	100 325-mg tablets bottle
T_003	Tylenol	$7.19	

We have illustrated the use of most clauses in the example except for **GROUP BY** and **HAVING**. These two clauses are closely tied to aggregate functions, which will be discussed in the next section.

8.3.2.2 Aggregate Functions

An aggregate function is a function that takes on a collection of values from a column and returns a single value that is a function of the input values. Some commonly used aggregate functions are as follows:

- **AVG( )** returns the average value
- **COUNT( )** returns the total number of rows
- **MAX( )** returns the maximum value
- **MIN( )** returns the smallest value
- **SUM( )** returns the total sum of the value
- **FIRST( )** returns the first value
- **LAST( )** returns the last value

The input to SUM and AVG must be a collection of numbers, but other operators can operate on a collection of non-numeric data types, such as string. To illustrate these functions, we have the following examples.

Example 8.18

Count the number of drugs that cost more than $5.99:

```
SELECT COUNT (*) FROM Drugs
WHERE Price > 5.99;
```

The corresponding result is

Count
4

Example 8.19

Count the number of unique drugs in the Drugs relation, not counting duplicates:

```
SELECT COUNT (DISTINCT Name) FROM Drugs;
```

The result is

Count
3

Example 8.20

Find the number of sales and the total sales in the Sales relation in the store with *Store_ID* = *S_001*:

```
SELECT COUNT (*) AS Count, SUM (Quantity) AS Sum FROM Drugs
WHERE Stored_ID = 'S_001';
```

The result of the query is a relation that consists of two columns, in which the **AS** clause is used to rename the returned columns:

Count	Sum
2	175

Example 8.21

Get the minimum, maximum, and average sales from the Sales relation:

```
SELECT MIN (Quantity) AS Min,
       MAX (Quantity) AS Max,
       AVG (Quantity) AS Avg
FROM Sales;
```

The result is a relation that consists of three columns. Again, the **AS** clause is used to rename the returned columns as follows:

Min	Max	Avg
25	300	158.3

Example 8.22

Query the number of transactions in each store and their total sales quantities:

```
SELECT Store_ID,
   COUNT (Drug_ID) AS Count
   SUM (Quantity) AS Sum
FROM Sales
GROUP BY Store_ID
ORDER BY Store_ID;
```

This query first groups the records in the Sales relation by store. Next, within each store, the number of sales and the total amount of sales are summarized using the aggregate functions **COUNT** and **SUM**. The result of this query is a relation with three columns, specifically, *Store_ID, Count,* and *Sum,* with the records sorted in ascending order by *Stored_ID*.

Store_ID	Count	Sum
S_001	2	175
S_002	1	300

Example 8.23

For stores with more than one transaction, find the number of store transactions and the corresponding sum of the total sales quantities:

```
SELECT Store_ID,
   COUNT (Drug_ID) AS Count
   SUM (Quantity) AS Sum
FROM Sales
   GROUP BY Store_ID
   HAVING COUNT (Drug_ID) >1;
```

The result is similar to the preceding example query, except that the **HAVING** clause filters the store with less than two transactions:

Store_ID	Count	Sum
S_001	2	175

By now, we have learned how to use **SELECT-FROM** in the main query. SQL provides a mechanism for nested queries as follows.

8.3.2.3 Nested Subqueries

A subquery is a **SELECT-FROM-WHERE** expression that is nested within another query, such as the **SELECT, INSERT, UPDATE,** and **DELETE** statements. The following example illustrates the use of subquery in a **SELECT** query.

Example 8.24

Find the phone number of the stores that have sold "Tylenol":

```
SELECT Store_ID, Phone_Number FROM Stores
WHERE Store_ID IN
   (SELECT Store_ID FROM Sales
   WHERE Drug_ID IN
         (SELECT Drug_ID FROM Drugs
         WHERE Name LIKE "%Tylenol%"));
```

This nested subquery has three layers. First, the most inner query selects the *Drug_ID* of drugs that contain "Tylenol" in their name. The query result is the set (T_001, T_003). The next query selects the *Store_ID* of the stores that sold drugs with the *Drug_ID* just retrieved. The set operator, **IN**, is used. Finally, the outer query selects both *Store_ID* and *Phone_Number* from the Stores relation for the specific *Store_ID*. The final query result is as follows:

Store_ID	Phone_Number
S_001	1-800-746-7287
S_002	1-800-925-4733

As seen from this example, a subquery is used to query across multiple tables. In some sense, this is similar to a JOIN command in SQL.

8.3.2.4 JOIN Query

The SQL **JOIN** command is used to query data from two or more tables, which is based on the relationship between certain columns in these tables. There are four types of **JOIN** operations:

- **JOIN**: This will return rows when there is at least one matching instance in both tables. In some database systems, it is called "INNER JOIN."
- **LEFT JOIN**: This will return all rows from the left table, even if there are no matches in the right table. Alternatively, the LEFT OUTER JOIN clause is used in some databases.
- **RIGHT JOIN**: This will return all rows from the right table, even if there are no matches in the left table. The alternative clause, RIGHT OUTER JOIN, is used in some databases.
- **FULL JOIN**: This will return rows when there is a match in one of the tables.

Let us use the following examples to illustrate the use of JOIN command.

Example 8.25

JOIN (INNER JOIN): Find the names of the drugs being sold:

```
SELECT Name
FROM Drugs
JOIN Sales
ON Drugs.Drug_ID = Sales.Drug_ID;
```

The result is a relation with one column, the heading is *Name*:

Name
Tylenol
Tylenol
Claritin

Note: Both the Drugs and Sales relations have the column *Drug_ID*. To avoid ambiguity, we need to explicitly specify where the *Drug_ID* column is queried from. Often, we can use the alias for each table. For example, the above query can be rewritten as follows:

```
SELECT d.Name
FROM Drugs d
JOIN Sales s
ON d.Drug_ID = s.Drug_ID;
```

This is also equivalent to

```
SELECT d.Name
FROM Drugs d, Sales s,
WHERE d.Drug_ID = s.Drug_ID;
```

The use of the alias *d* indicates that the *Name* is queried from the Drugs relation. Otherwise, the relational database will throw the "ambiguous" error.

Example 8.26

LEFT JOIN (LEFT OUTER JOIN): Find all the drugs and their sales information, if any:

```
SELECT d.*, s.Quantity, s.Store_ID, s.Transaction_ID
FROM Drugs d
LEFT JOIN Sales s
ON d.Drug_ID = s.Drug_ID
ORDER BY d.Drug_ID;
```

Given this query, we define Drugs as the left relation. Therefore, all the records from the Drugs relation will be retrieved. The result of the query is a relation that is composed of seven columns. For the records that have no corresponding record in the Sales relation, the NULL value applies. For the drugs with *Drug_ID* being T_002, T_003, and C_001, the *Quantity, Store_ID,* and *Transaction_ID* have NULL values.

Drug_ID	Name	Price	Description	Quantity	Store_ID	Transaction_ID
T_003	Tylenol	$8.99		NULL	NULL	NULL
T_002	Tri_Vitamin	$14.99	1500-400-35 solution 50-mL bottle	NULL	NULL	NULL
T_001	Tylenol	$17.99	100 325-mg tablets bottle	25	S_001	ST_001
T_001	Tylenol	$17.99	100 325-mg tablets bottle	300	S_002	ST_002
C_001	Claritin	$35.99	10 tablets	NULL	NULL	NULL

Example 8.27

RIGHT JOIN (RIGHT OUTER JOIN): Find all the sales and their drug information, if any:

```
SELECT s.Transaction_ID, s.Store_ID, s.Quantity, d.*
FROM Sales s
RIGHT JOIN Drugs d
ON s.Drug_ID = d.Drug_ID
```

Similar to the **LEFT JOIN** clause, the returned query for the drug with the ID being C_002 has NULL values.

Transaction_ID	Quantity	Store_ID	Drug_ID	Name	Price	Description
ST_001	25	S_001	T_001	Tylenol	$17.99	100 325-mg tablets bottle
ST_002	300	S_002	T_001	Tylenol	$17.99	100 325-mg tablets bottle
ST_003	150	S_001	C_002	NULL	NULL	NULL

Example 8.28

FULL JOIN: Query all drug details as well as their sales information per store:

```
SELECT d.*, s.Quantity, s.Store_ID, s.Transaction_ID
FROM Drugs d
FULL JOIN Sales s
ON d.Drug_ID = s.Drug_ID;
```

The **FULL JOIN** statement returns all the records from the left relation (Drugs) as well as all the rows from the right relation (Sales). If there are records in Drugs that do not have matches in the Sales relation, or if there are records in Sales that do not have matches in the Drugs relation, these records will be listed with NULL for the respective columns. The result of the query is shown as follows:

Drug_ID	Name	Price	Description	Quantity	Store_ID	Transaction_ID
T_003	Tylenol	$8.99		NULL	NULL	NULL
T_002	Tri_Vitamin	$14.99	1500-400-35 solution 50-mL bottle	NULL	NULL	NULL
T_001	Tylenol	$17.99	100 325-mg tablets bottle	25	S_001	ST_001
T_001	Tylenol	$17.99	100 325-mg tablets bottle	300	S_002	ST_002
C_001	Claritin	$35.99	10 tablets	NULL	NULL	NULL
C_002	NULL	NULL	NULL	150	S_001	ST_003

8.3.3 SQL Data Manipulation Language: Data Modification

Up until now, we have focused on the extraction of existing information from a database. In this section, we will learn how to add, remove, or change information using SQL commands.

8.3.3.1 INSERT

To insert data into a relation, we either specify a record to be inserted or write a query whose result is a set of records to be inserted.

The basic **INSERT** syntax is

```
INSERT INTO TableName [(columnList)] VALUES (dataValueList)
```

The optional *columnlist* statement specifies the list of columns within the table where the new records will be inserted. If it is not specified, SQL assumes that the list is in their original order specified in the **CREATE TABLE** command. Any missing data for specific columns must be declared as **NULL**, unless a **DEFAULT** value is specified for the column. The *dataValueList* statement is required to match the *columnlist* statement: the number of items in each list must be the same, the position of each item in each list must be the same, and data types in both lists must be matched perfectly.

Example 8.29

Add a new row into the *Sales* table when supplying data for all columns:

```
INSERT INTO Sales
VALUES ('ST_006,' 'C_002,' 'S_003,' 140);
```

This query inserts a new record into the Sales relation:

Transaction_ID	Drug_ID	Quantity	Store_ID
ST_006	C_002	140	S_003

The updated Sales relation now has the following information (the new record is highlighted in bold):

Transaction_ID	Drug_ID	Quantity	Store_ID
ST_001	T_001	25	S_001
ST_002	T_001	300	S_002
ST_003	C_002	150	S_001
ST_006	**C_002**	**140**	**S_003**

Example 8.30

Augment a new row into the Sales table, only supplying data for all mandatory columns:

```
INSERT INTO Sales (ST_ID)
VALUES ('ST_006');
```

or

```
INSERT INTO Sales
VALUES ('ST_006,' NULL, NULL, 0);
```

The query inserts a new record to the Sales relation, which has NULL for the *Drug_ID* and the *Store_ID* columns and 0 (default value) for the *Quantity* column as follows:

Transaction_ID	Drug_ID	Quantity	Store_ID
ST_006	NULL	0	NULL

The updated Sales relation now has the following information (the new record is highlighted in bold):

Transaction_ID	Drug_ID	Quantity	Store_ID
ST_001	T_001	25	S_001
ST_002	T_001	300	S_002
ST_003	C_002	150	S_001
ST_006	**NULL**	**0**	**NULL**

Example 8.31

Add a new row into the Drugs table for a new transaction, with the *ST_ID* being ST_003:

```
INSERT INTO Drugs
SELECT Drug_ID FROM Sales
WHERE ST_ID = "ST_003"
```

This query inserts a new record to the Drugs relation as follows:

Drug_ID	Name	Price	Description
C_002	NULL	NULL	NULL

The updated Drugs relation now has the following information (the new record is highlighted in bold):

Drug_ID	Name	Price	Description
T_001	Tylenol	$17.99	100 325-mg tablets bottle
T_003	Tylenol	$8.99	
T_002	Tri_Vitamin	$14.99	1500-400-35 solution 50-mL bottle
C_001	Claritin	$35.99	10 tablets
C_002	**NULL**	**NULL**	**NULL**

8.3.3.2 UPDATE

In some situations, we may want to change a value in a record without changing all values in the record. For this purpose, the **UPDATE** statement can be used.

The syntax of the **UPDATE** statement is

```
UPDATE TableName
SET columnName1 = dataValue1
    [, columnName2 = dataValue2...]
[WHERE searchcondition]
```

The *TableName* can be the name of the base table or an updatable view. The **SET** clause specifies the names of one or more columns that are to be updated with a new *dataValue.* The *dataValue* must be compatible with the data type for the corresponding column. The optional **WHERE** clause is used to filter the records to be updated based on the *searchCondition.* If the **WHERE** clause is omitted, all the records in the table will be updated.

Example 8.32

Give all drugs a 15% discount:

```
UPDATE Drugs
SET Price = Price*0.85;
```

Example 8.33

Apply a 15% discount to "Tylenol" drugs only:

```
UPDATE Drugs
SET Price = Price*0.85
WHERE Name = "Tylenol"
```

Similar to **INSERT**, a **SELECT-FROM** subquery may apply to specify the records to be updated.

8.3.3.3 DELETE

The **DELETE** statement is used to delete the whole record instead of some values of any particular column.

The **DELETE** syntax is applied as

```
DELETE FROM TableName [WHERE searchCondition]
```

The *TableName* is the name of the base table or an updatable view in which the records are to be deleted. Similar to the **UPDATE** clause, the optional **WHERE** clause is used to filter the records to be deleted based on the conditions. If it is omitted, then all the records in the relation will be removed. However, the relation still exists in the database. To remove the relation, the **DROP** statement is used, which is discussed in the next section.

Example 8.34

Delete all the records in the Sales relation:

```
DELETE FROM Sales
```

Example 8.35

Delete all sales with the *Drug_ID* being "C_002":

```
DELETE FROM Sales
WHERE Drug_ID = "C_002"
```

8.3.4 SQL Data Manipulation Language: Relation Modification

In a database, the created relation can be changed or removed using the **ALTER** and **DROP** statements.

8.3.4.1 ALTER TABLE

The SQL **ALTER TABLE** statement is used to add, edit, or modify a column from an existing table.

The basic syntax is

```
ALTER TABLE TableName
[ADD columnname datatype [NOT NULL][UNIQUE]]
[DROP columnname]
[ADD [constraint]]
[DROP [constraint]]
[ALTER [column] SET DEFAULT DefaultOption]
[ALTER [column] DROP DEFAULT]
```

The *TableName* is the table to be modified. This statement includes multiple optional statements, specifically, the **ADD** column with a specified *datatype* is to add a new column, the **DROP** statement is to drop the existing column, the **ADD/DROP** constraint (e.g., primary key, foreign key) are add or remove key relations, and the **ALTER** statement is to change the default value for the column.

Here we only show one simple example as follows:

Example 8.36

Add a column named "Transaction_Date" in the *Sales* relation:

```
ALTER TABLE Sales
ADD Transaction_Date date
```

8.3.4.2 DROP TABLE

Relations can easily be removed with the **DROP** statement. The simple syntax is

```
DROP TABLE TableName;
```

The **DROP** statement can also be used to remove the entire database as

```
DROP DATABSE DatabaseName;
```

In this section, we only briefly illustrate the relation modification. For advanced developer, please refer to specific database manual for comprehensive discussions on this topic.

8.4 Summary

This chapter reviews the history of SQL. An important notation technique, BNF, and its extension EBNF are introduced. Following the notation, basic SQL DDL statements are explained, including CREATE SCHEME, CREATE VIEW, CREATE DOMAIN, and CREATE TABLE. In addition, DML syntax, including the SELECT-FROM-WHERE clause, INSERT, UPDATE, DELETE, INSERT, DROP, ALTER, CREATE, and JOIN, are explained with examples. The use of aggregate function within the SELECT statement and the use of nest subqueries are also discussed with illustration examples.

EXERCISES

Review Exercises

1. SQL was initially developed by the World Wide Web Consortium (W3C). True or false?
2. DDL is used to modify tables within a database schema. True or false?
3. The SELECT statement is an example of a DML. True or false?
4. The referential integrity constraints for a database are manipulated by the TCL.
5. The {} character denotes a required element in the EBNF.
6. The following database elements that are modified by CREATE statements are:
   a. Domains
   b. Schemas
   c. Tables
   d. All of the above
   e. None of the above
7. Which of the following is a valid declaration to authorize username "Chandler" to access the Sales schema?
   a. CREATE AUTHORIZATION *chandler;*
   b. CREATE SCHEMA *sales;*
   c. CREATE SCHEMA *sales* AUTHORIZE *chandler;*

   d. INSERT AUTHORIZATION *chandler;*
   e. None of the above
8. Which of the following is not an optional statement in the CREATE VIEW clause?
   a. Query
   b. *columnlist*
   c. CHECK
   d. CASCADED
   e. LOCAL
9. Which of the following CREATE VIEW SQL code is not valid?
   a. CREATE VIEW AS SELECT * FROM *Book_Authors* WITH CHECK OPTION;
   b. CREATE VIEW *Author_M* AS SELECT * FROM *Book_Authors;*
   c. CREATE VIEW *Author_M* FROM *Book_Authors* WITH LOCAL CHECK OPTION;
   d. CREATE VIEW AS SELECT * FROM *Book_Authors* WITH CASCADED CHECK OPTION;
   e. All of the above are valid.
10. A domain essentially declares the data type of the entire table. True or false?
11. Declare a CREATE DOMAIN command to store the prices of computer spare parts. The price must be positive and can have, at most, two digits to the right and, at most, four digits to the left of the decimal point.
12. Declare a CREATE DOMAIN command to store the type of blood code named "Blood_Type_C." The type of blood code is a three-character field with default type "O+" and should not be NULL.
13. The CREATE TABLE statement is used to create new tables to hold entities within a database. True or false?
14. Suppose that we want to create a "Blood Donors" table that contains the "*Blood_Type_C*" domain from Review Exercise 12. To create a table with an attribute with a domain as a reference, we declare:
   a. CREATE TABLE *Blood_Donors* (*Donor_M* varchar(255), *Blood_Type* DOMAIN);
   b. CREATE TABLE *Blood_Donors* (*Donor_M* varchar(255), *Blood_Type Blood_Type_C*);
   c. CREATE TABLE *Blood_Donors* (*Donor_M* varchar(255), *Blood_Type* Varchar(3) NOTNULLFOREIGNKEY(*Blood_Type*)REFERENCE*Blood_Type_C*(*Blood_Type*));
   d. CREATE TABLE Blood_Donors (Donor_M varchar(255), Blood_Type DOMAIN Varchar(3));
   e. CREATE TABLE *Blood_Donors* (*Donor_M* varchar(255), *Blood_Type* AS *Blood_Type_C*);
15. The SELECT ALL FROM *Table_name;* is equivalent to
   a. SELECT ALL COLUMNS FROM *Table_name;*
   b. SELECT * FROM *Table_name;*
   c. SELECT FROM *Table_name;*
   d. FROM *Table_name* SELECT;
   e. None of the above. SELECT ALL FROM *Table_name;* is an erroneous statement.

16. The SELECT DISTINCT statement almost always returns less number of records as compared to the SELECT ALL statement. True or false?
17. The BETWEEN and statements are equivalent to the WHERE clause using the less than "<" and the greater than ">" commands. True or false?
18. Given the LIKE "Fac%" statement, the following column values will be returned except:
    a. Faculty
    b. Fac
    c. Fa
    d. All of the following will be returned.
    e. None of the following will be returned.
19. Given the LIKE "%201_" statement, the following column values will be returned except:
    a. Code2013
    b. 201334
    c. 2013
    d. All of the following will be returned.
    e. None of the following will be returned.
20. There is no need to declare the ORDER BY clause as ASC (ascending). True or false?
21. Given a table named *Faculty* with *Faculty_M* as the faculty name and *Faculty_R* as the faculty rank, write a query that will return an alphabetical listing of faculty members sorted by descending alphabetical faculty rank then by ascending faculty name.
22. Given the following JOINs on two tables with much more left table instances as compared to right table instances, which JOIN maximizes the COUNT statement?
    a. LEFT JOIN
    b. FULL JOIN
    c. JOIN
    d. RIGHT JOIN
    e. All join statements would return the same count.
23. Given the following JOINs on two tables with many more left table instances as compared to right table instances, which JOIN minimizes the AVG statement?
    a. LEFT JOIN
    b. FULL JOIN
    c. JOIN
    d. RIGHT JOIN
    e. All join statements would return the same count.
24. The following are valid aggregate functions except:
    a. AVG(Sales_Price)
    b. LAST(Name_Code)

c. COUNT(Sales_Price)
d. SUM(Supplier_Address)
e. All of the above are valid aggregate functions.

25. Suppose that a table named "Goods_Receipt" with "*Receipt_N*" as the goods receipt number and "*Receipt_D*" as the receipt data created in that order, where receipt number is numeric and is not null, is given. Which of the following is a valid INSERT statement?
    a. INSERT INTO *Goods_Receipt* VALUES('10-30-2013', '75774')
    b. INSERT INTO *Goods_Receipt(Receipt_D, Receipt_N)* VALUES('10-30-2013', '75774')
    c. INSERT INTO VALUES('10-30-2013', '75774') WHERE VALUE = *Goods_Receipt*
    d. INSERT INTO VALUES('10-30-2013', '75774') WHERE *Goods_Receipt(Receipt_D, Receipt_N)*
    e. None of the above
26. You can update a non-existent instance. True or false?
27. Given a table where *column_1* is required and *column_2* is optional, you can use the DELETE command to delete the value of *column_2*. True or false?

Practice Exercises

For practice questions 1 to 6, consider the following dataset:

Faculty

Fac_N	Fac_First_M	Fac_Last_M	Fac_Rank_C
13443	Iris	Martinez	ASCP
66234	Aura	Matias	PROF
45463	Lorelie	Grepo	ASTP
44556	Mickey	Mancenido	ASTP

Courses

Course_C	Course_M	Course_Units
IDOE	Design of Experiments	3
IBDM	Business Intelligence and Data Mining	3
MSCM	Supply Chain Management	3
SRES	Research Methods	1
SSIM	Simulation	4

Faculty_Courses

Course_C	Fac_N	Stud_Count	Sem_C
MSCM	13443	34	2012-1
MSCM	66234	32	2012-2
IDOE	44556	56	2012-1
SRES	13443	12	2012-1
IBDM	66234	40	2012-2
MSCM	44556	32	2013-1

1. Create a query to count the number of students that took Supply Chain Management. Name the column "Student_Count."
2. Create a query to handle the following desired query result:

Fac_Last_M	Num_Students
Mancenido	88
Martinez	46
Matias	72

3. Create a query for the following report:

Course Description	Rank	Last Name	Semester	Units
Business Intelligence and Data Mining	Prof	Matias	2012-2	3
Design of Experiments	ASTP	Mancenido	2012-1	3
Research Methods	ASCP	Martinez	2012-1	1
Supply Chain Management	PROF	Matias	2012-2	3
Supply Chain Management	ASCP	Martinez	2012-1	3
Supply Chain Management	ASCP	Mancenido	2013-1	3

4. The report in question 3 seems to lack the actual description of the faculty rank. Create a table named "Rank_Description" with the attributes:
   - *Fac_Rank_C*: A four-character attribute with default value INST and cannot be NULL. Additionally, this is the primary key of the table.
   - *Fac_Rank_M*: A variable character attribute that describes the rank.
   - *Min_Sal_Grade*: A numeric 1-digit number that can take values from 1 to 9, indicating the minimum salary grade of the rank.
5. Write an SQL command to populate the table in question 4 with the following data:

Fac_Rank_C	Fac_Rank_M	Min_Sal_Grade
INST	Instructor	1
ASTP	Assistant Professor	3
ASCP	Associate Professor	5
PROF	Professor	7
UPRF	University Professor	9

6. Write a nested query without join statements for the following report:

First Name	Last Name	Rank
Iris	Martinez	Associate Professor
Aura	Matias	Professor
Lorelie	Grepo	Assistant Professor
Mickey	Mancenido	Assistant Professor

For questions 7 to 12, consider the following tables:

Purchase_Orders

PO_N	PO_D	Supplier_N	PO_Pay_Terms
610557	2/27/2013	1335	NET30
610558	2/27/2013	2652	2/10NET20
610559	2/27/2013	1335	COD
610560	2/28/2013	1226	2/10NET20
610561	3/01/2013	2652	2/10NET20

Purchase_Order_Items

PO_N	Item_N	Item_Q	Item_Price
610557	36796	15	664.25
610557	36224	21	224.54
610559	36624	100	0.65
610560	36547	1	10887.10
610561	36869	224	336.65

7. The *PO_Pay_Terms* will be used in different tables; hence, a domain could be useful to handle this in the event of future changes. This attribute must contain a maximum of 10 characters; the default value should be "NET30," which corresponds to net 30 days; and should not be NULL. Write an SQL code to create a domain named "*Pay_Terms*" subject to the aforementioned specifications.
8. Create a "Suppliers" table with the following attributes:
   - Supplier_N: A four integer–digit primary key attribute used to uniquely identify a supplier and should not be NULL.
   - *Supplier_M*: A string that denotes the name of a supplier with a maximum of 255 characters.
   - *Supplier_Tier*: A single-digit number between 1 and 5 indicating the supplier's tier.
   - Default_Pay_Terms: An attribute that follows the domain "Pay_Terms" defined in question 6.
9. Populate the table generated in question 8 with the following information:

Supplier_N	Supplier_M	Supplier_Tier	Default_Pay_Terms
1335	MacLaren's Irish Pub	4	NET30
2652	Central Perk	3	NET30
1226	Beltway Coffee	Unknown	NET30

10. Create a query to for this report:

Supplier_M	PO_N	Item_N	Item_Q
MacLaren's Irish Pub	610557	36796	15
MacLaren's Irish Pub	610557	36224	21
Central Perk	610558	NULL	NULL
MacLaren's Irish Pub	610559	36624	100
Beltway Coffee	610560	36547	1
Central Perk	610561	36869	224

11. A round of supplier reviews has lapsed, and the Beltway Coffee supplier's master data need to be updated (see bold font).

Supplier_N	Supplier_M	Supplier_Tier	Default_Pay_Terms
1335	MacLaren's Irish Pub	4	NET30
2652	Central Perk	3	NET30
1226	Beltway Coffee	**3**	**2/10NET20**

    Create a single query to update the record.

12. A new attribute is needed for the *Purchase_Orders* relation to denote whether the purchase order instance is paid. Write the corresponding SQL code to add the following attributes:
    - Pay_Status: A three-character string indicating the status of the purchase order payment, with the default value "NYP" indicating not yet paid.

9

MySQL

Most relational database systems (RDBSs) adopt the basic syntax of Structured Query Language (SQL) data definition language (DDL) and data manipulation language (DML) discussed in Chapter 8, but may have some minor variations. In this chapter, a widely used open-source database system, MySQL, is introduced. First, the use of DDL and DML in MySQL is illustrated, followed by the discussions on the use of transaction control language (TCL) and data control language (DCL). Some unique features of MySQL, specifically, the utilities, the tools, and some SQL variations of MySQL, are also introduced. Since MySQL products are enhanced regularly, this chapter only covers the key aspects of this database system. When using a particular system version, it is recommended to consult the user manuals available online (http://www.mysql.com) for specific details.

9.1 Introduction

Currently, the most popular RDB management systems are Oracle (Oracle Corporation), SQL Server (Microsoft), DB2 (IBM), and MySQL (Oracle Corporation). As an open-source database server, MySQL is on the top of the list chosen by the users who are looking for a free or inexpensive database management system. Some notable advantages of MySQL include the following:

- *Ease of use*. MySQL is a high-performance but relatively simple database system and is much less complicated to set up and administer than larger systems such as Oracle and DB2.
- MySQL supports the SQL standard, which is the language of choice of all modern RDBS systems.
- *Capability*. In MySQL, many clients can connect to the server at the same time. Furthermore, developers can access MySQL interactively using several interfaces to enter queries and view the results. Examples of these interfaces are command-line clients, Web browsers, or X Window system clients. In addition, a variety of interfaces are available for programming languages, such as Perl and Java, just to name a few. MySQL also provides a number of connectors to support open database connectivity (ODBC), Java database connectivity (JDBC), and application platforms (e.g., .NET).
- *Connectivity and security*. MySQL is fully networked, and the database can be accessed from anywhere using the telecommunication network. MySQL has access control to ensure security issues. In addition, it supports encrypted connections using the AQ protocol.
- *Multi-platform*. MySQL runs on Unix, Windows, Linux, and Mac.

- *Small size*. MySQL has a modest distribution size, especially compared to the large disk space footprint of some commercial database systems.
- *Availability and cost*. MySQL was initially developed by Sun Microsystems as an open-source RDBS and was later acquired by Oracle Corporation. Currently, Oracle Corporation still supports MySQL as an open-source software, thus making it freely available under the terms of the general public license (GPL).

In the remaining of the chapter, we will explain how to work with MySQL.

9.2 Get Ready to Work with MySQL

Although MySQL runs across multiple operating systems (OSs), not all systems are equally good for supporting MySQL. To learn more about the supported systems, please check out the MySQL website (http://www.mysql.com).

Currently, some general MySQL tools are available as open distributions from the website:

- *MySQL Community Server* is a freely downloadable version of the open-source database. The supported platforms include Windows (×86, 32-bit, and 64-bit), Mac OS X, Linux, and Sun Solaris. There are different formats for download. It is recommended to download the binary distribution for easy installation. Binary distribution is also tuned for optimal and stable MySQL performance.
- *MySQL Cluster* is a real-time transactional database designed to serve intensive read and write workloads under high throughput conditions. It is available in both open-source and commercial edition versions.
- *MySQL Workbench* is a visual database design application that can be used to design, manage, and document database schema. It is also available in both open-source and commercial variants.
- *MySQL Connectors* is a list of available-for-download database drivers compatible with the ODBC and JDBC industry standards.
    - *ODBC Connector* is a standardized database driver for Windows, Linux, Mac OS, and Unix platforms.
    - *.NET Connector* is a standardized database driver for .NET platforms and development.
    - *J Connector* is a standardized database driver for Java platforms and development.
    - *C++ Connector* is a standardized database driver for C++ development.
    - *Python Connector* is a standardized database driver for Python platforms and development.
    - *C Connector* is a client library for C development (libmysql).
- *MySQL Installer* is a package that includes most MySQL tools such as MySQL Server, Workbench, Connectors, and MySQL for Excel and Visual Studio. However, unlike the tools listed above, the MySQL installer currently runs on Windows-based systems only. Note: Although the installer is 32-bit, it will run on both 32- and 64-bit binary OS versions.

Other than open distributions, MySQL has some commercial tools such as the MySQL Enterprise Edition and the MySQL Cluster Carrier Grade Edition (CGE).

In this chapter, the installation and configuration of MySQL using the MySQL installer for Windows is mainly illustrated. For Mac OS X, only the installation procedure of the MySQL Community Server is explained. Since the Workbench installation for Mac OS X is similar to Windows, we will not discuss it in detail. For learning purposes, the use of the command-line clients to practice SQL commands and the use of the Workbench tool to handle the database management are recommended.

9.2.1 MySQL Installation on Windows

To run MySQL on Windows systems, the Windows OS needs to be 2000, XP, Vista, Server 2008, Windows 7, Windows 8, or newer. Generally, one should install MySQL in Windows using an account that has administrator rights to have full access to the system environment. In addition, the system should support Transmission Control Protocol/Internet Protocol (TCP/IP) to run the MySQL Server.

There are different ways to install MySQL. New users are recommended to use the MySQL Installer, which provides wizards to install and configure the MySQL Server on Windows. The detailed procedure is outlined as follows:

Step 1: Download and launch the MySQL Installer. The process for starting the wizard depends on the content of the installation package being downloaded. If there is a setup.exe or .msi file present, just double-click on it to start the installation process.

Step 2: Once the installation is initiated, the welcome screen is shown, as seen in Figure 9.1. There are three options: Install MySQL Products, About MySQL, and

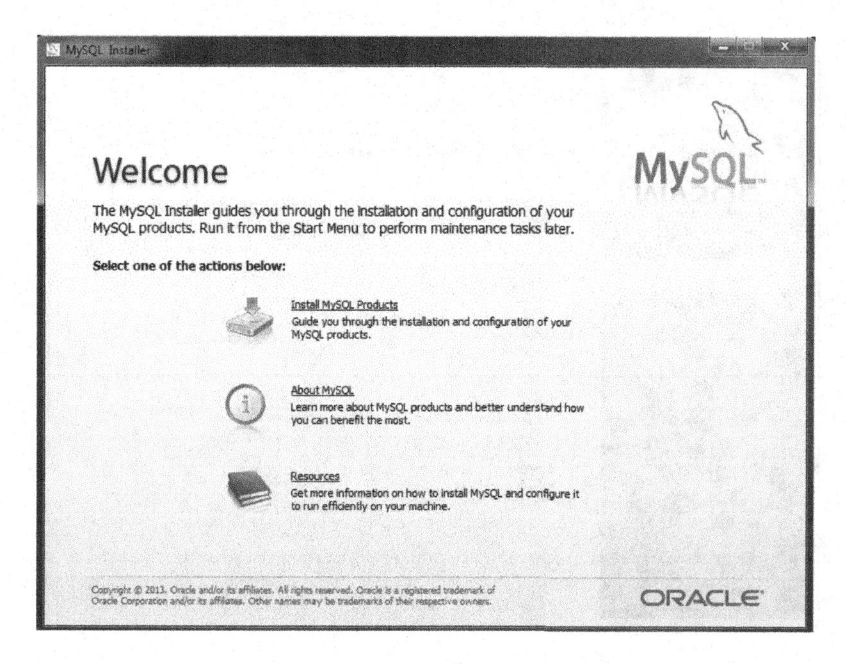

FIGURE 9.1
MySQL Installer Welcome screen.

Resources. If the Installer detects some MySQL components (e.g., connectors) available in the system, the installation will be launched as an update procedure.

To Install MySQL products for the first time, it is required to accept the license agreement before launching the installation process.

Step 3: Next, the setup type needs to be determined based on the installation preference. For the purposes of this chapter, we may choose the "Developer Default" setting. This includes MySQL Server, MySQL Workbench, MySQL Visual Studio plugin, MySQL Connectors, examples, and tutorial. In addition, the installation and data paths are defined here. The MySQL Installer will check the system for necessary external requirements (e.g., .NET framework for Visual Studio supportability) and then download and install missing components onto the system (Figure 9.2).

Step 4: Next, a comprehensive list of components to be installed will be popped up, as seen in Figure 9.3. Click Execute. This will start the installation.

Step 5: After the installation, the configuration will be initiated (Figure 9.4). This is to enable the TCP/IP networking component and specify the port number for the communication. If enabled, the MySQL client can access the server via the Internet. For example, let the IP address of the server be 149.169.43.22; the network-enabled communication is then set up over 149.169.43.22:3306 (assuming that the default port number 3306 is used). Otherwise, only localhost connections are allowed. The "Advanced Configuration" option provides additional logging options to configure. This includes defining file paths for the error log, general log, etc. (Figure 9.5).

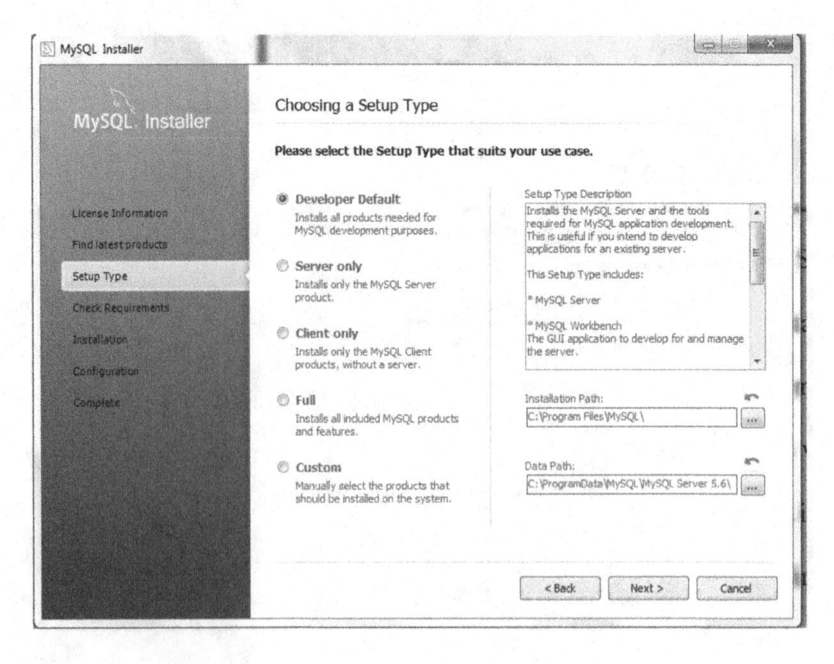

FIGURE 9.2
MySQL Installer: Choosing a Setup Type.

MySQL Installer

MySQL. Installer

License Information
Find latest products
Setup Type
Check Requirements
Installation
Configuration
Complete

Installation Progress

The following products will be installed or updated.

Product	Status	Progress	Notes
MySQL Server 5.6.10	To be installed		
MySQL Workbench CE 5.2.46	To be installed		
MySQL Notifier 1.0.3	To be installed		
MySQL For Excel 1.1.0	To be installed		
Connector/ODBC 5.2.4	To be installed		
Connector/C++ 1.1.2	To be installed		
Connector/J 5.1.23	To be installed		
Connector/NET 6.6.5	To be installed		
MySQL Documentation 5.6.10	To be installed		
Samples and Examples 5.6.10	To be installed		

Click [Execute] to install or update the following packages

< Back Execute Cancel

FIGURE 9.3
MySQL Installer: Installation Progress.

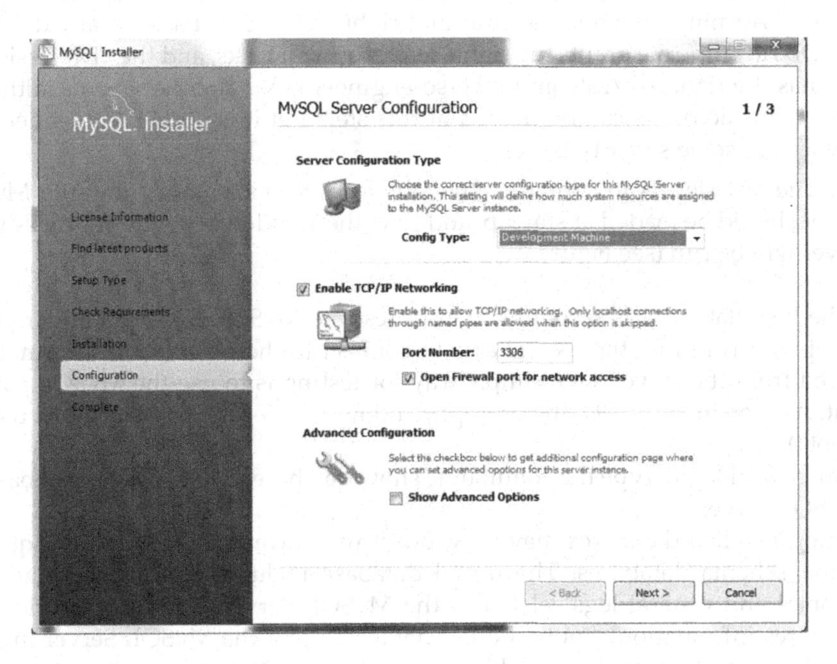

FIGURE 9.4
MySQL Installer: MySQL Server Configuration 1/3.

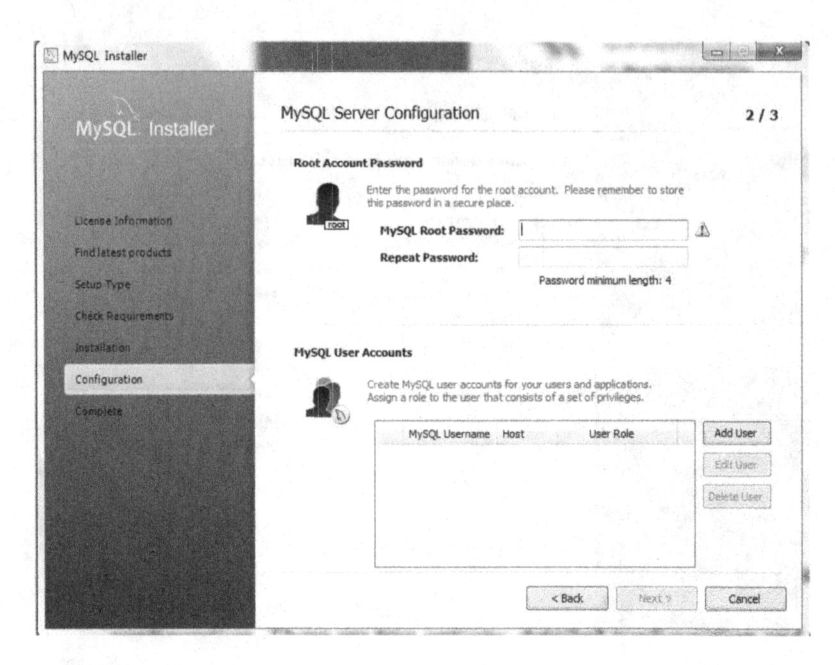

FIGURE 9.5
MySQL Installer: MySQL Server Configuration 2/3.

Step 6: Next, the password is required, with the username being "root" by default. At this point, it is optional to create additional users. There are several different pre-defined user roles that each have different permissions. For example, the "Backup Admin" user role has minimal rights needed to back up any database, the "DB Admin" role grants the rights to perform all tasks, and the "DB Designer" only has the right to create and reverse-engineer any database schema, although anonymous accounts can be created at this step, but it is not recommended as it may lead to some security issues.

Step 7: The last step is to configure the Windows Service name, how the MySQL Server should be loaded at startup, and how the Windows Service for the MySQL Server will be run (see Figure 9.6).

After the installation, it is recommended to test the MySQL Server to ensure that the server is up and running, the user is able to connect to the server, and information can be retrieved from the server. One simple way for testing is to use the MySQL command line client. As seen in Figure 9.7, the prompt window asks for the password (the username being "root").

Upon successful login, type the command "show databases;" to see what databases exist in the MySQL Server.

The list of installed databases may vary, but it will always include the mysql and the information_schema databases. The mysql database includes useful information such as help topics and general logs related to the MySQL Server. The information_schema database stores information about all other databases that the MySQL Server maintains. These databases contain read-only tables; thus, `INSERT`, `UPDATE`, or `DELETE` operations do not apply.

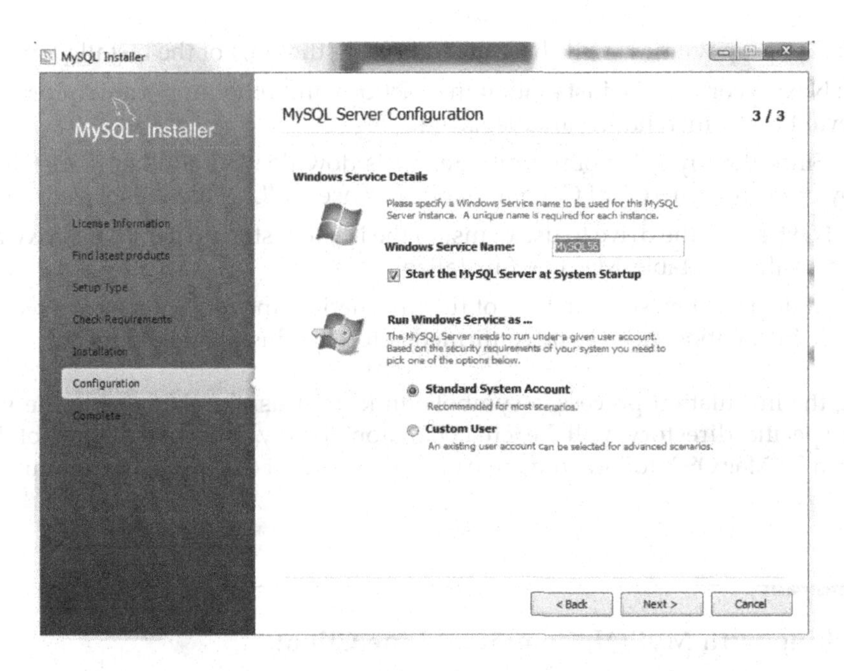

FIGURE 9.6
MySQL Installer: MySQL Server Configuration 3/3.

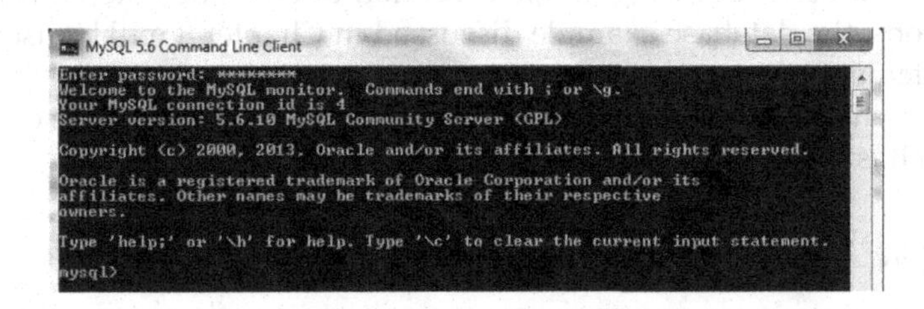

FIGURE 9.7
MySQL Command Line Client login.

9.2.2 MySQL Installation on Mac OS X

Unlike MySQL for Windows, the MySQL Community Server and the MySQL Workbench need to be downloaded and installed separately for a Mac OS. The open distributions for the MySQL Server are available in two different forms: the native package installer format and the tar package format. In this section, we will learn about the use of the native package installer, which uses the native Mac OS X installer to walk through the installation. Note that administrator privileges are required to perform the installation.

Step 1: Download and open the MySQL package installer, which is provided on a disk image (.dmg). This includes the main MySQL installation package, the MySQLStartupItem.pkg installation package, and the MySQL.prefPane. Double-click the disk image to open it.

Step 2: Double-click the MySQL Installer package. It will be named according to the version of MySQL being downloaded.

Step 3: A pop-up window will show up, indicating the start of the installation.

Step 4: Next, a copy of the installation instructions and other important information relevant to the installation are displayed.

Step 5: Since the MySQL Community Server is downloaded and being installed, a copy of the relevant GNU General Public License (GPL) is then displayed.

Step 6: Next, select the drive to use to install the MySQL startup item. The drive must have a valid, bootable, Mac OS X installed.

Step 7: In the last step, some details of the installation, including the space required for the installation, need to be confirmed before the installation launches.

During the installation process, a symbolic link from usr/local/mysql to the version/platform specific directory will be created automatically. The installation of MySQL Workbench for Mac OS X follows the same procedure as that of MySQL Community Server.

9.3 Working with MySQL Command Line Client

To better practice SQL command, it is recommended to start with MySQL Command Line Client. Note that, in MySQL, each command needs to end with a semicolon. The system recognizes this character and triggers the execution of the code entered before it. In the following sections, the database example discussed in Chapter 8 will be used to illustrate the MySQL statements (Figure 9.8).

Before we demonstrate the use of MySQL, we will first understand the data types used in MySQL which is discussed in the following.

9.3.1 Data Types

MySQL supports standard data types, including numbers, string values, temporal values such as dates and times, and NULL values (some are defined in Table 8.2) with some variations.

9.3.1.1 Numeric Data Types

MySQL specifies numbers as integers (with no fractional part) or floating point values (with a fractional part). Integers can be specified in decimal or hexadecimal format. The general numeric types supported by MySQL are summarized in Table 9.1.

FIGURE 9.8
MySQL Command Line Client: show databases.

TABLE 9.1

Numeric Types and Ranges in MySQL

Type	Meaning	Range
`tinyint`	A very small integer	Signed values: −128 to 127 (-2^7 to 2^7-1)
`smallint`	A small integer	Unsigned values: 0 to 255 (0 to 2^8-1)
`mediumint`	A medium-sized integer	Signed values: −32,768 to 32,767 (-2^{15} to $2^{15}-1$)
`int`	A standard integer	Unsigned values: 0 to 65,535 (0 to $2^{16}-1$)
`bigint`	A large integer	Signed values: −8,388,608 to 8,388,607 (-2^{23} to $2^{23}-1$)
`float (m, d)`	A single-precision floating-point number with M digits and D decimals	Unsigned values: 0 to 16,777,215 (0 to $2^{24}-1$)
`double (m, d)`	A double-precision floating-point number with M digits and D decimals	Signed values: −2,147,683,648 to 2,147,483,647 (-2^{31} to $2^{31}-1$)
`decimal (m, d)`	A floating-point number with M digits and D decimals	Unsigned values: 0 to 4,294,967,295 (0 to $2^{32}-1$)

9.3.1.2 String (Character) Data Type

A string is a sequence of character values, such as 'Phoenix,' or 'patient seeks treatment.' Either single or double quotes can be used to surround a string value. However, the American National Standards Institute (ANSI) SQL standard specifies single quotes, so statements written using single quotes would be more portable to other database engines. The string character types are summarized in Table 9.2. There are several escape sequences within strings, and these can be used to indicate special characters (see Table 9.3). Each sequence begins with a backslash character (\) to signify a temporary escape from the usual rules for character interpretation. Note that a NUL byte is not the same as the NULL value; NUL is a zero-valued byte, whereas NULL is the absence of a value. We will discuss more on NULL in MySQL in the later sections of this chapter.

9.3.1.3 Date and Time (Temporal) Data Types

Dates and times are values such as `'2013-07-15'` or `'17:34:43'`. MySQL also supports combined date/time values, such as `'2013-07-15 17:34:43'`. Take special note of the fact that

TABLE 9.2

String (Character) Type and Maximum Size

Type	Meaning	Maximum Size
`char (m)`	A fixed-length character string	M bytes
`varchar (m)`	A variable-length character string	M bytes
`tinyblob`	A very small binary large object (BLOB)	2^8-1 bytes
`blob`	A small BLOB	$2^{16}-1$ bytes
`mediumblob`	A medium-sized BLOB	$2^{24}-1$ bytes
`longblob`	A large BLOB	$2^{32}-1$ bytes
`tinytext`	A very small text string	2^8-1 bytes
`text`	A small text string	$2^{16}-1$ bytes
`mediumtext`	A medium-sized text string	$2^{24}-1$ bytes
`longtext`	A large text string	$2^{32}-1$ bytes
`enum ('value1'...)`	An enumeration; column values may be assigned one enumeration member	65,525 members
`set ('value1'...)`	A set; column values may be assigned multiple set members	64 members

TABLE 9.3
String Escape Sequences

Sequence	Meaning
`\0`	NUL (ASCII 0)
`\'`	Single quote
`\"`	Double quote
`\b`	Backspace
`\n`	Newline (linefeed)
`\r`	Carriage return
`\t`	Tab
`\\`	Backslash
`\Z`	Ctrl-Z (Windows EOF character)

TABLE 9.4
Date and Time

Type	Meaning	Range
`date`	A date value, in "YYYY-MM-DD" format	"1000-01-01" to "9999-12-31"
`time`	A time value, in "hh:mm:ss" format	"–838:59:59" to "838:59:59"
`datetime`	A date and time value, in "YYYY-MM-DD hh:mm:ss" format	"1000-01-01 00:00:00" to "9999-12-31 23:59:59"
`timestamp`	A timestamp value, in YYYYMMDDhhmmss format	19700101000000 to some time in the year 2037
`year`	A year value, in YYYY format	1901 to 2155 for YEAR(4)

MySQL follows the ANSI SQL standard (known as the ISO 8601 format) and represents Sdates in year-month-day order, which is YYYY-MM-DD. Although we can display date values any way we want by using the `DATE _ FORMAT( )` function, the default display format lists the year first, and input values must be specified with the year first (Table 9.4).

9.3.1.4 NULL Value

The NULL data type is a "typeless" value and is usually used to mean "no value," "unknown value," "missing value," "out of range," "not applicable," "none of the above," and so on. The user can insert NULL values into tables, retrieve them from tables, and test whether a value is NULL. However, the user cannot perform arithmetic operations on NULL values; if one tries, the result is still NULL.

9.3.2 Practice Data Definition Language in MySQL

9.3.2.1 Statements for Database Operations

MySQL provides several database-level statements: `USE` for selecting a default database, `CREATE DATABASE` for creating databases, `DROP DATABASE` for removing them, and `ALTER DATABASE` for modifying global database characteristics.

9.3.2.1.1 Creating A Database

The `CREATE DATABASE` statement is used to create a database, with syntax as follows:

```
CREATE DATABASE db_name;
```

When creating a database, the following constraints must be considered: the name must conform to standards, the database must not already exist, and the creator must have sufficient privileges to create it.

Example 9.1

Create a database named healthcaresystem:

```
mysql> create database healthcaresystem;
Query OK, 1 row affected (0.00 sec)
```

9.3.2.1.2 Selecting A Database

The `USE` statement selects a database to make it the default (current) database for a given connection to the server. The user must have access privileges for the database in question; otherwise, the user cannot select it. The syntax for selecting a database is

```
USE db_name;
```

Example 9.2

Choose the database named healthcaresystem:

```
mysql> use healthcaresystem;
Database changed
```

Note: When a connection to the server terminates, any notion of what the default database is also disappears. That is, if we connect to the server again, it does not remember what database we had selected previously. We will have to select the database again.

9.3.2.1.3 Altering Databases

The `ALTER DATABASE` statement makes changes to a database's global characteristics or attributes. For example,

```
ALTER DATABASE db_name DEFAULT CHARACTER SET charset;
```

The `charset` statement should be the name of a character set supported by the server, such as latin1 (cp1252 West European) or binary. Please refer to the MySQL online manual for a comprehensive list of character sets supported by the server.

Example 9.3

Change the default character set from latin1 to latin2:

```
mysql> alter database healthcaresystem character set latin2;
Query OK, 1 row affected (0.00 sec)
```

9.3.2.1.4 Dropping Databases

Dropping a database is just as easy as creating one, assuming that we have sufficient privileges. The syntax for deleting databases is

```
DROP DATABASE db_name;
```

However, the DROP DATABASE statement is something that the user should use with caution. It removes the database and all the tables within it. After the database is dropped, it is gone forever. If only some or all of the tables within the database need to be removed while keeping the database, the DROP TABLE statement should be used instead.

Example 9.4

Remove the database named healthcaresystem:

```
mysql> drop database healthcaresystem;
Query OK, 0 rows affected (0.11 sec)
```

9.3.2.2 Statements for Relation Operations

In general, MySQL follows the ANSI SQL standard to operate on tables. These include CREATE TABLE for creating tables, DROP TABLE for removing the table from the database, and ALTER TABLE for changing table characteristics. Note that, in the previous illustrative examples, the query results show "0 rows affected" since the table is empty with no records yet.

9.3.2.2.1 Creating Tables

To create a table, the CREATE TABLE statement is used.

Example 9.5

Create the Drugs table from Chapter 8:

```
mysql> CREATE TABLE Drugs (
    ->         Drug_ID char (20) NOT NULL,
    ->         Name    varchar (255),
    ->         Price           decimal (10, 2),
    ->         Description     varchar (255),
    ->         CHECK (Price > 0),
    ->         PRIMARY KEY (Drug_ID))
    -> ;
Query OK, 0 rows affected (0.04 sec)
```

Note: In this example, the decimal data type (Table 9.1) instead of the numeric data type is used for the Price column.

9.3.2.2.2 Altering Table Structure

ALTER TABLE is a versatile statement in MySQL. The basic use of this statement is to rename tables, add or drop columns, change column types, and add or drop constraints.

Example 9.6

Rename the Drugs relation with a new table name New_Drugs:

```
mysql> alter table drugs rename to new_drugs;
Query OK, 0 rows affected (0.10 sec)
```

Example 9.7

Change the data type of the column Price from decimal to float:

```
mysql> alter table drugs modify price float(10,2);
Query OK, 0 rows affected (0.20 sec)
Records: 0  Duplicates: 0  Warnings: 0
```

Example 9.8

Remove the Primary key constraint from the Drugs relation:

```
mysql> alter table drugs drop primary key;
Query OK, 0 rows affected (0.16 sec)
Records: 0  Duplicates: 0  Warnings: 0
```

9.3.2.2.3 Dropping Tables

Dropping a table is much easier than creating it because there is no need to specify anything about its contents. For example,

Example 9.9

Delete the Drugs relation:

```
mysql> drop table drugs;
Query OK, 0 rows affected (0.11 sec)
```

In MySQL, if the user is not sure whether a table exists, but the user wants to drop it if it does, `IF EXISTS` can be used in the statement. For example,

Example 9.10

Drop the relation named "Sales" by using `IF EXISTS`.

```
mysql> drop table if exists sales;
Query OK, 0 rows affected, 1 warning (0.00 sec)
```

Note: If the Sales relation does not exist, without using `IF EXISTS`, MySQL will throw an error message:

```
mysql> drop table sales;
ERROR 1051 (42S02): Unknown table 'healthcaresystem.sales'
```

9.3.3 Practice Data Manipulation Language in MySQL

Similar to DDL, MySQL adopts SQL DML for data manipulations. These include `INSERT INTO` for adding new records, `UPDATE` for changing the records, `SELECT` for querying the records, and `DELETE` for removing the records. The concept of nested query also applies.

9.3.3.1 Adding Records to a Table

The `INSERT INTO` statement is used to add new records into an existing table.

Example 9.11

Add the four records into the Drugs relation.

Drug_ID	Name	Price	Description
T_001	Tylenol	$17.99	100 325mg Tablets Bottle
T_003	Tylenol	$8.99	
T_002	Tri_Vitamin	$14.99	1500-400-35 Solution 50ml Bottle
C_001	Claritin	$35.99	10 Tablets

```
mysql> INSERT INTO Drugs VALUES ('T_001', 'Tylenol', 17.99, '100 325mg Tablets Bottle');
Query OK, 1 row affected (0.00 sec)

mysql> INSERT INTO Drugs VALUES ('T_003', 'Tylenol', 8.99, NULL);
Query OK, 1 row affected (0.00 sec)

mysql> INSERT INTO Drugs VALUES ('T_002', 'Tri_Vitamin', 14.99, '1500-400-35 Solution 50ml Bottle');
Query OK, 1 row affected (0.00 sec)

mysql> INSERT INTO Drugs VALUES ('C_001', 'Claritin', 35.99, '10 Tablets');
Query OK, 1 row affected (0.00 sec)
```

Note: An easy way to populate multiple rows into a table is to create a text file (e.g., Notepad, and Microsoft Word is NOT recommended) containing all the `INSERT` statements. The user can copy the statement, and in the command-line window, go to Edit | Paste; a series of executions are triggered to have the records added into the table one by one. Please take note of the difference between symbols ' ' (MySQL) and ‘ ’ (text editor). Inappropriate use of the symbols will throw an error message "Error 1064 (42000): You have an error in your SQL syntax."

9.3.3.2 Querying Tables

To query table(s), the `SELECT` statement is used. We have previously learned the `USE` statement for selecting the database. If the user does have access to a database, the tables can be queried directly without selecting the database explicitly. This is done by referring to table names with the database name. For example,

Example 9.12

Retrieve the contents of the Drugs table in the healthcaresystem database without selecting the database first:

```
mysql> select * from healthcaresystem.drugs;
+---------+-------------+-------+----------------------------------+
| Drug_ID | Name        | Price | Description                      |
+---------+-------------+-------+----------------------------------+
| C_001   | Claritin    | 35.99 | 10 Tablets                       |
| T_001   | Tylenol     | 17.99 | 100 325mg Tablets Bottle         |
| T_002   | Tri_Vitamin | 14.99 | 1500-400-35 Solution 50ml Bottle |
| T_003   | Tylenol     |  8.99 | NULL                             |
+---------+-------------+-------+----------------------------------+
4 rows in set (0.00 sec)
```

It is, however, much more convenient to refer to tables without having to specify a database qualifier. Thus, the `USE` statement is recommended before executing queries. Another benefit of the `USE` statement is that the user can issue the statements to switch among databases as long as the access privileges are granted.

Note: By default, MySQL returns the results in descending order over the first column Drug_ID.

9.3.3.3 Updating Tables

When the values of existing records in a table need to be changed, the `UPDATE` statement is used.

Example 9.13

Update the price of all drugs in the Drugs relation with 15% discount:

```
mysql> update drugs set price=price*0.85;
Query OK, 4 rows affected, 4 warnings (0.02 sec)
Rows matched: 4  Changed: 4  Warnings: 4
```

Note: The MySQL `UPDATE` statement has a notable difference from standard SQL. That is, if the user accesses a column from the table to be updated, MySQL `UPDATE` uses the current value (after the update) of the column instead of the original value of the column. For example, in the statement:

```
UPDATE table_1 SET col1 = col1 + 1, col2 = col1;
```

col2 will be set with the new col1 value after the update. As a result, both two columns have the same values.

9.3.3.4 *Deleting Records*

The DELETE statement is used to remove records from an existing table.

Example 9.14

Remove all records from the Drugs relation:

```
mysql> delete from drugs;
Query OK, 4 rows affected (0.04 sec)
```

Note: The DELETE statement will only delete records. Even if the user has all the records removed from the relation, the empty table still exists in the database.

9.3.4 MySQL Transaction Control Language

A transaction consists of a sequence of query and/or update statements. By default, a connection to the MySQL Server begins with the AUTOCOMMIT mode enabled, which automatically commits every SQL statement being executed. When the user wants to issue a sequence of DML statements and commit them or roll them back all together, the user needs to switch AUTOCOMMIT off and end each transaction with the COMMIT or the ROLLBACK command when appropriate. The COMMIT statement will commit the current transaction and make the update performed by the transaction permanently in the database. After the transaction is committed, a new transaction is automatically started. The ROLLBACK statement causes the current transaction to be rolled back. That is, it undoes all the updates performed by the SQL statements in the transaction. Thus, the database state is restored to what it was before the first statement of the transaction was executed. The following examples show two transactions, in which the first one is committed, while the second one is rolled back.

Example 9.15

Use of commit. In this example, we first create a relation named "Drugs." Next, we insert a new record into the relation. The COMMIT statement indicates that it is the end of the transaction.

```
mysql>  CREATE TABLE Drugs (
    ->          Drug_ID char (20) NOT NULL,
    ->          Name    varchar (255),
    ->          Price           decimal (10, 2),
    ->          Description     varchar (255),
    ->          CHECK (Price > 0),
    ->          PRIMARY KEY (Drug_ID))
    -> ;
Query OK, 0 rows affected (0.10 sec)

mysql> -- Do a transactioin with autocommit turned on (this is by Default);
mysql> insert into drugs values ('T_001', 'Tylenol', 17.99, '100 325mg Tablets Bottle');
Query OK, 1 row affected (0.00 sec)

mysql> commit;
Query OK, 0 rows affected (0.00 sec)

mysql> select * from drugs;
+---------+---------+-------+--------------------------+
| Drug_ID | Name    | Price | Description              |
+---------+---------+-------+--------------------------+
| T_001   | Tylenol | 17.99 | 100 325mg Tablets Bottle |
+---------+---------+-------+--------------------------+
1 row in set (0.00 sec)
```

Note: In MySQL, the "--" (double-dash) character is used as a comment syntax. Take note that the second dash should be followed by at least one whitespace or control character (such as a space, tab, newline, and so on).

Example 9.16

Use of rollback. In this example, we first turn off `AUTOCOMMIT`. Next, we insert three new records into the relation then delete the new record that we added in Example 9.15. The `ROLLBACK` statement undoes the three new insert commands and one delete; thus, the relation is restored to the original state after Example 9.15.

```
mysql> -- Do a transaction with autocommit turned off;
mysql> SET autocommit=0;
Query OK, 0 rows affected (0.05 sec)

mysql> insert into drugs values ('T_003', 'Tylenol', 8.99, NULL);
Query OK, 1 row affected (0.00 sec)

mysql> insert into drugs values ('T_002', 'Tri_Vitamin', 14.99, '1500-400-35 Solution 50ml Bottle');
Query OK, 1 row affected (0.00 sec)

mysql> insert into drugs values ('C_001', 'Claritin', 35.99, '10 Tablet');
Query OK, 1 row affected (0.00 sec)

mysql> delete from drugs where Drug_ID='T_001';
Query OK, 1 row affected (0.04 sec)

mysql> select * from drugs;
+---------+-------------+-------+----------------------------------+
| Drug_ID | Name        | Price | Description                      |
+---------+-------------+-------+----------------------------------+
| C_001   | Claritin    | 35.99 | 10 Tablet                        |
| T_002   | Tri_Vitamin | 14.99 | 1500-400-35 Solution 50ml Bottle |
| T_003   | Tylenol     |  8.99 | NULL                             |
+---------+-------------+-------+----------------------------------+
3 rows in set (0.00 sec)

mysql> -- Now we undo the last three inserts and one delete;
mysql> rollback;
Query OK, 0 rows affected (0.00 sec)

mysql> select * from drugs;
+---------+---------+-------+---------------------------+
| Drug_ID | Name    | Price | Description               |
+---------+---------+-------+---------------------------+
| T_001   | Tylenol | 17.99 | 100 325mg Tablets Bottle  |
+---------+---------+-------+---------------------------+
1 row in set (0.00 sec)
```

9.3.5 MySQL Data Control Language

The account information of MySQL is stored in the tables of the MySQL database. There are a number of administrative statements available in MySQL. For example, `CREATE USER` is used to add new users to the system, `GRANT` is used to offer the privileges to existing users, `REVOKE` is used to revoke privileges from existing users, and `DROP USER` is used to remove a user from the system. Only the system administrator has the authority to use these statements. Here, we only use one simple example to illustrate the DCL in MySQL as follows:

Example 9.17

First, create a new user named 'Joel,' and full privilege to the database named "healthcaresystem" is also granted. Next, revoke the privileges of "Joel" and delete 'Joel' from the system.

```
mysql> -- Let's first create a new user named joel;
mysql> create user 'joel'@'localhost';
Query OK, 0 rows affected (0.00 sec)

mysql> use healthcaresystem;
Database changed
mysql> -- Now, we grant the privilege to healthcaresystem database to joel;
mysql> grant all on healthcaresystem to 'joel'@'localhost';
Query OK, 0 rows affected (0.00 sec)

mysql> -- Now, we revoke the privilege from joel;
mysql> revoke select on healthcaresystem from 'joel'@'localhost';
Query OK, 0 rows affected (0.00 sec)

mysql> -- We can also remove the user from the system;
mysql> drop user 'joel'@'localhost';
Query OK, 0 rows affected (0.00 sec)
```

9.3.6 MySQL Utilities

To better study MySQL, there are some important utilities we need to know, for example, SHOW, DESCRIBE, and HELP. We will discuss briefly in this section.

9.3.6.1 `SHOW` Statement

MySQL provides a `SHOW` statement that displays information about databases and the corresponding tables within them. The `SHOW` command is helpful for keeping track of the contents of the databases and for reminding the users about the structure of the tables.

Example 9.18

List the current databases managed by the server:

```
mysql> show databases;
+--------------------+
| Database           |
+--------------------+
| information_schema |
| healthcaresystem   |
| mysql              |
| performance_schema |
| sakila             |
| test               |
| world              |
+--------------------+
7 rows in set (0.00 sec)
```

Example 9.19

Obtain a listing of tables in the healthcaresystem database:

```
mysql> show tables from healthcaresystem;
+----------------------------+
| Tables_in_healthcaresystem |
+----------------------------+
| drugs                      |
+----------------------------+
1 row in set (0.00 sec)
```

Note: Alternatively, if the `USE` statement is used prior to this command, the simple type "show tables" will get the same results.

Example 9.20

Display information about the columns in the Drugs table within the healthcaresystem database:

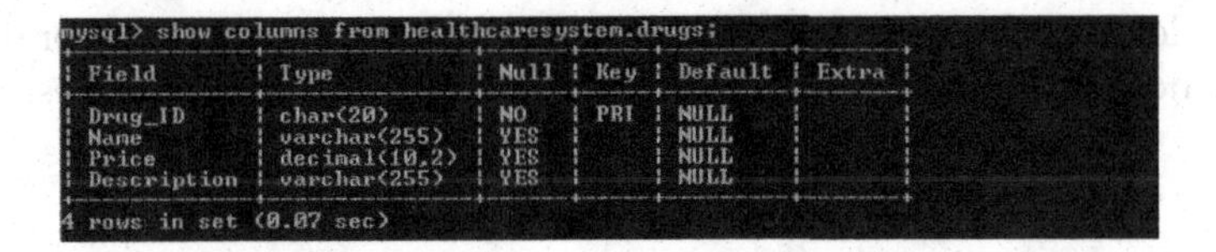

```
mysql> show columns from healthcaresystem.drugs;
+-------------+---------------+------+-----+---------+-------+
| Field       | Type          | Null | Key | Default | Extra |
+-------------+---------------+------+-----+---------+-------+
| Drug_ID     | char(20)      | NO   | PRI | NULL    |       |
| Name        | varchar(255)  | YES  |     | NULL    |       |
| Price       | decimal(10,2) | YES  |     | NULL    |       |
| Description | varchar(255)  | YES  |     | NULL    |       |
+-------------+---------------+------+-----+---------+-------+
4 rows in set (0.07 sec)
```

Example 9.21

List the tables in the MySQL database containing "help" in the name:

```
mysql> show tables from mysql like '%help%';
+-------------------------+
| Tables_in_mysql (%help%) |
+-------------------------+
| help_category           |
| help_keyword            |
| help_relation           |
| help_topic              |
+-------------------------+
4 rows in set (0.00 sec)
```

9.3.6.2 `DESCRIBE` Statement

The `DESCRIBE` statement provides information about the columns in a table. It is an alternative to the `SHOW COLUMN FROM` clause.

Example 9.22

Display the information about the columns in the Drugs table in the healthcaresystem database using the `DESCRIBE` statement:

```
mysql> describe drugs;
+-------------+---------------+------+-----+---------+-------+
| Field       | Type          | Null | Key | Default | Extra |
+-------------+---------------+------+-----+---------+-------+
| Drug_ID     | char(20)      | NO   | PRI | NULL    |       |
| Name        | varchar(255)  | YES  |     | NULL    |       |
| Price       | decimal(10,2) | YES  |     | NULL    |       |
| Description | varchar(255)  | YES  |     | NULL    |       |
+-------------+---------------+------+-----+---------+-------+
4 rows in set (0.01 sec)
```

9.3.6.3 `HELP` Statement

The `HELP` statement searches the help tables from the MySQL database for the given string and displays the result of the search. Note that the search string provided by the user is not case sensitive.

Example 9.23

List the information related to "show tables:"

```
mysql> help 'show tables';
Name: 'SHOW TABLES'
Description:
Syntax:
SHOW [FULL] TABLES [{FROM | IN} db_name]
    [LIKE 'pattern' | WHERE expr]

SHOW TABLES lists the non-TEMPORARY tables in a given database. You can
also get this list using the mysqlshow db_name command. The LIKE
clause, if present, indicates which table names to match. The WHERE
clause can be given to select rows using more general conditions, as
discussed in http://dev.mysql.com/doc/refman/5.6/en/extended-show.html.

This statement also lists any views in the database. The FULL modifier
is supported such that SHOW FULL TABLES displays a second output
column. Values for the second column are BASE TABLE for a table and
VIEW for a view.

If you have no privileges for a base table or view, it does not show up
in the output from SHOW TABLES or mysqlshow db_name.

URL: http://dev.mysql.com/doc/refman/5.6/en/show-tables.html
```

So far, we have learned the use of command-line clients to practice SQL commands. Next, we will learn another MySQL tool, Workbench, and its basic functionalities.

9.4 Working with MySQL Workbench

The MySQL Workbench is a unified visual tool for database architects, developers, and administrators. MySQL Workbench provides the platform for data modeling and SQL development, as well as comprehensive administration tools for server configuration and user administration. Figure 9.9 illustrates the three main functionalities of the MySQL Workbench. Each will be briefly introduced in the following sections.

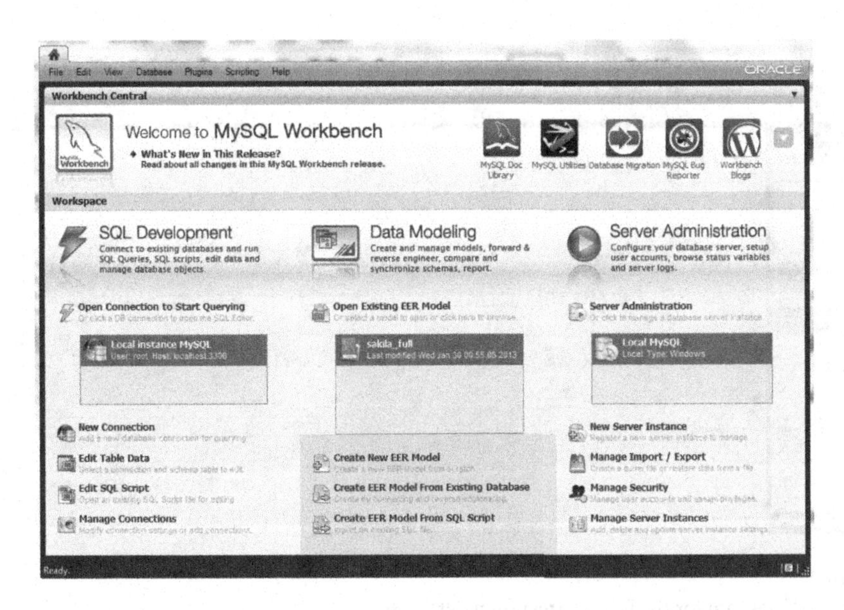

FIGURE 9.9
Launching MySQL Workbench.

9.4.1 Data Modeling

The data modeling module enables the users to create models of the database schema graphically, reverse- and forward-engineer a schema and a live database, and edit all aspects of the database using the comprehensive Table Editor. The Table Editor provides easy-to-use facilities for editing tables, columns, indexes, triggers, partitioning, options, inserts and privileges, routines, and views. We will use the following example to show how to work with data modeling in MySQL Workbench.

Example 9.24

Use of data modeling to create the healthcaresystem database from scratch.

Step 1: Click on the Add Diagram icon in the Model Overview window. A new Enhanced Entity-Relational (EER) diagram tab will then appear. We can create any number of EER diagrams just as we may create any number of schemas.

Step 2: Click the EER diagram tab to navigate to the canvas used to graphically manipulate database objects. The vertical toolbar is on the left side of this page. When using the toolbar, we are able to create relations and columns for each relation, and define the relationship constraints between the relations (see Figure 9.10).

Step 3: Click on the MySQL model tab. We will then see the physical schemas panel. The MySQL database schema has automatically populated the three relations created in the EER diagram.

Step 4: If we want to grant privileges to a given user, we will work with the Schema Privileges panel, where we can add new and assign roles to users.

Step 5: Next, click on the Database | Forward engineer. We will be asked to connect to the server and populate the database schema to the server. A database called "healthcaresystem" with three relations is created.

Next, we can use the SQL development tools to populate the data to the database.

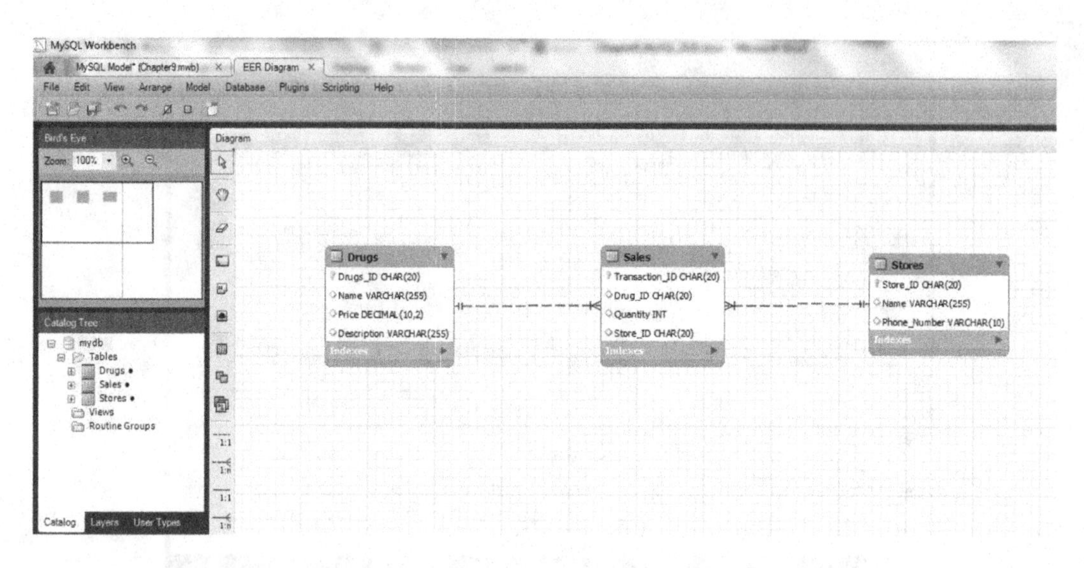

FIGURE 9.10
Creating an EER diagram for the healthcaresystem database.

9.4.2 SQL Development

The MySQL Workbench provides the capability to create and manage connections to database servers, configure connection parameters, and execute SQL queries on the database connections using the built-in SQL editor. Since we have extensively discussed the use of SQL commands in MySQL Command Line Client, we only show one screenshot on the use of Workbench for populating data to the Drugs relation (Figure 9.11).

The easiest way is to navigate to the database and right-click on the relation. The Table Editor is used to edit data and commit changes using a simple grid format. For example, "edit the table" is chosen to create new records, "alter the table" is chosen to make changes to an existing table, and "drop the table" is chosen to remove the table.

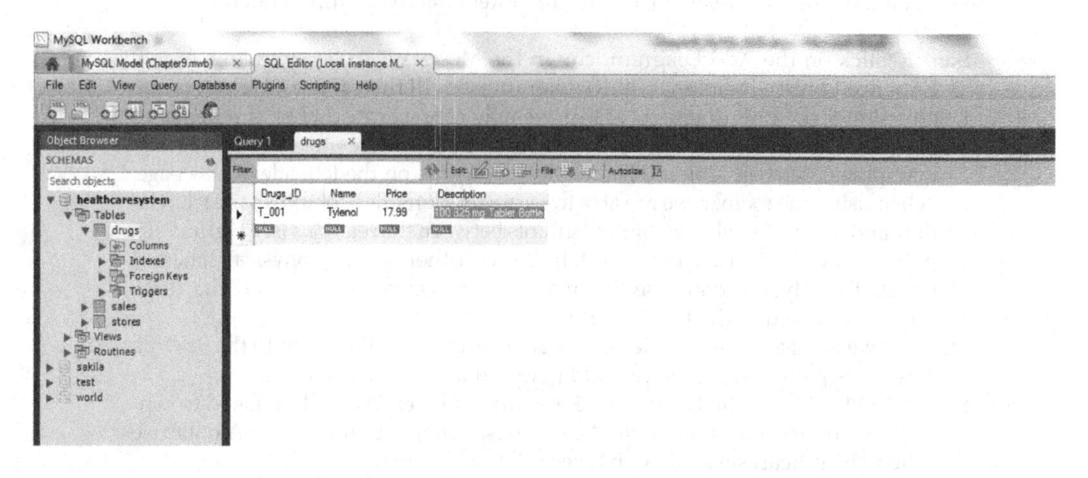

FIGURE 9.11
MySQL Workbench: SQL development.

The visual SQL editor allows the users to build, edit, and run queries; create and edit data; and view and export results. Multiple queries can be executed simultaneously, and results can be viewed on individual tabs in the Results window. The export facility allows the users to export results data to common formats, while the History panel provides a complete session history of queries and statements showing what queries were run and when. The users can easily retrieve, review, rerun, append, or modify previously executed SQL statements. In addition, the SQL Snippet panel allows the users to save and easily reuse common Selects, DML, and DDL code. The most common queries can be placed in snippets so that they can quickly be retrieved and reused later by simply double-clicking in the snippet list.

9.4.3 Server Administration

A server instance is created to provide a way of connecting to a server to be managed. The server administration tool enables the users to create and administer these server instances (Figure 9.12).

The administrator functionality is grouped into several tabs:

- *Management*:
  - *Server Status*: This includes CPU Utilization, Memory Usage, Connection Usage, Traffic, Query Cache Hit Rate, and Key Efficiency.
  - *Startup/Shutdown*: This allows the users to start and stop the MySQL Server and view the startup message log.
  - *Status and System Variables*: This displays system and status variables.
  - *Server Logs*: This displays server log file entries.
- *Configuration*: This allows the users to view and edit the MySQL configuration file (my.ini or my.cnf) using the GUI.
- *Security*: This allows the users to create user accounts and assign roles and privileges.
- *Connection*: This displays connections to the MySQL Server.
- *Data Export/Restore*: This is an important utility to import different data files (e.g., MS Excel, .csv) into MySQL and export MySQL schema objects/tables to an SQL file.

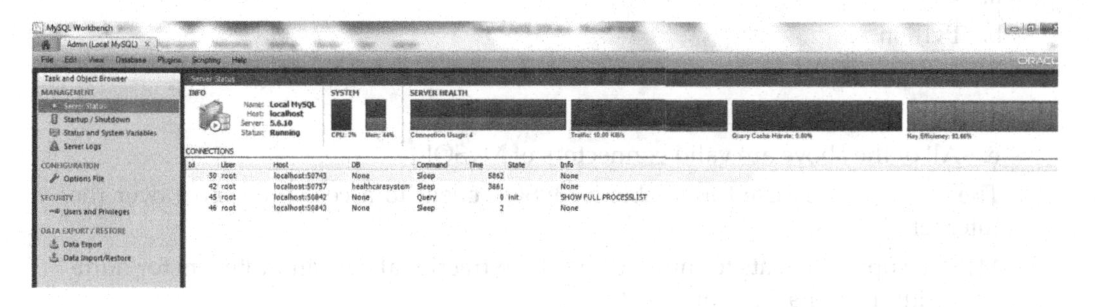

FIGURE 9.12
MySQL Workbench: server administration.

9.5 Summary

MySQL is one of the top choice open-source RDBS. This chapter briefly discusses how to install and configure the MySQL Server. Two important MySQL tools are introduced: the MySQL Command Line Client and the MySQL Workbench. To serve the purpose of learning SQL, the MySQL Command Line Client tool is the first focus to demonstrate (1) how to use DDL and DML commands in MySQL, (2) how to manage and control MySQL transactions using TCL and DCL, and (3) how to use some important MySQL utilities including the SHOW, DESCRIBE, and HELP statements. To serve the purpose of learning database management, data modeling, and MySQL Workbench, a GUI tool is then briefly reviewed.

We only cover the basics of MySQL. Advanced readers are referred to the online manuals for more information.

EXERCISES

1. MySQL is an open-source database management system. True or false?
2. Oracle Corporation originally established MySQL. True or false?
3. The following are advantages of MySQL except:
   a. MySQL runs on Linux and Unix.
   b. MySQL supports SSL protocol.
   c. MySQL can connect to .NET applications.
   d. MySQL has many features at a trade-off of large installation size.
   e. All of the above are advantages of MySQL.
4. The MySQL __________ is a MySQL version that handles high-volume transactions per time period.
5. Which of following are functions of the MySQL Workbench?
   a. Design a database schema.
   b. Normalize data stores.
   c. Document the database schema.
   d. All of the above are valid functions.
   e. Only a and c are valid.
6. The following are valid connectors of MySQL except:
   a. C++
   b. Python
   c. J
   d. C
   e. All of the above are valid connectors of MySQL.
7. The __________ is used to enable a MySQL client to access the server over the Internet.
8. MySQL supports floats for numbers with no fractional part and integers for numbers with fractions. True or false?
9. Escape sequences are started using the backslash (\) character. True or false?

10. MySQL follows the "MM-DD-YYYY" format, which is the ISO 8601 standard format for dates. True or false?
11. The NULL value is interpreted as a value of zero when performing arithmetic operations in MySQL. True or false?
12. When creating databases in MySQL, which of the following need to be considered?
    a. The user must have sufficient privileges to create a database.
    b. The database name must conform to standards.
    c. The database name must be unique.
    d. All of the above.
    e. None of the above.
13. MySQL can set a default database to access using the `USE` command even after restarting MySQL. True or false?
14. The `IF EXISTS` command can be used to avoid error messages in MySQL for the `DROP` statement. True or false?
15. By default, the AUTOCOMMIT is enabled when a new connection is started. True or false?
16. Given a database for the purchasing, purchasing_records, write a code in MySQL to list all tables within the database.
17. The __________ and __________ are two alternative commands to list the columns of a table in MySQL.
18. The Table Editor can accomplish the following except:
    a. Edit databases
    b. Modify tables
    c. Edit columns
    d. Edit user privileges
    e. All of the above

10

Object-Based Database Systems

Relational database systems (RDBS) originated from the IBM R project in the 1970s and gained great popularity in business applications. As of today, RDBS products are still the dominant choice of database systems for information management problems. However, with the advancement of computer software and hardware, several limitations of RDBS products began to attract a lot of attention in the late 1980s. First, structured query language (SQL)-92 only supports numbers and strings, but several applications began to include complex data objects such as geographic points, images, and digital signal data. Second, relational tables are flat and do not support nested structures efficiently. Third, RDBS do not take advantage of object-oriented (OO) approaches, which have been widely accepted in industry software engineering practices. This chapter introduces some improvements and advancements made to RDBS, specifically OODBMS and object RDBMS (ORDBMS).

10.1 Object-Oriented Database Management System

10.1.1 Object-Oriented Database Concepts

An OO database is an extension of the OO programming language (OOPL) technique for persistent data management. It is a system that integrates database capabilities with OOPL capabilities.

As in OOPL, an object in an OO database typically has two components: its state (value) and its behavior (operations), except that it usually has a complex data structure as well as several specific operations defined by the developer. Objects in an OOPL exist only during program execution and are, hence, called "transient" objects. On the other hand, an OO database can extend the existence of objects so that they can be stored permanently. As a result, the objects persist beyond program termination and can be retrieved later and shared to other programs. These types of objects are called "persistent" objects.

One major advantage of an OO database over a relational database is that it preserves the direct relationship between real-world and the database objects. Additionally, it keeps the integrity and identity of objects consistently, and these objects can be easily identified and operated upon. In addition, complex data structures are often used to organize the information for better describing the objects. In contrast, relational databases represent the complex object structures over many scattered relations and tuples (records), which result only in an implicit mapping between the real-world entities and its database representations.

To learn OO databases, some important concepts are reviewed first as follows.

10.1.1.1 Objects and Identities

Each real-world entity is modeled as a database object. The object needs a unique identity to be stored in the system. This unique identity is implemented via a unique, system-generated object identifier, termed as "OID." An OID is used internally by the system instead of the developers to identify each object uniquely and create and manage inter-object references. Using OIDs over keys for object identification presents at least one advantage, that is, the developers do not need to be concerned about selecting the appropriate keys for the objects since the OID is generated by the system. On the other hand, an OID is criticized as having little semantic meaning to the users. Interestingly, Cattell (1991) argues that, although the keys are important for the user, they present some difficulties. For example, short keys (e.g., social security number [SSN]) may not contain significant semantic information, whereas longer keys (first and last names, book titles) tend to be extremely inefficient if used as foreign keys.

10.1.1.2 Complex Objects

A complex object is an object that includes the set of attributes associated to it. The value of an attribute can be an object or a set of objects. This characteristic enables an arbitrary complex object to be defined from other objects. Complex objects are built by applying constructors to simpler objects, which are data types such as integers, real numbers, character strings, Booleans, and other basic data types supported by the system. These constructors are composed of several other types: sets, tuples, lists, bags, and arrays. Typically, the set constructor is used to define a set of objects identified by OIDs, for example, $\{OID_1, OID_2, \ldots OID_n\}$. The tuple constructor is used to represent the property of the objects, for example, $\{a_1: OID_1, a_2: OID_2, \ldots a_n: OID_n\}$, where a_1 is an attribute name for the object. The list constructor is similar to a set constructor except that the OIDs in a list are ordered. The bag constructor is similar to a set constructor except that all elements in a set must be distinct, whereas a bag can have duplicate elements. The array constructor is a one-dimensional array of elements. The main difference between a list and an array is that a list can have an arbitrary number of elements, whereas an array has a maximum size.

10.1.1.3 Encapsulation

Encapsulation is derived from the concepts of abstract data types and the information hidden in OOPL. Each object consists of two elements that can be accessed and manipulated by other objects: an interface and an implementation. By definition, the interface element is the specification of the set of operations that can be invoked on the object, and it is the only visible part to the users. On the other hand, the implementation element contains the data of the object and the methods that provide the implementation of each operation.

Let us revisit the Patient relation in the healthcaresystem database to illustrate the encapsulation concept. In a relational database, a set of patients is represented by a table in which each row represents each patient. This relation can be queried using SQL data manipulation language statements (e.g., `SELECT-FROM`). There exists a clear separation between the query statements (operations) and the data in a relational database. Moreover, the structures of the database objects are visible to the users and the external programs. Hence, the SQL statements are generic and may be applied to any relation in the database. In an OO database, however, the entity Patient is defined as an object consisting of a data component (similar to the table structure defined in a relational database) and an operational component, for example, GetPatientHistory. Both the data and the operational components are stored in a database. The external users of the Patient object are only presented with information on the interface of the object (e.g., the name and the parameters of each operation), whereas the internal data structure and the associated methods are hidden from the users. As a result, the encapsulation implements the idea of revealing only pertinent information to users.

10.1.1.4 Classes and Inheritance

In an OO database, all objects sharing the same set of attributes and methods are grouped together into classes. A class can be defined as another instance of one or more existing classes and will inherit the attributes and methods of such classes. This class is often referred to as a "subclass," whereas the classes from which it has been defined are referred to as a "superclass." For example, we define a Person class as

```
class Person {
            attribute string FirstName;
            attribute string LastName;
            attribute date DateOfBirth;
            attribute string Address;
            attribute string SSN;
            short Age( )
}
```

In this class, the FirstName, LastName, SSN, Address, and DateOfBirth are stored as attributes, and the Age function is a method to calculate the Age from the value of the DateOfBirth attribute and the current date. To illustrate the concept of a subclass, let us assume a new class, Patient, which has similar attributes to that of the Person class to be created. The Patient class can be defined as

```
class Patient {
            attribute string FirstName;
            attribute string LastName;
            attribute date DateOfBirth;
            attribute string Address;
            attribute string SSN;
            short Age( )
            string GetPatientHistory( )
            string ReactionToTreatment( )
}
```

It is observed that the Patient class has the same attributes and an Age function as compared to the Person class, with two added functions: GetPatientHistory and ReactionToTreatment. Given a Person class defined previously, the Patient class can be redefined as

```
class Patient extends Person {
            string GetPatientHistory( )
            string ReactionToTreatment( )
}
```

10.1.1.5 Overloading, Overriding, and Late Binding

Overloading, overriding, and late binding are also known as "polymorphism of operations." This concept allows different methods to be associated with a single operation name, leaving the system to determine which method should be used to execute a given operation.

10.1.2 Object-Oriented Database Design and Modeling

The idea of implementing OO concepts into a database design was inspiring. However, the practical implementation of this concept is not an easy task, mainly due to the lack of existing standards. In 1991, a consortium of OODBMS vendors, called "Object Data Management Group" (ODMG), was formed to work toward standardizing OO database management. In 1993, the first standard, ODMG 1.0, was released, and the final version, ODMG 3.0 Standard, was released in 2001. The ODMG standards outlined three components: object modeling, object definition language (ODL), and object query language (OQL), which are explained in the following sections.

10.1.2.1 Object Modeling

The object model is designed to be the foundation for object database systems upon which the ODL and OQL are defined. It supports the concepts discussed previously, including the representation of complex objects, encapsulation, and inheritance. Furthermore, object modeling has some common concepts with relational modeling (e.g., entity-relational model) but with some fundamental differences. The key comparisons are provided as follows, and the key terms in an OO data model and its counterpart in the relational model are summarized in Table 10.1.

10.1.2.1.1 Object versus Entity and Tuple

An OODM object resembles the entity and the tuple in the relational models, with additional characteristics such as behavior, inheritance, and encapsulation. These characteristic make OO modeling much more natural than relational modeling.

TABLE 10.1

Summary of OO Data Models

OO Data Model	Description	Corresponding Relational Model Element
Class	An entity that has a well-defined role in real-world application, which is described by its state, behavior, and identity.	Entity set
Object	An instance of a class that encapsulates data and behavior.	Entity
State	Consists of an object's properties and the values of the properties.	Attributes
Operation	A function that is provided by all the instances of a class.	None or not available
Method	Implements the operations on the objects such as query and update.	None or not available
Association	A named relationship between or among object classes, for example, unary and binary.	Degree of relationship
Multiplicity	A specification that indicates how many objects participate in a given relationship.	Cardinality
Class diagram	Illustrates the static structure of an OO data model: the classes, their structure, and interrelationship.	E-R diagram

10.1.2.1.2 Class versus Entity Set and Table

The concept of a class can be associated with the relational models' concepts of entity sets and tables. However, the concept of a class is more powerful in that it allows the description not only of the data structure, but also of the behavior (represented by method) of the class objects.

10.1.2.1.3 Encapsulation and Inheritance

Classes are organized in class hierarchies. An object belonging to a class inherits all the properties of its superclasses. Additionally, encapsulation means that the data representation and the methods' implementation are hidden from other objects and the end user, whereas in a relational model, data are directly accessible from the external environment.

10.1.2.1.4 Object ID

Object ID is not supported in the relational model.

10.1.2.1.5 Relationship

The main property of any data model is found in its representation of relationships among the data components. The relational model uses a value-based relationship approach. This means that a relationship among entities is established through a common value in one or several of the entity attributes. In contrast, the object model uses the object ID, which is identifier based, to establish relationships among objects, and such relationships are independent of the state of the object. The relationships in an object model can either be inter-class references or class hierarchy inheritance.

Example 10.1

Present the healthcaresystem database learned in Chapter 5 using an object model.

The solution is shown in Figure 10.1. As seen, the Patient class has five attributes—FirstName, LastName, DateOfBirth, Address, and SSN—and three methods—Age( ), GetPatientHistory( ), and ReactionToTreatment( ). The TreatmentPlan class has two attributes—Drug_Description and TreatmentDate—and one method—PatientOutcome( ). The Drug class has three attributes—Drug_ID, Name, and Description—and one method—Sales( ). Each Patient may have multiple TreatmentPlans, and each TreatmentPlan may be applied to multiple Patients. In each TreatmentPlan, multiple drugs may be prescribed, and each Drug may be prescribed in multiple TreatmentPlans. In the next section, we will learn the use of ODL to define the classes and their relationships.

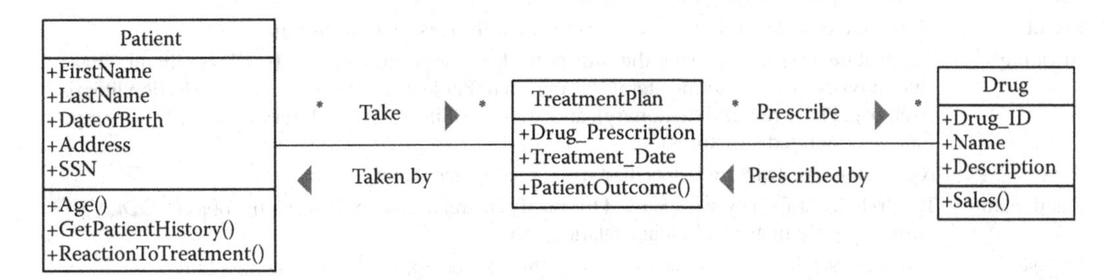

FIGURE 10.1
Object data model for part of the healthcaresystem database.

10.1.2.2 Object Definition Language

Similar to SQL-92 data definition language, which specifies a logical schema for a relational database, object definition language allows the users to specify a logical schema for an OO database. ODL is designed to support the semantic constructs of the ODMG object model and is independent of any particular programming language. Table 10.2 summarizes the keywords used to define a class, extent, attribute, struct, and relationship in ODL. An operation is specified without the use of a keyword; a parenthesis is simply added after its name.

In the example shown in Figure 10.1, the ODL for the three classes are as follows. Since each Patient can take multiple TreatmentPlans, the keyword **set** is used after the **relationship** keyword. Another keyword is **extent**, which is simply used to refer to the class set.

```
class Patient {
(       extent Patients)
        attribute string FirstName;
        attribute string LastName;
        attribute date DateOfBirth;
        attribute string Address;
        attribute string SSN;
        relationship set <TreatmentPlan> takes inverse
        TreatmentPlan::taken_by
        short Age( )
        string GetPatientHistory( )
        string ReactionToTreatment( )
}


class TreatmentPlan {
(       extent TreatmentPlans)
        attribute string Drug_Prescription;
        attribute date Treatment_Date;
        relationship set <Patient> taken_by inverse Patient::takes
        relationship set <Drug> prescribes inverse Drug::prescribed_by
        string PatientOutcome( )
}
```

TABLE 10.2
Some Important ODL Keywords

Keywords	Description
Class	A class is specified using the "class" keyword.
Extent	The extent of a class is the set of all instances of the class within the database.
Attribute	An attribute is specified using the "attribute" keyword and can either be OID or literal. The literal types include atomic literals (string, char, Boolean, float, short, long), collection literals (set, bag, list, array, and dictionary), and structural literals (Date, Interval, Time, Timestamp, and user-defined structures).
Struct	A user-defined structure is specified using the "struct" keyword.
Relationship	The "relationship" keyword is used to specify the relationship between the objects. ODL supports only unary and binary relationships.
Inverse	The "inverse" keyword is used to specify the relationship in the reverse direction.

```
class Drug {
(       extent Drugs)
        attribute string Drug_ID;
        attribute string Name;
        attribute string Description;
        relationship set <TreatmentPlan> prescribed_by inverse
        TreatmentPlan::prescribes
        string SalesTransaction( )
}
```

Let a user-defined struct Size for the patient class be as follows:

```
struct Size {
        float weight;
        float height;
        short BMI;
}
```

The Patient class can then be modified with the added reference to the user-defined struct Size:

```
class Patient {
(       extent Patients)
        attribute string FirstName;
        attribute string LastName;
        attribute date DateOfBirth;
        attribute string Address;
        attribute string SSN;
        attribute SIZE patient_size;
        relationship set <TreatmentPlan> takes inverse
        TreatmentPlan::taken_by
        short Age( )
        string GetPatientHistory( )
        string ReactionToTreatment( )
}
```

Compared to the example from Chapter 5, fundamental differences between an OO database schema and a relational database schema are clearly observed. In the following section, the query language for an OO database, OQL, is introduced.

10.1.2.3 Object Query Language

OQL is the query language designed to work closely with the programming language for which an ODMG binding is defined. Hence, an OQL query embedded into any OOPL can return objects that match the type system of that language. The OQL syntax for queries is similar to the syntax of SQL-92, with additional features for ODMG concepts, such as object identity, complex objects, operations, inheritance, polymorphism, and relationships. In an OQL query, the developer needs to query from the **extent** of the class. For example, if one wants to query for patient information, the query is executed on the Patients class.

Example 10.2

Query from one class: Retrieve the Address of the patient whose name is "Joe Smith":

```
SELECT p.Address
FROM Patients p
WHERE p.FirstName = "Joe" and p.LastName = "Smith";
```

Example 10.3

Including an operation in a query: Retrieve the name of all patients who are 40 years or older:

```
SELECT p.FirstName, p.LastName
FROM Patients p
WHERE p.age >= 40;
```

Example 10.4

Use of aggregate functions: Find the number of patients being treated with the corresponding code:

```
SELECT count (*)
FROM Patients p
```

Example 10.5

Query from multiple classes: Retrieve the name of the drug used in treating the patient named "Joe Smith" without duplicates:

```
SELECT DISTINCT d.Name
FROM Drugs d, Patients p, Treatments t
        d.prescribed_by t
        p.takes t
WHERE p.FirstName = "Joe" and p.LastName = "Smith"
```

Example 10.6

Use of subquery: Retrieve the name of the patient who has a "Positive" response to the treatment and is over 40 years:

```
SELECT x.FirstName, x.LastName
FROM (
  SELECT p FROM Patients where p.ReactionToTreatment = "Positive") as x
WHERE x.age >= 40
```

Note that only a subset of OQL queries are illustrated above. Please refer to ODMG 3.0 Standard (2000) for more standard OQL features.

10.1.3 Summary of OODBMS

In summary, OODBMS yield several benefits over relational database systems. Most of these benefits are based on the previously explored OO concepts. Database systems such as hierarchical, network, and relational systems have been very successful in traditional business applications. However, new business operations require more complex structures for objects, new data types for storing images or large textual items, and the need to define nonstandard application-specific operations. OO databases were proposed to meet these needs. These databases offer the flexibility to handle some of the requirements without being limited by the data types and query languages available in traditional database systems.

However, OODBMS is not without limitations. First, OODBMS vendors have discovered the difficulty of tying the database design closely to the application design. Maintaining and evolving an OODBMS-based information system has also been an arduous undertaking. In addition, the OO data model lacks the mathematical foundations present in relational models, which make an OODBMS not well understood as compared to RDBS.

Inspired by the OO concept and noting some complementary features of OO and relational models, there has been an effort to combine the two methods in developing large information systems, which produced the ORDBMS.

10.2 Object RDBMS

10.2.1 Object-Relational Model

The object-relational (OR) data model enhances a relational model with objects. The main task of the model is to provide a flexible framework for organizing and manipulating database objects that correspond to real-world objects.

To achieve this, the OR model extends the relational model by incorporating attributes that are atomic in nature as well as those that have a structured data type. Similar to relational databases, an OR database is composed of a group of tables made of rows, and all rows may consist of values from structurally identical specific data types instead of only atomic types. Like an OO database, the structured types are the types built from atomic types via the use of constructors. This extension enables an OR database to have nested relations. Let us use the following example to illustrate the concept.

Example 10.7

Design a nested relation schema for the Patient relation that incorporates an attribute called "TreatmentPlan," representing all the treatments that the patient has had.

For simplicity, let the Patient relation have three original attributes: Name, Address, and SSN. Furthermore, let the TreatmentPlan have two attributes: Drug_Description and TreatmentDate. An example of the design is shown in Figure 10.2. As seen, Name, Address, and SSN have atomic values, whereas TreatmentPlan has a relation as a structured type. It includes two attributes, Drug_Description and Treatment_Date, which are atomic in nature.

The detailed schema for this example, thus, is written as

```
Patient (Name (FirstName, LastName), Address, SSN, TreatmentPlan (Drug_
Description, Treatment_Date))
```

Since the drug "Claritin" was prescribed to two patients, it appears twice in the relation. Based on the relational database design principle, this violates Boyce–Codd normal form (BCNF) which we have discussed in the earlier chapter. To address this issue, the OR database introduces reference idea, which allows the tuples to point to other tuples to avoid duplicates (shown in Figure 10.3).

Name		Address	SSN	TreatmentPlan	
FirstName	LastName	Flamingo Dr. Chandler, AZ	322-18-9533	Drug_Description	Treatment_Date
Joe	Smith			Claritin	02/25/2013
				Claritin	04/10/2013
FirstName	LastName	Riggs Rd. Gilbert, AZ	905-35-5217	Drug_Description	Treatment_Date
Tyler	O'Neil			Claritin	02/25/2013
				Claritin	05/01/2013

FIGURE 10.2
Nested relation for Patients and associated TreatmentPlan.

Name		Address	SSN	TreatmentPlan
FirstName	LastName	Flamingo Dr. Chandler, AZ	322-18-9533	
Joe	Smith			
FirstName	LastName	Riggs Rd. Gilbert, AZ	905-35-5217	
Tyler	O'Neil			

Drug_Description	Treatment_Date

FIGURE 10.3
Illustration of the reference concept used in the OR database.

The schemas of the two relations, thus are:

```
TreatmentPlans (Drug_Description, Treatment_Date)
Patient (Name (FirstName, LastName), Address, SSN, TreatmentPlan
({*TreatmentPlans}))
```

We use the Patient (... {*TreatmentPlans}) statement to describe that in Patient relation, in which a reference to a single tuple with a relation schema named TreatmentPlans exists. Note that this is different from decomposing relations as is done in relational database design. The TreatmentPlan attribute is still an attribute in Patient relation since the structured type is defined in a separate relation and is referenced in the Patient relation.

Note: This example does not discuss the use of methods as part of the OR schema. In practice, the SQL-99 standard and OR database allow the same ability as the ODL in defining methods associated with any relation.

10.2.2 Object-Relational Query Language

The OR database adopts the query-centric approach as done in a relational database. All data handling uses declarative SQL statements. Since an OR database supports structured data types, the query statements are capable to deal with complex objects (e.g., Name) instead of atomic data types only (e.g., integer, varchar). Indeed, SQL-99 adds an extensive type system to the earlier SQL standards and allows structured types and type inherence. All these extensions can serve for OR database queries.

Example 10.8

Create Patient relation, that is,

```
Patient (Name (FirstName, LastName), Address, SSN, TreatmentPlan
({*TreatmentPlans}))
```

```
TreatmentPlans (Drug_Description, Treatment_Date)
```

First, let the following two structured types be defined as

```
CREATE TYPE PersonName AS (
          FirstName varchar (20),
          LastName varchar (20))


CREATE TYPE TreatmentPlan AS (
          Drug_Description varchar (255),
          Treatment_Date date)
```

Next, the Patient relation is defined as

```
CREATE TABLE Patients (
Name PersonName NOT NULL,
Address varchar (255),
SSN char (9),
TreatmentPlan   TreatmentPlan NOT NULL,
Primary Key (SSN))
```

As seen, an OR table is structurally very similar to its relational counterpart, and the same data integrity and physical organization rules can be enforced over it. The major difference is specifying column types. In the OR table, atomic data types such as varchar and char are the same as in relational tables, but the other two columns (PersonName, TreatmentPlan) are completely new. From an OO perspective, these data types correspond to class names and can be user defined.

In terms of query, it is very similar to SQL. For example, if the user wants to get the information from the patient named "Joe Smith," the query can be written as

```
SELECT FROM Patients
WHERE Name = PersonName ("Joe Smith")
```

In this section, the basics of OR queries are briefly illustrated. Advanced readers can refer to SQL-99 for a comprehensive discussion on the OR query language features.

10.2.3 Summary of ORDBMS

In summary, ORDBMS synthesizes the features of RDBS with the best ideas of OODBMS. Although ORDBMS reuse the relational model as SQL interprets it, the OR data model is opened up in novel ways. New data types and functions can be implemented using general-purpose languages such as C and Java. In other words, ORDBMS allow the developers to embed new classes of data objects into the relational data model abstraction. Moreover, ORDBMS schemas have additional features that are not present in an RDBS schema. Several OO structural features such as inheritance and polymorphism are included in the OR data model. Second, ORDBMS adopt the RDBS query-centric approach to data management. All data access in an ORDBMS is handled with declarative SQL statements. There is no procedural, or object-at-a-time, navigational interface. ORDBMS persist with the idea of a data language that is fundamentally declarative and, therefore, mismatched with procedural OO host languages. This significantly affects the internal design of the ORDBMS engine, and it has profound implications for the interfaces that the developers use to access the database. Third, from a systems architecture point of view, ORDBMS are implemented as a central server program rather than as a distributed data architecture typical in OODBMS products. However, ORDBMS extends the functionality of RDBS products significantly, and an information system using an ORDBMS can be deployed over multiple machines.

10.3 Summary

This chapter gives a basic overview of two advanced database systems: the OO database and the OR database. Inspired by the success of OO programming, the OO database was designed to incorporate complex data objects (e.g., images, pictures). However, it has not been widely accepted due to the lack of mathematical rigor and the labor intensiveness of maintaining the system. As a result, most relational database vendors choose to develop an OR database. Some fundamentals such as data modeling and queries of both the two systems are briefly discussed in this chapter.

EXERCISES

1. The two components of an object are its current values and its methods. True or false?
2. Objects are transient objects, that is, they only exist when the program is closed. True or false?
3. The following are advantages of an OO database over a relational database except:
    a. It makes object identification easier.
    b. It provides an implicit mapping of real-world entities.
    c. It preserves a direct relationship between a real object and a database object.
    d. All of the above are valid advantages of an OO database.
    e. None of the above is a valid advantage.
4. The following are properties of OIDs except:
    a. An internally generated ID to act as a primary key for an entity
    b. No need for primary keys to uniquely identify each element
    c. Has less semantic meaning
    d. Used to create and manage inter-entity relationships
    e. All of the above are valid properties
5. A complex object can be composed of other complex objects. True or false?
6. In encapsulation, the ____________ element is a set of operations that is visible to the users, while the ____________ element is hidden.
7. A superclass can inherit the following from a subclass:
    a. Attributes
    b. Data points
    c. Methods
    d. All of the above
    e. Only a and c

8. The ___________ of operations concept allows different methods to be associated with a single operation.
9. An OODM object incorporates encapsulation as compared to an RDBS entity. True or false?
10. An operation in OODM is equivalent to a query in RDBS. True or false?
11. Using an ODL, an operation is specified using a keyword "define." True or false?
12. An OQL query can be integrated into any OOPL to query objects. True or false?
13. An OR database extends the relational model by treating atomic attributes as objects. True or false?
14. What is the main advantage of using an OQL over relational databases?

    For questions 14 and 15, consider the following schema:

```
Person (Person_M(First_M, Last_M), Person_Address(Number, Street,
City, State, Zip))
Background (Academic_Inst, Year_Completed, Degree)
Courses (Course_M, Course_Units)
```

15. Write an ODL code to create a table called "Faculty_Member" that inherits all the attributes from a person schema with academic backgrounds. Additionally, the new table should include a Faculty_N integer attribute to act as a primary key and a new attribute called "Specialization" to denote a faculty's academic specialization.
16. Write an ODL code to create a table named "Faculty_Courses" from these schemas. This table should list down all courses as well as the faculty members teaching the courses, the number of students for the courses, and the semester and year that the courses are taught.

Section III

Windows Application Development

Microsoft Visual Studio is a collection of tools that developers can use to design, develop, debug, and deploy Windows applications and Web applications. It is based on the .NET framework and includes several different programming languages such as Visual Basic, Visual C#, Visual C++, Visual F#, JScript, and ASP.NET. Figure III.1 shows the basic components of Visual Studio, where common language runtime manages the execution of a programming project writing in any language; .NET framework class library is a library of pre-defined class objects including Windows forms, controls, and more; and ADO.NET is an object model to enable the access to data from diverse sources.

Section III of this book covers Visual Basic programming language (VB.NET) and Visual Basic for Application (VBA). VB.NET is an object-oriented programming language working with objects. Example objects are form objects, button objects, text box objects, and

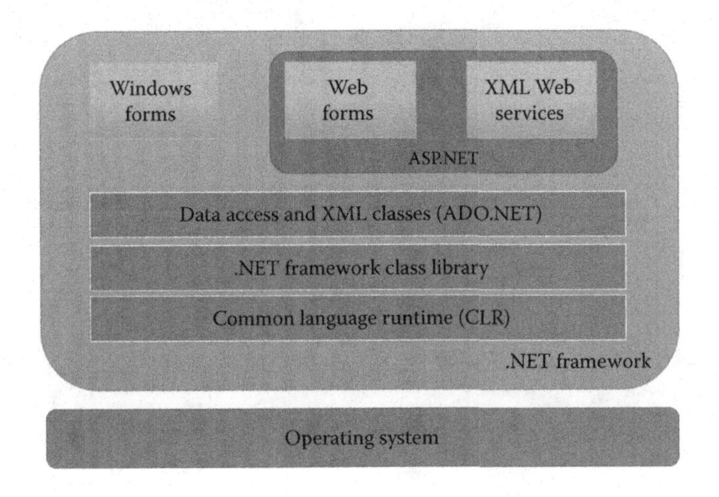

FIGURE III.1
NET framework.

label objects, just to name a few. VB is also known as an event-drive programming language in that the developers may write the program code to respond to the events controlled by end users. Examples include "click a button." The properties and events of the objects are covered in Chapter 11, followed by Visual Basic programming in Chapter 12. Next, the application of ADO.NET in VB.NET for Windows application with database connectivity is covered in Chapter 13. Chapters 14 and 15 focus on VBA, explaining the objects in VBA and covering the programming basics of the use of ADO.NET in Excel application, respectively.

11

Windows Forms and Controls in Microsoft Visual Studio

Visual Studio 2013 is an integrated development environment that integrates and manages software development tools in a single environment, including various programming languages (e.g., Visual Basic, Visual C++, and Visual C#), database development, Windows application development and Web application development. We use Visual Basic in this text. Using Visual Studio 2013, we can develop a Windows-based application that runs on a computer and is used by one user at a time, or a Web-enabled application that can be used by multiple users via the Internet. Chapters 11 to 13 focus on developing a Windows-based application. Chapter 16 describes the development of a Web-enabled application. A Windows-based application consists of a graphical user interface (GUI), which is described in this chapter; functions implemented through Visual Basic programming, which is described in Chapter 12; and connection with database, which is described in Chapter 13. The following sections describe Windows forms and controls in the GUI of a Windows-based application.

11.1 Visual Basic Development Environment

At the first time of using Visual Studio 2013, we get the window shown in Figure 11.1 when we start Visual Studio 2013. We select Visual Basic for Development Settings, and then click on the Start Visual Studio button. This brings up the Start Page shown in Figure 11.2, which allows us to select either starting a new project or opening an existing project. We select Start a New Project, which brings up the New Project window shown in Figure 11.3. We use the New Project window to name our new project "Healthcare Information System." The New Project window has Windows Forms Application set as default, and we take this default application type for our project. After we click the OK button in the New Project window, the Visual Basic development environment, as shown in Figure 11.4, is brought up. This development environment has the menus and shortcut icons at the top and several windows that assist the development. On the left side, there are two tabs for Toolbox and Data Sources. The Toolbox tab contains controls and other tools that we can use to create a GUI. The Data Sources tab initially has no data sources and will contain databases and other types of data sources when they are added to the project. If the Toolbox tab is closed, it can be brought up again by selecting Toolbox in the View menu. If the Data Source tab is closed, it can be brought up again by selecting Other Windows in the View menu and then Data Sources in the pop-up menu. In the Design window, there is the Design View of Form1, which is a default form. The Solution Explorer window gives the list of components in the project. At the beginning, the list includes only three components in the project: My Project, App.config, and Form1.vb. The Properties window shows properties of the selected object in the Design window, which is Form1 in Figure 11.4.

FIGURE 11.1
Starting window of Visual Studio 2013.

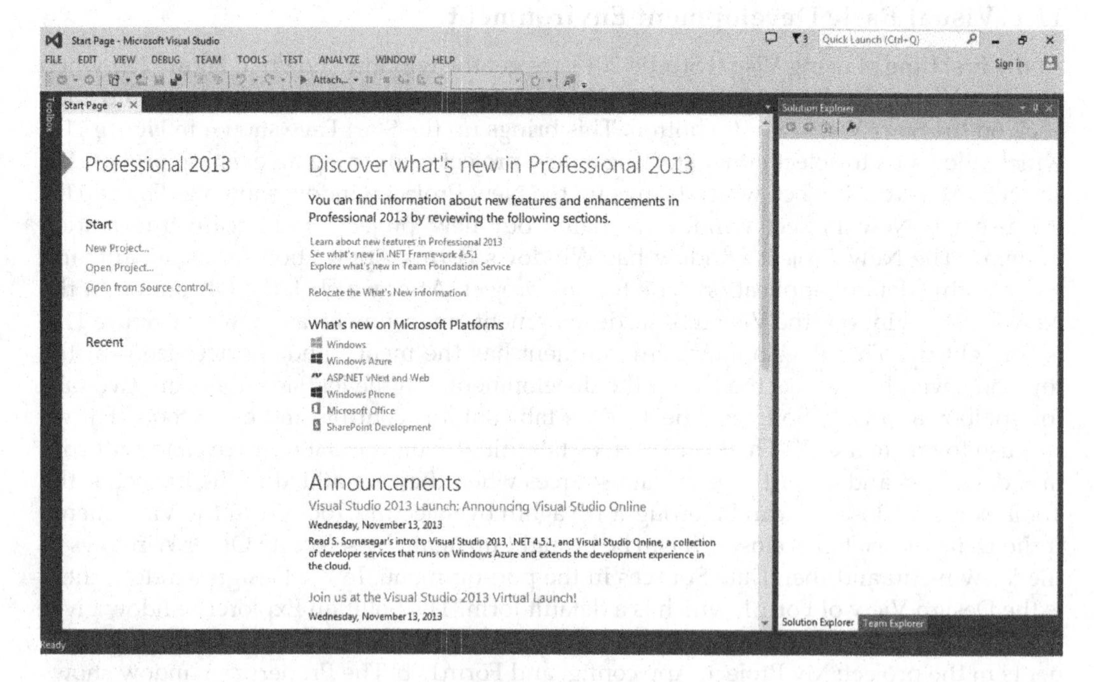

FIGURE 11.2
Start Page window.

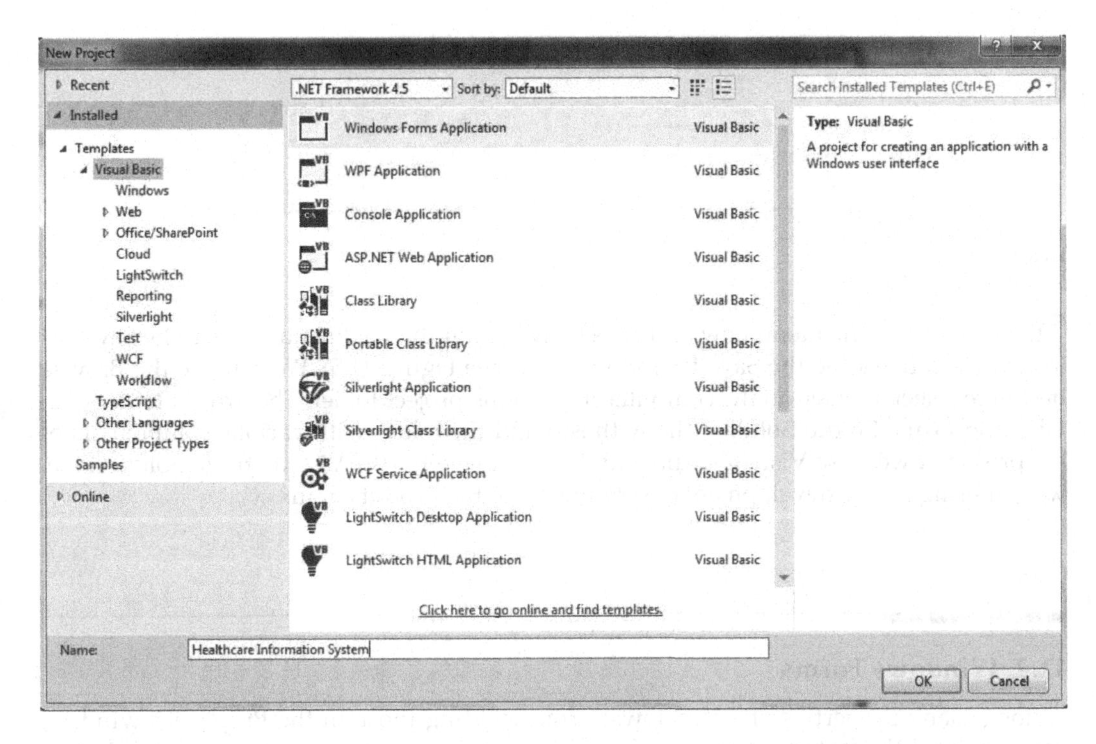

FIGURE 11.3
New Project window.

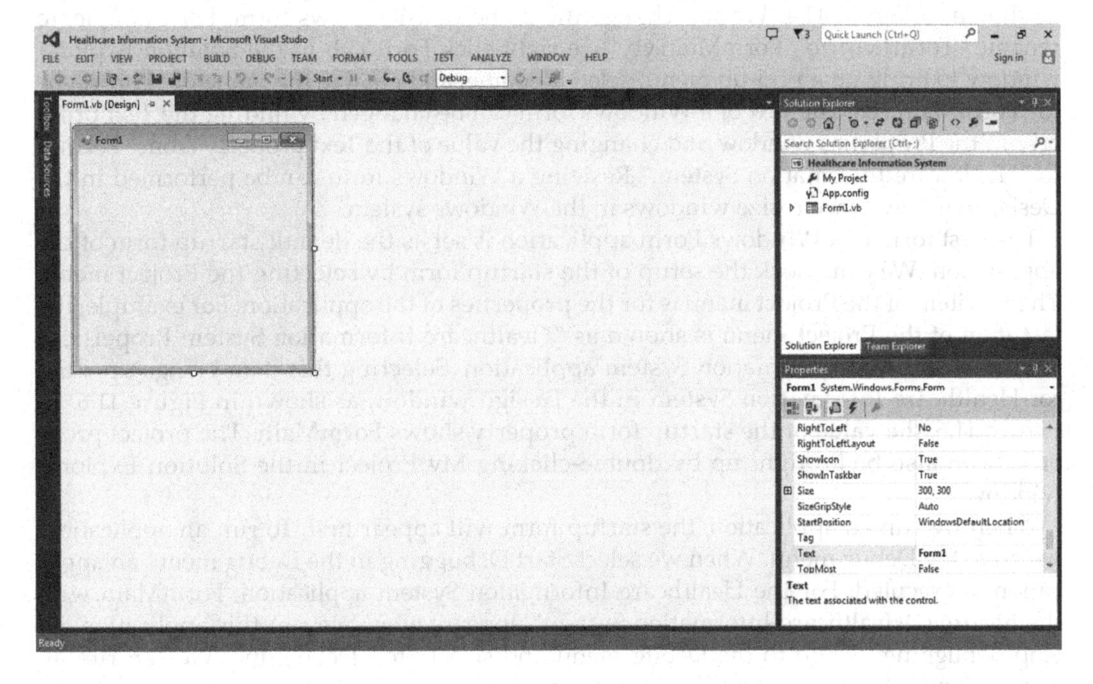

FIGURE 11.4
Visual Basic development environment.

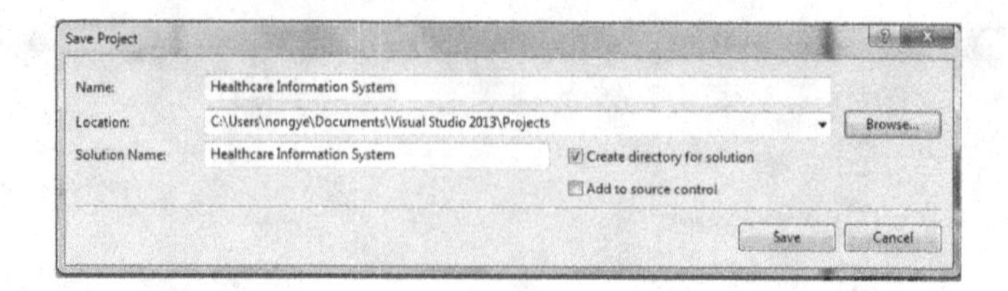

FIGURE 11.5
Save Project window.

If we go to the File menu now and select Save All in the File menu or click the Save All icon at the top, we see the Save Project window (see Figure 11.5). We can use the Browse button to select a folder on the computer to keep the project folder. The project folder contains the Visual Studio Solution file with .sln and the folder with various components of the project. If we close Visual Studio and then double-click the Visual Studio Solution file, we can bring up the development environment for the project again.

11.2 Windows Forms

Among many properties of a Windows form including those in the Properties window, the name and the text of a Windows form often need to be changed. For example, Form1 in Figure 11.4 has the name of Form1.vb and the text of Form1. The name appears in the Solution Explorer window, in the tab of the Design window, and in the Properties window, as shown in Figure 11.4. The text appears as the heading at the top of the Windows form, as shown in Figure 11.4. We can change the name of a Windows form. For example, to rename "Form1.vb" to "FormMain.vb," we right-click Form1.vb in the Solution Explorer window to bring up a pop-up menu, select Rename, and change the name of "Form1.vb" to "FormMain.vb." The text of a Windows form can be changed by finding the Text property in the Properties window and changing the value of the Text property from "Form1" to "Healthcare Information System." Resizing a Windows form can be performed in the Design window as we resize windows in the Windows system.

The first form of a Windows Form application is set as the default startup form of the application. We can check the setup of the startup form by selecting the Project menu. The last item of the Project menu is for the properties of the application. For example, the last item of the Project menu is shown as "Healthcare Information System Properties" for the Healthcare Information System application. Selecting this item brings up a tab for Healthcare Information System in the Design window, as shown in Figure 11.6. In Figure 11.6, the value of the startup form property shows FormMain. The project properties can also be brought up by double-clicking My Project in the Solution Explorer window.

When we run an application, the startup form will appear first. To run an application, we go to the Debug menu. When we select Start Debugging in the Debug menu, an application is executed. For the Healthcare Information System application, FormMain with the heading "Healthcare Information System" appears when we run this application. To stop debugging, we go to the Debug menu and select Stop Debugging. We can run an application without opening Visual Studio by going into the bin folder inside the project folder, opening the Debug folder in the bin folder, and double-clicking the application's

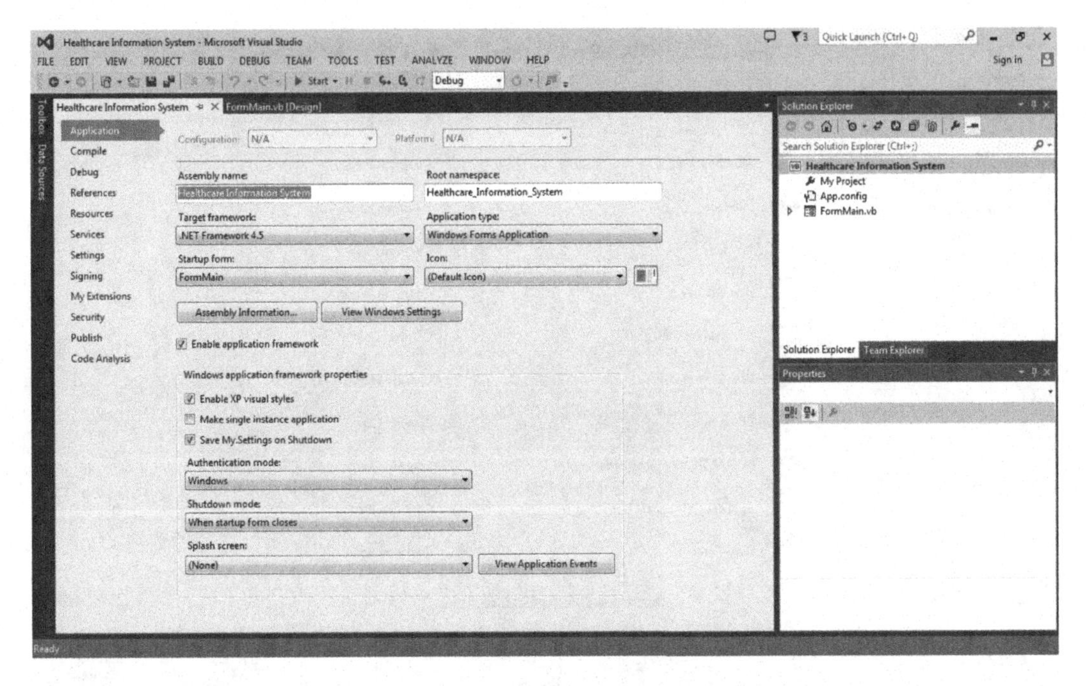

FIGURE 11.6
Properties window of a Windows Form application.

executable file in the Debug folder. We can move the application's executable file in the Debug folder to the desktop to create an icon on the desktop and then double-click the icon to execute the application.

We can add more Windows forms to our application. To add a Windows form, we select the Project menu and select Add Windows Form in this menu. This brings up the Add New Item window with Windows Form selected, as shown in Figure 11.7. We name the new Windows form, for example, FormEmployers, and click the Add button. Now, we see FormEmployers added in the Solution Explorer window. We change the Text property of FormEmployers to Employers so that the heading shows Employers.

With two Windows forms in the Healthcare Information System application and FormMain as the startup form, we want to close FormMain and open FormEmployers when we click on FormMain, and then we want to close FormEmployers and open FormMain when we click on FormEmployers. This requires adding code to both Windows forms. First of all, we change from the Design View to the Code View of FormMain by either double-clicking on FormMain in the Design View or selecting Code in the View menu. Figure 11.8 shows the Code View of FormMain, which presents the code of FormMain. The top portion of the Code View has two boxes with drop-down arrows. The left box has a list of objects. The right box has a list of events for the selected object in the left box. The code of FormMain contains a default subroutine that begins with the keywords, Private Sub, and ends with the keywords, End Sub. The keywords, Private Sub, are followed by the name of the Windows form and the Load event that triggers the execution of the subroutine. The Load event means the loading of the Windows form. This subroutine is executed when the FormMain_Load event occurs. Hence, Visual Basic programming in Visual Studio 2013 is an event-driven programming to have the code written and executed in response to events.

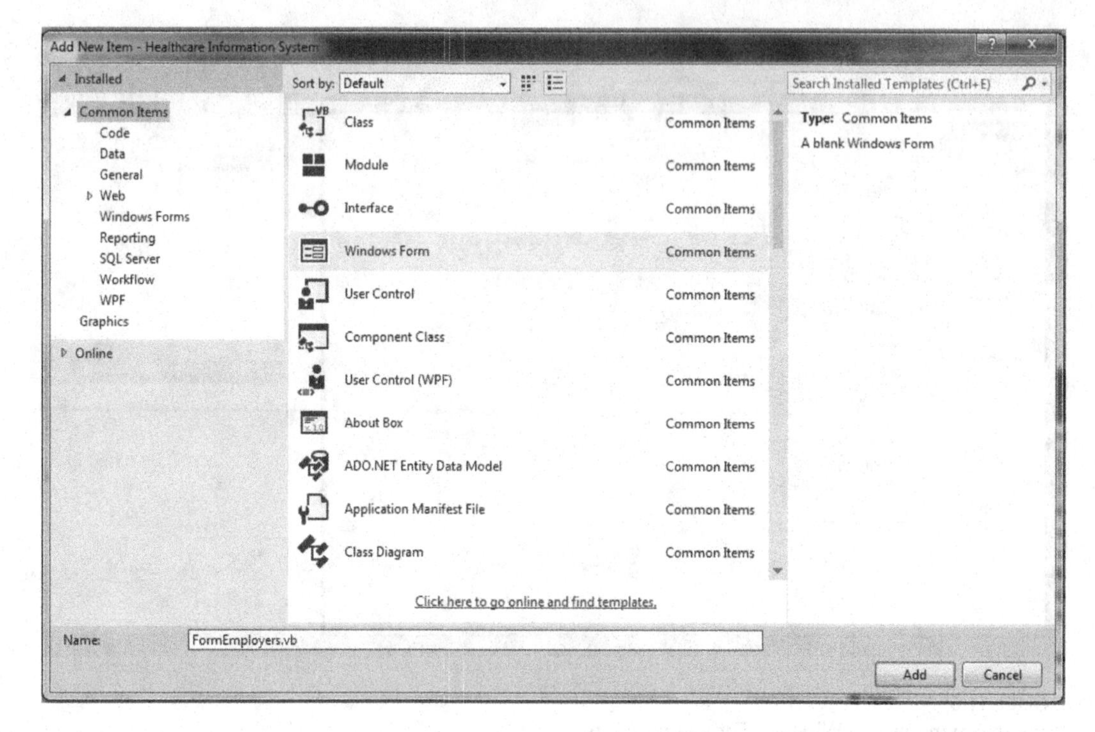

FIGURE 11.7
Add New Item window.

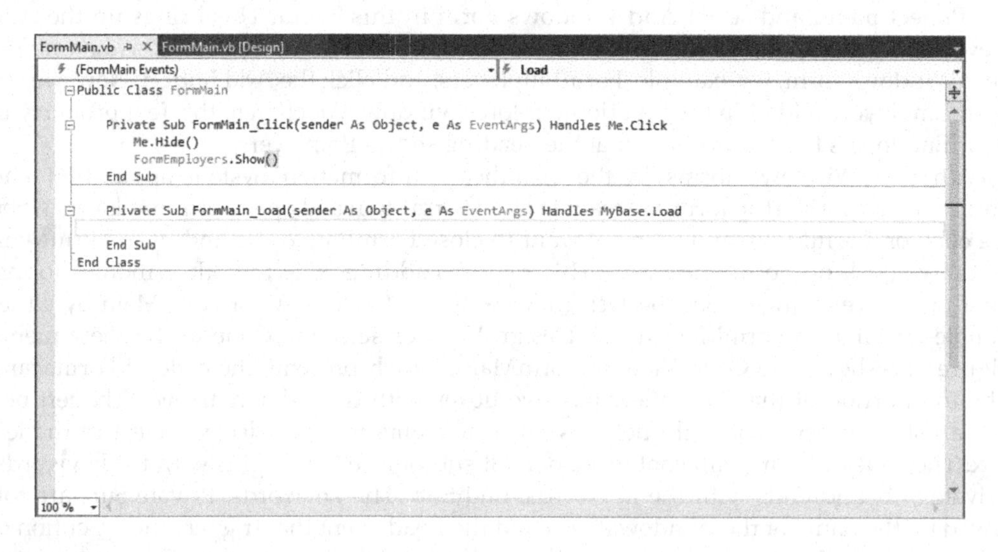

FIGURE 11.8
Code View of a Windows form.

Many events may occur on a Windows form. When we select the Click event from the list of events, another subroutine is added to the code of FormMain for FormMain_Click. This new subroutine is executed when we run the application and click on FormMain. We add the following code in the subroutine for FormMain_Click:

```
Me.Hide()
FormEmployers.Show()
```

The keyword, Me, denotes this form, that is, FormMain. Hide() is a function to hide this form. The code FormEmployers.Show() shows FormEmployers. We add a similar code to FormEmployers by double-clicking FormEmployers.vb in the Solution Explorer window, opening the Code View of FormEmployers, and adding a subroutine for FormEmployers_Click and the following code in this subroutine:

```
Me.Hide()
FormMain.Show()
```

With the above codes of FormWelcome and FormEmployers, we can click on FormMain to switch to FormEmployers and click on FormEmployers to switch to FormMain when we run the application.

Example 11.1

Add FormPatients, FormProviders, FormInsuranceCompanies, FormPharmacies, and FormUpdates, with Patients, Providers, Insurance Companies, Pharmacies, and Updates as the value of the Text property for these forms, respectively, to the Healthcare Information System application.

The steps in adding these forms are given as follows:

1. Select Add Windows Form in the Project menu, which brings up the Add New Item window.
2. Change the name of the Windows form to FormPatients in the Add New Item window, and click the Add button.
3. Change the value of the Text property in the Properties window to Patients.
4. Click the Save All icon to save the change.
5. Repeat Steps 1 to 4 to add FormProviders, with Providers as the value of the Text property.
6. Repeat Steps 1 to 4 to add FormInsuranceCompanies, with Insurance Companies as the value of the Text property.
7. Repeat Steps 1 to 4 to add FormPharmacies, with Pharmacies as the value of the Text property.
8. Repeat Steps 1 to 4 to add FormUpdates, with Updates as the value of the Text property.

11.3 GUI Controls

Example 11.2 shows how to add GUI controls to a Windows form.

Example 11.2

Add the GUI controls, as shown in Figure 11.9, including one label, six radio buttons, and one button, to FormMain. Add code so that checking one of the six radio buttons and then clicking on the Next button switches the Windows form from FormMain to the Windows form corresponding to the selected radio button.

By double-clicking FormMain.vb in the Solution Explorer window, we bring the Design View of FormMain in the Design window. After clicking on the Toolbox tab, we see several categories of tools, including All Windows Forms, Common Controls, and so on. Expanding Common Controls gives us the list of common controls, as shown in Figure 11.10.

We select Label from the list of common controls, drag and drop it to FormMain, and use the Properties window to set the value of the Text property to "Select one of the following." We can change the font and the size of this label by setting the Font and the Size properties in the Properties window. Other appearance features of this label can also be changed using the Properties window.

We then drag and drop RadioButton from the list of common controls to FormMain and use the Properties window to set the value of the Text property to Employers and the value of the Name property to RadioButtonEmployers. We add the other five radio buttons for Patients, Healthcare Providers, Insurance Companies, Pharmacies, and Updates in the same way, and set the name of these five radio buttons to RadioButtonPatients, RadioButtonProviders, RadioButtonCompanies, RadioButtonPharmacies, and RadioButtonUpdates, respectively. No matter how many radio buttons we add to this Windows form, the radio button function lets the user check only one of these radio buttons.

We now drag and drop Button to FormMain and use the Properties window to set the value of the Text property to Next and the value of the Name property to ButtonNext. There are many other types of common controls that give other types of GUI functions for a developer to choose from.

Double-clicking ButtonNext brings up the Code View of FormMain and adds a subroutine for the event ButtonNext_Clicked. The code that we enter into this subroutine is shown in Figure 11.11. The code hides FormMain, examines which of the radio buttons is checked, and shows the Windows form corresponding to the checked radio button.

Example 11.3 shows how to take the user's input using TextBox and displays an output message using MessageBox.

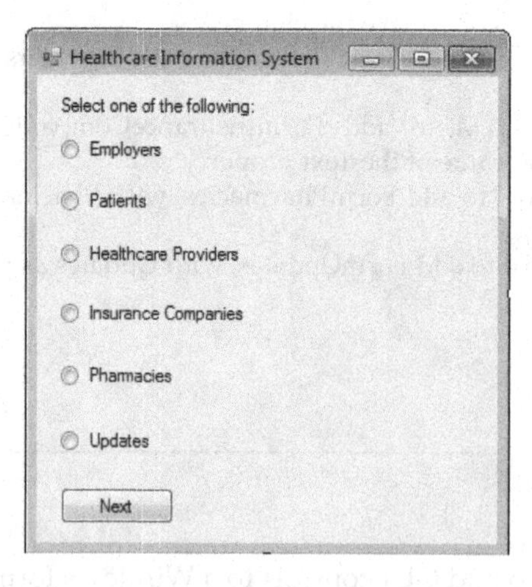

FIGURE 11.9
GUI of FormMain in the Healthcare Information System.

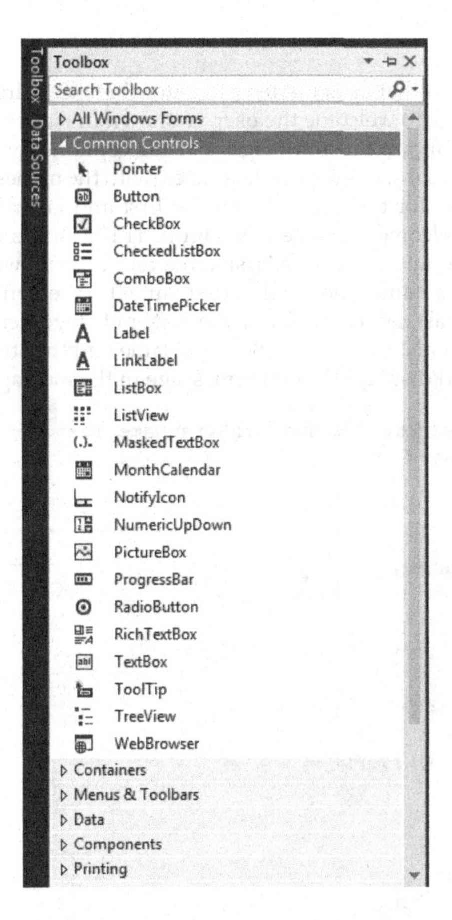

FIGURE 11.10
List of common controls in the Toolbox tab.

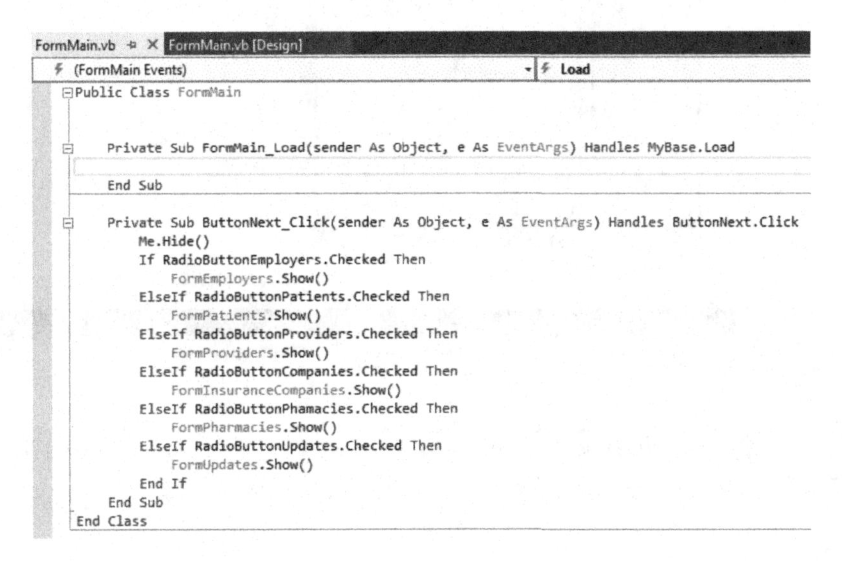

FIGURE 11.11
Code for the Click event of ButtonNext in FormMain of the Healthcare Information System.

Example 11.3

Create a Windows form that asks the user to enter the user's first and last names and then displays a message to welcome the user, as shown in Figure 11.12a.

The design of the Windows form in Figure 11.12a contains two labels to prompt the user to enter the first and last names, two textboxes (with the names of TextBoxFirstName and TextBoxLastName) for the user to enter the first and last names, and a button to click to display the welcome message box. Figure 11.12b shows the code for the Click event of the Welcome button. TextBoxFirstName.Text returns what the user enters in the text box for the first name, and TextBoxLastName.Text returns what the user enters in the text box for the last name. The following code puts together the values in the Text property of TextBoxFirstName and TextBoxLastName and the string "Welcome," using the string operator and displays the entire message in the message box.

```
MessageBox.Show("Welcome," & TextBoxFirstName.Text & " " &
TextBoxLastName.Text)
```

(a)

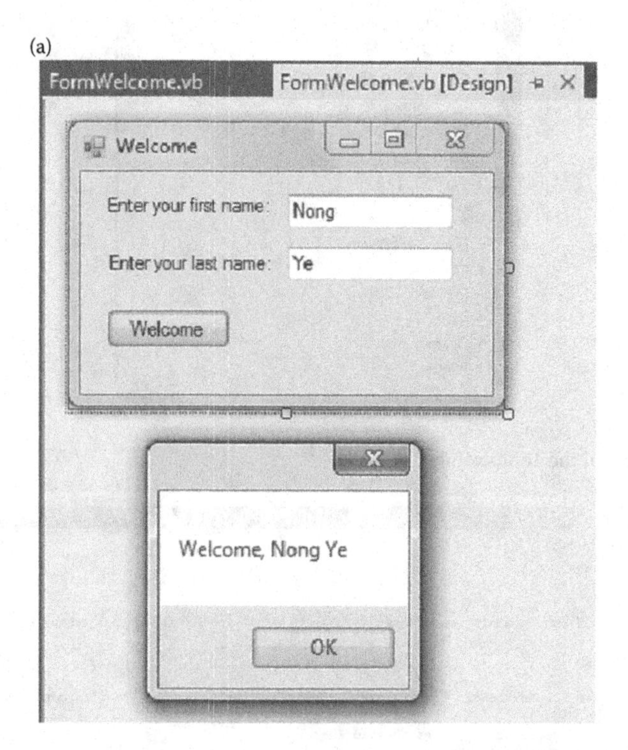

(b)

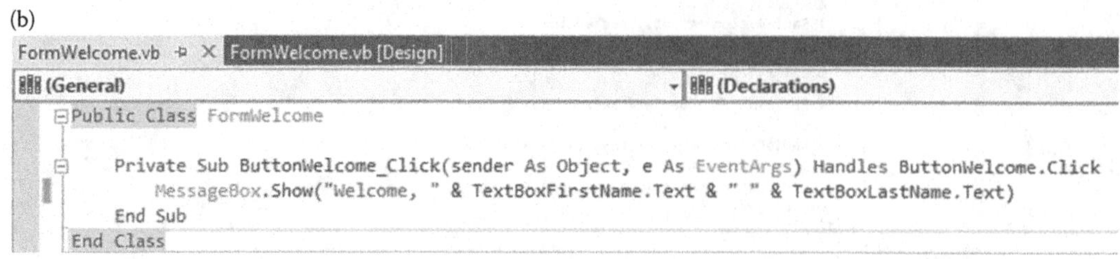

FIGURE 11.12
Design View and Code View of FormWelcome for Example 11.3. (a) Design View of FormWelcome and the message box to welcome a user. (b) Code View of FormWelcome.

11.4 GUI of the Healthcare Information System Application

Figure 7.13 shows all the tables in the healthcare database implemented in Access. Chapter 7 gives some example queries for the healthcare database. Chapter 26 gives the complete list of queries for employers, patients, healthcare providers, insurance companies, and pharmacies. The Healthcare Information System application lets a user select and view data from the queries and update data in the tables through GUI. Figure 11.13 shows the GUI design of the Healthcare Information System application, which includes a hierarchy of Windows forms. The startup Windows form at the top level of the hierarchy on the left side of Figure 11.13, FormMain, organizes data from the queries into several categories for employers, patients, providers, companies, and pharmacies using five radio buttons. Checking one of the five radio buttons takes the user to one of the five Windows forms at the middle level of the hierarchy—FormEmployers, FormPatients, FormProviders, FormCompanies, and FormPharmacies. Each of the five Windows forms gives a list of data from the list of queries in Chapter 26 for the selected category (employers, patients, providers, companies, or pharmacies). Selecting a data item in the list takes the user to the Windows form at the bottom level of the hierarchy, which presents the data. FormMain also contains a radio button for updating data in the tables. Checking the radio button for Updates takes the user to a Windows form at the middle level of the hierarchy, which includes a list of the tables. Selecting a table takes the user to the Windows form at the bottom level of the hierarchy, which presents the data in the selected table. The user can change the table data directly in the Healthcare Information System application. Note that Query 8 and Query 9 for insurance companies are action queries of updating and deleting data. Updating and deleting data in a table will be done on the table data by taking the route through FormUpdates. Hence, there are no radio buttons in FormCompanies corresponding to Query 8 and Query 9 for insurance companies.

Example 11.4

Complete the GUI of the Healthcare Information System application that has been started in Example 11.2 to create all the Windows forms, the GUI elements of each Windows form, and the code for making transitions among these Windows forms, as shown in Figure 11.13, except the Windows forms at the bottom level of the hierarchy, which present data from the queries and the tables.

Figures 11.9 and 11.11 give the Design View and the Code View of FormMain at the top level of the hierarchy. Figures 11.14 to 11.19 give the Design View and the Code View of the Windows forms at the middle level of the hierarchy for the Healthcare Information System application.

11.5 Execution of a Windows-Based Application Outside Visual Studio

We can execute a Windows-based application without launching Visual Studio by double-clicking the executable application file in the project folder of the application. For example, we first go to the bin folder in the Healthcare Information System application folder and then open the Debug folder in the bin folder. When we double-click the executable application file, Healthcare Information System.exe, in the Debug folder, the Healthcare

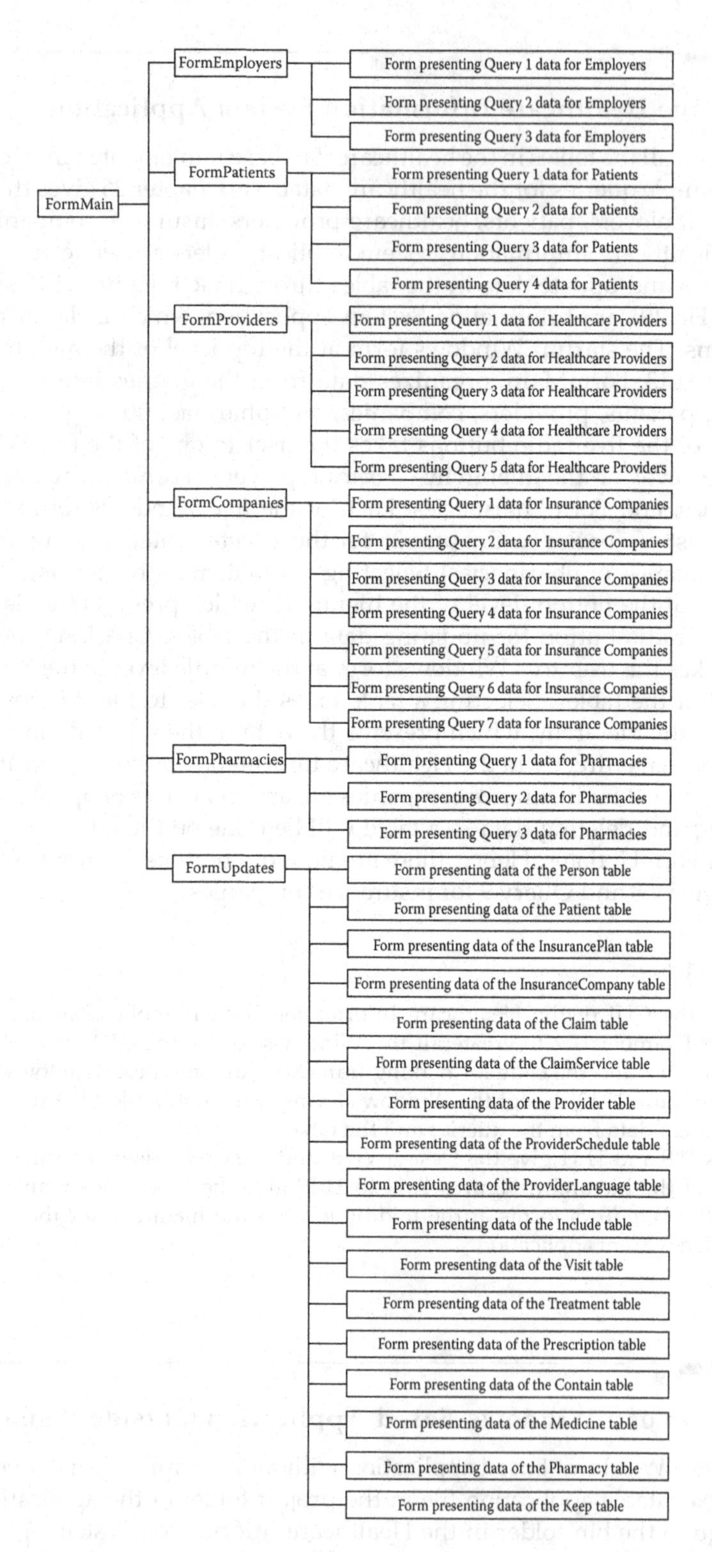

FIGURE 11.13
Hierarchy of Windows forms in the Healthcare Information System application.

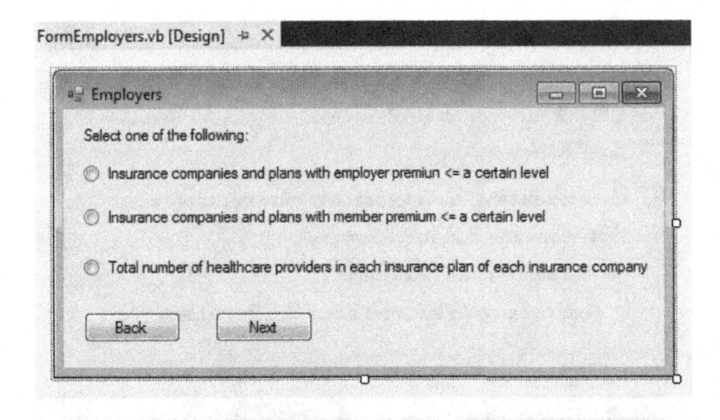

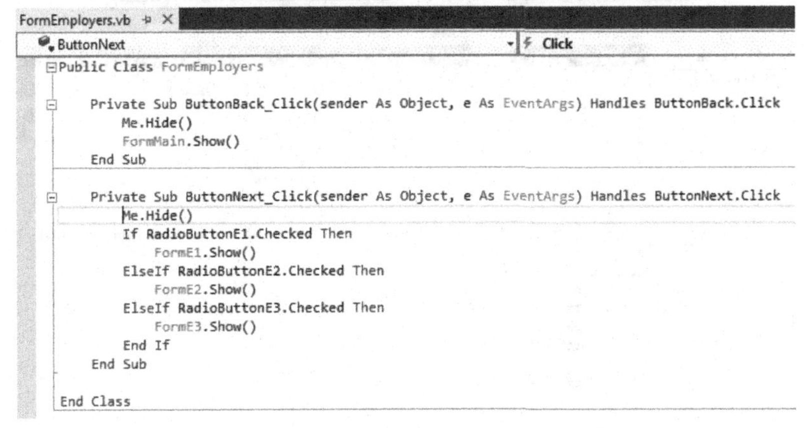

FIGURE 11.14
Design View and Code View of FormEmployers in the Healthcare Information System application.

Information System application is executed. All files in the Debug folder can be moved to another location on the computer, for example desktop, so that the user can double-click on the icon of the executable file on the desktop to execute the application.

11.6 Summary

This chapter describes the following:

- Setting up the Visual Basic development environment in Visual Studio 2013
- Changing the properties of a Windows form, adding a Windows form, setting up the startup Windows form, and adding code to make transitions among Windows forms
- Adding common controls, such as label, radio button, button, and text box, to a Windows form to create a GUI of a Windows form
- Adding event-driven programming code

The next chapter gives more details of Visual Basic programming in Visual Studio 2013.

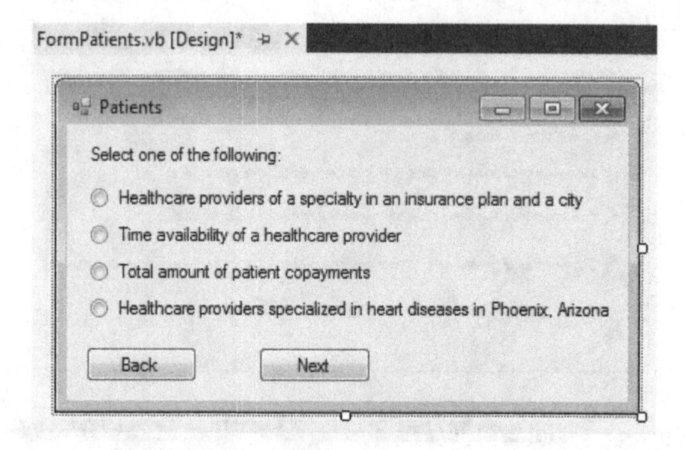

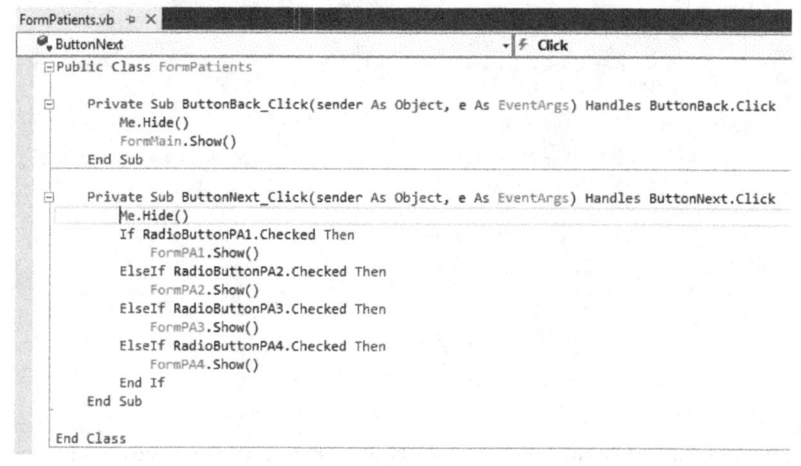

FIGURE 11.15
Design View and Code View of FormPatients in the Healthcare Information System application.

EXERCISES

1. Create a GUI of the movie theater application using Visual Studio 2013 to include all the Windows forms, which are similar to those at the top and the middle levels in the hierarchy of Windows forms for the Healthcare Information System application shown in Figure 11.13. The list of the Windows forms is given as follows:
   - *FormWelcome at the top level*: Contains five radio buttons for movies, producers, customers, statistics, and updates, and the Next button
   - *FormMovies at the middle level to appear after a user selects the radio button for movies in FormWelcome*: Contains five radio buttons for Query 2, Query 3, Query 4, Query 8, and Query 9 in the movie theater database from Exercise 2 in Chapter 7, and the Back and the Next buttons
   - *FormProducers at the middle level to appear after a user selects the radio button for producers in FormWelcome*: Contains one radio button for Query 7 in the movie theater database from Exercise 2 in Chapter 7, and the Back and the Next buttons

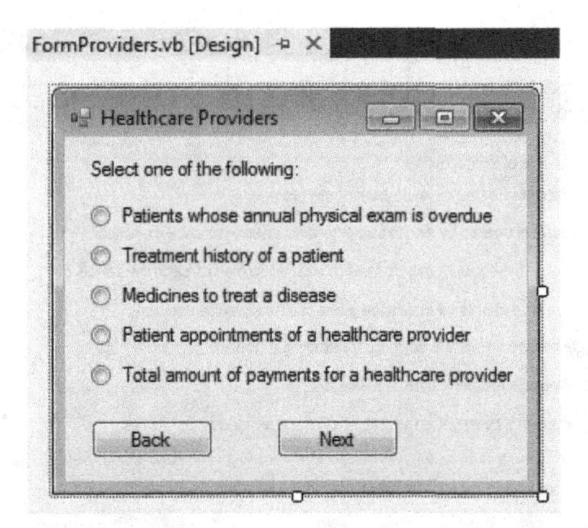

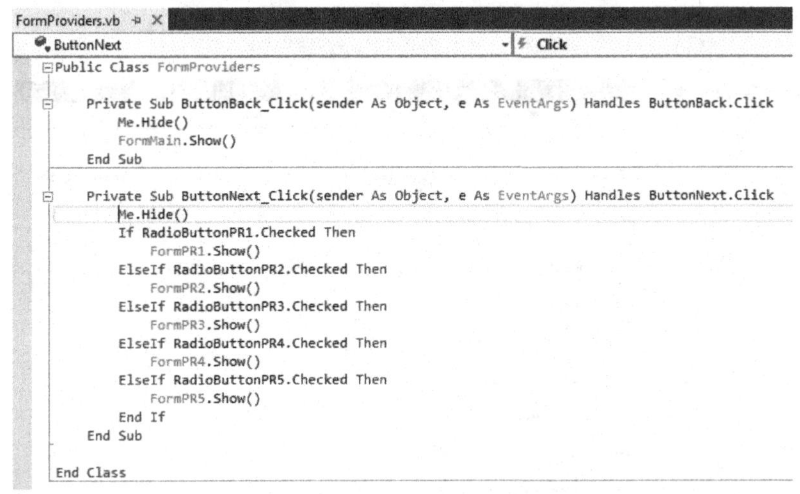

FIGURE 11.16
Design View and Code View of FormProviders in the Healthcare Information System application.

- *FormCustomers at the middle level to appear after a user selects the radio button for customers in FormWelcome*: Contains two radio buttons for Query 6 and Query 10 in the movie theater database from Exercise 2 in Chapter 7, and the Back and the Next buttons
- *FormStatistics at the middle level to appear after a user selects the radio button for statistics in FormWelcome*: Contains two radio buttons for Query 1 and Query 5 in the movie theater database from Exercise 2 in Chapter 7, and the Back and the Next buttons
- *FormUpdates at the middle level to appear after a user selects the radio button for updates in FormWelcome*: Contains the radio buttons for all the tables in the movie theater database from Exercise 1 in Chapter 7, and the Back and the Next buttons

Add code to the Click event of the Back and the Next buttons in the above Windows forms to make transitions among these Windows forms.

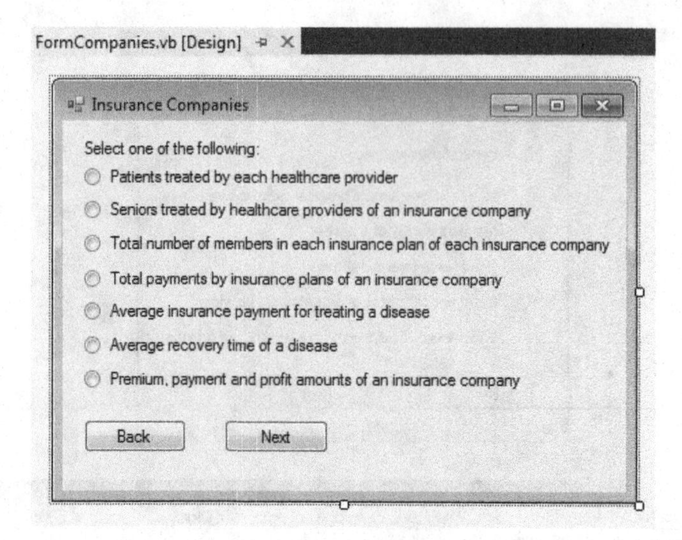

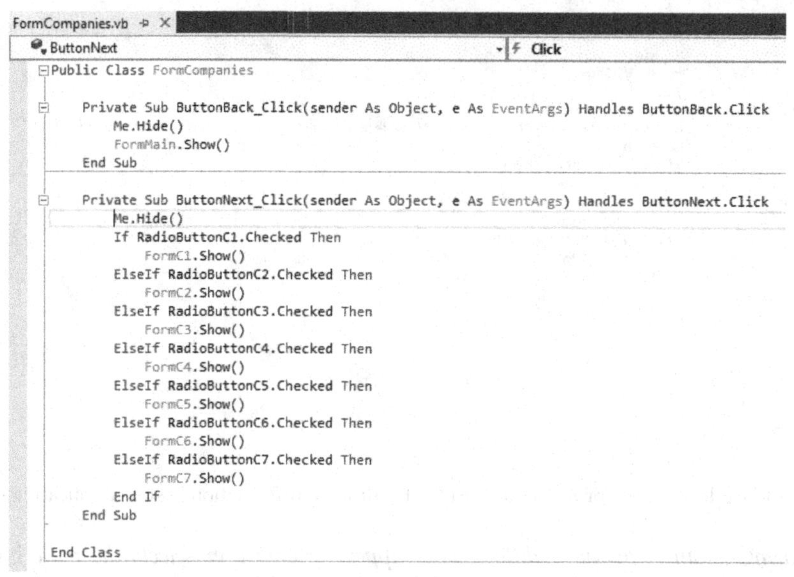


```
Public Class FormCompanies

    Private Sub ButtonBack_Click(sender As Object, e As EventArgs) Handles ButtonBack.Click
        Me.Hide()
        FormMain.Show()
    End Sub

    Private Sub ButtonNext_Click(sender As Object, e As EventArgs) Handles ButtonNext.Click
        Me.Hide()
        If RadioButtonC1.Checked Then
            FormC1.Show()
        ElseIf RadioButtonC2.Checked Then
            FormC2.Show()
        ElseIf RadioButtonC3.Checked Then
            FormC3.Show()
        ElseIf RadioButtonC4.Checked Then
            FormC4.Show()
        ElseIf RadioButtonC5.Checked Then
            FormC5.Show()
        ElseIf RadioButtonC6.Checked Then
            FormC6.Show()
        ElseIf RadioButtonC7.Checked Then
            FormC7.Show()
        End If
    End Sub

End Class
```

FIGURE 11.17
Design View and Code View of FormCompanies in the Healthcare Information System application.

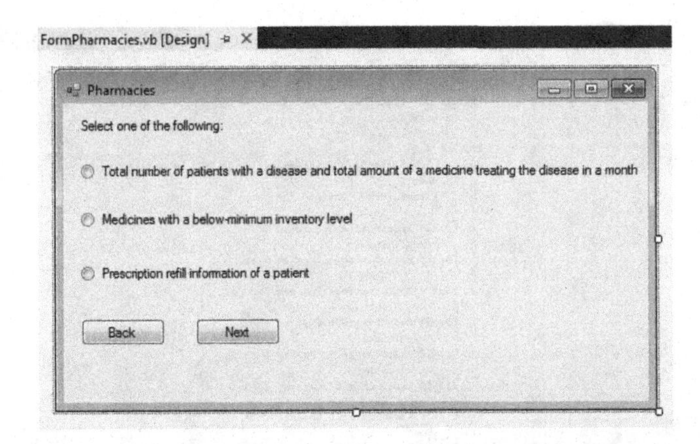

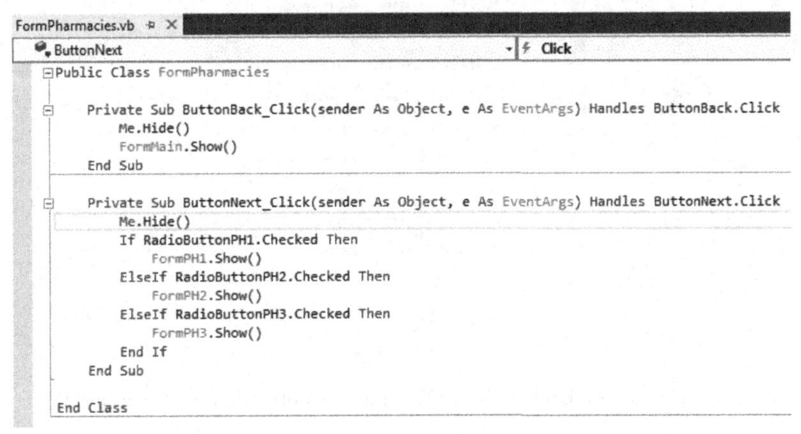

FIGURE 11.18
Design View and Code View of FormPharmacies in the Healthcare Information System application.

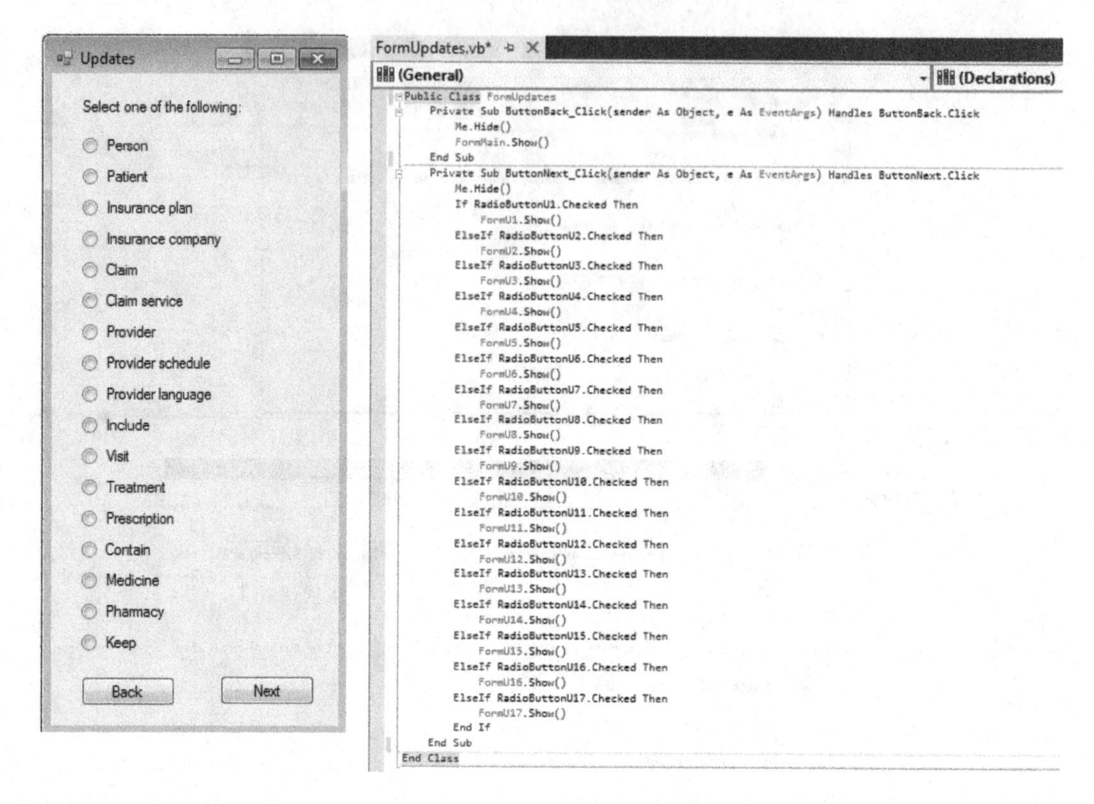

FIGURE 11.19
Design View and Code View of FormUpdates in the Healthcare Information System application.

12

Visual Basic Programming in Microsoft Visual Studio

In Chapter 11, we add some Visual Basic programming code to subroutines for events of objects. This chapter describes various elements used to put together statements in Visual Basic code, including variables, constants, control structures, and operators used in expressions. We give examples of Visual Basic code that use these elements.

12.1 Variables and Data Types

A variable must be first declared before being used in a Visual Basic statement. The syntax for the variable declaration is

```
Dim <variable1, …, variableN> As <data type> [= <initial value>]
```

where Dim and As are the keywords. The keywords of Visual Basic appear in blue color in Visual Studio. The initial value in the square brackets is optional. We use square brackets to indicate an optional element. Some commonly used data types are as follows:

- Integer
- Long
- Double
- String
- Char
- Boolean
- Date
- DateAndTime

where Long is used for a large integer, and Double is used for a value with decimals.

Example 12.1

Declare two variables, *i* and *j*, as Integer, with an initial value of 1, and declare the variable, note, as String.

The declaration of the variables is given as

Dim *i*, *j* as Integer = 1
Dim note as String

The syntax of declaring an array is

```
Dim <array1, …, arrayN> (<size of the 1st dimension>, …, <size of the mth
dimension>) As <data type>
```

Example 12.2

Declare the variable, students, as a one-dimensional array that keeps names of up to 100 students, and declare the variable, schoolstudents, as a two-dimensional array that keeps names of up to 100 students in each of 9 to 12 grades.

The declaration of the two arrays are given as

Dim students (99) as String
Dim schoolstudents (3, 99) as String

Note that the size of the array for students is 99 because the index of an element in the array goes from 0 to 99. Hence, schoolstudents (3, 99) has four rows for four grades (9 to 12 grades) and 100 columns.

The ith element of the students array is referred to as students(i). The element in the ith row and the jth column of the schoolstudents matrix is referred to as schoolstudents(i, j).

Table 12.1 lists a number of functions to check and convert data types. Example 12.3 illustrates the use of some conversion functions.

Example 12.3

Create Visual Basic statements that use an InputBox function to take the user's input of two numbers, perform the division of the two numbers, and use a MessageBox to display the integer result of the division.

The Visual Basic statements are given as

```
Dim Number1, Number2, Result as Integer
Number1 = InputBox("Enter Number1:", "Division")
Number2 = InputBox("Enter Number2:", "Division")
Result = CInt(Number1/Number2)
MessageBox.Show("The division result is:" & Result, "Division")
```

Figure 12.1 shows the two InputBoxes and the MessageBox.

TABLE 12.1

List of Functions to Check and Convert Data Types

Function	Description
IsNumeric(<object name>)	Return True if the object is numeric; False if otherwise.
IsDate(<object name>)	Return True if the object is a date; False if otherwise.
IsArray(<object name>)	Return True if the object is an array; False if otherwise.
CInt(<object name>)	Convert the object to an integer.
CLng(<object name>)	Convert the object to a long integer.
CDbl(<object name>)	Convert the object to a real value.
CStr(<object name>)	Convert the object to a string.
CBool(<object name>)	Convert the object to a Boolean value.
CDate(<object name>)	Convert the object to a date.

The syntax of the InputBox function is

```
<variable> = InputBox(<prompt>, [title], [default response],
[x coordinate of the InputBox on the screen], [y coordinate of the
InputBox on the screen])
```

The syntax of the MessageBox is

```
MessageBox.Show(<message>, [title], [type], [icon])
```

The types of MessageBox are MessageBoxButtons.OKCancel, MessageBoxButtons.YesNoCancel, and MessageBoxButtons.AbortRetryIgnore, to include different sets of responses for the user to select. The icons of MessageBox are MessageBoxIcon.Warning, MessageBoxIcon.Information, MessageBoxIcon.Error, and MessageBoxIcon.Question to show four different icons in the MessageBox.

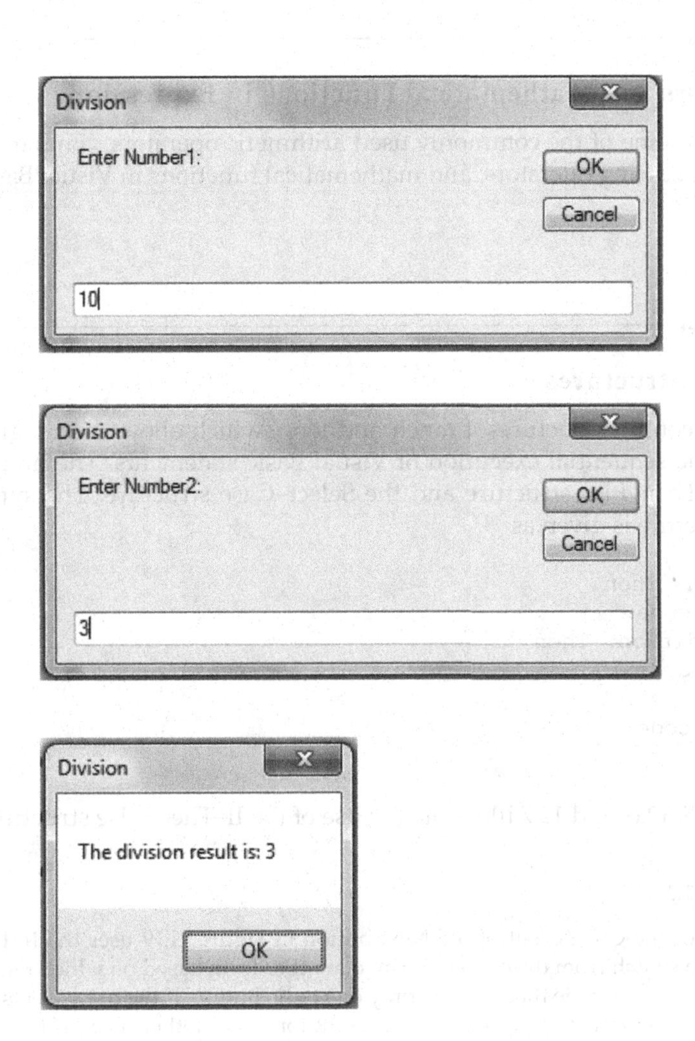

FIGURE 12.1
Examples of InputBoxes and MessageBox.

12.2 Constants

A constant can be declared for a commonly used value, for example, sale tax rate. The syntax of declaring a constant is

```
Const <constant name> As <data type> [= <value>]
```

Example 12.4

Declare a constant for the sale tax rate, which is equal to 9%.

The declaration of this constant is given as

```
Const tax As Double = 0.09
```

12.3 Operators and Mathematical Functions in Expressions

Table 12.2 gives some of the commonly used arithmetic operators, comparison operators, logical operators, string operators, and mathematical functions in Visual Basic, along with examples.

12.4 Control Structures

There are two control structures, branch and loop, which allow the program control to change from the sequential execution of Visual Basic statements. The branch structures include the If–Then–Else structure and the Select–Case structure. The syntax of the If–Then–Else structure is given as

```
If <condition> Then
       <body code>
[Elseif <condition> Then
       <body code>]
[Else
       <body code>]
End If
```

Examples 12.5, 12.6, and 12.7 illustrate the use of the If–Then–Else structure.

Example 12.5

The code for the Click event of the Next button in Figure 11.19 uses the If–Then–Else structure to switch from the current form to another form based on which radio button is selected. Create a code that handles only one radio button. If the user selects the radio button, the program switches from the current form to another form. If the user does not select the radio button, the program does nothing.

TABLE 12.2

Operators and Some Mathematical Functions

Operator or Function	Description	Example
+	Addition	$a + b$
–	Subtraction	$a - b$
*	Multiplication	$a * b$
/	Division	a / b
^	Exponential	a ^ b (a to the power of b)
Mod	Modulo	a Mod b
=	Equal	$a = b$
<>	Not equal	$a <> b$
>	Greater than	$a > b$
>=	Greater than or equal to	$a >= b$
<	Less than	$a < b$
<=	Less than or equal to	$a <= b$
AND	Logical and	a AND b
OR	Logical or	a OR b
NOT	Logical not	NOT a
XOR	Logical exclusive or	a XOR b
&	Concatenate two strings	a & b
String.Length	Return the length of a string	String.Length(a)
Math.Abs	Return the absolute value	Math.Abs(a)
Math.Sqrt	Return the square root	Math.Sqrt(a)
Math.Pow	Return the power	Math.Pow(a, b) (return a to the power of b)
Math.Log	Return the natural logarithm	Math.Log(a)
Math.Log10	Return the base 10 logarithm	Math.Log10(a)
TimeOfDay	Return the current time	TimeOfDay
DateString	Return the current date	DateString
Year	Return the year of a date as an integer	Year(DateString)
Month	Return the month of a date as an integer	Month(DateString)

The code is given as

```
If RadioButtonU1.Checked Then
        Me.Hide()
        FormU1.Show()
End If
```

Example 12.6

Create a code that is similar to the code in Example 12.5 to handle one radio button. If the user selects the radio button, the program switches from the current form to another form. If the user does not select the radio button, the program displays a message to prompt the user to select the radio button.

The code is given as

```
If RadioButtonU1.Checked Then
        Me.Hide()
        FormU1.Show()
```

```
Else
        MessageBox("Please select the radio button.")
End If
```

Example 12.7

Create a code that is similar to the code in Example 12.5 but handles two radio buttons. If the user selects the radio button, the program switches from the current form to another form. If the user does not select the radio button, the program does nothing.

The code is given as

```
If RadioButtonU1.Checked Then
        Me.Hide()
        FormU1.Show()
Elseif RadioButtonU2.Checked Then
        Me.Hide()
        FormU2.Show()
End If
```

The syntax of the Select–Case structure is given as

```
Select Case <test expression>
Case Is <expression>
        <body>
...
[Case Is <expression>
        <body code>]
Case Else
        <body code>
End Select
```

Example 12.8 illustrates the use of the Select–Case structure.

Example 12.8

Write a Visual Basic code for Example 12.6 using the Select–Case structure.

The code is given as

```
Select Case RadioButtonU1.Checked
Case True
        Me.Hide()
        FormU1.Show()
Case Else
        MessageBox("Please select the radio button.")
End Select
```

The loop structures include the Do–Loop–While, the Do–While–Loop, the Do–Loop–Until, the Do–Until–Loop, and the For–Next structures. The syntax and example of each loop structure are given as follows:

The syntax of the Do–Loop–While structure is

```
Do
<body code>
Loop While (<condition>)
```

This loop structure goes into the body of the loop and then checks the condition. If the condition is true, the program goes back into the body of the loop; otherwise, the program gets out of the loop.

The syntax of the Do–While–Loop structure is

```
Do While (<condition>)
<body code>
Loop
```

This loop structure first checks the condition. If the condition is true, the program goes into the body of the loop; otherwise, the program gets out of the loop.

The syntax of the Do–Loop–Until structure is

```
Do
<body code>
Loop Until (<condition>)
```

This loop structure goes into the body of the loop and then checks the condition. If the condition is true, the program gets out of the loop; otherwise, the program goes back into the body of the loop.

The syntax of the Do–Until–Loop structure is

```
Do Until (<condition>)
<body code>
Loop
```

This loop structure first checks the condition. If the condition is true, the program gets out of the loop; otherwise, the program goes back into the body of the loop.

The syntax of the For–Next structure is

```
For <counter> = <first value> To <last value> [Step <increment value>]
<body code>
Next
```

This loop structure goes into the body of the loop for a number of times from the first value to the last value of the counter.

The above five loop structures are illustrated in Examples 12.9, 12.10, 12.11, 12.12, and 12.13, respectively. The code in each of the five examples implements the same function that takes two input integers, *a* and *b*, from the user, computes base *a* to the power of *b*, and displays the result.

Example 12.9

Create a Visual Basic code for the power function using the Do–Loop–While structure.

The code is given as

```
Dim a, b As Integer
Dim result As Integer = 1
a = InputBox("Enter the base number.")
b = InputBox("Enter the power number.")
Do
        result = result * a
        b = b - 1
```

```
Loop While (b > 0)
MessageBox.Show("The result is:" & result)
```

Example 12.10

Create a Visual Basic code for the power function using the Do–While–Loop structure.
The code is given as

```
Dim a, b As Integer
Dim result As Integer = 1
a = InputBox("Enter the base number.")
b = InputBox("Enter the power number.")
Do While (b > 0)
          result = result * a
          b = b - 1
Loop
MessageBox.Show("The result is:" & result)
```

Example 12.11

Create a Visual Basic code for the power function using the Do–Loop–Until structure.
The code is given as

```
Dim a, b As Integer
Dim result As Integer = 1
a = InputBox("Enter the base number.")
b = InputBox("Enter the power number.")
Do
          result = result * a
          b = b - 1
Loop Until (b = 0)
MessageBox.Show("The result is:" & result)
```

Example 12.12

Create a Visual Basic code for the power function using the Do–Loop–Until structure.
The code is given as

```
Dim a, b As Integer
Dim result As Integer = 1
a = InputBox("Enter the base number.")
b = InputBox("Enter the power number.")
Do Until (b = 0)
          result = result * a
          b = b - 1
Loop
MessageBox.Show("The result is: " & result)
```

Example 12.13

Create a Visual Basic code for the power function using the For–Next structure.
The code is given as

```
'i is the counter for the For-Next loop
Dim a, b, i As Integer
Dim result As Integer = 1
a = InputBox("Enter the base number.")
b = InputBox("Enter the power number.")
```

```
For i = 1 to b
        result = result * a
Next
MessageBox.Show("The result is:" & result)
```

The first line of the above node is a comment line that starts with '.

12.5 Summary

This chapter gives the syntax and examples of the various elements used to put together a Visual Basic code, including variable declaration; constant declaration; and some of the commonly used arithmetic, comparison, logical, and string operators, mathematical functions, and branch and loop control structures. The next chapter describes how to embed an Access database in a Windows-based application.

EXERCISES

1. Create a Windows form containing a currency converter that takes an amount in one currency and converts the amount to another currency.
2. Create a Windows form containing a simple calculator that can perform addition, subtraction, multiplication, and division.
3. Create a Window form that lets a user enter an integer number, computes and displays the factorial of this number.
4. Create a Visual Basic code that sorts and rearranges values in an array of 100 integers to a nondecreasing order of these values.

13

Database Connection in Microsoft Visual Studio

This chapter describes how to connect to an Access database in a Windows-based application so that we can present data from tables and queries in the Access database in the Windows-based application. We first introduce how to add an Access database as a data source to an application. We then bring data from tables and simple Select queries to an application. We also describe other capacities such as handling Crosstab and Select queries with a parameter, displaying related data from parent and child tables, and writing Visual Basic code to retrieve, add, and delete data of an Access database.

13.1 Adding an Access Database as a Data Source

We can select Add New Data Source in the Project menu or open the Data Sources windows and click the Add New Data Source icon to start the procedure of adding an Access database as a data source in a Windows-based application. Figure 13.1 shows the procedure, which includes the following steps in the Data Source Configuration Wizard:

1. Select Database in the Choose a Data Source Type window.
2. Select Dataset in the Choose a Database Model window.
3. Click the New Connection button in the Choose Your Data Connection window.
4. Select Microsoft Access Data File and click the Continue button in the Choose Data Source window.
5. Use the Browse button to select an Access database file on the computer (e.g., Healthcare Access Database.accdb), click the Test Connection button to see a message showing "Test connection succeeded," and click the OK button, all in the Add Connection window, which shows Microsoft Access Database File labeled as "OLB DB."
6. Click the Next button in the Choose Your Data Connection window, which now shows the selected Access database file.
7. Click the Yes button in the message box to copy the database file to the project folder.
8. Click the Next button in the Save the Connection String to the Application Configuration File window showing that the connection to Healthcare Access Database.accdb is saved as "Healthcare_Access_DatabaseConnectionString."
9. Select both Tables and Views, expand Tables to see the list of tables, expand Views to see the list of queries, and click the Finish button, all in the Choose Your Database Objects window, which shows Healthcare_Access_DatabaseDataSet as the Dataset name.

(a)

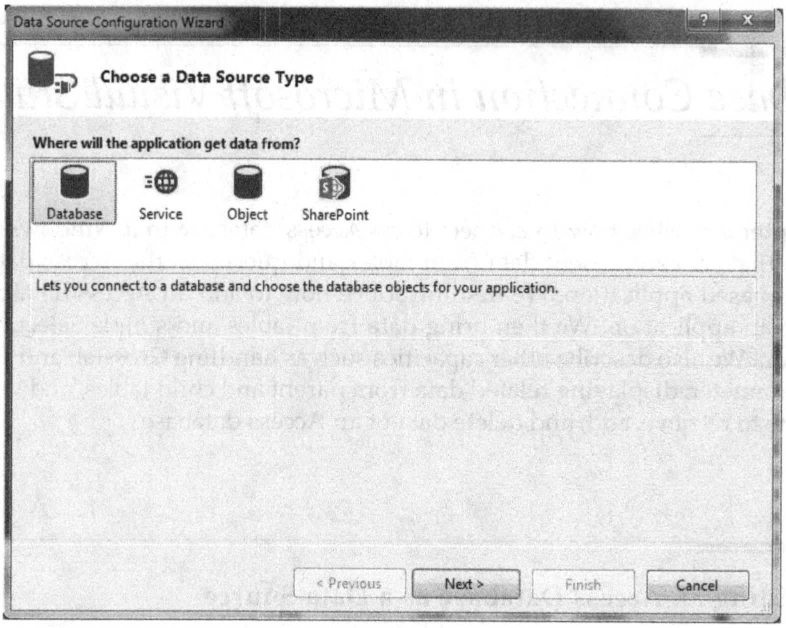

Data Source Configuration Wizard

Choose a Data Source Type

Where will the application get data from?

Database Service Object SharePoint

Lets you connect to a database and choose the database objects for your application.

< Previous Next > Finish Cancel

(b)

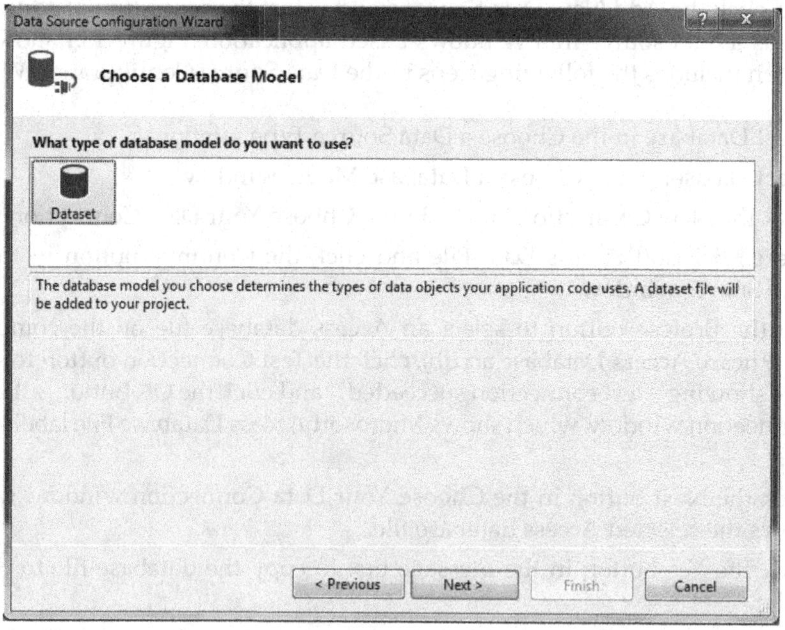

Data Source Configuration Wizard

Choose a Database Model

What type of database model do you want to use?

Dataset

The database model you choose determines the types of data objects your application code uses. A dataset file will be added to your project.

< Previous Next > Finish Cancel

FIGURE 13.1
Procedure of adding an Access database as a data source in a Windows-based application. (a) Choose a data source type. (b) Choose a database model.

(c)

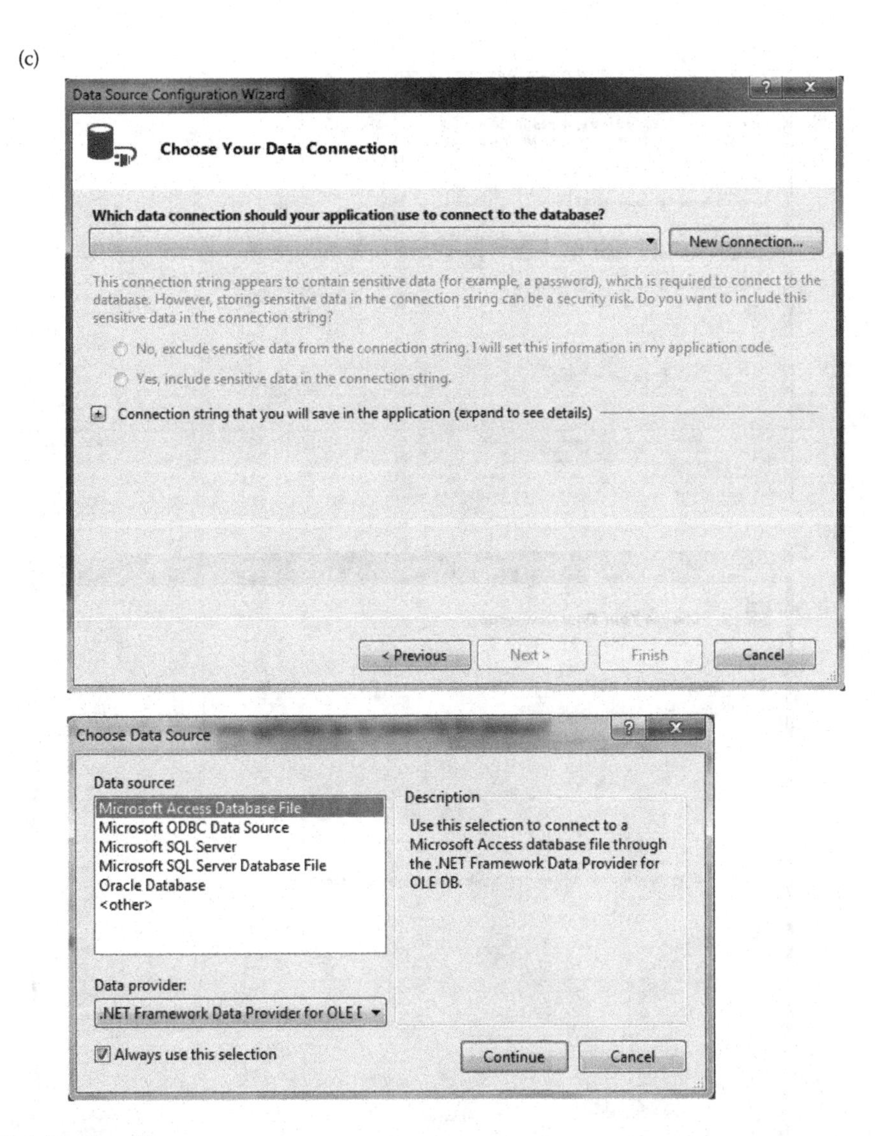

FIGURE 13.1 (Continued)
Procedure of adding an Access database as a data source in a Windows-based application. (c) Choose a data connection.

(d)

Add Connection

Enter information to connect to the selected data source or click "Change" to choose a different data source and/or provider.

Data source:

Microsoft Access Database File (OLE DB) Change...

Database file name:

C:\Users\nongye\Documents\My\yenong\publ Browse...

Log on to the database

User name: Admin

Password:

Save my password

Advanced...

Test Connection OK Cancel

(e)

Data Source Configuration Wizard

Choose Your Data Connection

Which data connection should your application use to connect to the database?

Healthcare Access Database.accdb New Connection...

This connection string appears to contain sensitive data (for example, a password), which is required to connect to the database. However, storing sensitive data in the connection string can be a security risk. Do you want to include this sensitive data in the connection string?

No, exclude sensitive data from the connection string. I will set this information in my application code.

Yes, include sensitive data in the connection string.

Connection string that you will save in the application (expand to see details)

< Previous Next > Finish Cancel

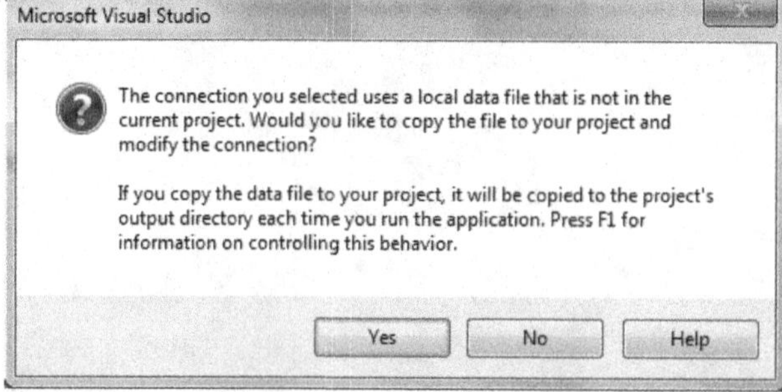

FIGURE 13.1 (Continued)
Procedure of adding an Access database as a data source in a Windows-based application. (d) Add connection. (e) Choose a data connection.

(f)

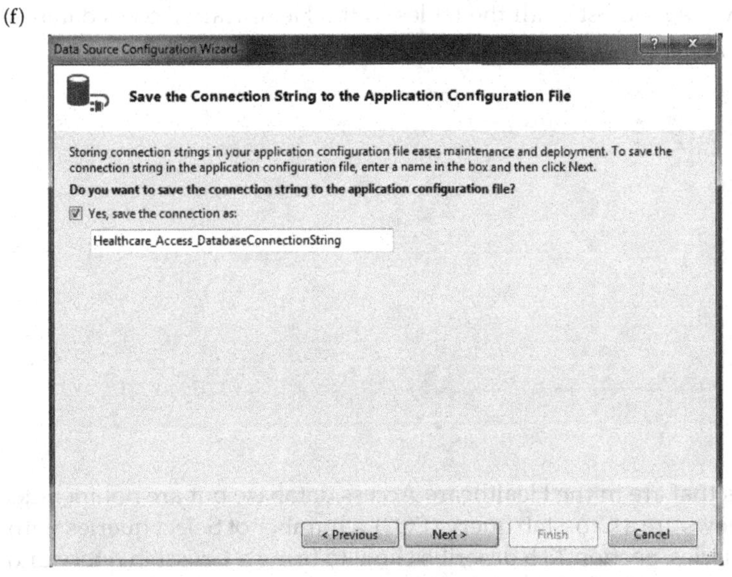

(g)

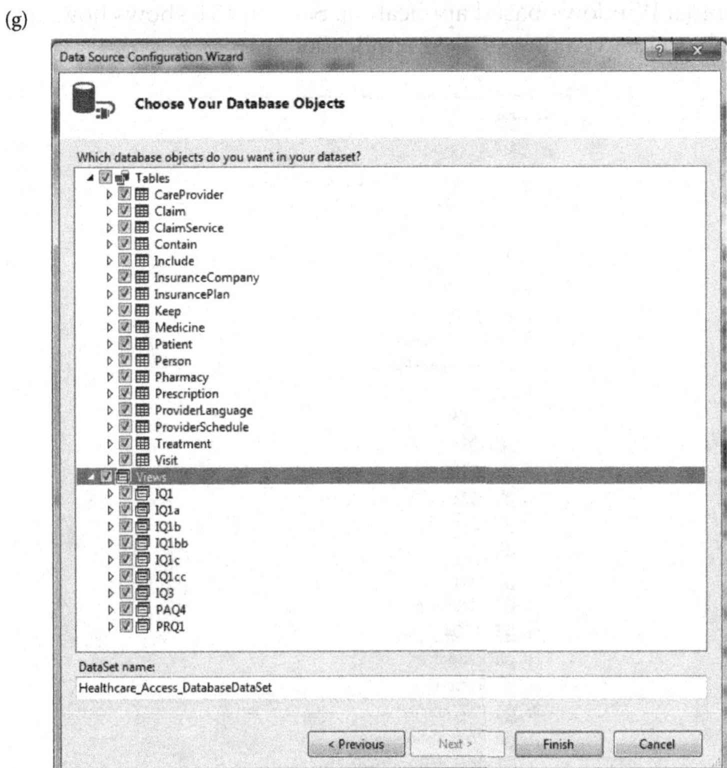

FIGURE 13.1 (Continued)
Procedure of adding an Access database as a data source in a Windows-based application. (f) Save the connection string. (g) Choose database objects.

In Step 9, we see the list of all the tables in the Healthcare Access database and a list of the following, but not all, queries in the Healthcare Access database:

- IQ1
- IQ1a
- IQ1b
- IQ1bb
- IQ1c
- IQ1cc
- IQ3
- PAQ4
- PRQ1

The queries that are in the Healthcare Access database but are not included in the list of queries for Views are a Crosstab query (EQ3), a number of Select queries with parameter(s), and action queries. Section 13.5 describes how to bring a Crosstab query and a query with parameter(s) into a Windows-based application. Section 13.6 shows how to update data of an Access database in a Windows-based application.

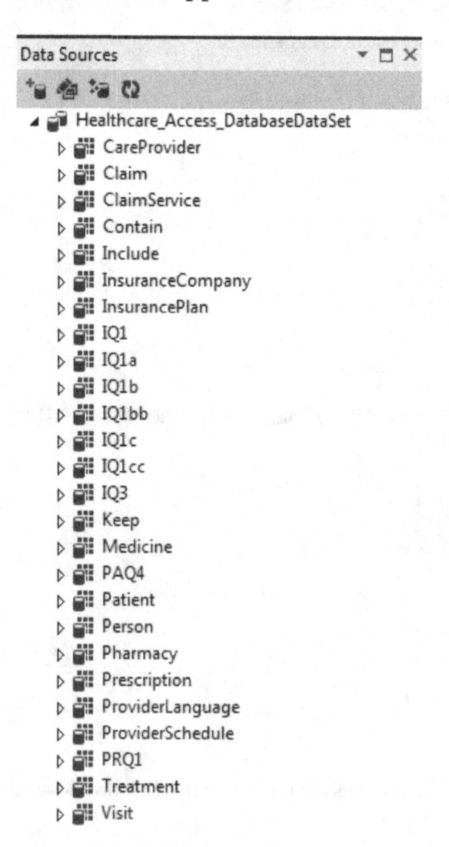

FIGURE 13.2
Tables and views in the data set for the Healthcare Access database.

When we open the Data Sources window, we see the list of the Healthcare_Access_DatabaseDataSet, which includes all the tables and the queries in the Healthcare Access database, excluding Crosstab queries, Select queries with parameter(s), and action queries, as shown in Figure 13.2.

13.2 Presenting Data from Tables and Simple Select Queries

Data from a table can be brought to a Windows form by dragging the table from Healthcare_Access_DatabaseDataSet in the Data Sources window and dropping it in the Windows form, which creates a DataGridView to present data from the table.

Example 13.1

When the user selects the Person radio button in FormUpdates of the Healthcare Information System application, the program brings up FormU1 to present data from the Person table. Add the Person table to FormU1.

Figure 13.3a shows the Design View of FormU1 after we drag the Person table from Healthcare_Access_DatabaseDataSet in the Data Sources window and drop it to FormU1, which adds the Toolstrip and the DataGridView. The Toolstrip shows a number indicating the selected record, the total number of records in the DataGridView, navigation buttons to go through records back and forth, the Add New button to add a new record, the Delete button to delete a selected record, and the Save Data button. The DataGridView is a graphical user interface (GUI) control (named "PersonDataGridView" in the Properties window) on FormU1 for presenting data from the Person table in Healthcare_Access_DatabaseDataSet. Figure 13.3b shows data of the Person table presented in the DataGridView when we execute the Healthcare Information System application.

A query available in Healthcare_Access_DatabaseDataSet in the Data Sources window can be dragged and dropped to a Windows form to present the query result in the Windows form.

Example 13.2

Add the query, PRQ1, to FormPR1 in the Healthcare Information System application to present the query result showing patients whose annual physical exam is overdue.

Figure 13.4 shows the Design View of FormPR1 after we drag the query, PRQ1, from Healthcare_Access_DatabaseDataSet in the Data Sources window and drop it to FormPR1.

13.3 Architecture of Database Connection

As seen in Figure 13.3, five objects related to the database connection are added to a window below the Design View window of FormU1 when we drag and drop the Person table in Healthcare_Access_DatabaseDataSet to FormU1. These five objects are as follows:

- Healthcare_Access_DatabaseDataSet
- PersonBindingSource

(a)

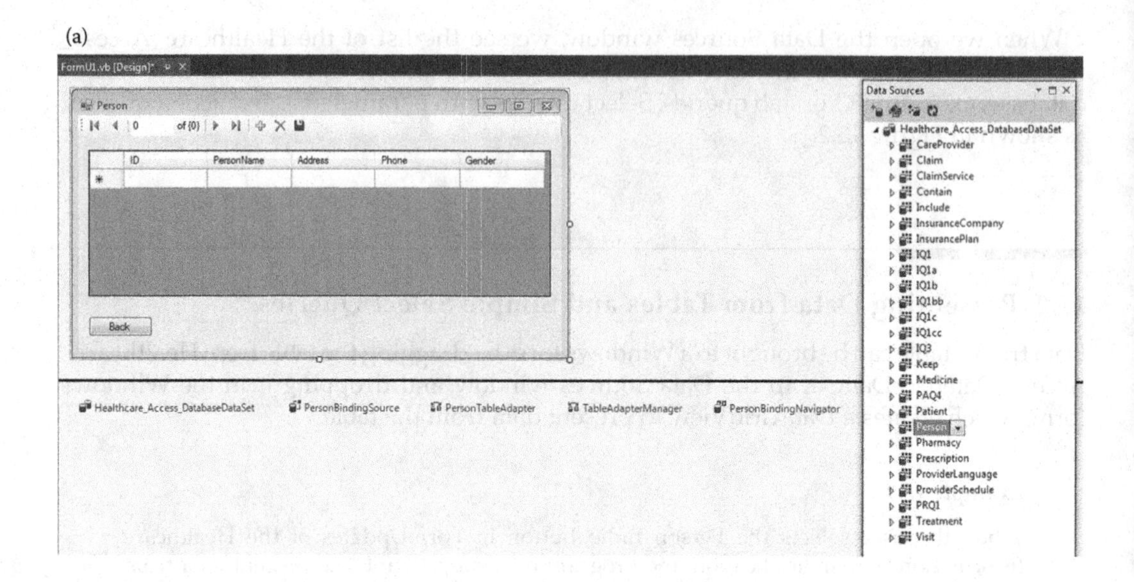

(b)

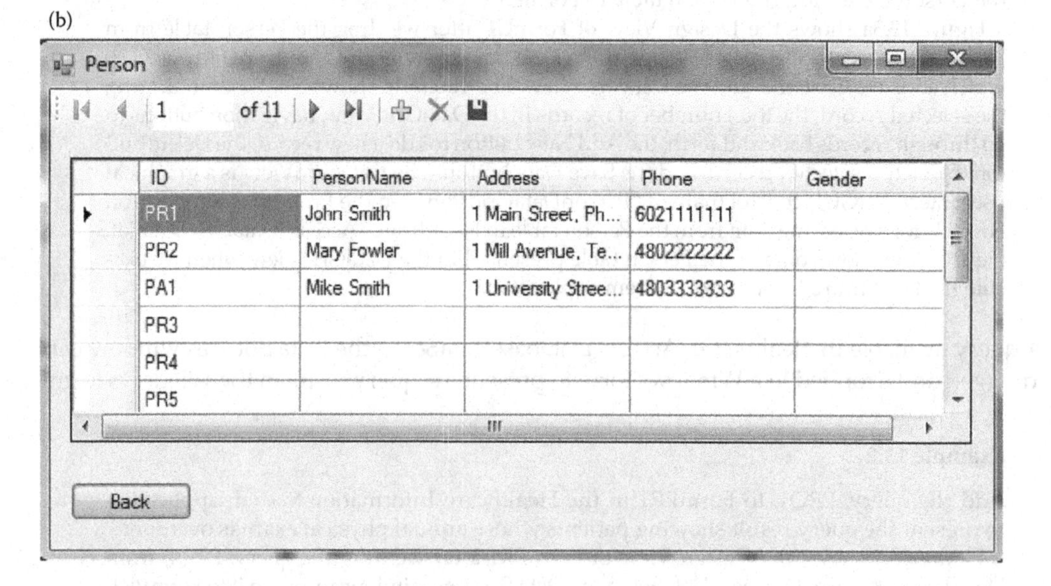

FIGURE 13.3
FormU1 to present the data of the Person table. (a) Design View of the FormU1 in the Healthcare Information System application to present data of the Person table. (b) Data of the Person table presented when the application is executed.

- PersonTableAdapter
- TableAdapterManager
- PersonBindingNavigator

Similarly, five objects are added to a window below the Design View window of FormPR1 when we drag and drop the query, PRQ1, in Healthcare_Access_DatabaseDataSet to FormPR1, as shown in Figure 13.4. These five objects are as follows:

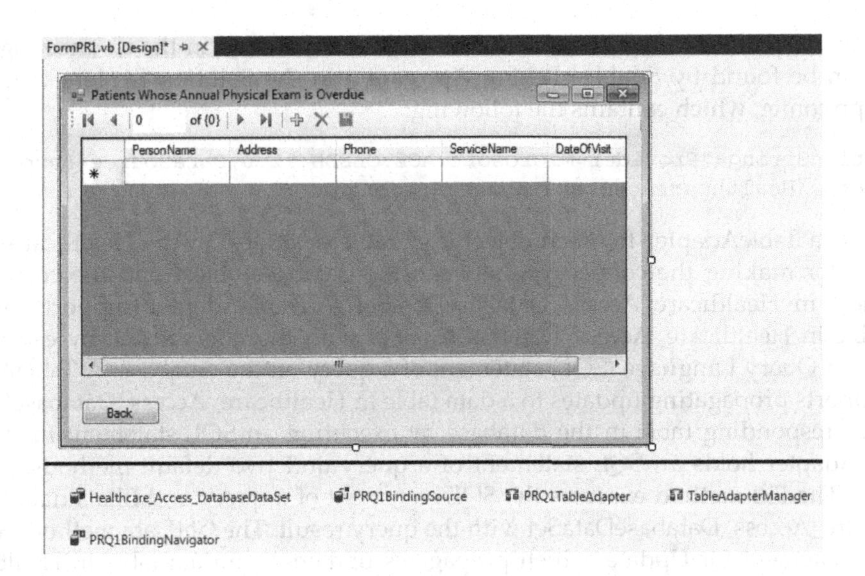

FIGURE 13.4
Design View of Form PR1, showing the query, PRQ1, to present patients whose annual physical exam is overdue.

- Healthcare_Access_DatabaseDataSet
- PRQ1BindingSource
- PRQ1TableAdapter
- TableAdapterManager
- PRQ1BindingNavigator

Figure 13.5 shows how these objects work together to enable the connection between a database and a data presentation in a DataGridView on a Windows form. When we make the connection to the Healthcare Access database, as described in Section 13.1, Healthcare_Access_DatabaseDataSet is created to include data tables that keep data in objects (e.g., tables and queries) of the Healthcare Access database in the memory for the Healthcare Information system application. That is, the data tables in Healthcare_Access_DatabaseDataSet are the memory storage of data from the Healthcare Access database for the Healthcare Information System application.

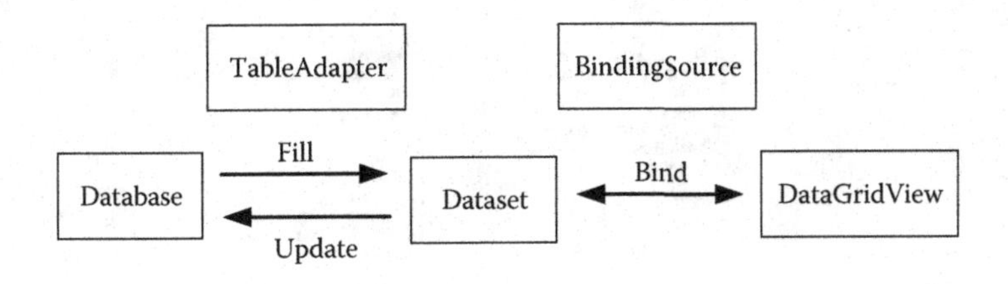

FIGURE 13.5
Architecture of database connection.

The path and other information of the connection are stored in the ConnectingString, which can be found by double-clicking App.config in the Solution Explore window to open App.config, which contains the following:

```
ConnectionString="Provider=Microsoft.ACE.OLEDB.12.0;DataSource=&quot;|Data
Directory|\HealthcareAccessDatabase.accdb&quot;"
```

There is a TableAdapter for each object (e.g., table or query) in the Healthcare Access database for making the connection between the database object and the corresponding object in Healthcare_Access_DatabaseDataSet. A TableAdapter supports filling a data table in Healthcare_Access_DatabaseDataSet with the query result by executing a Structured Query Language (SQL) statement of a query on the database. A TableAdapter also supports propagating updates to a data table in Healthcare_Access_DatabaseDataSet to the corresponding table in the database by executing an SQL statement for Update. A TableAdapter holds an SQL statement of a query and two default methods: Fill and GetData. The Fill method executes the SQL statement of a query and fills a data table in Healthcare_Access_DatabaseDataSet with the query result. The GetData method executes an SQL statement for Update, which propagates updates to a data table in Healthcare_Access_DatabaseDataSet back to the corresponding table in the database.

A TableAdapter and its Fill and GetData methods can be viewed by double-clicking Healthcare_Access_DatabaseDataSet in the Solution Explorer window to bring up the Design View window of Healthcare_Access_DatabaseDataSet. The Design View window of Healthcare_Access_DatabaseDataSet can also be brought up by selecting a TableAdapter below the Design View of a Windows form containing a DataGridView, right-clicking on the right arrow button at the top-right corner of the TableAdapter, and selecting Edit Queries in DataSet Designer. Figure 13.6 shows a part of the Design View of Healthcare_Access_DatabaseDataSet, which includes the TableAdapters for all the tables in the Healthcare Access database, relationships of the tables, and certain queries. Right-clicking a TableAdapter in the Design View of Healthcare_Access_DatabaseDataSet brings up a pop-up menu. Selecting Configure in the pop-up menu brings up the TableAdapter

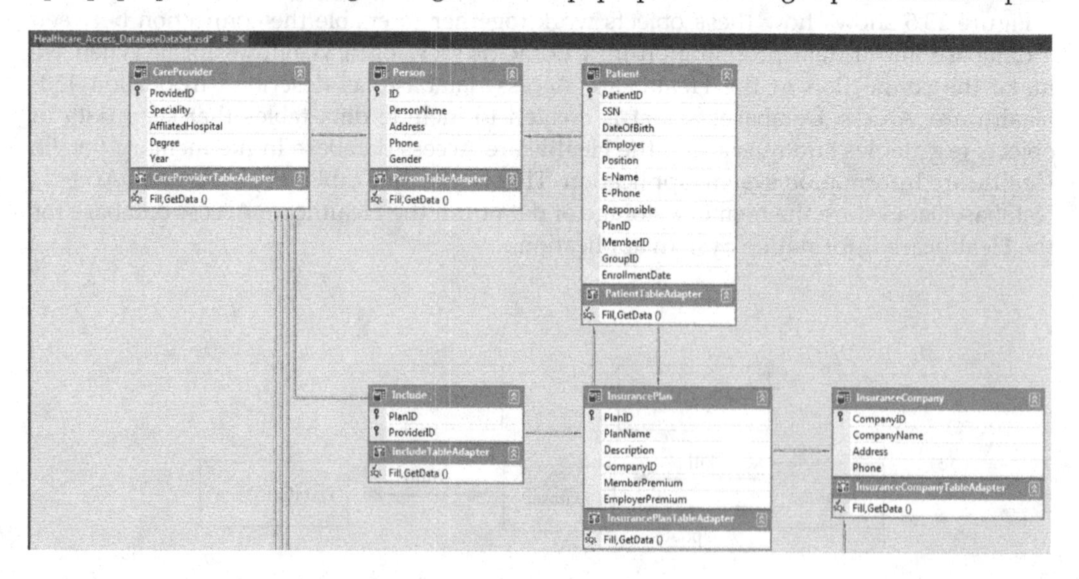

FIGURE 13.6
Part of the Design View of Healthcare_Access_DatabaseDataSet.

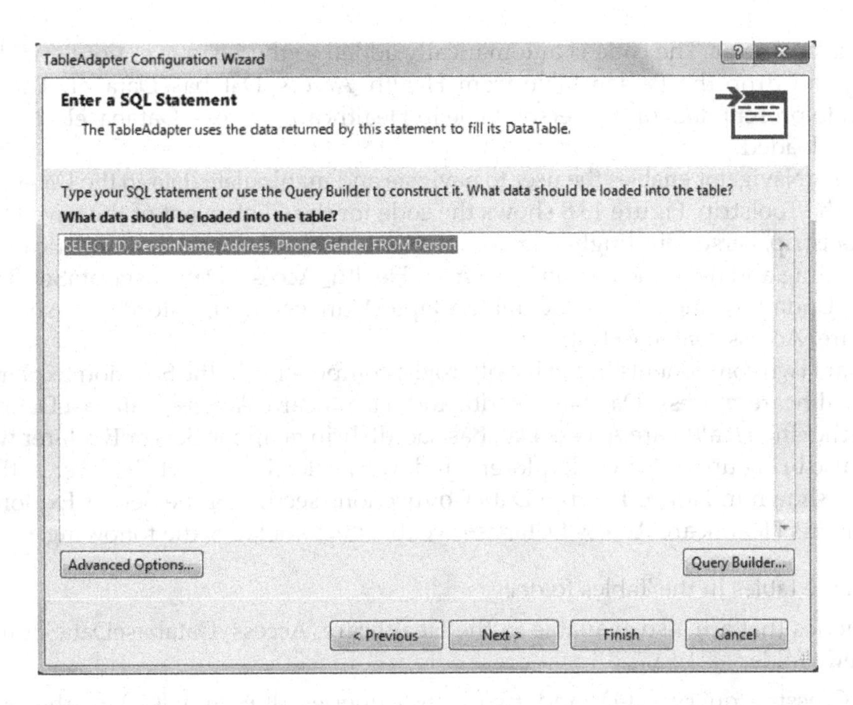

FIGURE 13.7
SQL statement of a Select query held by PersonTableAdapter for the Person table in the Healthcare Access database.

Configuration Wizard (see Figure 13.7), which shows the SQL statement of a query held by the TableAdapter.

Each DataGridView has a BindingSource as the value of the Data Source property of the DataGridView to bind the DataGridView to the corresponding data table in Healthcare_Access_DatabaseDataSet for bringing data in the data table in Healthcare_Access_DatabaseDataSet to the DataGridView. When a Windows form containing a DataGridView is loaded, the subroutine for the Load event of the Windows form object contains the code to execute the Fill method of the TableAdapter, which is to execute the SQL statement of the query held by the TableAdapter and fill the DataGridView with the query result. Figure 13.8 shows the code for the Load event of FormU1 to execute the Fill method of the

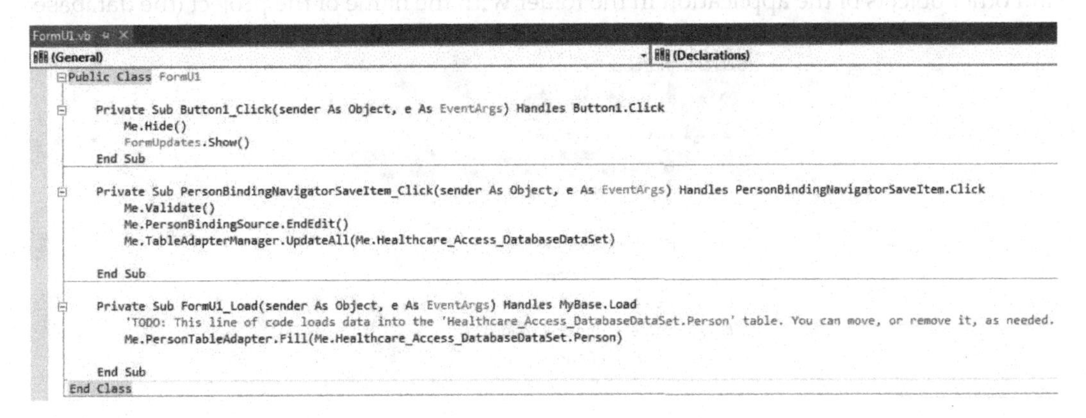

FIGURE 13.8
Code of FormU1 in the Healthcare Information System application.

PersonTableAdapter. The code is automatically added to the Load event of FormU1 when we drag and drop the Person table from Health_Access_DatabaseDataSet. The code is executed to present data of the Person table in Healthcare_Access_DatabaseDataSet when FormU1 is loaded.

A BindingNavigator enables the user to navigate and manipulate data in the DataGridView through the Toolstrip. Figure 13.8 shows the code for the Click event of the Save Data icon in the Toolstrip, PersonBindingNavigatorSaveItem_Click. The code is automatically added when we drag and drop the Person table from Health_Access_DatabaseDataSet. The code uses the UpdateAll method of the TableAdapterManager to update the Person table in Healthcare_Access_DatabaseDataSet.

There are two components in the list of project components in the Solution Explorer window: Healthcare Access Database.accdb and Healthcare_Access_DatabaseDataSet.xsd. Double-clicking Healthcare Access Database.accdb brings up the Server Explorer window. We can also bring up the Server Explorer window by selecting Server Explorer in the View menu. As shown in Figure 13.9, the Data Connections section of the Server Explorer window contains Healthcare Access Database.accdb, which includes the following:

- All the tables in the Tables folder
- Queries that are also available in the Healthcare_Access_DatabaseDataSet in the Views folder
- A Crosstab query (EQ3) and two action queries (IQ8 and IQ9) in the Stored Procedures folder
- Queries with parameter(s) in the Functions folder

Hence, the Server Explorer window shows all the objects of the Healthcare Access database in Healthcare Access Database.accdb, whereas the Data Sources window shows the data tables and views in HealthcareAccessDatabaseDataSet. Because there are no data tables in Healthcare_Access_DatabaseDataSet for the queries in the Stored Procedures and the Functions folders shown in the Server Explorer, we cannot directly add the queries in these two folders to a Windows form.

There are two copies of the Healthcare Access database in the project folder of the Healthcare Information System application, with one copy in the Debug folder of the bin folder (the database file in the execution environment) and the other copy along with forms and other objects of the application in the folder with the name of the project (the database

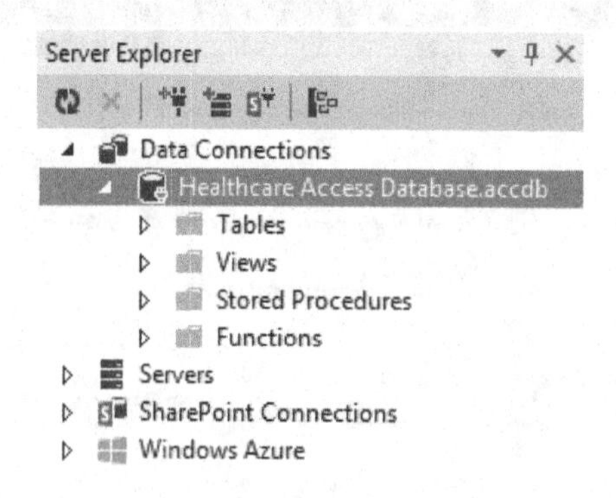

FIGURE 13.9
Server Explorer window.

file in the design environment). When we execute the Healthcare Information System application, Healthcare_Access_DatabaseDataSet is bound to the copy of the Healthcare Access database in the execution environment, and any database changes made by the application are made to this copy of the Healthcare Access database. However, when we open the Healthcare Information System application file in Visual Studio to design the application, the copy of the Healthcare Access database in the design environment is used.

13.4 Displaying Related Data in Parent and Child Tables

In the Healthcare Access database, data in the Person table (the parent table) are related to data in the CareProvider table (the child table) through the common data fields (ID in the Person table and ProviderID in the CareProvider table). Data in the parent and child tables can be displayed in a way to relate them together. Example 13.3 shows the procedure of displaying related data in parent and child tables and the display of related data.

Example 13.3

With data of the Person table added to FormU1 in Example 13.1, add data of the CareProvider table to FormU1 so that data of the Person table and data of the CareProvider table are displayed in a synchronized way.

The following steps are taken:

1. Double-click FormU1.vb in the Solution Explorer window to bring up FormU1 in the Design View.
2. Double-click the Person table in the Data Sources window to expand the Person table and bring up a list of the data fields in the Person, the CareProvider, and the Patient tables, which are related to the Person table (see Figure 13.10a).
3. Drag and drop the CareProvider table in the expanded list for the Person table to FormU1 (see Figure 13.10a). Note that dragging and dropping the CareProvider table to FormU1 adds CareProviderBindingSource and CareProviderTableAdapter to FormU1 (see Figure 13.10a).

Figure 13.10b shows the display of related data from the Person and the CareProvider tables. Note that only one record of the CareProvider table corresponding to the selected record in the Person table is displayed. If the selected record in the Person table (e.g., the record with the ID value of PA1) does not have a corresponding record in the CareProvider table, nothing is displayed in the CareProvider table. Hence, when we use the Toolstrip to navigate one record to another record in the Person table, the record displayed in the CareProvider table changes accordingly.

Because the ID field of the Person table and the ProviderID field of the CareProvider table are the common data fields and have the same value, we do not need to display the ProviderID field. We can delete the ProviderID field by selecting the CareProvider table in FormU1 and clicking the right arrow at the top-right corner of the CareProvider table to bring up a pop-up menu (see Figure 13.10a). Selecting Edit Columns in the menu brings up the Edit Columns window, as shown in Figure 13.11. By selecting ProviderID in the Selected Columns area and clicking the Remove button, we remove ProviderID from the CareProvider table on FormU1.

In Figure 13.10a, the CareProvider table is in the list of data tables in the Data Sources window and is also embedded in the expanded list of data fields and child table(s) for

(a)

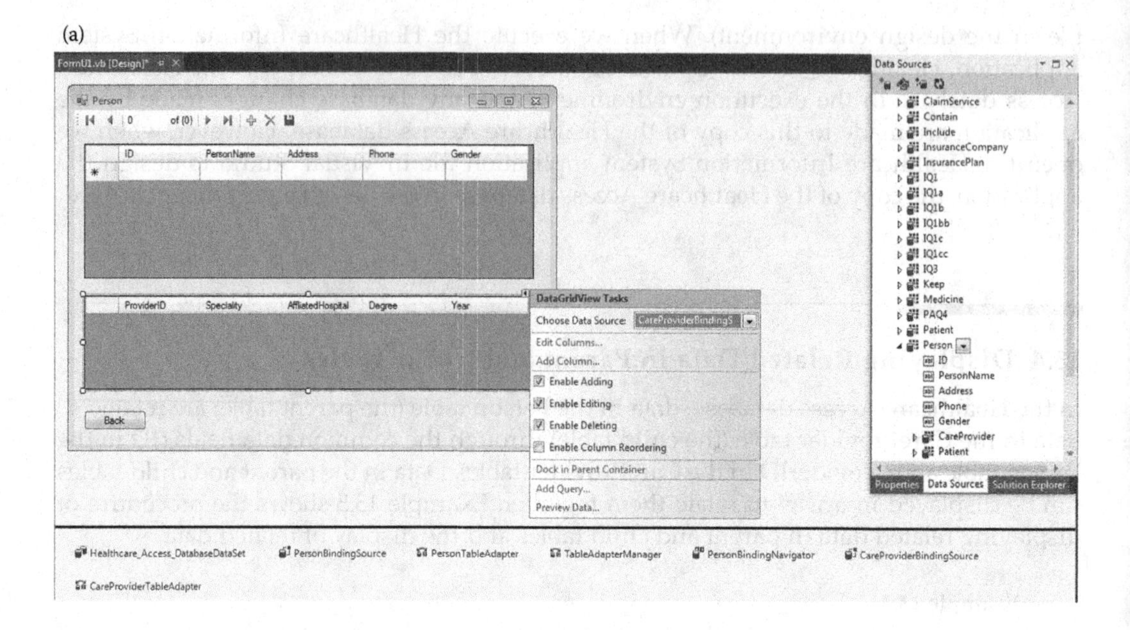

(b)

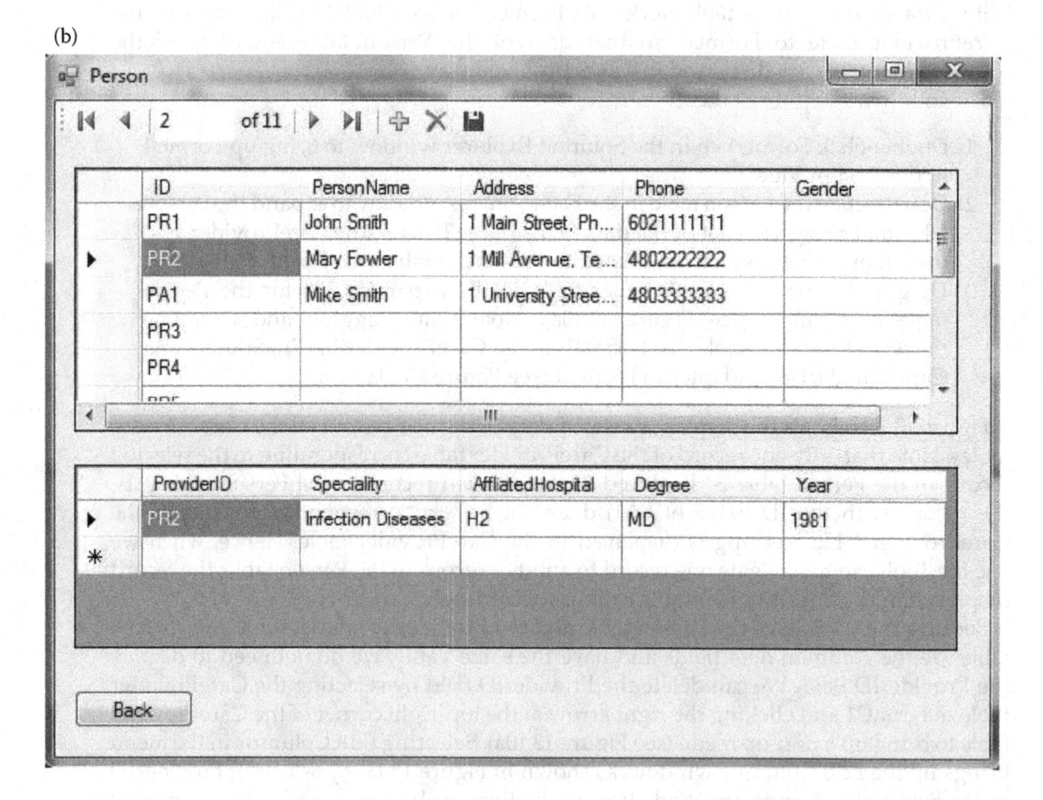

FIGURE 13.10
Use of the Data Sources window to display related data in parent and child tables. (a) Design View of FormU1 with related data from the Person table and the CareProvider table. (b) Display of related data from the Person table and the CareProvider table when the application is executed.

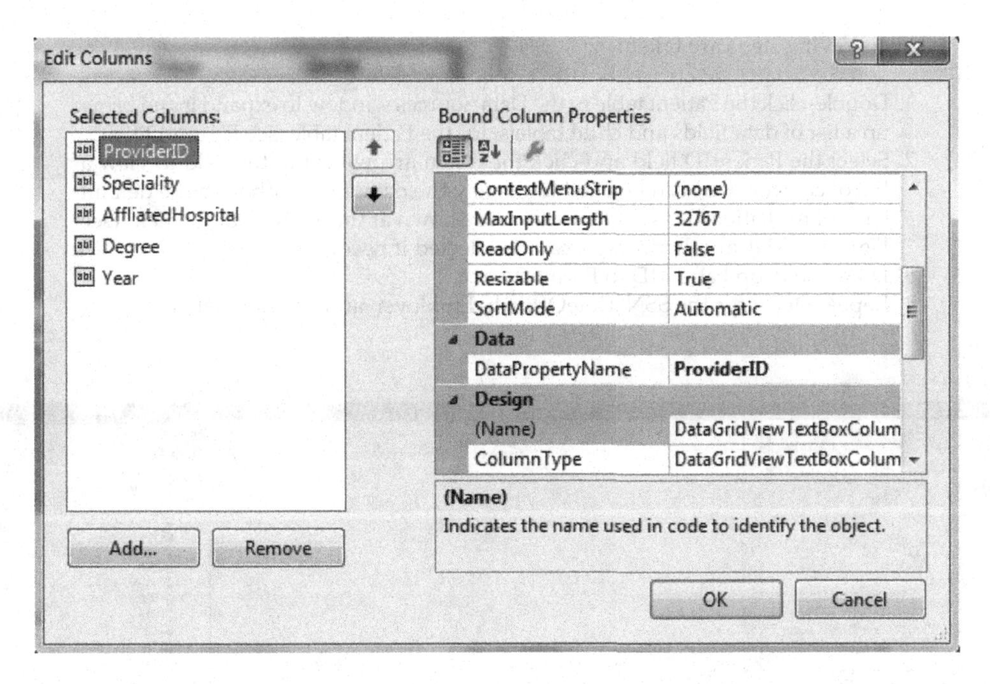

FIGURE 13.11
Edit Columns window.

the Person table. To display related data of parent and child tables, we have to drag and drop the child table embedded in the parent table. For example, if we drag and drop the CareProvider table in the list of data tables in the Data Sources window (not the embedded child table) to FormU1, data of the Person and the CareProvider tables in FormU1 will not be related. That is, the selected record of the Person table and the displayed record of the CareProvider table are not necessarily for the same person and do not necessarily have the same value of the common data fields.

13.5 Displaying Data Fields Using Controls Other Than DataGridView

In Examples 13.1 and 13.3, all records and data fields of the Person table are displayed in the DataGridView control on FormU1 after we drag and drop the Person table to FormU1. Using the Edit Columns window shown in Figure 13.11, we can display only the selected data fields of a table in the DataGridView control. We can also display the selected data fields and only one record of a table using a control for an individual data field rather than the DataGridView control. Example 13.4 shows the procedure for this.

Example 13.4

Add the PatientID, SSN, DateOfBirth, Employer, and Position fields of the Patient table to FormU2 in the Healthcare Information System application using the TextBox control to display each data field.

The following steps are taken:

1. Double-click the Patient table in the Data Sources window to expand it and bring up a list of data fields and child table(s) for the Patient table (see Figure 13.12a).
2. Select the PatientID field and click the down arrow next to the field to view a list of controls that can be used to display the data field. Although the default control for PatientID is set to TextBox as shown at the left side of the field (see Figure 13.12a), another control can be selected if needed.
3. Drag and drop PatientID to FormU2.
4. Repeat Step 3 for the SSN, DateOfBirth, Employer, and Position fields.

(a)

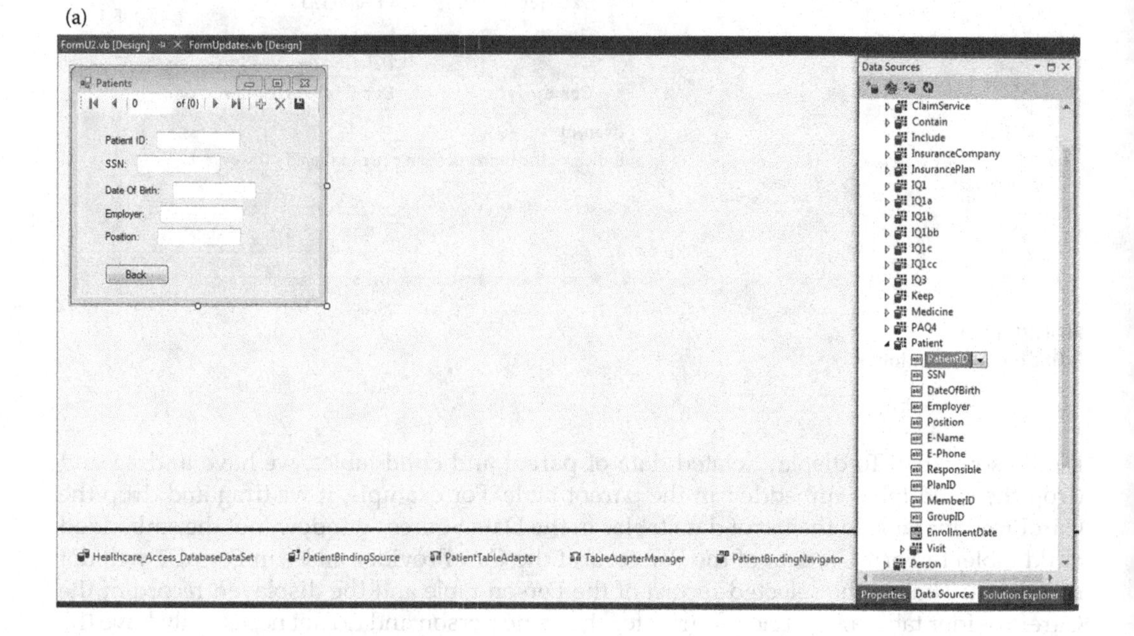

(b)

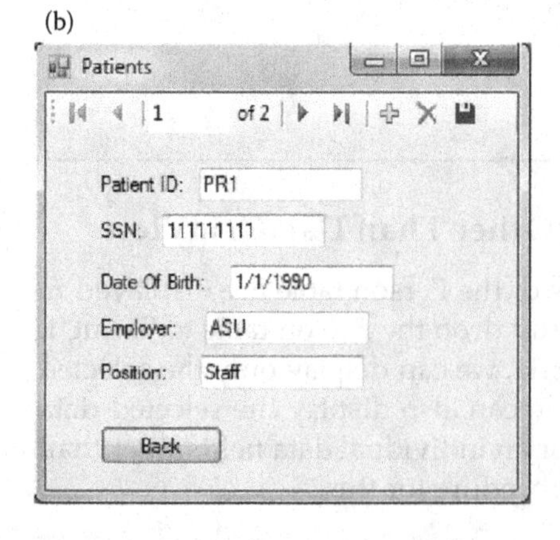

FIGURE 13.12
Display of individual data fields and one record at a time from the Patient table in the Healthcare Information System application. (a) Design View of FormU2 and the Data Sources window. (b) Display of individual data fields in one record from the Patient table.

Figure 13.12b shows the display of one record containing the values for these data fields from the Patient table. We can navigate from one record to another record of the Patient table using the Toolstrip.

13.6 Handling Crosstab and Select Queries with Parameters

EQ3 in the Healthcare Access database is a Crosstab query and is taken as a stored procedure after we add the Healthcare Access database as a data source to the Healthcare Information System application. As a stored procedure, EQ3 is not included in Healthcare_Access_DatabaseDataSet and cannot be directly dragged and dropped to a Windows form to display the query result. To display the query result of a Crosstab query in an Access database such as EQ3, we can create a regular Select query that embeds the Crosstab query. The regular Select query is included in Healthcare_Access_DatabaseDataSet of the database in a Windows-based application. Example 13.5 illustrates how to include a Crosstab query in an Access database in a Windows-based application.

Example 13.5

Create a Select query, EQ3-Select, which embeds EQ3 in the Healthcare Access database; update Healthcare_Access_DatabaseDataSet in the Healthcare Information System application; and add EQ3-Select to FormE3.

Figure 13.13a shows the Design view of EQ3, the Crosstab query, in the healthcare Access database. Figure 13.13c shows the query result of EQ3. We open the healthcare Access database in the project folder of the Healthcare Information System application, and create EQ3-Select by adding EQ3 to the table pane and putting EQ3.* in the data field. Figure 13.13b shows the Design View of EQ3-Select. The query result of EQ3-Select is the same as that shown in Figure 13.13c. To update Healthcare_Access_DatabaseDataSet in the Healthcare Information System application, we click the Configure Data Source with Wizard icon in the Data Sources window, select Continue with Wizard in the Choose DataSet Editor window, check Views in the Choose Your Database Objects window, and click the Finish button. Figure 13.14 shows the updated Healthcare_Access_DatabaseDataSet, which includes EQ3-Select. We drag and drop EQ3-Select from Healthcare_Access_DatabaseDataset to FormE3 to display the query result.

All the other queries in the Healthcare Access database but not in the Healthcare_Access_DatabaseDataSet in the Healthcare Information System application are queries that request parameter input(s) from the user. Such queries with parameter(s) are shown as functions in the Server Explorer window under Healthcare Access Database.accdb. Because a query with parameters is not included in the DataSet of an Access database, we need to create the query using the Query Builder in Visual Studio. Example 13.6 shows how to create a query with a parameter input from the user using the Query Builder in Visual Studio.

Example 13.6

Create EQ1 using the Query Builder in Visual Studio and display the query result of EQ1 in FormE1 of the Healthcare Information System application.

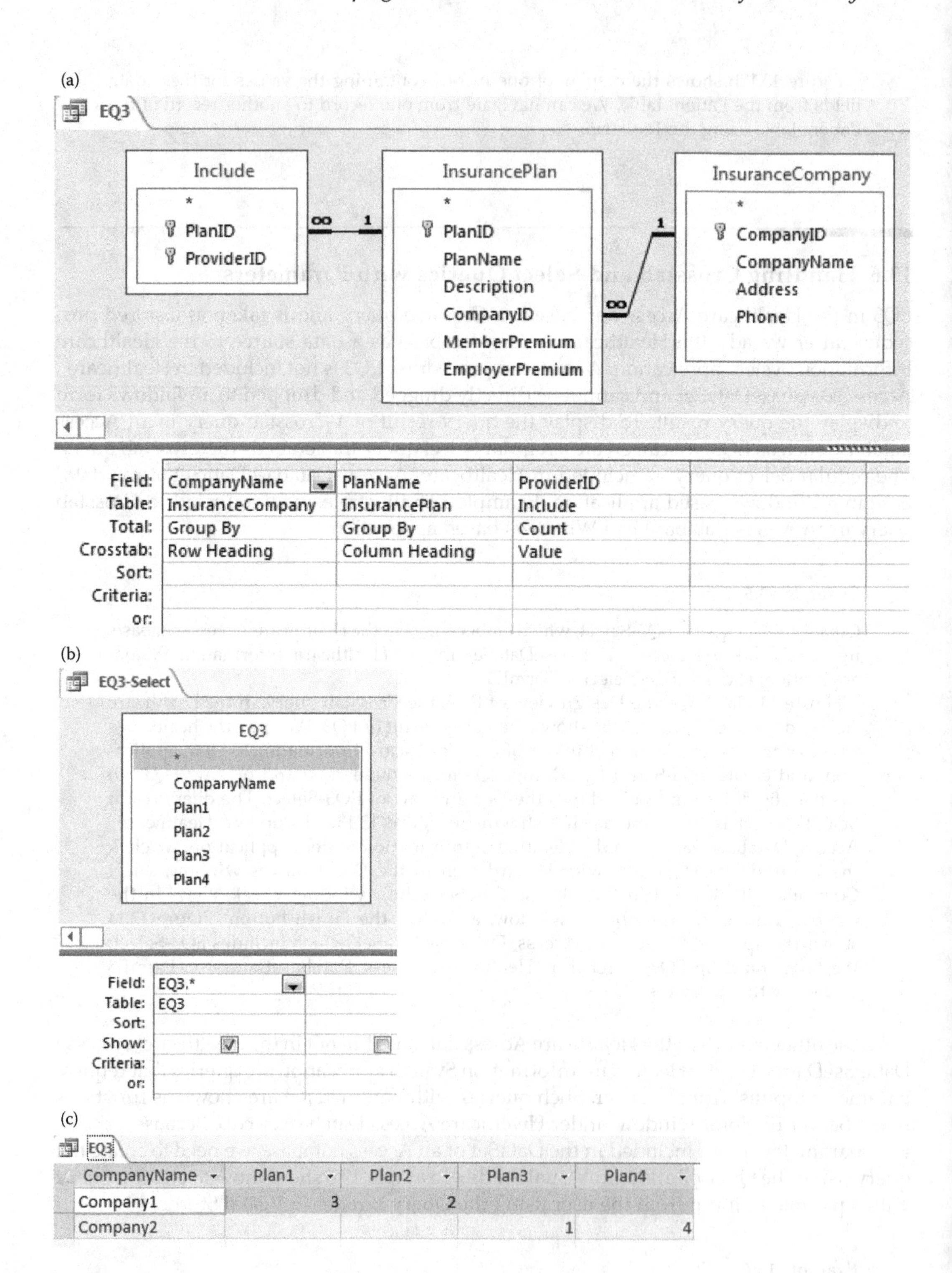

FIGURE 13.13
Creation of a Select query that embeds a Crosstab query. (a) Design View of EQ3, a Crosstab query. (b) Design View of EQ3-Select. (c) Query result of EQ3 and EQ3-Select.

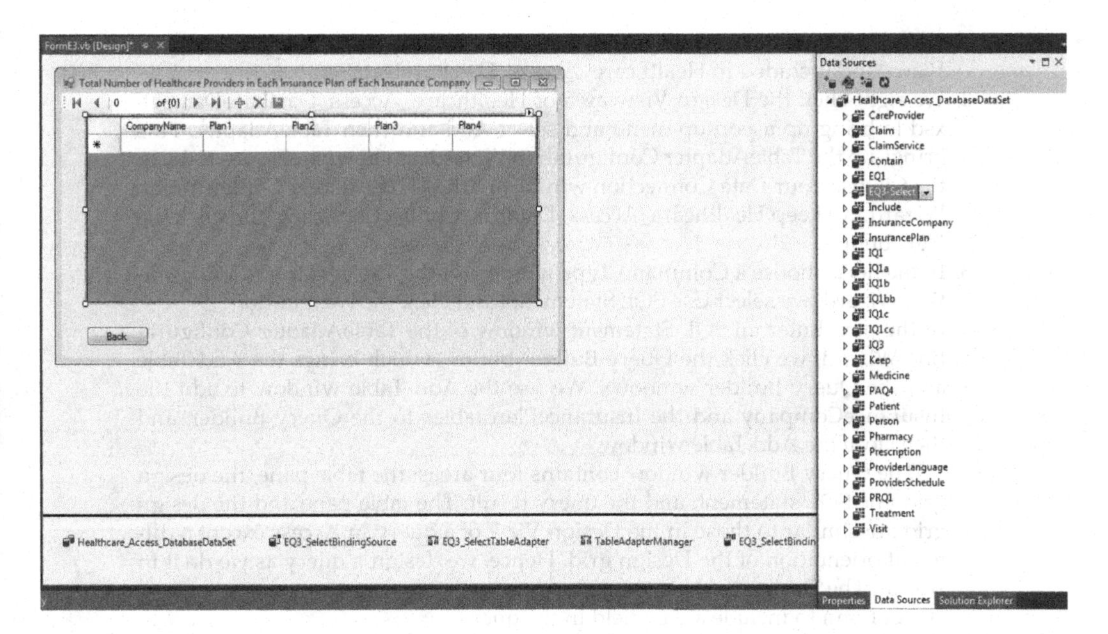

FIGURE 13.14
Healthcare_Access_DatabaseDataSet in the Data Sources window including EQ3-Select and the Design View of FormE3.

Figure 13.15 shows the Design View of EQ1 in the healthcare Access database. The procedure of creating EQ1 in Visual Studio includes the following steps:

1. We double-click Healthcare_Access_DatabaseDataSet.xsd in the Solution Explorer window to bring up the Design View of Healthcare_Access_DatabaseDataSet.xsd. The Design View of Healthcare_Access_DatabaseDataSet.xsd contains all the

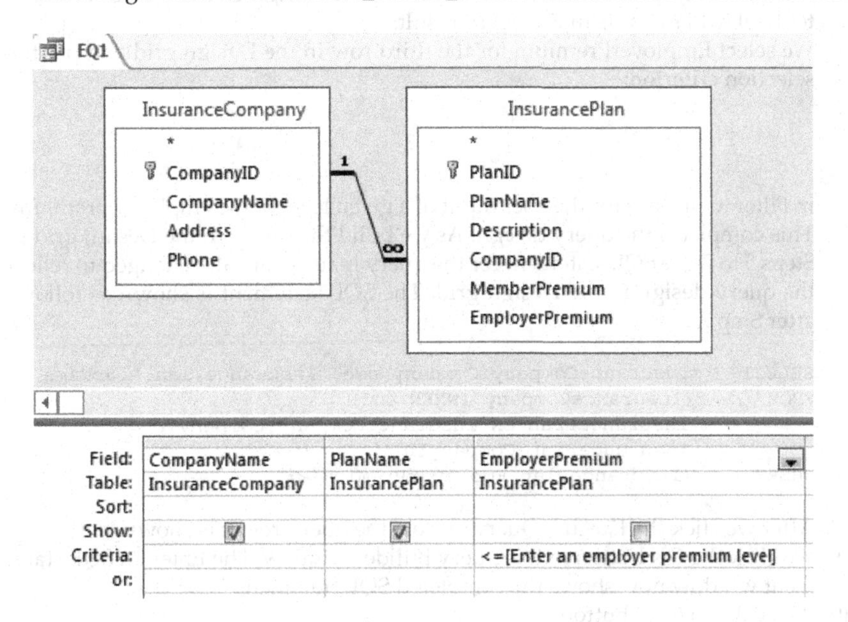

FIGURE 13.15
The Design View of EQ1 in the healthcare Access database.

tables, relationships of the tables, and individual queries that are taken as Views and included in Healthcare_Access_DatabaseDataSet.

2. We right-click the Design View area of Healthcare_Access_DatabaseDataSet.xsd to bring up a pop-up menu and select Add and then TableAdapter. This brings up the TableAdapter Configuration Wizard, as shown in Figure 13.16. In the Choose Your Data Connection window of the TableAdapter Configuration Wizard, we keep Healthcare_Access_DatabaseConnectionString and click the Next button.
3. In the next Choose a Command Type window of the TableAdapter Configuration Wizard, we select Use SQL Statements and click the Next button.
4. In the next Enter an SQL Statement window of the TableAdapter Configuration Wizard, we click the Query Builder button, which brings the Add Table and the Query Builder windows. We use the Add Table window to add the InsuranceCompany and the InsurancePlan tables to the Query Builder, and then close the Add Table window.

   The Query Builder window contains four areas: the table pane, the design grid, the SQL statement, and the query result. The table pane and the design grid are similar to those in the Design View of a query in Access except a different orientation of the Design grid. Hence, we design a query as we do it in Access. The Design grid has the following:
   - Column to include a data field in the query
   - Alias to give a display name for the column
   - Output to indicate whether we want to display the column in the query result
   - Sort Type and Sort Order to determine how to sort the query result
   - Filter to specify the selection criterion.

   Right-clicking the table pane brings up a pop-up menu. Selecting Add Group By in the pop-up menu adds Group By in the Design grid.
5. We select CompanyName for the first row in the Design grid and check Output to display CompanyName in the query result.
6. We select PlanName for the second row in the Design grid and check Output to display PlanName in the query result.
7. We select EmployerPremium for the third row in the Design grid and put the selection criterion:

   <= ?

   in Filter to prompt for the user input of a given level of the employer premium. This completes the query design. As we build the query in the Design grid in Steps 5 to 7, the SQL statement for the query is automatically changed to reflect the query design in the Design grid. The SQL statement is shown as follows after Step 7:

```
SELECT    InsuranceCompany.CompanyName, InsurancePlan.PlanName
FROM      (InsuranceCompany INNER JOIN
          InsurancePlan ON InsuranceCompany.CompanyID =
          InsurancePlan.CompanyID)
WHERE     (InsurancePlan.EmployerPremium <= ?)
```

8. After we click the Execute Query button, the query result is shown.
9. We click the OK button in the Query Builder window. The Enter an SQL statement window now shows the completed SQL statement.
10. We click the Next button.

(a)

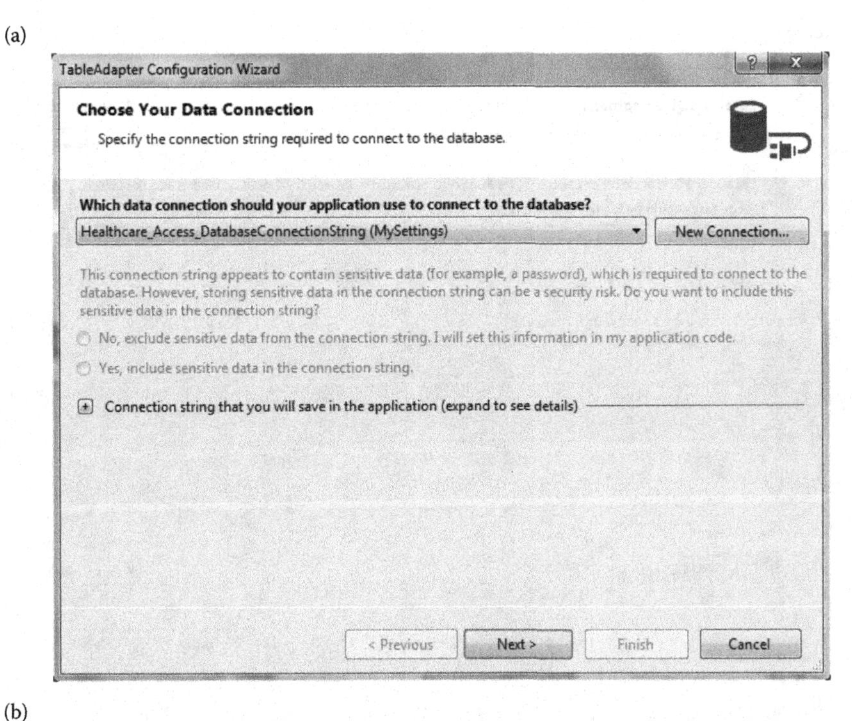

(b)

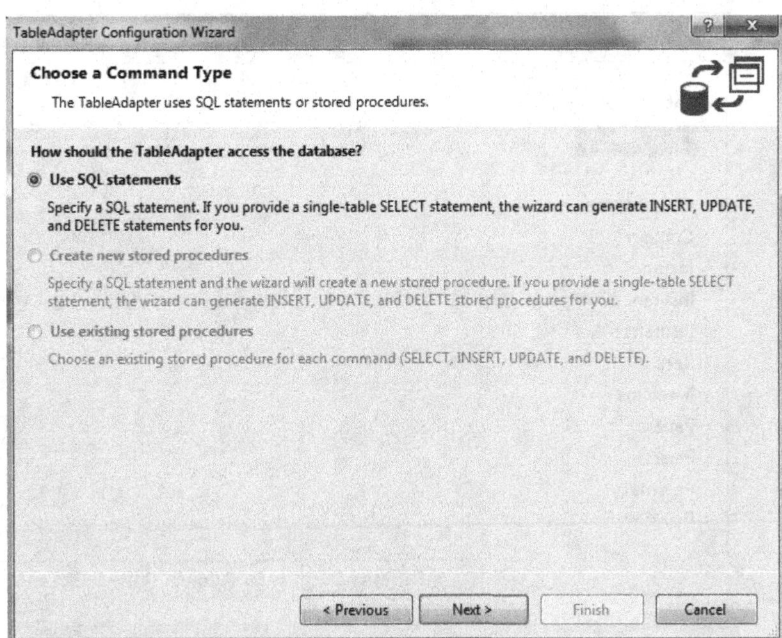

FIGURE 13.16
Procedure of using the Query Builder in Visual Studio to create EQ1, a query with a parameter input. (a) Choose a data connection. (b) Choose a command type.

(c)

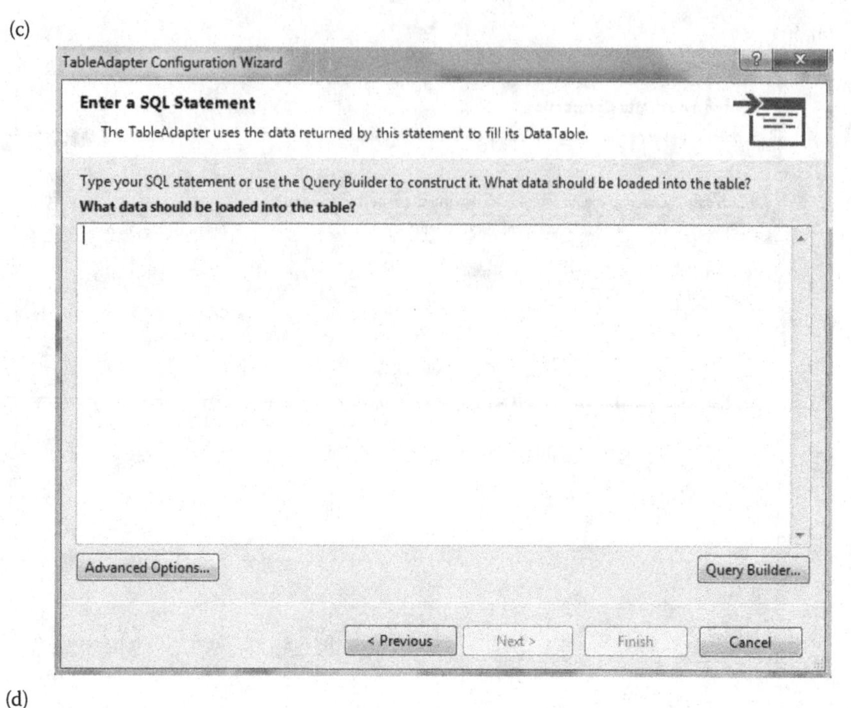

(d)

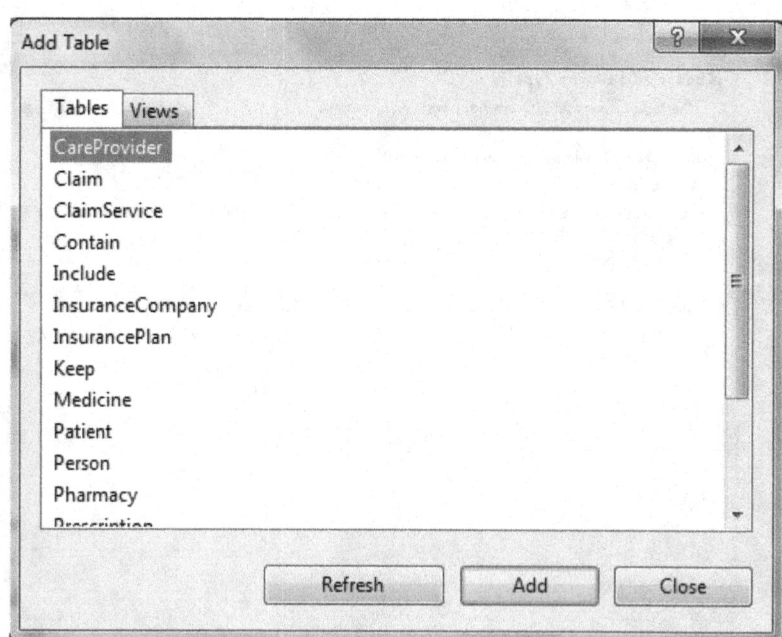

FIGURE 13.16 (Continued)
Procedure of using the Query Builder in Visual Studio to create EQ1, a query with a parameter input. (c) Enter an SQL statement. (d) Add table.

(e)

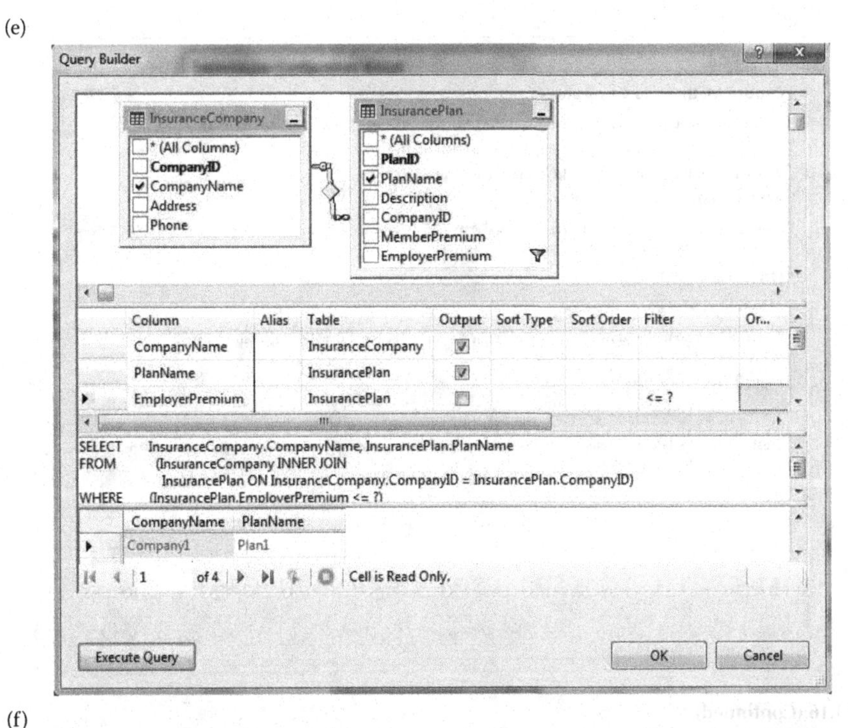

(f)

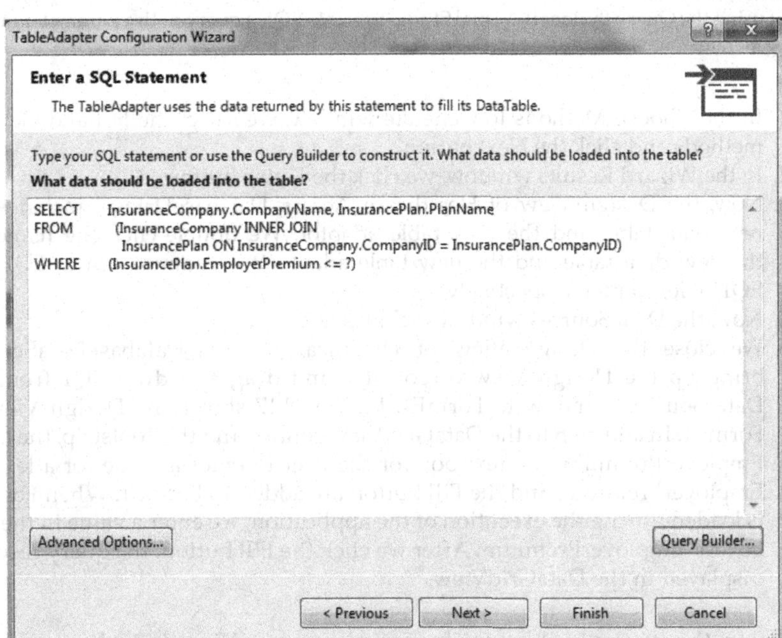

FIGURE 13.16 (Continued)
Procedure of using the Query Builder in Visual Studio to create EQ1, a query with a parameter input. (e) Query builder. (f) Enter an SQL statement.

(g)

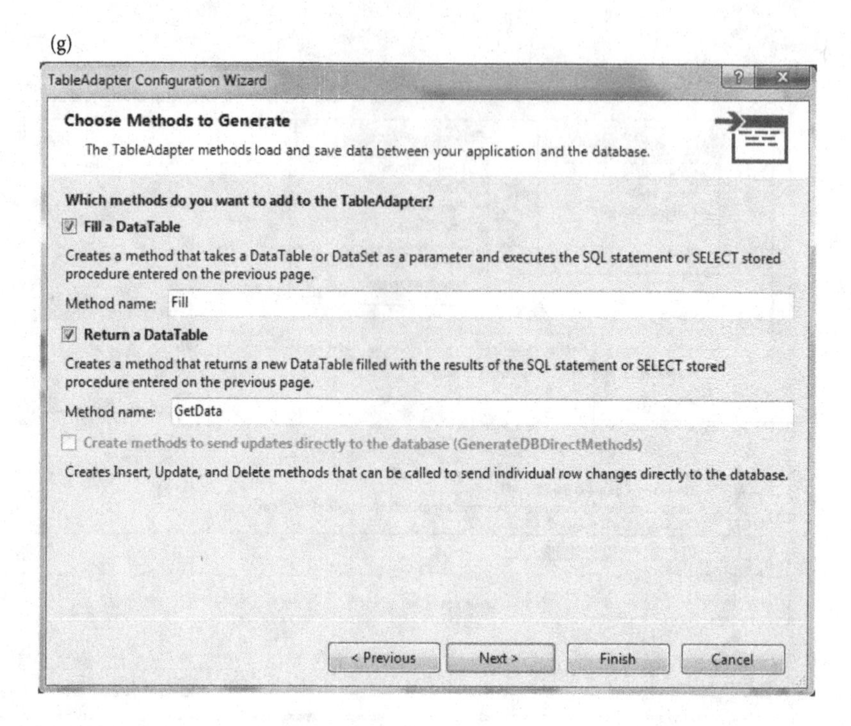

FIGURE 13.16 (Continued)
Procedure of using the Query Builder in Visual Studio to create EQ1, a query with a parameter input. (g) Choose methods to generate.

11. In the Choose Methods to Generate window, we select the Fill and GetData methods and click the Next button.
12. In the Wizard Results window, we click the Finish button.
13. Now, the Design View of Healthcare_Access_DatabaseDataset includes the new data table and the new table adaptor. We double-click the names of the new data table and the new table adaptor to rename them to EQ1 and EQ1TableAdapter, respectively.
14. Now, the Data Sources window includes EQ1.
15. We close the Design View of Healthcare_Access_DatabaseDataSet.xsd, bring up the Design View of FormE1, and drag and drop EQ1 from the Data Sources window to FormE1. Figure 13.17 shows the Design View of FormE1. In addition to the DataGridView control and the Toolstrip, the label, EmployerPremium, the text box for the user to enter a value for a level of EmployerPremium, and the Fill button are added to FormE1. When FormE1 is loaded during the execution of the application, we enter a value in the text box for EmployerPremium. After we click the Fill button, the query result is displayed in the DataGridView.

When using the query builder to design a query in Visual Studio, some features of the query design in Access do not work in Visual Studio. For example, in Access, the asterisk (*) is used to match any string, whereas in Visual Studio, the percentage (%) should be used to match any string. In Access, a pair of double quotes is used to enclose a string, whereas in Visual Studio, a pair of single quotes is used to enclose a string. Hence, the selection criterion such as "PA*" in Access should be 'PA%' in Visual Studio.

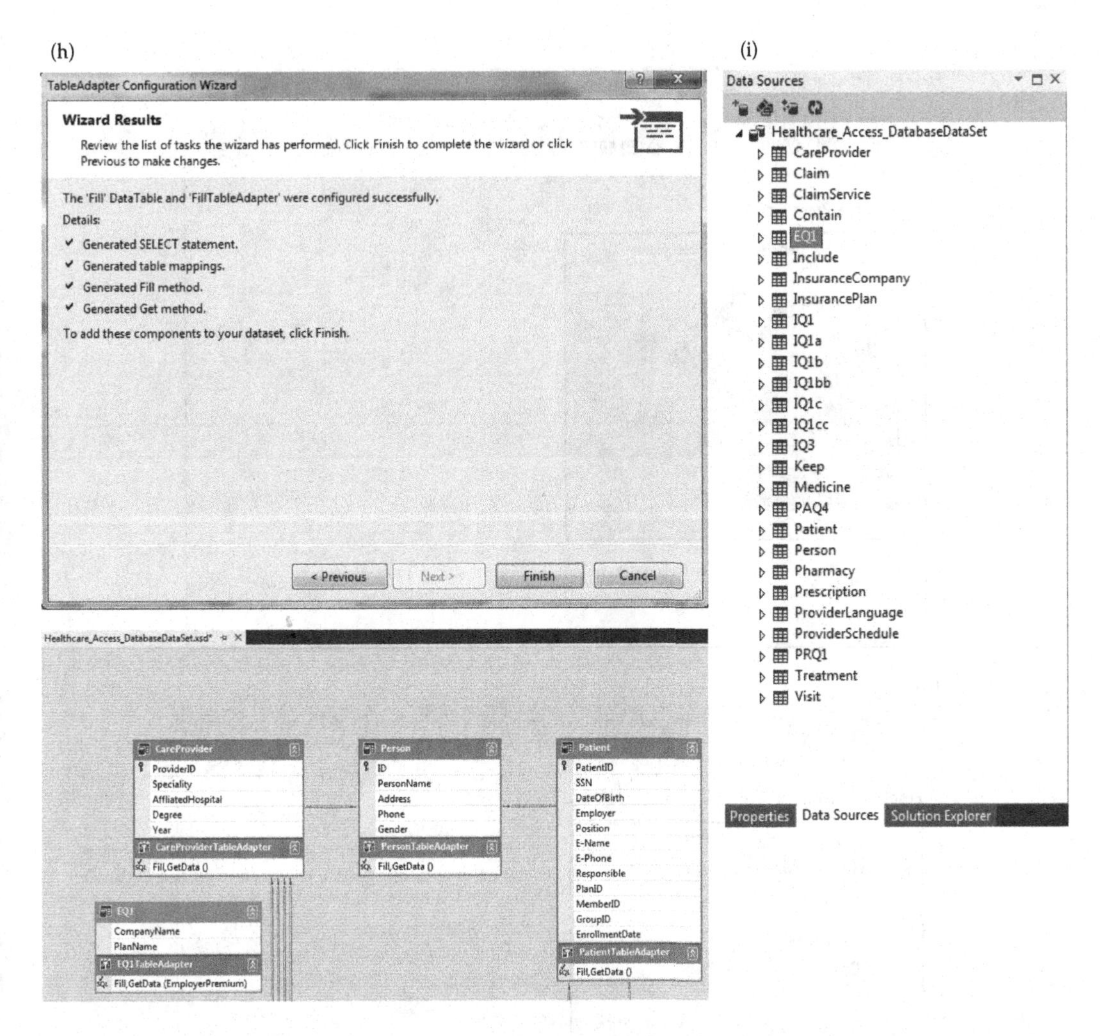

FIGURE 13.16 (Continued)
Procedure of using the Query Builder in Visual Studio to create EQ1, a query with a parameter input. (h) Wizard results. (i) Data sources.

In Access, brackets with a prompt, e.g., [Enter a provider name], are used to create a parameter input in a query design. In Visual Studio, the question mark (?) is used in a place where a parameter input is needed (see the query design in Example 13.6). Some time-related functions such as Year() and Month() in Access also do not work in Visual Studio. In Visual Studio, the function datepart() can be used to extract a specific part of a date/time. Examples of extracting the year, quarter, month, day, day of year, week, hour, minute, and second of the current date.time using datepart() in Visual Studio are given below:

Year: datepart('yyyy', now())
Quarter: datepart('q', now())
Month: datepart('m', now())

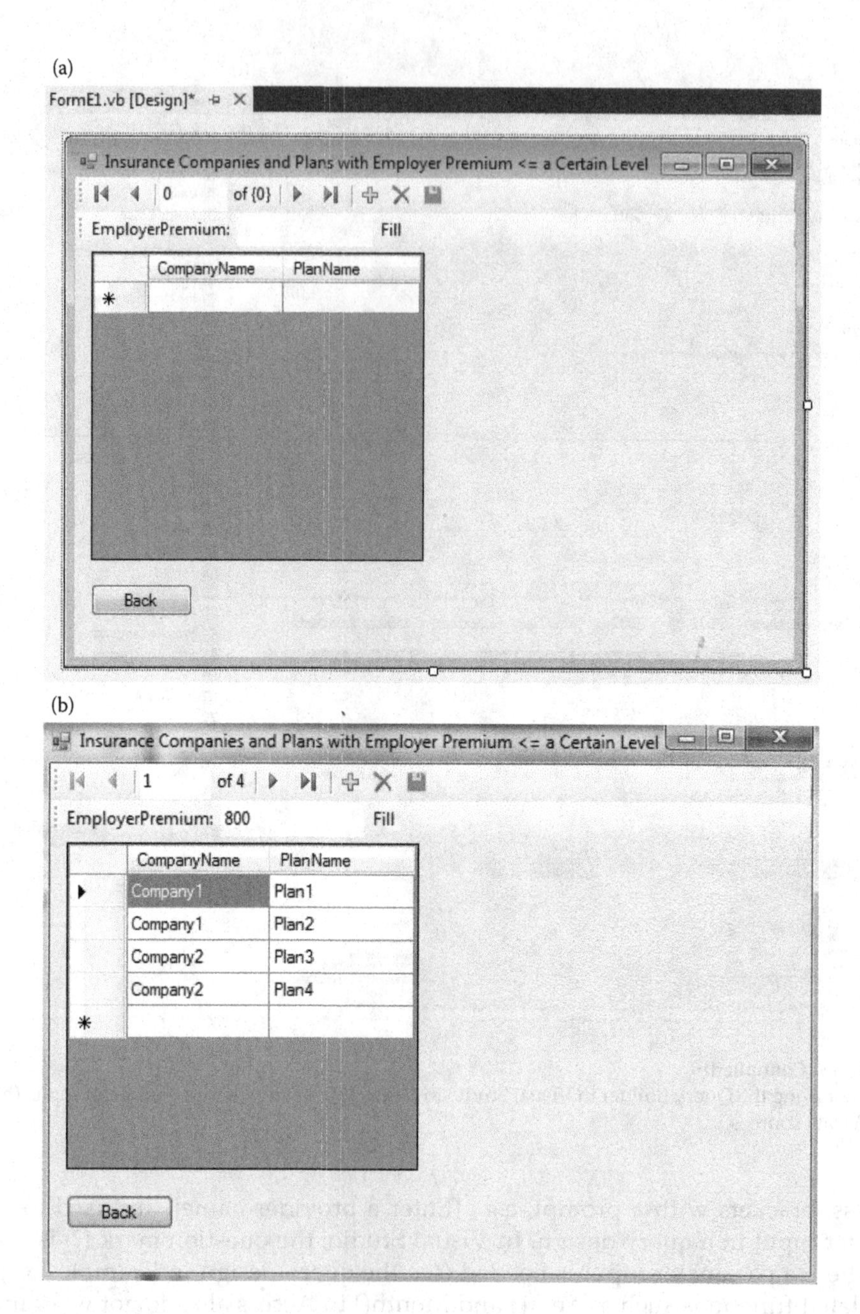

FIGURE 13.17
Design View of FormE1 in the Healthcare Information System application and the display of the query result. (a) Design View of FormE1 showing the query result of EQ1, a query with a paramater input. (b) The display of the query result when FormE1 is executed.

Day: datepart('d', now())
Day of year: datepart('y', now())
Week: datepart('ww', now())
Hour: datepart('h', now())
Minute: datepart('n', now())
Second: datepart('s', now())

13.7 Updating Data to an Access Database

The Toolstrip associated with a DataViewGrid control contains the following:

- Add new button for the user to add a new row of data in the DataGridView control
- Delete button for the user to delete a row of data in the DataGridView control
- Save Data button to save changes in the DataGridView control to the DataSet and update the database using the DataSet

Figure 13.18 shows the code for the Save Data button in FormU3, which is automatically generated when we drag and drop the InsurancePlan table from Healthcare_Access_DatabaseDataSet to FormU3. In the subroutine for the Click event of the Save Data button, the first line of code:

```
Me.Validate()
```

validates and saves changes made in the DataGridView control to the Healthcare_Access_DatabaseDataSet. The second line of code:

```
Me.InsurancePlanBindingSource.EndEdit()
```

stops editing to Healthcare_Access_DatabaseDataSet. The third line of code:

```
Me.TableAdapterManager.UpdateAll(Me.Healthcare_Access_DatabaseDataSet)
```

executes the Update SQL statement held by TableAdapterManager to propagate the content of the Healthcare_Access_DatabaseDataSet to the healthcare Access database in the Debug folder of the bin folder. Hence, the user can use the Toolstrip to update data to a database.

```
Private Sub InsurancePlanBindingNavigatorSaveItem_Click(sender As Object, e As EventArgs) Handles InsurancePlanBindingNavigatorSaveItem.Click
    Me.Validate()
    Me.InsurancePlanBindingSource.EndEdit()
    Me.TableAdapterManager.UpdateAll(Me.Healthcare_Access_DatabaseDataSet)
End Sub
```

FIGURE 13.18
Code for the Save Data button in the Toolstrip.

13.8 Visual Basic Code to Retrieve, Add, and Delete Data of an Access Database

Instead of using the DataGridView control and the Toolstrip, we can use other Windows controls and code to add a row of data, delete a row of data, update a value, and then update a database with changes as shown in Figure 13.19. Figure 13.19a shows a Windows form with the InsurancePlan table dragged and dropped from Healthcare_Access_DatabaseDataset, along with the Toolstrip. On the left side below the InsurancePlan table on the Windows form, GUI controls including labels, textboxes for PlanID, PlanName, Description, CompanyID, MemberPremium, and EmployerPremium, and the button for Enter the New Plan, are added. Figure 13.19b gives the Visual Basic code to be executed for adding a new data record to the InsurancePlan table in Healthcare_Access_DatabaseDataSet and in the healthcare Access database, when the user enters the values for PlanID, PlanName, Description, CompnanyID, MemberPremium, and EmployerPremium in the textboxes and clicks the button for Enter the New Plan. The comments in the code explain the purpose of each code block. In the middle part below the InsurancePlan table on the Windows form, GUI controls, including the labels, three textboxes for PlanID, MemberPremium, and EmployerPremium and the button for Update Premiums, are added. Figure 13.19c gives the Visual Basic code to retrieve a data record with the value of PlanID entered by the user and update the value of MemberPremium and the value of EmployerPremium, when the user enters the parameter inputs and click on the button for Update Premiums. On the right side below the InsurancePlan table on the Windows form, GUI controls, including the labels, a textbox and a button for Delete the Plan, are added. Figure 13.19d gives the Visual Basic code to delete the data record with the value of PlanID entered by the user, when the user enters the parameter input and clicks on the button for Delete the Plan.

Example 13.7 shows how to add the login function to the Healthcare Information System application.

Example 13.7

Add the login function to the Healthcare Information System application as follows. When the user selects the radio button for Updates and clicks on the Next Button on the Windows form, FormMain, the user is taken to the Windows form, FormLogin, which allows the user to enter a username and a password. The application checks to see if the username and the password matches any pair of username and password kept by the application. If there is a match, the user is taken to FormUpdates, which allows the user to access the list of the tables in the healthcare Access database for updating data of the tables. If there is no match, a message is displayed to tell the user that the access is denied.

We first use Access to add a new table, Login, to the healthcare Access database. Figure 13.20a shows a data record with a username and a password stored in the Login table. We then open the Healthcare Information System application in Visual Studio. Since we add the Login table in the healthcare Access database, we need to update the healthcare Access database embedded in the Healthcare Information System application. To update the healthcare Access database in the Healthcare Information System application, we use the Solution Explorer window to delete the existing healthcare Access database and the existing Hearlthcare_Access_DatabaseDataSet, and then use the Data Sources window to add a new data source using the updated healthcare

(a)

Insurance Plans

	PlanID	PlanName	Description	CompanyID	MemberPremium	EmployerPremium
*						

Enter information of a new plan:

PlanID:

PlanName:

Description:

CompanyID:

MemberPremium:

EmployerPremium:

Enter the New Plan

Update member premium and employer premium of a plan:

PlanID:

MemberPremium:

EmployerPremium:

Update Premiums

Delete a plan:

PlanID:

Delete the Plan

Back

(b)

```
Private Sub ButtonEnter_Click(sender As Object, e As EventArgs) Handles ButtonEnter.Click
    Dim NewPlanRow As Healthcare_Access_DatabaseDataSet.InsurancePlanRow
    'The following code takes values from the textboxes
    NewPlanRow = Healthcare_Access_DatabaseDataSet.InsurancePlan.NewInsurancePlanRow
    NewPlanRow.PlanID = TextBoxPlanID.Text
    NewPlanRow.PlanName = TextBoxPlanName.Text
    NewPlanRow.Description = TextBoxDescription.Text
    NewPlanRow.CompanyID = TextBoxCompanyID.Text
    NewPlanRow.MemberPremium = CDbl(TextBoxMemberPremium.Text)
    NewPlanRow.EmployerPremium = CDbl(TextBoxEmployerPremium.Text)
    Healthcare_Access_DatabaseDataSet.InsurancePlan.Rows.Add(NewPlanRow)
    'The following code is same as the code for the Save Data button in the Toolstrip
    Me.Validate()
    Me.InsurancePlanBindingSource.EndEdit()
    Me.TableAdapterManager.UpdateAll(Me.Healthcare_Access_DatabaseDataSet)
    'The following code clears the textboxes
    TextBoxPlanID.Text = ""
    TextBoxPlanName.Text = ""
    TextBoxDescription.Text = ""
    TextBoxCompanyID.Text = ""
    TextBoxMemberPremium.Text = ""
    TextBoxEmployerPremium.Text = ""
End Sub
```

FIGURE 13.19
Design View of FormU3 in the Healthcare Information System application and the code for the three buttons for updating the InsurancePlan table in the Healthcare Access database. (a) Design View of FormU3. (b) Code for the Enter the New Plan button.

(c)

```
Private Sub ButtonUpdate_Click(sender As Object, e As EventArgs) Handles ButtonUpdate.Click
    Dim PlanRow As Healthcare_Access_DatabaseDataSet.InsurancePlanRow
    'The following code finds the row with the PlanID value entered by the user in the textbox and takes premium values
    PlanRow = Healthcare_Access_DatabaseDataSet.InsurancePlan.FindByPlanID(TextBoxPlanIDUpdate.Text)
    PlanRow.MemberPremium = TextBoxMemberPremiumUpdate.Text
    PlanRow.EmployerPremium = TextBoxEmployerPremiumUpdate.Text
    'The following code is same as the code for the Save Data button in the Toolstrip
    Me.Validate()
    Me.InsurancePlanBindingSource.EndEdit()
    Me.TableAdapterManager.UpdateAll(Me.Healthcare_Access_DatabaseDataSet)
    'The following code clears the textboxes
    TextBoxPlanIDUpdate.Text = ""
    TextBoxMemberPremiumUpdate.Text = ""
    TextBoxEmployerPremiumUpdate.Text = ""
End Sub
```

(d)

```
Private Sub Button2_Click(sender As Object, e As EventArgs) Handles Button2.Click
    Dim PlanRow As Healthcare_Access_DatabaseDataSet.InsurancePlanRow
    'The following code finds the row with the PlanID value entered by the user in the textbox and delete the row
    PlanRow = Healthcare_Access_DatabaseDataSet.InsurancePlan.FindByPlanID(TextBoxPlanIDDelete.Text)
    PlanRow.Delete()
    'The following code is same as the code for the Save Data button in the Toolstrip
    Me.Validate()
    Me.InsurancePlanBindingSource.EndEdit()
    Me.TableAdapterManager.UpdateAll(Me.Healthcare_Access_DatabaseDataSet)
    'The following code clears the textbox
    TextBoxPlanIDDelete.Text = ""
End Sub
```

FIGURE 13.19 (Continued)
Design View of FormU3 in the Healthcare Information System application and the code for the three buttons for updating the InsurancePlan table in the Healthcare Access database. (c) Code for the Update Premiums button. (d) Code for the Delete the Plan button.

Access database. Next, we add the Windows form, FormLogin, to the Healthcare Information System application. We add GUI controls, including the labels, two textboxes for username and password, the button for Login, and the button for Back as shown in Figure 13.20b. We also drag and drop the Login table from Healthcare_Access__DatabaseDataSet to FormLogin. This adds five objects to FormLogin: HealthcareAccessDatabaseDataSet, LoginBindingSource, LoginTableAdapter, TableAdapterManager, and LoginBindingNavigator (see Figure 13.20b). Although we need LoginTableAdapter in the Visual Basic code for the Login button, we do not really need the Login table and the Toolstrip on FormLogin. Hence, we select the Login table on the Windows form and press the Delete button on the keyboard to delete the Login form. We right-click on LoginBindingNavigator and select Delete in the pop-up menu to delete LoginBindingNavigator and the Toolstrip. We now double-click the Login button in the Design view of FormLogin to add the Visual Basic code (see Figure 13.20c) to check the match of the entered username and password with any pair of username and password stored in the Login table of the healthcare Access database. We double-click the Back button in the Design view of FormLogin to add the code for this button (see Figure 13.20c). Note that in the code for the Login button, the underline, __, is a line breaker to break one line of code into code written in two lines. At last, we update the Visual Basic code (see Figure 13.20d) for FormMain so that when the user selects the radio button for Updates and clicks on the Next button, the user is taken to FormLogin instead of FormUpdates.

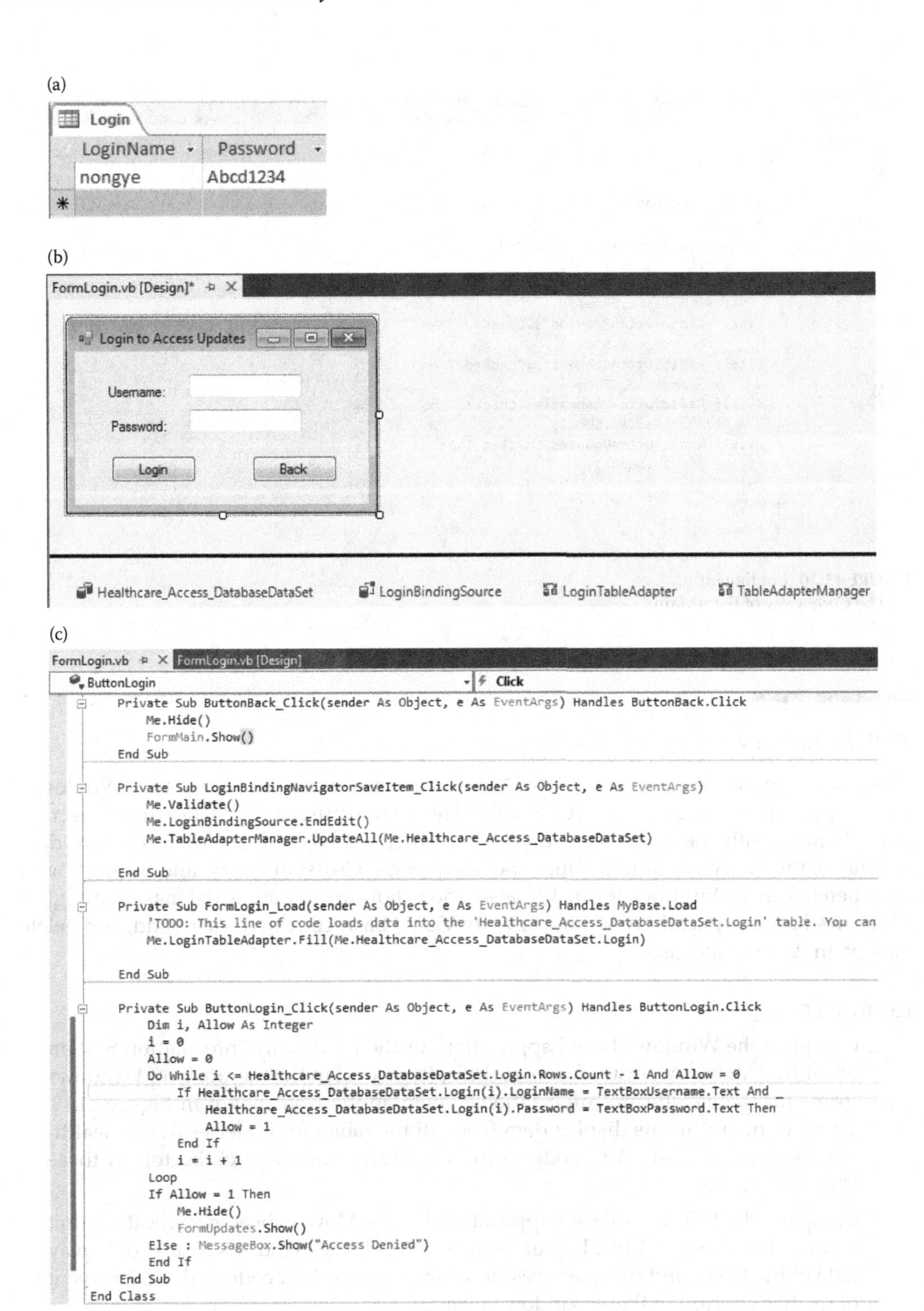

FIGURE 13.20
(a) The Login table added to the healthcare Access database and a data record stored in the Login table. (b) The Windows form, FormLogin, added to the Healthcare Information System application. (c) The Code view of FormLogin.

(d)

```
FormMain.vb  FormMain.vb [Design]
ButtonNext                                   Click
Public Class FormMain

    Private Sub ButtonNext_Click(sender As Object, e As EventArgs) Handles ButtonNext.Click
        Me.Hide()
        If RadioButtonEmployers.Checked Then
            FormEmployers.Show()
        ElseIf RadioButtonPatients.Checked Then
            FormPatients.Show()
        ElseIf RadioButtonProviders.Checked Then
            FormProviders.Show()
        ElseIf RadioButtonCompanies.Checked Then
            FormCompanies.Show()
        ElseIf RadioButtonPharmacies.Checked Then
            FormPharmacies.Show()
        ElseIf RadioButtonUpdates.Checked Then
            FormLogin.Show()
        End If
    End Sub
End Class
```

FIGURE 13.20 (Continued)
(d) The Code view of FormMain.

13.9 Summary

This chapter describes how to add an Access database as a data source in a Windows-based application using Visual Studio 2013. The architecture of the database connection is explained. With the established database connection, we show how to display data of a table, the query result of a simple Select query, a Crosstab query, and a query with parameter(s) in a Windows form. We also show how to update a database through a Windows-based application and how to write Visual Basic code to retrieve, add, and delete data of an Access database.

EXERCISES

1. Complete the Windows-based application for the Healthcare Information System by adding all the Windows forms at the bottom level in the hierarchy of Windows forms for the Healthcare Information System application shown in Figure 11.13. These Windows forms display data from all the tables and queries in the healthcare Access database. Add code to the Click event of the Back button in these Windows forms.
2. Complete the Windows-based application for the Movie Theater application that is started in Exercise 1 in Chapter 11 by adding all the Windows forms to display data of the tables and the query results of the queries. Add code to the Click event of the Back button in these Windows forms.

14

Working with VBA in Excel

VBA, which stands for Visual Basic for Applications, is a programming language developed by Microsoft Corporation. VBA is embedded in several Microsoft Office products, including Word, Excel, PowerPoint, and Visio to augment their capabilities. Other products, such as the Arena system simulation software, also provide an interface to handle VBA to further enhance their functionalities. While VBA is implemented in many applications, this chapter will primarily focus on Microsoft Excel 2013 to illustrate VBA fundamentals. After familiarizing with VBA's object models in Excel, this chapter will cover how to work with VBA, which includes Visual Basic Editor (VBE), an editor environment for VBA development; Windows Forms and Controls in VBA; various VBA procedures; and some VBA programming essentials. Lastly, the use of VBA functions and the integration of Excel's built-in worksheet functions into VBA are discussed.

14.1 Introduction to VBA

Previous chapters of the book (Chapters 11 and 12) introduced both the VB.NET and the VB programming languages. Although VBA and VB share a lot in common for syntax, VB is used to develop stand-alone applications, whereas VBA is more suited to be embedded in an existing application (e.g., Excel) and interfaces with the application. Taking Excel as an example, one notable advantage of VBA is that it can automate some highly repetitive tasks in Excel. It can also enhance the application by executing custom-built functions that are not handled in Excel. This benefit, however, does not come with ease. This chapter will cover how to write programs in VBA to fully leverage its benefits. One notable disadvantage of VBA though is that it is a moving target. As Microsoft continuously upgrades Office products, any VBA program may need modifications to be compatible with a newer version of the application.

Before we delve into VBA details, let us first create a simple macro in Excel.

Example 14.1

For the Excel spreadsheet shown in the following, we want to create a macro to copy the drug information for "Tylenol" into a separate spreadsheet within the same workbook. We can accomplish this task without a macro; however, it is tedious and the procedure is composed of repetitive tasks. In this example, we will demonstrate the use of a Macro to automate the task.

Recording a Macro

Step 1: Open the workbook entitled Chapter14_Macro.xlsx, as shown in Figure 14.1.

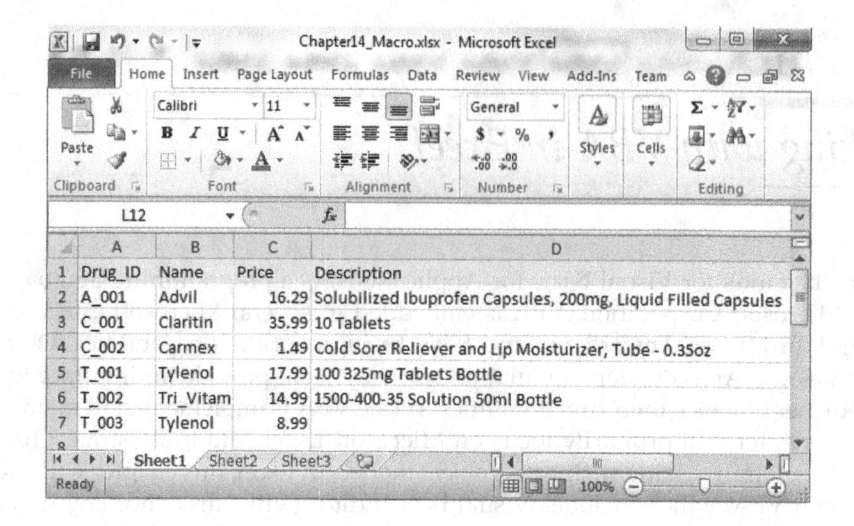

FIGURE 14.1
Example Excel Spreadsheet.

Step 2: Click View → Macro → Record Macro. The following screen is then shown. Please specify the name for the macro (e.g., GetTylenolInfo), provide the required shortcut key (e.g., Ctrl + Shift + T), add a description if needed, and click OK (Figure 14.2).

Step 3: After clicking OK, Excel now records any steps made by the user. Now, it is ready to record the manual steps for the macro. Within the current worksheet ("Sheet1"), first select the range of cells named "Name." Then, click Data → Filter. Next, click on the Filter symbol at the cell "Name," uncheck "(Select All)," check "Tylenol," then click OK. The screenshot of the filter is shown in Figure 14.3.

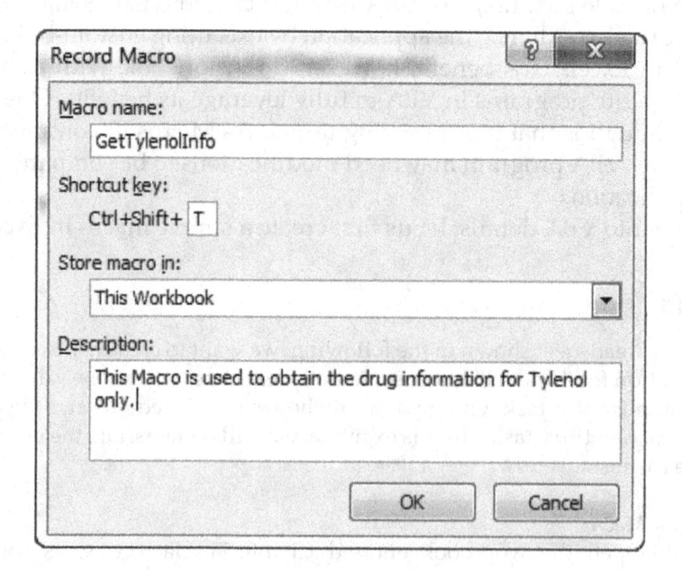

FIGURE 14.2
Start recording a macro.

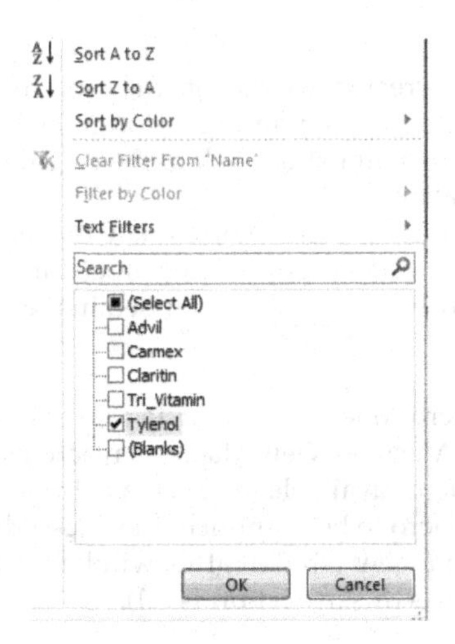

FIGURE 14.3
Use Filter in Excel.

Step 4: After clicking OK, the spreadsheet now contains only the information related to "Tylenol." Copy the rows containing the "Tylenol" information using the Ctrl + C command.

Step 5: Add a new spreadsheet using the Insert New Spreadsheet option, and paste the information as seen in Figure 14.4 using the Ctrl + V command starting from cell A1.

Step 6: Click View → Macro → Stop Recording. Since this is a macro-enabled Excel, please save the Excel workbook with extension .xlsm (Excel macro-enabled workbook).

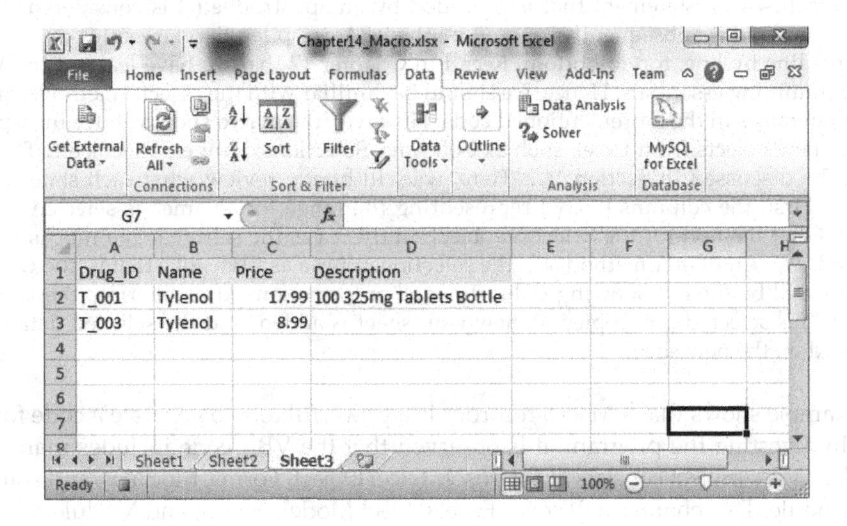

FIGURE 14.4
Paste filtered information on a new spreadsheet.

Testing the Macro

To test the macro we just created, we execute the procedure as follows:

Step 1: Click the spreadsheet containing the original drug information. Data → Unclick Filter. This will get us back to the original spreadsheet listing all the drug information.

Step 2: Press Ctrl + Shift + T. This shortcut key is used to trigger the macro to automate the Filter–Select–Copy–Paste procedure. There are other ways to trigger the macro, which will be discussed in the following sections.

Examining the Macro

To take a look at the macro code just created, use the following steps:

Step 1: Click View → Macro → View Macro. A macro window will then pop up, listing all the macros available in Excel in a drop-down list.

Step 2: Choose the macro to be examined and click Edit Macro. The following code is shown in a new window (this window is called "Code window," and we will discuss this in Section 14.3.1).

```
Sub GetTylenolInfo()
'GetTylenolInfo Macro
'This Macro is used to obtain the drug information for Tylenol only.
'
'Keyboard Shortcut: Ctrl + Shift + T
'
   Columns("B:B").Select
   Selection.AutoFilter
   ActiveSheet.Range("$B$1:$B$16").AutoFilter Field:=1,
   Criteria1:="Tylenol"
   Range("A1:D7").Select
   Selection.Copy
   Sheets.Add After:=Sheets(Sheets.Count)
   ActiveSheet.Paste
   Columns("D:D").EntireColumn.AutoFit
End Sub
```

The GetTylenolInfo( ) macro (also known as a "sub-procedure") consists of several statements. Any statement that is preceded by an apostrophe (') is considered as a comment and is subsequently ignored by the VBA compiler. Excel executes the statements line by line, top to bottom. Recall in Chapter 12 that we have learned the VB programming essentials. Hence, we should be familiar with the objects, methods, and dot operators in the aforementioned code. However, the macro shown above incorporates new objects from Excel, such as Columns, Selections, Ranges, Sheets, etc. These will be discussed in Section 14.2. Here, we will briefly review what each statement does. First, the columns ("B:B") representing the range for "Name" is selected. The AutoFilter method for the Selection object, in this case, the selected columns, is triggered. The AutoFilter method sets the selection criteria as "Tylenol," that is, the worksheet will be filtered showing only "Tylenol" information. After filtering, the range ("A:D") is selected and copied. A new worksheet is added, and the selected data are pasted on the new sheet.

This example shows that a Macro recorder is a powerful tool to generate a code for a VBA project. In digesting the program, it is observed that the VBA code includes many objects. Generally, a VBA project consists of Microsoft Excel Object, Forms, Modules, and sometimes, a Class Module. This chapter will cover Excel Object Model, Forms, and Modules. Advanced readers may refer to VB programming resources for creating and using Class Modules.

14.2 Excel Object Models

All Excel-based VBA projects start with an Excel Application as an object and are composed of other objects. These objects, in turn, contain more sub-objects. Simply put, the Excel Object Model involves an object hierarchy that can be presented as a tree structure, with the root of the tree being the Excel Application object. A basic object hierarchy in an Excel Application is shown in Figure 14.5.

Figure 14.5 comes across an important concept—collections. By definition, a collection is a group of objects of the same type and is, itself, an object. In an Excel Application object, some commonly used collections are Workbooks, which is a collection of all currently opened Workbook objects; Worksheets, which is a collection of all Worksheet objects contained in a particular Workbook object and usually contains items such as range and formulas; Sheets, which is a collection of all sheets contained in a particular Workbook object, consists not only of worksheets, but also of other types of sheets, such as Chart sheet; and Charts, which is a collection of all Chart objects contained in a particular Workbook object, which takes up an entire worksheet but not charts that are inserted as part of a worksheet.

Following the tree structure, the Excel Object Model hierarchy can be explained as follows: the Excel Application object contains multiple Workbooks collection objects, and within each Workbook, there may be more than one Workbook object. Within each Workbook object there are Worksheets collection objects, Sheets collection objects, and Charts collection objects. Worksheets may contain multiple Worksheet objects, whereas Charts may contain multiple Chart objects. As for Sheets collection, it may contain Worksheet and/or Chart objects. Each Worksheet object contains multiple Range objects.

The developers can work with an entire collection of objects. However, more often, the developer works with a specific object within the collection. In this case, referring to an object is important. The following examples illustrate how to reference an object in an Excel Application.

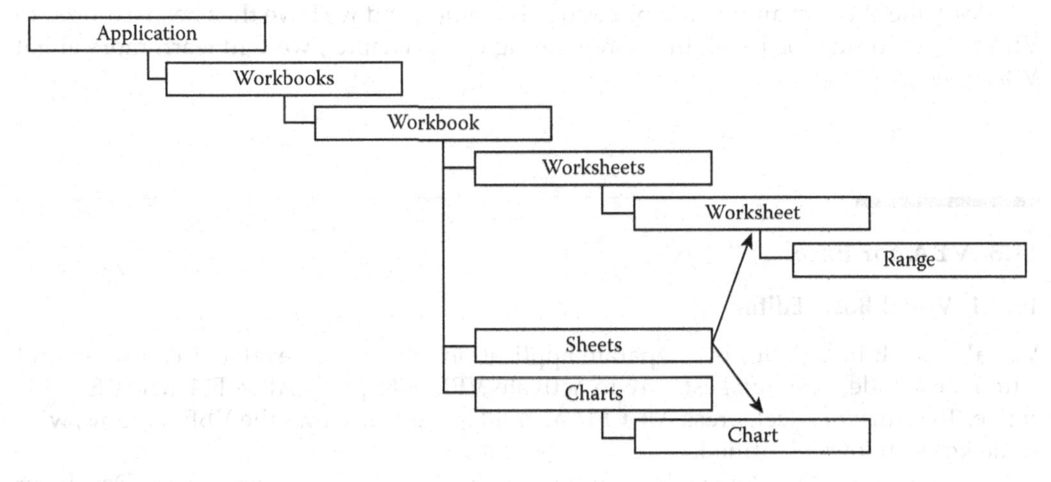

FIGURE 14.5
Basic Excel Object Model hierarchy.

Example 14.2

Get to the Workbook object named "Chapter14.xlsm" and use the Application object:

```
Application.Workbooks("Chapter14.xlsm")
```

Example 14.3

Get the Worksheet object named "Sheet1" under the workbook named "Chapter14.xlsm":

```
Application.Workbooks("Chapter14.xlsm").Worksheets("Sheet1")
```

or

```
Application.Workbooks("Chapter14.xlsm").Worksheets(1)
```

if Sheet1 is the first worksheet in the collection.

Example 14.4

Get to the value of cell "A1" on the Worksheet object named "Sheet1" from the workbook named "Chapter14.xlsm":

```
Application.Workbooks("Chapter14.xlsm").Worksheets("Sheet1").
Range("A1").value
```

Clearly, dot operators are extensively used here. Take a look at Example 14.4, the Application code is referring to an application object. The Workbooks code, on the other hand, is a Workbooks collection object belonging to the Application object. "Chapter14.xlsm" is used to identify the specific Workbook object within the Workbooks collection. For the workbook named "Chapter14.xlsm," the Worksheets object is a collection object belonging to it, and "Sheet1" is used to identify the specific Worksheet object within the Worksheets collection. This Worksheet object contains a Range object. The value property of the Range object ("A1") is obtained.

Indeed, the above examples are pieces of VBA codes, and we have thus been exposed to VBA programming for Excel. In the remaining of the chapter, we will learn more about VBA basics.

14.3 VBA for Excel

14.3.1 Visual Basic Editor

Visual Basic Editor (VBE) is a separate application where the developer can write and edit a VBA code. The quickest way to activate VBE is to press Alt + F11 when Excel is active. To return to Excel, press Alt + F11 again. Figure 14.6 shows the VBE window, with some key windows identified.

Alternatively, the Developer tab can be used to navigate to VBE. However, the Developer tab is hidden by default. To include it first in the ribbon, use the following steps: File → Options → Customize Ribbon → select the Developer tab.

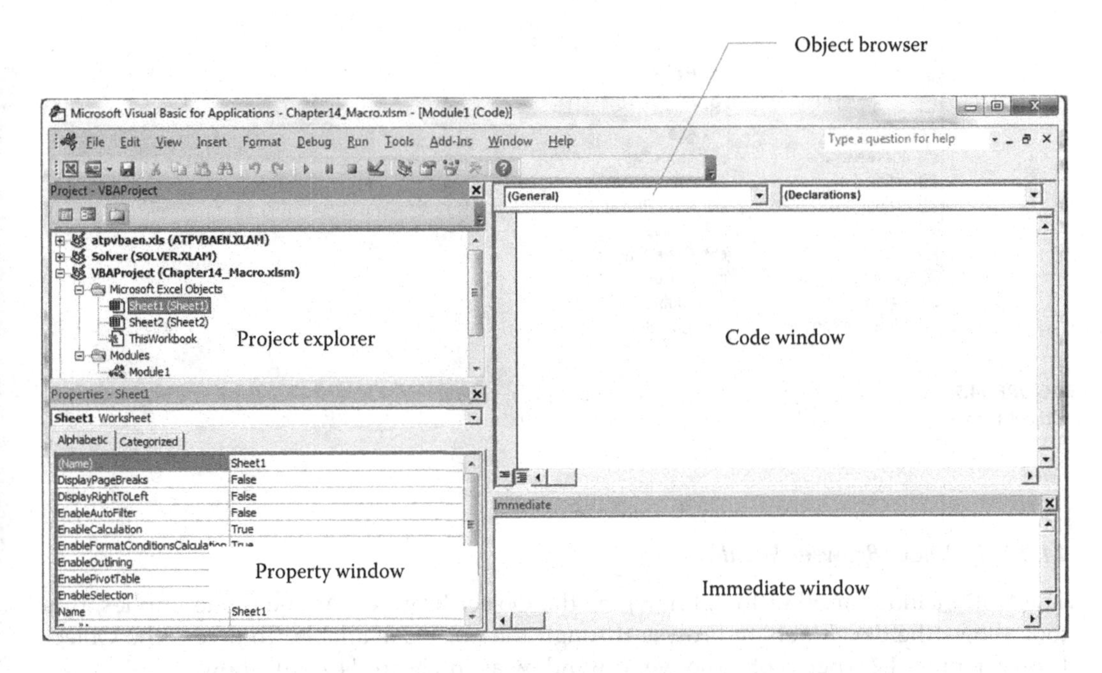

FIGURE 14.6
VBE window.

14.3.1.1 Project Explorer Window

The Project Explorer window displays a tree diagram that consists of every workbook currently open in Excel (including add-ins and hidden workbooks). For each workbook, the Microsoft Excel objects are displayed. These include Sheets, Modules, Class Modules, and UserForms.

14.3.1.2 Code Window

A Code window (sometimes known as a "Module window") displays the VBA code of a particular project. Every object in a project has an associated Code window. The macro recorded in Example 14.1 is stored in a module, and the VBA code can be directly modified in the code window.

14.3.1.3 Property Window

The Property window has the same functions as that in VB.NET. It represents the associated properties of the selected object. Each control, such as Form, Label, Command Button, etc., has its own unique set of properties. When we click over different objects, a different list appears in the property window in response to the selected control.

14.3.1.4 Immediate Window

The Immediate window is most useful for executing VBA statements directly and for debugging the code.

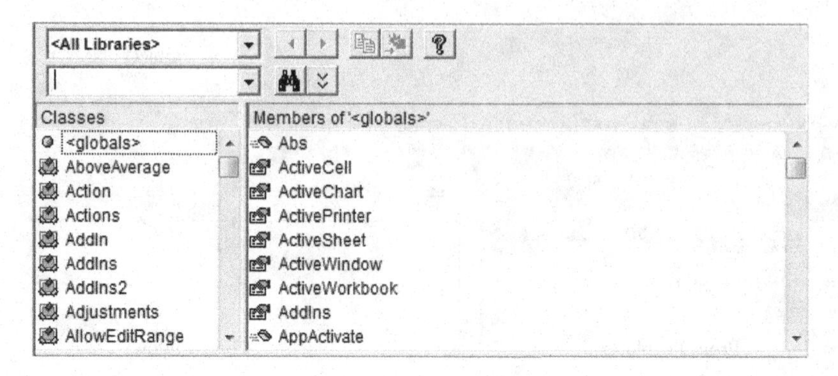

FIGURE 14.7
Object Browser.

14.3.1.5 Object Browser Window

The VBE includes another tool, known as the Object Browser. As the name implies, this tool allows the developers to browse through the objects available. To access the Object Browser, press F2 when VBE is active. A window as in Figure 14.7 will show.

The drop-down list at the top left contains a list of all currently available object libraries. To browse through VBA objects, select the VBA Project option from the drop-down list.

For all the windows discussed above, if they are not shown when VBE is launched, please choose View and activate the windows as needed.

14.3.2 Graphical User Interface: Forms and Controls

Recall that the main objects used to help a user interact with the computer are called "Windows controls." Chapter 11 covers various Windows forms and controls. The Windows forms and controls of VBA are similar to those in VB.NET, with some variations. For example, there are two ways to access controls with Excel.

14.3.2.1 Controls in Excel

In Excel, we can add or position controls in a document. In this case, the graphical user interface is built upon the controls on the Excel spreadsheet. This VBA project may not contain any Forms object. To use various controls with Excel, first click the Developer tab on the Ribbon, then click Insert in the Control section (see Figure 14.8).

This would display the list of controls available in Microsoft Excel. The controls appear in two sections: Form Controls and ActiveX Controls. If we are working on a spreadsheet in Excel, it is recommended to use the controls in the ActiveX Controls section. In other cases, Form controls may be used. Every control in the Form Controls or the ActiveX Controls section has a specific name. The ones shown in Figure 14.8 are the most widely used and are summarized in Table 14.1.

Other than the ActiveX controls listed above, one can always click the More Controls button to add more controls available in the computer system.

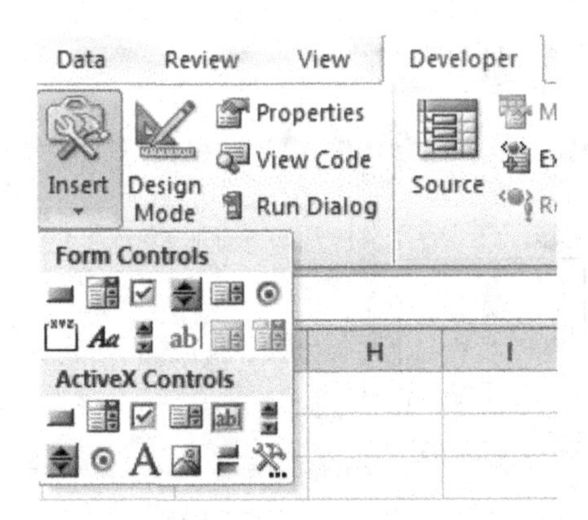

FIGURE 14.8
Adding controls to Excel documents.

TABLE 14.1
Controls with Excel or VBA

Form controls	Name	ActiveX controls	Name
	Command button		Command button
	Combo box		Combo box
	Check box		Check box
	Spin button		Spin button
	List box		List box
	Option button		Option button
	Scroll bar		Scroll bar
	Frame		Text box
Aa	Label		Label
			Image
			Toggle button
			More Controls button

14.3.2.2 Controls with Forms

An alternative way to work with a control is with Windows forms, in which case the VBA project will then have the Forms object. If the VBE is activated, we can add or position controls on the UserForm by right clicking Microsoft Excel Objects → Insert → UserForm.

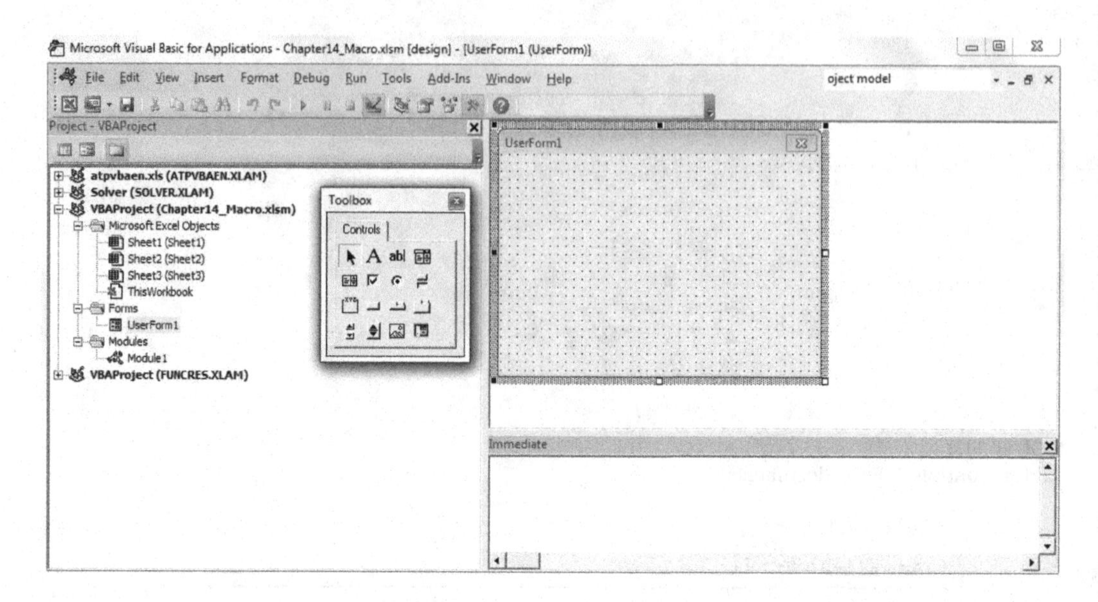

FIGURE 14.9
Adding controls to Windows forms.

A UserForm window, as shown in Figure 14.9, along with a toolbox equipped with various controls will appear. The controls in the toolbox are the same as those in the form control section discussed previously. Since this approach is very similar to what is covered in VB.NET programming, please use Chapter 11 as a guideline to work with Windows Form and Controls. These include: (1) set the properties of the controls using the Property window, (2) change the properties of the controls in running time, and (3) use the control's event procedure. Please remember that the VBA code for the forms and controls will reside in the Forms object.

14.3.3 Modules: Sub and Function Procedure

In a VBA project, the Module object is a very important object. Modules are mainly used to store procedures that can be accessed either from within that module or from anywhere in the project. The VBA code usually resides in the procedure and there are two most common types of procedures: Subs and Functions.

A Sub procedure is a group of VBA statements that performs an action within Excel. Most of the macros in VBA are Sub procedures. A Sub procedure can be considered as a command, that is, execute the Sub procedure and something happens. Every Sub procedure starts with the keyword Sub and ends with an End Sub statement.

Example 14.5

Here is an example of a Sub procedure:

```
Sub ShowMessage()
   MsgBox("Hi there")
End Sub
```

This example shows a procedure named ShowMessage. A set of parentheses follows the procedure's name. In most cases, these parentheses are empty. However, we may pass arguments to Sub procedures from other procedures. In this case, the arguments will be listed between the parentheses.

A Function procedure is a group of VBA statements that takes in one or more arguments, performs a calculation, and returns a single value. Excel includes many worksheet functions, for example, SUM and VLOOKUP. These pre-defined functions can be used in formulas. The same goes for the Function procedures written by the developers using VBA. Every Function procedure starts with the keyword Function and ends with an End Function statement.

Example 14.6

Here is an example of a Function procedure:

```
Function Perimeter(radius)
   Perimeter = 2 * radius * WorksheetFunction.Pi()
End Function
```

This function, named "Perimeter," takes one argument named "radius," which is enclosed in parentheses. Functions can have any number of arguments or none at all. When the function is executed, it returns a single value, Perimeter, which is the perimeter of the circle for a given radius. In this example, a worksheet function Pi is called, and it returns the value of Pi (3.14) to calculate the perimeter.

There are two ways to execute a Function procedure: execute it from another procedure (a Sub or another Function procedure) or use it in a worksheet formula. Unlike the Sub procedure, the Excel Macro recorder cannot be used to record a Function procedure. The function procedure has to be coded by the developers manually.

14.3.4 Modules: Variables

Other than a Sub procedure and a Function procedure, modules are also used to store variables, constants, and declarations to be accessed from anywhere within the project. VBA's main purpose is to manipulate data. Some data, such as worksheet ranges, reside in objects. Other data are stored in variables. A variable is simply a named storage location in the computer's memory. We assign a value to a variable using the equals sign operator ("="). A variable's value may change over the course of the procedure. Sometimes, we may need to refer to a value or string that never changes, we use constant. The variables and constants in a VBA project are used in a similar manner. Hence, we will only focus on the discussion of variables in this chapter.

Naming variables in VBA is rather flexible, although there are a few rules: (1) the name can use letters, numbers, and some punctuation characters; however, the first character must be a letter; (2) no space or period is allowed in a variable's name; (3) no special characters such as #, $, %, &, and ! are allowed in the name; (4) VBA-reserved words, such as Sqr, Sub, With, For, etc., are not allowed to be used as variable names.

VBA has a variety of built-in data types (e.g., Integer, Date) that are the same as those in VB.NET. When a variable is declared without explicitly being assigned to a data type, the default data type "variant" is used. Data stored as a variant will change its type depending on the code, and VBA will automatically handle the conversion; however,

TABLE 14.2
Variable's Scope

Scope	How the Variable Is Declared
Procedure only	By using a Dim or a Static statement in the procedure that uses the variable
Module only	By using a Dim statement before the first Sub or Function statement in the module
All procedures in all modules	By using a Public statement before the first Sub or Function statement in a module

remember that this comes with the cost of speed and memory. To force the developers to declare all the variables, the following statement can be added at the beginning of the VBA Module:

```
Option Explicit
```

Keep in mind that the Option Explicit statement only applies to the module in which it resides. If there is more than one VBA module in a project, such statement needs to be added for each module. To ensure that the Option Explicit statement is inserted automatically whenever a new VBA module is created, the Require Variable Definition option can be turned on by choosing Tools → Options.

Recall that a workbook can have any number of VBA modules, and a VBA module can have any number of Sub and Function procedures. A variable's scope determines which modules and procedures can use the variables, and this information is summarized in Table 14.2.

14.3.4.1 Procedure-Only Variables

Variables declared with this scope can be used only in the procedure in which they are declared. When the procedure ends, the variable no longer exists, and Excel frees up its memory. If we execute the procedure again, the variable is re-initiated, but its previous value is lost. The most common way to declare a procedure-only variable is with a Dim statement placed between a Sub statement and an End Sub statement (or between a Function and an End Function statement). The Dim keyword is short for "dimension," which simply means that we are setting aside memory for a particular variable. We usually place Dim statements immediately after the Sub or Function statement and before the procedure's code. The following example shows some procedure-only variables declared by using Dim statements:

```
Sub MySub( )
       Dim x As Integer, y As Integer
       Dim DOB As Date
       Dim Name As String
       Dim Price As Long
       Dim Description
       ' ... [The procedure's code goes here.] ...
End Sub
```

Notice that the last Dim statement in the example does not declare a data type; it declares only the variable itself. The effect is that the variable Description is a "variant." Also, please note that VBA does not allow us to declare a group of variables to be a particular data type by separating the variables with commas. For example, although valid, the following statement does not declare all the variables as integer:

```
Dim x, y As Integer
```

Instead, only y is declared to be an integer; x is declared as variant.

If a variable is declared with procedure-only scope, other procedures in the same module can use the same variable name, but each instance of the variable is unique to its own procedure. In general, variables declared at the procedure level are the most efficient variables because VBA frees up the memory that they use when the procedure ends.

14.3.4.2 Module-Only Variables

To make a variable available to all procedures in a module, the variable needs to be declared before the module's first Sub or Function statement. This is done in the Declaration section, where the Option Explicit statement is located. With this declaration in place, the variable has a module-only scope and, thus, can be used in any procedures within the module, and the variable retains its value from one procedure to another.

14.3.4.3 Public Variables

To make a variable available not only within the module, but also to all procedures in the VBA project, the variable needs to be declared at the same location as the module-level variable, but with the Public keyword. Here is an example:

```
Public Price As Long
```

The Public keyword makes the Price variable available to any procedure in the workbook. Often, for a large VBA project, a better way to manage public scope variables is to pull all public declarations in a separate module.

14.3.4.4 Static Variables

Often, when a procedure ends, all of the variables are reset. Static variables are a special case because they retain their values even when the procedure ends (the project is still executing). A static variable is declared at the procedure level using the Static keyword. A static variable may be useful, for example, if we want to track the number of times that we execute a procedure, a static variable can be used here, and it increases by one each time the procedure is executed. Since static variables are procedure level, their values are not available to any other procedures in the same module and in other modules in the workbook.

14.4 Programming with VBA and Excel

14.4.1 Excel: Range Object

In a typical VBA Excel project, much of the programming focuses on the Range object.

14.4.1.1 Range

A Range object represents a range contained in a Worksheet object. It can be as small as a single cell or all cells in a worksheet (A1:IV 65536, or 16,777,216 cells). We can refer to a Range object with a code as follows:

```
Range("A:D")—Columns A to D
```

or if the range has a variable name (created by selecting Insert → NameBox, and entering the name of the range), we use

```
Range("Drug_List")
```

Unless Excel is informed otherwise, it assumes that the Range object is from the active worksheet. If anything other than a worksheet is active (such as a chart sheet), the range reference fails, and an error message will show. Certainly, a range outside the active sheet can be referenced by qualifying the Range object with a worksheet name and a workbook name. For example:

```
Worksheets("Sheet1").Range("A1:C5")
```

This statement refers to the range on "Sheet1" for the active workbook. Alternatively,

```
Workbooks("Healthcaresystem.xlsm").Worksheets("Sheet1").Range("A1:C5")
```

can be used to refer to a range from a different workbook.

14.4.1.2 Cells

Rather than using the VBA Range keyword, a range can be referenced via the Cells property. It may be confusing, but note that Cells is a property that VBA evaluates instead of an object. To use the Cells property appropriately, two arguments need to be provided: rows and columns. For example, to refer to cell C2 on "Sheet1," the expression is as follows:

```
Worksheets("Sheet1").Cells(2, 3)
```

The Cells property can also be used to refer to a multi-cell range such as:

```
Range(Cells(1, 1), Cells(10, 10))
```

This expression refers to a 100-cell range that extends from cell A1 (row 1, column 1) to cell J10 (row 10, column 10). It is equivalent to

```
Range("A1:J10")
```

The advantage of using the Cells property to refer to ranges is that we can use actual numbers as the cell's argument, which may make looping an easier task.

14.4.1.3 Offset

The Offset property provides another handy means for referring to ranges. When we record a macro to handle multiple cells, often, the Offset property is used. This property, which operates on a Range object and returns another Range object, lets us refer to a cell, which is a particular number of rows and columns away from the current Range cell. Like the Cells property, the Offset property takes two arguments. The first argument represents the number of rows to offset, while the second argument represents the number of columns to offset. The following expression refers to a cell one row below cell A1 and two columns to the right of cell A1. In other words, this refers to cell C2. This example is illustrated as follows:

```
Range("A1").Offset(1, 2)
```

The Offset property can also use negative arguments. A negative row offset refers to a row above the range. A negative column offset refers to a column to the left of the range.

14.4.1.4 Some Useful Properties and Methods for the Range Object

The Range object has dozens of operable properties. Some Range properties are read-only properties, which means that we cannot change their current values. Some Range properties are both readable and writable. Table 14.3 summarizes some of the more commonly used Range properties. For complete details, consult the Help system in VBE.

In addition, some commonly used Range object methods are listed in Table 14.4.

14.4.2 VBA Functions and Excel Functions

To develop a powerful application, we need to integrate VBA with Excel seamlessly. By seamlessly, we mean to utilize the data stored in the Excel worksheet and the variables declared and used in the module. Seamless integration also means to take advantage of

TABLE 14.3
Some Useful Range Object Properties

Property	Description	Example
Value	Represents the value contained in a cell. It is a read–write property.	Worksheets("Sheet1"). Range("A1").Value
Text	Returns a string representing the text as displayed in a cell. It is a read-only property.	Worksheets("Sheet1"). Range("A1").Text
Count	Returns the number of cells in a Range object. It is a read-only property.	Range("A1:C3").Count
Column	Returns the column number of a single cell range. It is a ready-only property.	Sheets("Sheet1").Range("A1"). Column
Row	Returns the row number of a single cell range. It is a read-only property.	Sheets("Sheet1").Range("A1").Row
Address	Displays the cell address for a Range object in an absolute notation (e.g., $A$1). It is a ready-only property.	Range(Cells(1, 1), Cells(5, 5)). Address
HasFormula	Returns True if all cells in the Range object contains a formula, False if all cells in the Range object contains no formula, and Null if mixture.	Range("A1:A2").HasFormula
Formula	Represents the formula in a cell. It is a read–write property.	Range("A1").Formula = "=SUM(A2:A12)"
NumberFormat	Represents the number format of the Range object. It is a read–write property.	Columns("A:A"). NumberFormat="0.00%"

TABLE 14.4
Some Useful Range Object Methods

Method	Description	Example
Select	Selects a range of cells. Note that we need to ensure that the worksheet is activated.	Sheets("Sheet1").Activate Range("A1:C12").Select
Copy	This method is applicable to the Range object.	Sub CopyRange() Range("A1:A12").Select Selection.Copy Range("C1").Select ActiveSheet.Paste End Sub
Paste	This method is applicable to the Worksheet object.	
Clear	Deletes the contents of a range and all the cell formatting.	Columns("D:D").Clear
Delete	Deletes a range; Excel will shift the remaining cells around to fill up the range that we deleted.	Rows("1:1").Delete

existing Excel built-in functions, VBA built-in functions, and the functions that we write in VBA.

14.4.2.1 VBA Functions

VBA provides numerous built-in functions. Some of these functions take input arguments, while some do not. For example, the Date(') function returns the current date and, thus, requires no input argument, whereas the Len(myString) function returns the length of the string and requires a string argument (e.g., myString). The best source to learn the properties of a particular VBA function is the Excel Visual Basic Help system. To query the details of a particular function, type the function name into a VBA Module, move the cursor anywhere in the text, and press F1.

What deserves more discussion here is that, among the VBA built-in functions, a few go beyond simply performing the tasks and returning a single value. If used appropriately, these functions can greatly enhance the functionalities of the application. A few examples of these built-in functions are discussed as follows:

- MsgBox: The MsgBox command displays a dialog box containing a message and a button. For example,

```
MsgBox("Hi, there")
```

This will return the following window:

- InputBox: The InputBox command displays a simple dialog box that asks the user for some input. The function returns whatever the user enters into the dialog box. For example,

  ```
  Dim UserName As String

  UserName = InputBox("Please enter your UserName.")
  ```

  When we execute the VBA, the following window will pop up. Let us type "Joe Smith" in the input box, and the variable UserName will be assigned with the value being "Joe Smith."

  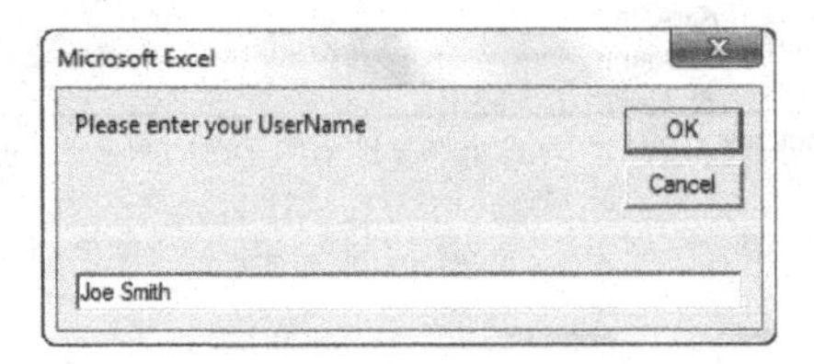

- Shell: The Shell command executes another program. The function returns the task ID (a unique identifier) of the other program (or an error if the function cannot start the other program). The following is an example:

  ```
  Dim ApplicationID As String
  ApplicationID = Shell("C:\Program Files (x86)\Internet Explorer\
  iexplore.exe")
  MsgBox(ApplicationID)
  ```

  When we execute the program, the Shell function will launch Internet Explorer with the path stated as "C:\Program Files (x86)\Internet Explorer\iexplore.exe," and the ApplicationID will be returned.

14.4.2.2 Excel Functions

Although VBA offers a collection of built-in functions, we take note of the important concept that the various Excel's worksheet functions can be incorporated into the particular VBA project. VBA makes Excel's worksheet functions available through the WorksheetFunction object, which is contained in the Application object. Therefore, any statement that uses a worksheet function must use the Application and/or the WorksheetFunction qualifier.

For example, the following three expressions will work exactly as running Excel's native functions. To calculate the sum of Range("A2:A12") and write it back to Cells(1, 1) (the range of A1):

```
Cells(1, 1) = Application.WorksheetFunction.Sum(Range("A2:A12"))
Cells(1, 1) = WorksheetFunction.Sum(Range("A2:A12"))
Cells(1, 1) = Application.Sum(Range("A2:A12"))
```

The first statement states the Application.WorksheetFunction, the second statement uses the WorksheetFunction qualifier, and the third statement omits the Worksheet but uses the Application qualifier.

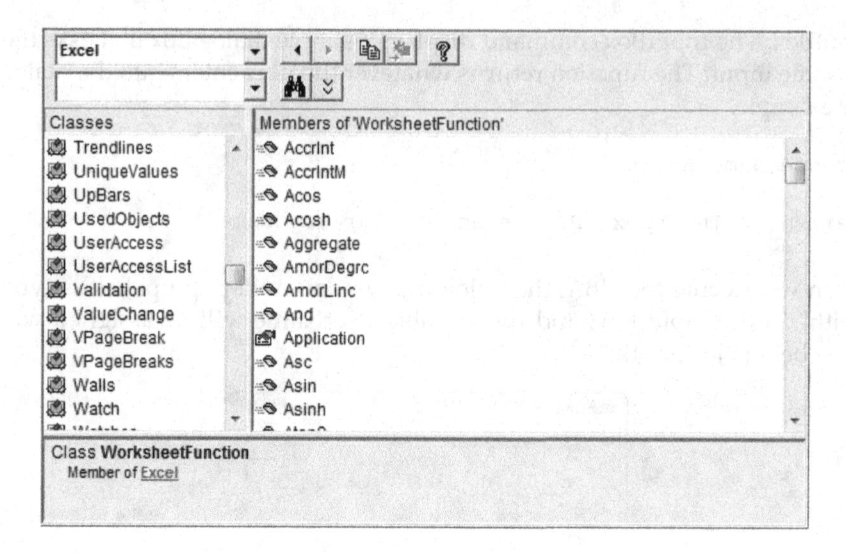

FIGURE 14.10
Comprehensive list of WorksheetFunction.

The WorksheetFunction object contains the worksheet functions available to the VBA procedures. To see the list of these functions, please use the following steps to display a complete list of worksheet functions available in VBA:

Step 1: In VBE, press F2. The Object Browser appears.

Step 2: In the Project/Library drop-down list, select Excel.

Step 3: In the list labeled "Classes," select WorksheetFunction.

The members of the list show all the worksheet functions that can be used in a VBA project (Figure 14.10).

In a VBA project, we may get confused with VBA's built-in functions and Excel's functions. A good practical rule is to first check whether VBA has the functions to meet the need. If not, we then check out Excel's functions. If all fail, we may have to write our own function by using VBA.

14.4.3 Some Important Events in Excel

Similar to VB programming, where the event-handling procedure is an important procedure for Windows forms and controls, there are also several important events in Excel that are very helpful for running a VBA project. These events reside in the Code window of an Excel Object Module (not the standard VBA Module).

A list of some useful workbook-related events is provided in Table 14.5, and a list of some worksheet-related events is provided in Table 14.6.

One example of using a workbook-related event is as follows:

When the workbook is opened, some initializations (e.g., initialize variables, set the constant) can be done under this event procedure:

TABLE 14.5
Partial List of Workbook-Related Events

Workbook Event	Description
Activate	The workbook is activated.
Open	The workbook is opened.
WindowActivate	The workbook window is activated.
BeforeClose	The workbook is closed.
BeforeSave	The workbook is saved.
Deactivate	The workbook is deactivated.

TABLE 14.6
Partial List of Worksheet-Related Events

Worksheet Event	Description
Activate	The worksheet is activated.
BeforeDoubleClick	A cell in the worksheet is double-clicked.
BeforeRightClick	A cell in the worksheet is right-clicked.
Calculate	The worksheet is recalculated.
Change	A change is made to a cell in the worksheet.
Deactivate	The worksheet is deactivated.

```
Private Sub Workbook_Open()
      ' Our code in initialization goes here.
End Sub
```

A good example of using a worksheet-related event is as follows:

Let us assume that the user is asked to input data to the Excel spreadsheet in Cells(1, 1) and that the data entered must be greater than 100. A piece of VBA codes can be written under the event of Worksheet_Change( ).

```
Private Sub Worksheet_SelectionChange()
      If Cells(1, 1) < 100 Then
          MsgBox("Sorry. The number is less than 100!")
      End If
End Sub
```

By now, we have touched upon some basics of VBA. Next, we will work on a mini project using VBA.

14.5 Mini Project

14.5.1 Introduction

An industrial engineering student would want to query a subset of the top Institute of Science Index (ISI)-indexed industrial engineering (IE) journals. Consider the data from the "Chapter14MiniProjectRawData.xlsx" Excel data sheet. The data consist of the name of the journal; its publisher; its corresponding IE field specialization; and its 2012 ISI impact factor, which is used to measure a journal's relative importance. A simple Excel VBA-enabled spreadsheet could be used to handle this scenario.

14.5.2 Spreadsheet Specifications

The specifications of the VBA enabled spreadsheet are as follows:

- Upon opening the spreadsheet, an overview page as in Figure 14.11 is shown to the user.
- A "command button" control with the caption "Execute Search Query" would act as a starting point to query the journals. Upon clicking the button, the following UserForm, as presented in Figure 14.12, is shown to the user.
    - The publisher option acts like a filter to return only journals that matched the selection of the user. Additionally, the publisher option is a "combo box"

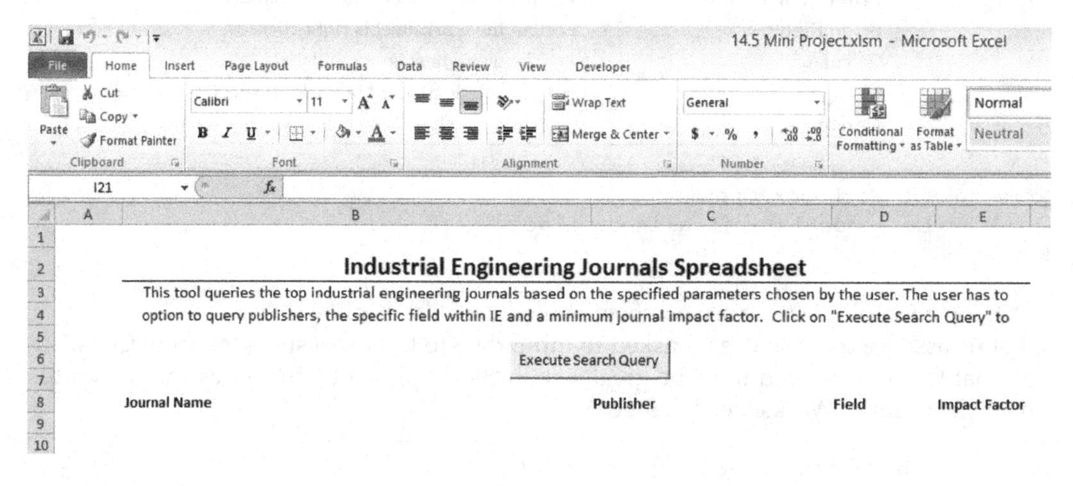

FIGURE 14.11
Sample Welcome screen of the VBA spreadsheet.

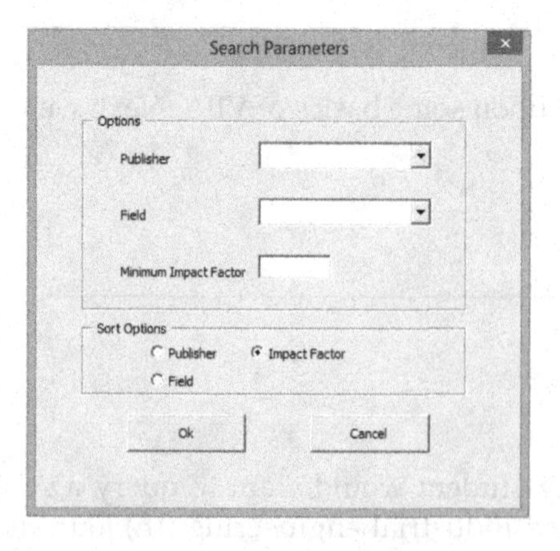

FIGURE 14.12
Sample Search Parameters form.

control where the drop-down list is the list of unique publishers queried from the data.

- The filed option is similar to the publisher option, which acts like a filter to return only journals that matched the selection of the user. Furthermore, the field option is a "combo box" control where the drop-down list is the list of unique fields queried from the data.
- The minimum impact factor is a text box that accepts only numeric values. This serves as a filter that queries journals that meet the minimum requirement impact factor provided by the user.
- The Sort Options panel is an "option button" control that determines how the data are to be sorted. By default, the data are sorted by descending impact factor. Additionally, the user may choose to sort by publisher or by field alphabetically.

14.6 Summary

This chapter gives a nutshell on developing a VBA project in Excel. First, a Macro recorder tool is demonstrated on how to generate a VBA code. By digesting the VBA code, the use of object and its related properties and methods is noted. This chapter then briefly overviews some objects in an Excel VBA project. Since we have extensively studied Windows forms objects and control objects in the earlier chapters, this chapter mainly focuses on Excel objects. Specifically, the Range object is discussed in detail. The module concepts, including the sub procedure and the function procedure, are then explained. Given the uniqueness of VBA in Excel, some important VBA functions, Excel functions, and Excel-related event procedures are reviewed, which, together, could help develop a strong application.

EXERCISES

1. The Record Macro feature of Microsoft Excel can be accessed using the Insert → Macro → Record Macro navigation option. True or false?
2. A shortcut key is a required step in recording a macro and must be unique across macros. True or false?
3. Which of the following Excel file types allows the execution permission of macros by default?
   a. *.xls
   b. *.xlsx
   c. *.xlxm
   d. xlsm.*
   e. None of the above
4. Which of the following ways are valid procedures in inspecting a recorded macro?
   a. View → Macros → View Macro → Edit Macro
   b. Developer → Macros → View Macro → Edit Macro
   c. Insert → Macros → View Macro → Edit Macro

   d. All of the above are valid ways to inspect a macro.
   e. Only a and b are valid ways to inspect a macro.
5. A workbook is composed of the following objects except:
   a. Worksheets
   b. Workbooks
   c. Sheets
   d. Charts
   e. All of the above are valid objects within a workbook.
6. The following are examples of valid Excel collections except:
   a. Charts
   b. Sheets
   c. Worksheets
   d. Workbooks
   e. All of the above are valid collections.
7. Write a VBA code to refer to the value of cell "A10" of the worksheet named "DataSheet" within the Excel file "Medicines.xlsm" using the appropriate VBA object hierarchy.
8. Write a VBA code to refer to the values of a 10 × 1 array starting from the "A1" cell. However, the name of the worksheet is unknown, but on the other hand, we know that it is the eighth sheet within the "Medicines.xlsm" Excel file.
9. The __________ window displays a tree diagram for every workbook in Excel.
   a. Project Explorer
   b. Code
   c. Property
   d. Immediate
   e. None of the above
10. The Object Browser has the following properties:
    a. Shows all VBA classes, objects, and their corresponding members
    b. Can be accessed using the F3 key
    c. Can also be accessed using the View → Object Browser from the Excel menu
    d. All of the above are true.
    e. None of the above is true.
11. VBA controls are accessed through the Excel Developer tab. True or false?
12. User-defined forms can be added to an Excel spreadsheet using the following except:
    a. VBE: Insert → UserForm
    b. Right-click on the VBA project within the Project Browser: Insert → UserForm
    c. Excel: Insert → UserForm
    d. All of the above are true.
    e. Only a and b are true.

13. The two common types of subroutines in VBA are Procedures and Functions. True or false?
14. A function can have multiple input arguments or could have none at all. True or false?
15. VBA variables have the following properties except:
    a. An equals sign operator is used to assign a value to a variable.
    b. The first character to name a variable should be a letter.
    c. Reserved words cannot be named as a variable.
    d. The default type of a variable is the variant type.
    e. A variable's range determines which modules can see the variable.
16. A procedure variable exists only within the procedure even if the procedure ends. True or false?
17. Using the code: "Dim string1,string2 As String," "string2" is declared as a string while "string1" is declared as a:
    a. String
    b. Variant
    c. Dim
    d. Integer
    e. Error, and will not compile
18. The primary advantage of the Cell property over the Range property is that it can be used within a loop. True or false?
19. Using the Cell property, write a code to refer to the values of a 10 × 2 array starting from the "A1" cell of the "Drugs" worksheet within the "Medicines.xlms" Excel file.
20. Using the Offset property, write a code to refer to the value of a cell two rows to the left and one cell below from cell "D2" of the "Drugs" worksheet within the "Medicines.xlms" Excel file.
21. From the selected values from Review Exercise 19, write the VBA code to select and copy the values.
22. From Review Exercise 21, paste the copied values into a new sheet called "DrugsCopy" starting from cell "A1."
23. Built-in Excel functions are accessed using the Application.WorksheetFunction command. True or false?
24. The following are valid events in Excel except:
    a. Workbook_Activate()
    b. UserForm_Initialize()
    c. Button_Click()
    d. All of the above are valid events.
    e. None of the above is a valid event.

15

Database Connectivity with VBA

After learning the fundamentals of Visual Basic for Applications (VBA) in Chapter 14, this chapter will cover how to use ADO.NET in Excel VBA to connect to external data sources. These include databases such as a Microsoft Access database or a MySQL database. In addition, connecting to Web sources such as uniform resource locator (URL) and extensible markup language (XML) documents is also briefly discussed.

15.1 ActiveX Data Object

ActiveX Data Object (ADO) is an object model that enables the developers to access data stored in a variety of data sources. Commonly, the data source could be a database, but it could also be a text file, an Excel spreadsheet, or an XML file. Since different data sources use different protocols, there is no single set of classes that allows the developers to accomplish this task universally. ADO.NET is a set of object-oriented libraries that support the communication of Excel VBA to the right data source using the right protocols. In ADO.NET, the libraries are called "data providers" and are usually named after the protocol or data source type that they allow the developers to interact with. Table 15.1 lists the data providers that are included in the ADO.NET framework.

15.2 ADO.NET Objects in Data Provider Libraries

ADO.NET includes many objects that can be used to work with data. The five core objects are Connection, Command, DataReader, Dataset, and DataAdapter. In VBA, the Connection, Command, and Recordset (instead of Dataset) objects are commonly used. We will briefly discuss the objects and illustrate their uses through some examples.

15.2.1 Connection Object

The Connection object is used to create an open connection to a database via which the developers can access and manipulate the records within the database. The connection object needs to specify the database server, database name, username, password, and other parameters that are required for initiating the connection.

TABLE 15.1

ADO.NET Data Providers

Provider Name	Data Source Description
OLEDB data provider	For data sources exposed to an OLEDB interface, that is, Access
ODBC data provider	For data sources exposed by using ODBC
SQL data provider	For interacting with the Microsoft SQL Server
Oracle data provider	For Oracle databases. The .NET Framework Data Provider for Oracle supports Oracle client software version 8.1.7 and later.

Example 15.1

Create a connection object to connect to a MySQL database named "healthcaresystem." The syntax is as follows:

```
cntMyConnection.ConnectionString="DRIVER={MySQL ODBC 5.2 Driver};
Server=localhost; Database=healthcaresystem;Uid=root; pwd=12345678;
OPTION=3;"
```

Since we want to connect to MySQL database, parameters such as the database driver, server name, database name, user ID, and password need to be specified and assigned to the ConnectionString property of the object named "cntMyConnection."

Example 15.2

Create a connection object to connect to the Access database named "healthcaresystem." The syntax is as follows:

```
cntMyConnection.ConnectionString="Provider=Microsoft.Jet.OLEDB.4.0; Data
Source=c:/healthcaresystem.accdb"
```

When connecting to an Access database, only the database provider (Microsoft Jet Engine) and the data source (note that there is a space between the two) that specifies the location of the database are needed to define the ConnectionString property of cntMyConnection.

The connection object is then used by command objects to ensure that the commands are applied to the right database.

15.2.2 Command Object

The command object is used to execute a single query on a database. To execute a set of commands that work on a group of data, the DataAdapter object is used. The query can perform actions like creating, adding, retrieving, deleting, or updating records. The data queried will be returned as a Recordset object, that is, the retrieved data can be manipulated by the properties, collections, methods, and events of the Recordset object.

Example 15.3

Create a command object based on the cntMyConnection, that is, the connection object created in Example 15.1. The syntax is

```
mysqlcommand.CommandText="Select * from drugs"
mysqlcommand.ActiveConnection=cntMyConnection
mysqlcommand.Execute
```

15.2.3 Recordset Object

The Recordset object is used to hold a group of records from a database. It consists of a set of records and columns. When Recordset is first opened, the current record pointer will point to the first record, and the Beginning of File (BOF) property is true and the End of File (EOF) property is false. If there are no records, the BOF and the EOF are true.

Example 15.4

An example of using the Recordset object to retrieve information from the Drugs table based on the connection created previously is as follows:

```
src="SELECT * FROM Drugs"
rstFirstRecordset.Open Source:=src, ActiveConnection:=cntMyConnection
```

In this example, the rstFirstRecordset is the Recordset object. The query command is assigned to a variable src. The ActiveConnection parameter is set to cntMyConnection (Example 15.1).

With the basic understanding of ADO.NET objects used in VBA, some examples are used to demonstrate the applications in detail.

15.3 Illustrative Example 1: Connecting to MySQL

Let us use the healthcaresystem MySQL database created in Chapter 9 to demonstrate how to connect VBA to an external database using ADO.NET.

Before starting, we need to consider the following:

1. Be sure that the MySQL database is installed on the computer and runs appropriately.
2. Download the database driver for MySQL from http://www.mysql.com. As discussed in Chapter 9, there is a wide variety of connectors available. In this example, we need to download the open database connectivity (ODBC) connector, which is the standardized database connection driver. Do make sure that the correct driver is downloaded (e.g., Windows vs. Mac, 32 bit vs. 64 bit). After installing the driver, a better way to check the database driver is to go to Control Panel → Administrative Tools → Data Source (ODBC). As shown in Figure 15.1, the system has the MySQL ODBC 5.2a Driver installed for the MySQL database. This information will then be used in the VBA project.
3. Activate Excel Visual Basic Editor, choose Tools → References → Microsoft ADOs 2.8 Library (see Figure 15.2). There may be a number of ADO object libraries available. We may choose any versions of the libraries that would provide the needed database connection capability.

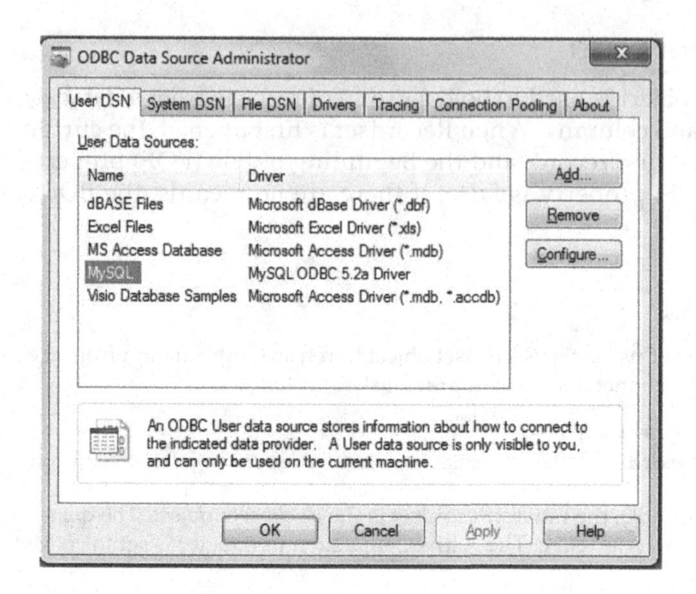

FIGURE 15.1
ODBC Data Source Administrator.

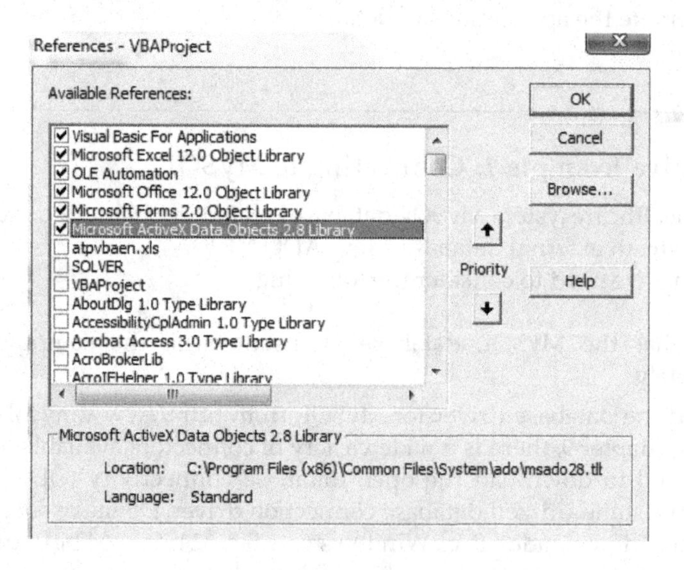

FIGURE 15.2
Adding ADO Library Reference to a VBA project.

Now, we will add a module to the VBA project. In the code window, we type the following:

```
1. Sub querydatafromMySQL()
2.        Dim cntMyConnection As ADODB.Connection
3.        Set cntMyConnection=New ADODB.Connection
4.        cntMyConnection.ConnectionString="DRIVER={MySQL ODBC 5.2a
          Driver};Server=localhost;Database=healthcaresystem;Uid=
          root; pwd=12345678; OPTION=3;"
```

```
5.          cntMyConnection.Open
6.          Dim rstFirstRecordset As ADODB.Recordset
7.          Set rstFirstRecordset=New ADODB.Recordset
8.          Dim src as String
9.          src="SELECT * FROM Drugs"
10.         rstFirstRecordset.Open Source:=src, ActiveConnection:=
              cntMyConnection
11.         Worksheets("DatabaseConnection").Activate
12.            Dim column as Integer, row as Integer
13.            With ActiveSheet
14.                For column=1 to rstFirstRecordset.Fields.Count
15.                    .cells(1, column)=rstFirstRecordset(column -1).
                         name
16.                Next column
17.                row=2
18.                While Not rstFirstRecordset.EOF
19.                    For column=1 to rstFirstRecordset.Fields.Count
20.                        .Cells(row, column)=rstFirstRecordset
                                (column -1).Value
21.                    Next column
22.                        rstFirstRecordset.MoveNext
23.                        row=row + 1
24.                    Wend
25.                End With
26.                Set rstFirstRecordset=Nothing
27.                cntMyConnection.Close
28.                Set cntMyConnection=Nothing
29. End Sub
```

This example retrieves data from the Drugs table in a MySQL database named healthcaresystem. This Drugs table has four columns: Drugs_ID, Name, Price, and Description. In this example, we use both Connection and Recordset objects. Let us go through the codes in detail.

Lines 1, 29: This is a Sub Procedure with Sub–End Sub.

Lines 2 to 3: Declare an object of Connection class named cntMyConnection. Next, a new instance of the object is created.

Line 4: Specifies the details of the database connection string, including the database driver, in this example, MySQL ODBC 5.2a (this information is gathered from Data Source [ODBC]) is used; the name of the server, localhost, and connection details including the username and the password are also included. We also include OPTION=3 in the MySQL connection string. This option setting allows us to direct the MySQL server to behave in a specific manner for the duration of each connection. The "3" setting corresponds to the field length, which implies that there is no need to optimize the column width, and for the row, simply return the matching rows. We can find more information about this setting at the MySQL Reference Manual.

If a remote database is to be connected, the syntax is as follows:

```
DRIVER={MySQL ODBC 5.2 Driver};Server=myserverAddress;Database=myDatabase;
Uid=myUsername; pwd=myPassword; OPTION=3;"
```

Furthermore, if the transmission control protocol (TCP)/internet protocol (IP) port needs to be specified when connecting to a remote database, the syntax is as follows:

```
DRIVER={MySQL ODBC 5.2 Driver};Server=myserverAddress;Port=3306;
Database=myDatabase; Uid=myUsername; pwd=myPassword; OPTION=3;"
```

Line 5: The Open method of the cntMyConnection object is called. This opens the connection to the database.

Lines 6 to 7: Similar to lines 2 to 3, a new object from the Recordset class named "rstFirstRecordset" is first declared then instantiated.

Lines 8 to 9: Declare src as a string data type. An SQL statement is then assigned to src.

Line 10: The Open method of the rstFirstRecord object is triggered. There are two parameters: the SQL statement, which is src (line 9), and the active connection to the database, cntMyConnection, which is established in line 5.

Line 11: This is a statement for the Excel object. It makes the worksheet named "DatabaseConnection" active.

Line 12: Declare two integer variables: column representing the columns and row representing the rows. These two variables are used as increment in the following loop statements.

Lines 13, 25: For the active sheet, we use the With–End With Statement. The advantage of using this statement pair is that, when the Cells property is called, we do not need to spell out the name of the worksheet every time.

Lines 14 to 16: Use the For Loop here to get the field names from the "Drugs" table. The property rstFirstRecordset.Fields.Count is used to indicate the number of columns retrieved. The property rstFirstRecordset(column).name is used to indicate the name of the specific column. This information is written back to the Excel worksheet in the first row, from the first column to the rstFirstRecordset.Fields.Count column. In this example, we move from column 1 to column 4. Note that, in the Excel object, the index starts with 1, whereas in the ADO.NET object, the index starts with 0. As a result, to retrieve the name held by rstFirstRecordset, we will use (column –1) to list the name for columns 1 to 4.

Line 17: Assign the row to be 2, that is, from now on, the data retrieved from the Recordset will be populated to the spreadsheet from the second row onward.

Lines 18, 23, 24: Use the While–End Loop to retrieve the records one by one. The rstFirstRecordset.EOF property is used to check if all the records (rows) are retrieved from the Recordset. The row variable has an increment of one in each iteration.

Lines 19 to 21: Similar to lines 14 to 16, for the specific row, data are retrieved for each column and written onto the worksheet.

Line 22: The method MoveNext of the RecordSet object is called. This will move the cursor from the current row to the next row.

Lines 26 to 28: These statements are used to release the system resources. As we are done with the database transaction, we will set the Recordset and the connection object to Nothing to release the computer resource.

In this example, we demonstrate the query from the database. To update the database, for example, performing INSERT, DELETE, and UPDATE operations, the statement in line 9 needs to be changed in response to the appropriate SQL commands. Be cautious when using the INSERT statement—the data types need to be matched.

For example, a new drug information is added to the worksheets ("new_drugs") on Range ("A1:A4"). The command statement to add the Drug_ID information is defined as follows:

```
src="INSERT INTO Drugs (Drug_ID) Values (' & worksheets ("new_drugs").
Range("A1") & ')"
```

In this case, the Drug_ID is defined as a character data type. If Drug_ID is defined as a numeric data type, the corresponding statement is as follows:

```
src="INSERT INTO Drugs (Drug_ID) Values ( & worksheets("new_drugs").
Range("A1") & )"
```

The difference between the two statements is the use of the single quote pair ' '. In the first example, Drug_ID is a character data type, and it should be encoded with ' ' when being inserted into the table, whereas in the second example, the numeric data type does not need a single quote pair for the data variable.

15.4 Illustrative Example 2: Connecting to an Access Database

There are two ways to connect Excel and Access. The easiest approach is to go to Data → From Access. Following the procedure, we will be able to populate the data from the Access database. However, more often than that, a VBA project using ADO.NET can be developed to connect to an Access database programmatically and dynamically. In this case, the procedure is similar to that in MySQL, with a couple of issues deserving special attention. First, Microsoft is making it confusing to us in terms of using 32-bit versus 64-bit Windows operating system, and 32-bit versus 64-bit Office product. If we are working on an older version of MS Office, we need to modify the codes in lines 4 and 5 as follows:

```
ConnectionString="Provider=Microsoft.Jet.OLEDB.4.0; Data
Source=myAccessdatabase"
cntMyConnection.open ConnectionString
```

If we are working on a newer version of Office (e.g., Access 2013), the syntax will be as follows:

```
ConnectionString="Driver={Microsoft Access Driver (*.mdb, *.accdb)};
DBQ=myAccessdatabase;"
cntMyConnection.open ConnectionString
```

The main difference is that it is "Driver" instead of "Provider" following the ODBC connection syntax, and the database query (DBQ) parameter is used in replacing "Data Source." Thus, the correct Access Driver needs to be downloaded and registered as that in MySQL.

15.5 Illustrative Example 3: Connecting to XML

XML stands for extensible markup language, also known as a semi-structured database. It is becoming a standard platform for data exchange over the Internet; thus, many applications are providing the interface to XML, including Microsoft Office products. More details of XML will be discussed in Chapters 17 and 18. Here, we will learn how Excel interacts with an XML document.

Importing XML to a worksheet is a relatively simple procedure: choose Data → From Other Sources → From XML Import. Following the wizard, the data in XML can be mapped to the cells in a worksheet. On the other hand, exporting data from Excel cells to an XML document may need some VBA programming. Simply exporting an Excel worksheet to an XML document may generate a messy XML file that is difficult to read. The following example shows how to use ActiveX data objects database (ADODB) objects to query from an Excel worksheet and save the data into an XML document. Here, we need to use the document object model (DOM) object, which will be explained in Chapter 18. As shown in Figure 15.3, references to the Microsoft ADO Library as well as the Microsoft XML Library need to be added.

```
1. Sub ExporttoXML()
2.     Dim xmlMyConnection As ADODB.Connection
3.     Set xmlMyConnection=New ADODB.Connection
4.     Dim xmlMyRecordset As ADODB.Recordset
5.     Set xmlMyRecordset=New ADODB.Recordset
6.     Dim xmlMyXML As DOMDocument
7.     Set xmlMyXML=New DOMDocument
8.     Dim xmlMyWorkbook As String
9.     xmlMyWorkbook=Application.ThisWorkbook.FullName
10.    xmlMyConnection.Open "Provider=Microsoft.Jet.OLEDB.4.0;" & _
```

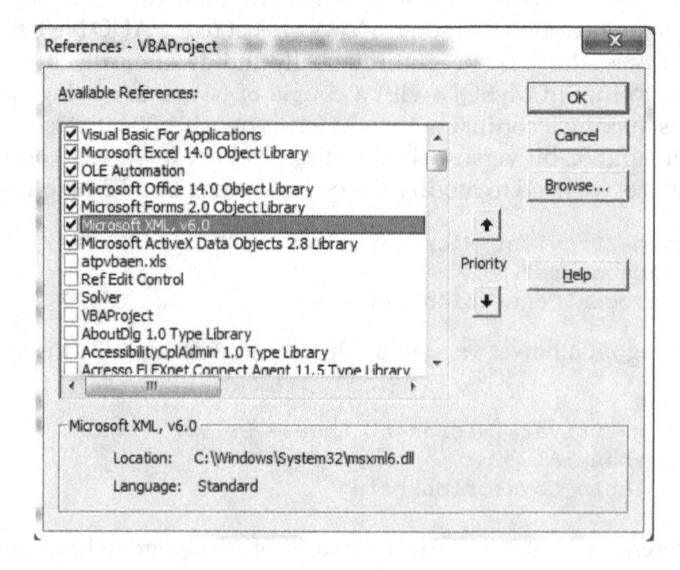

FIGURE 15.3
Adding references to a VBA project.

```
            "Data Source=" & xmlMyWorkbook & ";" & _
            "Extended Properties=excel 8.0;" & _
            "Persist Security Info=False"
11.     xmlMyRecordset.Open "Select * from [Sheet1$]",
        xmlMyConnection,            adOpenStatic
12.     If Not xmlMyRecordset.EOF Then
13.     xmlMyRecordset.MoveFirst
14.     xmlMyRecordset.Save xmlMyXML, adPersistXML
15.     End If
16.     xmlMyXML.Save (ThisWorkbook.Path & "\Output.xml")
17.     xmlMyRecordset.Close
18.     Set xmlMyConnection=Nothing
19.     Set xmlMyRecordset=Nothing
20.     Set xmlMyXML=Nothing
21. End Sub
```

Lines 1, 21: This is a Sub Procedure with Sub–End Sub.

Lines 2 to 5: Initiate two ADODB objects: the xmlMyConnection object and the xmlMyRecordset object.

Lines 6 and 7: Create an XML DOM Document object named xmlMyXML (refer to Chapter 18 for DOM concepts).

Lines 8 and 9: Get the name of the Excel workbook and assign it to a string variable xmlMyWorkbook.

Line 10: The open method for the connection object named xmlMyConnection is triggered. Provider and Data Source information about the source data (the Excel workbook) need to be specified.

Line 11: Execute the open method for xmlMyRecordset. Other than the connection parameter and the status of the connection, we specify the "query"–"Select * from [Sheet1$]", which will search for all the data contained in Sheet1 and assign the set of records to the Recordset.

Lines 12 to 15: Use the If–Then–End If loop to read the Recordset. The EOF property is used to check if the Recordset has records. If yes, the cursor is moved to the first record, and all the records are saved to the DOM object.

Line 16: Save the XML file as "output.xml" in the same directory as the Excel workbook.

Lines 17 to 20: Release all the system resources.

Let us use the Excel in Figure 15.4 as the example. The exported output.xml is shown as follows. Please refer to Chapter 18 for more details on the XML document.

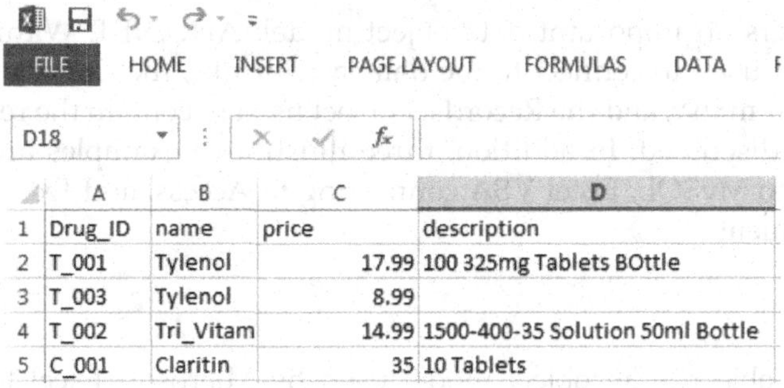

FILE HOME INSERT PAGE LAYOUT FORMULAS DATA

D18

	A	B	C	D
1	Drug_ID	name	price	description
2	T_001	Tylenol	17.99	100 325mg Tablets BOttle
3	T_003	Tylenol	8.99	
4	T_002	Tri_Vitam	14.99	1500-400-35 Solution 50ml Bottle
5	C_001	Claritin	35	10 Tablets

FIGURE 15.4
Exporting Excel to an XML example.

```
<xml xmlns:s="uuid:BDC6E3F0-6DA3-11d1-A2A3-00AA00C14882"
xmlns:dt="uuid:C2F41010-65B3-11d1-A29F-00AA00C14882"
xmlns:rs="urn:schemas-microsoft-com:rowset" xmlns:z="#RowsetSchema">
<s:Schema id="RowsetSchema">
 <s:ElementType name="row" content="eltOnly">
   <s:AttributeType name="Drug_ID" rs:number="1" rs:nullable="true"
rs:maydefer="true" rs:writeunknown="true">
       <s:datatype dt:type="string" dt:maxLength="255"/>
     </s:AttributeType>
    <s:AttributeType name="name" rs:number="2" rs:nullable="true"
rs:maydefer="true" rs:writeunknown="true">
       <s:datatype dt:type="string" dt:maxLength="255"/>
     </s:AttributeType>
    <s:AttributeType name="price" rs:number="3" rs:nullable="true"
rs:maydefer="true" rs:writeunknown="true">
       <s:datatype dt:type="float" dt:maxLength="8" rs:precision="15"
rs:fixedlength="true"/>
     </s:AttributeType>
    <s:AttributeType name="description" rs:number="4" rs:nullable="true"
rs:maydefer="true" rs:writeunknown="true">
       <s:datatype dt:type="string" dt:maxLength="255"/>
     </s:AttributeType>
     <s:extends type="rs:rowbase"/>
    </s:ElementType>
</s:Schema>
<rs:data>
     <z:row Drug_ID="T_001" name="Tylenol" price="17.989999999999998"
description="100 325mg Tablets Bottle"/>
     <z:row Drug_ID="T_003" name="Tylenol" price="8.9900000000000002"/>
     <z:row Drug_ID="T_002" name="Tri_Vitamin" price="14.99"
description="1500-400-35 Solution 50ml Bottle"/>
     <z:row Drug_ID="C_001" name="Claritin" price="35" description="10
Tablets"/>
</rs:data>
</xml>
```


15.6 Summary

This chapter covers an important data object model: ADO.NET. Within ADO.NET, the Connection object used to connect to the database (or file), the Command object used to hold the query command, and the Recordset object used to contain the retrieved data from the database are discussed. In addition, three illustrative examples are provided: Excel VBA connecting to MySQL, Excel VBA connecting to Access, and Excel VBA connecting to the XML document.

EXERCISES

1. An OLEDB object is an object model used by Microsoft Excel to connect to a Microsoft Access Database. True or false?
2. ADO.NET libraries are usually named by their data source type. True or false?

3. When using VBA, the following objects are used in a VBA database connection except:
   a. Dataset
   b. Connection
   c. Command
   d. Recordset
   e. All of the above are used in VBA
4. When using a connection object, the following are commonly included in the declaration except:
   a. Password user account accessing the database
   b. Username of the user accessing the database
   c. Database server trough an IP address or a local drive directory
   d. All of the above are true
   e. None of the above is true
5. A DataAdapter is an example of a command object. True or false?
6. In VBA, a Recordset supports the database disconnected mode. True or false?
7. A Recordset object is a two-dimensional object composed of records and column attributes. True or false?
8. Write an instruction to find the currently installed MySQL connection driver version within a Windows PC.
9. Write an instruction to activate the Microsoft ADOs Library in Excel 2010.
10. A loop is typically used to scroll through Recordset objects. True or false?
11. Given a list called "DrugRecordset" queried from a table consisting of 14 drug records, how many times will the subroutine PrintData be executed in the following code?

```
While DrugRecordset.BOF
        Call PrintData
        DrugRecordset.MoveNext
Wend
```

12. The RecordsetObject.EOF is used to check if the pointer is positioned at the start of a Recordset to start the retrieval of data. True or false?

Section IV

Web Application Development

Section IV of this book covers the development of Web applications and Web services. A Web application is a network application that runs using the server–client protocol for Web. It is becoming increasingly popular due to the ubiquity of the Internet, Web browsers, and Web servers. In Chapter 16, the application of ASP.NET in Microsoft Visual Studio 2013 to develop Web applications is covered. To learn Web services, an important backbone standard, extensible markup language (XML) is described. Chapters 17 and 18 discuss the fundamentals of XML, followed by discussions on developing Web services in Microsoft Visual Studio.

16

Web Applications in Microsoft Visual Studio

A Web application is a network application involving a hypertext transfer protocol (HTTP) server and HTTP clients. Internet Explorer is a common HTTP client software package for the Windows environment. Internet information server (IIS) is a common HTTP server software package for the Windows environment. IIS is integrated in Visual Studio 2013. An HTTP server is a software system that hosts and stores Web pages. An HTTP client is a software system that reads and displays Web pages. Hypertext markup language (HTML) is used to construct the appearance of a Web page, while there is program code such as Visual Basic code to provide the back-end functions of a Web page. An HTTP server and an HTTP client work together through request–response cycles. In each request–response cycle, the HTTP client sends a request for an HTML file using universal resource locator (URL) to locate the HTTP server and the HTML file held by the HTTP server, and the HTTP server receives the request and responds to send the requested file to the HTTP client.

In this chapter, we first show how to start building a Web application project in Visual Studio 2013. We then introduce the use of HTML and Visual Basic code to construct a Web page. We also show how to incorporate an Access database into a Web application.

16.1 Starting a Web Application in Microsoft Visual Studio

Example 16.1 shows how to start building a Web application for the Healthcare Information System–Web using Visual Studio 2013.

Example 16.1

Create a Web application, the Healthcare Information System–Web, including a Web form that is similar to the Windows form in the Healthcare Information System shown in Figure 11.9.

After starting Visual Studio 2013, we see Start Page. We close Start Page and select New Website in the File menu. In the New Website window (see Figure 16.1), we select Visual Basic for Templates, select ASP.NET Empty Website, and click the Browse button, which brings up the Choose Location window. In ASP.NET, ASP stands for active server pages. In the Choose Location window, we select File System on the left side of the window so that we can navigate folders in the file system and select a folder in which a project folder is created. For example, we can enter: C:\Users\nongye\Documents\My\yenong\Healthcare Information System–Web in which yenong is an existing folder and Healthcare Information System–Web is the name that we give to our new project folder held in the yenong folder. We then click the Open button in the Choose Location window, which brings up a message telling us that the folder, C:\Users\nongye\Documents\My\yenong\Healthcare Information System–Web, does not exist and asking us if we want to create this folder. We click the Yes button, which brings us back to the New Website window. We click the OK button, which brings up

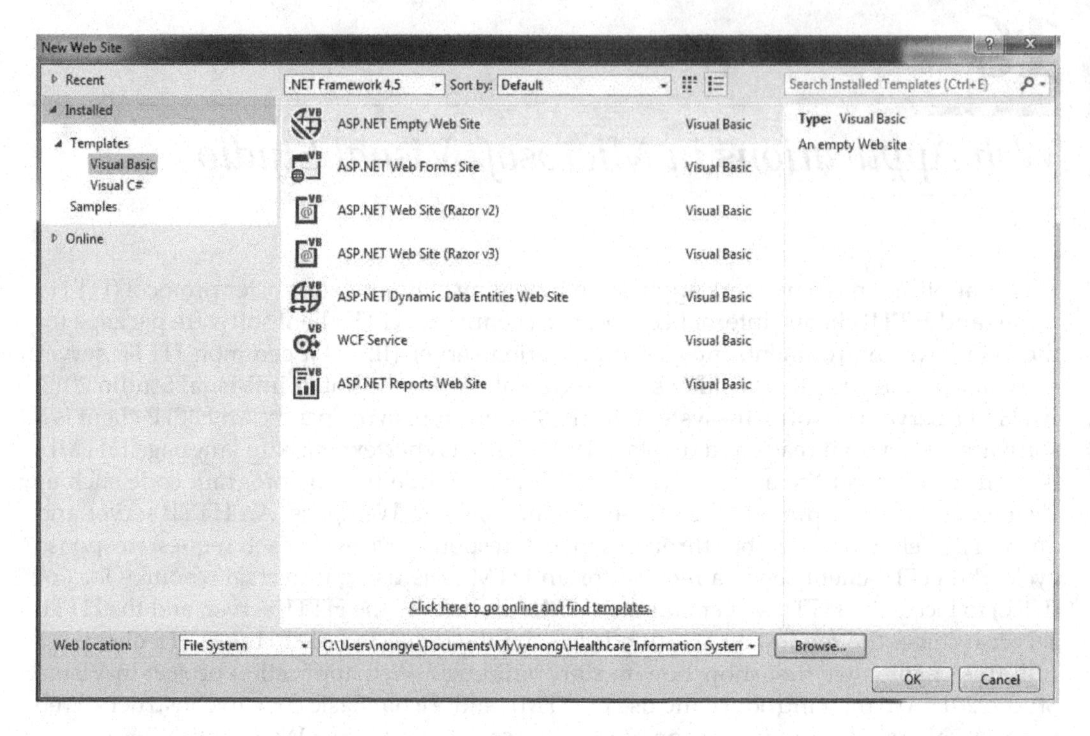

FIGURE 16.1
New Website window.

the same development environment that we use to develop a Windows-based application in Chapter 11. The development environment has the design area along with the Toolbox tab, the Solution Explorer window, the Properties window, etc.

Next, we select Add New Item in the Website menu, which brings up the Add New Item window (see Figure 16.2). We select Web Form, change the name of the Web file to FormMain.aspx, and click the Add button, which brings up the Design View of the Web form kept in the Web file. We select the Web form and use the Properties window to change the ID of the Web form to FormMain. We want to add to this main Web form for the Healthcare Information System–Web application the same graphical user interface controls as those in FormMain (see Figure 11.9) in the Windows-based application "Healthcare Information System." We drag and drop a Label from the Standard part of Toolbox to the Web form and use the Properties window to change the value for the Text property of the label. Next, we drag and drop a RadioButtonList from the Standard part of ToolBox to the Web form. Selecting the RadioButtonList bring up a right arrow button. Clicking the right arrow button and then selecting Edit Items in the pop-up menu brings up the ListItem Collection Editor window (see Figure 16.3). Using the Add button, we can add six items. We then select each ListItem, use ListItem Properties to change the text of ListItem, and click the OK button. We drag and drop a button from the Standard part of Toolbox to the Web form and use the Properties window to change the text of the button.

Figure 16.4 shows the Design View of the Web form. Switching from the Design tab to the Source tab at the bottom of the screen, we see the Source View of the Web form, which shows the Web page in HTML (see Figure 16.5). Going back to the Design View and double-clicking the Next Button brings up the Code View of the Web form (see Figure 16.6).

We right-click FormMain.aspx in the Solution Explorer and select Set as Start Page in the pop-up menu to set FormMain.aspx as the Start page of the Web application. To run

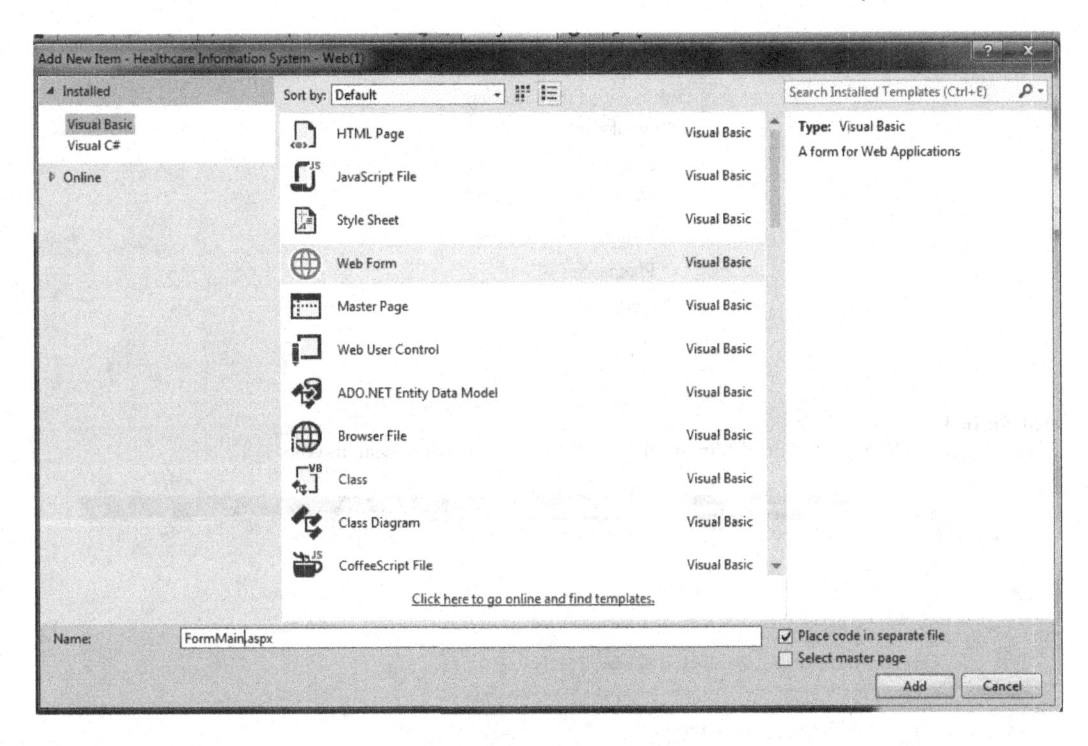

FIGURE 16.2
Add New Item window.

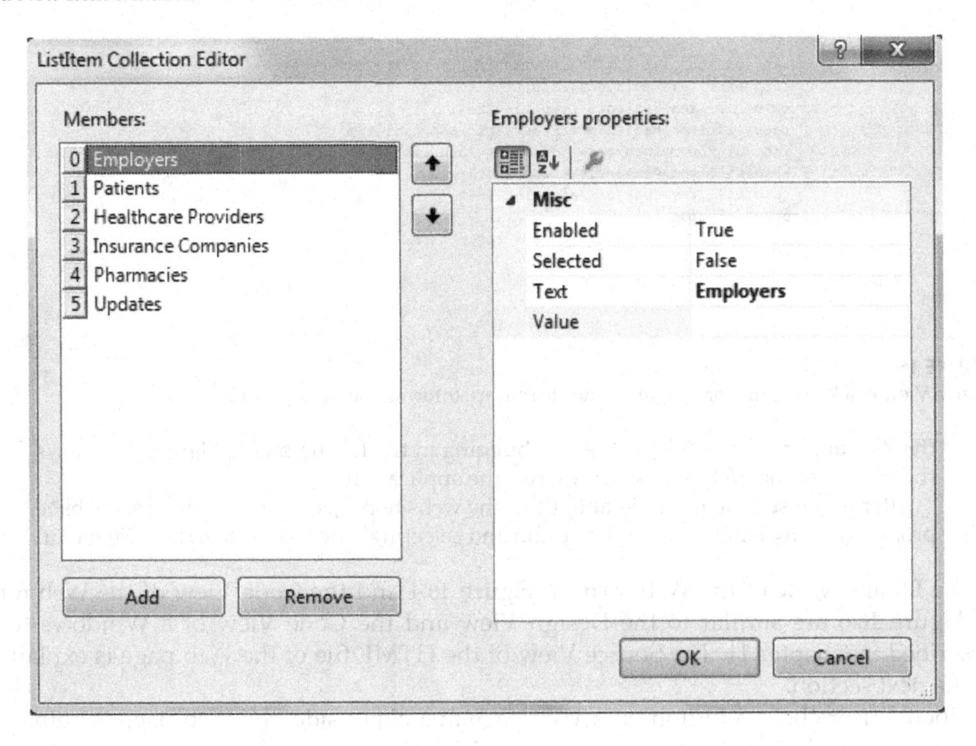

FIGURE 16.3
ListItem Collection Editor window.

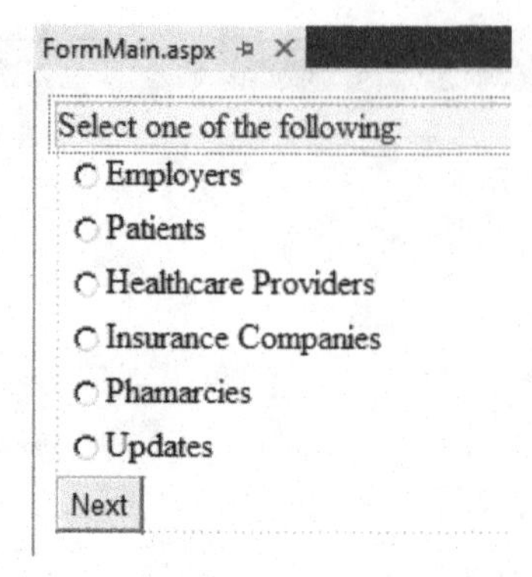

FIGURE 16.4
Design View of a Web form, FormMain, in the Healthcare Information System–Web.

```
FormMain.aspx
<%@ Page Language="VB" AutoEventWireup="false" CodeFile="FormMain.aspx.vb" Inherits="Main" %>

<!DOCTYPE html>

<html xmlns="http://www.w3.org/1999/xhtml">
<head runat="server">
    <title>Healthcare Information System</title>
</head>
<body>
    <form id="form1" runat="server">
    <div>

        <asp:Label ID="Label1" runat="server" Text="Select one of the following:"></asp:Label>

    </div>
        <asp:RadioButtonList ID="RadioButtonList1" runat="server">
            <asp:ListItem>Employers</asp:ListItem>
            <asp:ListItem>Patients</asp:ListItem>
            <asp:ListItem>Healthcare Providers</asp:ListItem>
            <asp:ListItem>Insurance Companies</asp:ListItem>
            <asp:ListItem>Phamarcies</asp:ListItem>
            <asp:ListItem>Updates</asp:ListItem>
        </asp:RadioButtonList>
        <asp:Button ID="ButtonNext" runat="server" Text="Next" />
    </form>
</body>
</html>
```

FIGURE 16.5
Source View of a Web form, FormMain, in the Healthcare Information System–Web.

the Web application, we select Start Debugging in the Debug menu. Figure 16.7 shows the display of the Web page when we run the application.

After we close Visual Studio and, thus, the website project, we can open the website project again by launching Visual Studio and selecting Open Website in the File menu.

The Design View of the Web form in Figure 16.4 and the Code View of the Web form in Figure 16.6 are similar to the Design View and the Code View of a Windows form described in Chapter 11. The Source View of the HTML file of the Web page is explained in the next section.

When a user clicks a button on a Web page at a client side, the Web page is automatically posted back for server-side processing. Many events may occur on a Web page at a client side, for example, mouse movement. For all controls on a Web page except the button

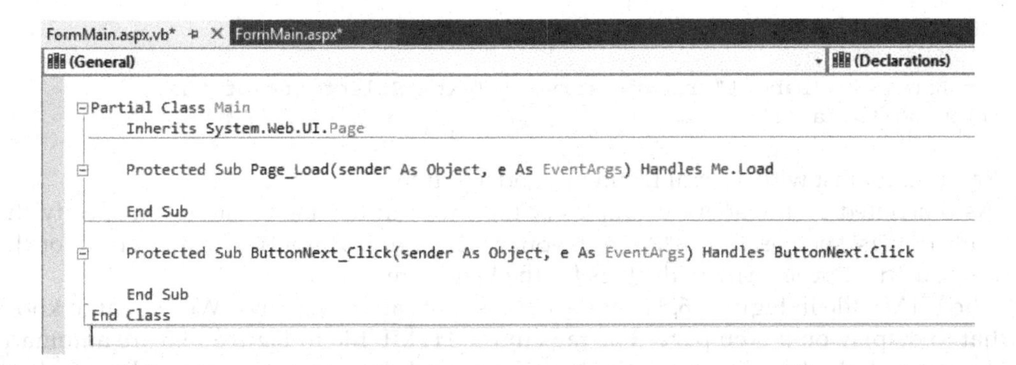

FIGURE 16.6
Code View of a Web form, FormMain, in the Healthcare Information System–Web.

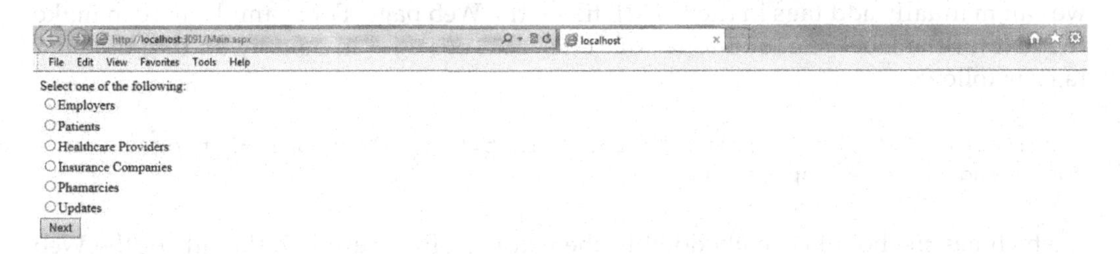

FIGURE 16.7
The display of the Web page for FormMain.aspx when the website application, Healthcare Information System–Web, is executed.

control, the AutoPostBack properties of controls are set to False because the performance of the Web application will be degraded if the Web page is posted back to the server for every event. We can use the Properties window to change the AutoPostBack property of a control to True for enabling the automatic post-back.

16.2 Hypertext Markup Language

Figure 16.5 shows the HTML file of FormMain.aspx. HTML uses tags to mark Web data. For example, in Figure 16.5, we have

```
<head runat="server"></head>
<body></body>
```

to mark two major parts of a Web page: head and body. The head in Figure 16.5 consists of a title marked by

```
<title></title>
```

The body in Figure 16.5 consists of the Web form marked by

```
<form id="FormMain" runat="server"></form>,
```

a label marked by

```
<asp:Label ID="Label1" runat="server" Text="Select one of the
following:"></asp:Label>,
```

a RadioButtonList with several ListItems, and a button.

As illustrated by the above examples of tags, we can use tag attributes to specify the details of tags, such as Text="Select one of the following:" in the tags for the label and id="FormMain" in the tags for the Web form.

The HTML file in Figure 16.5 is made of tags that can be read by a Web client to know what to display on a Web page. The tags in the HTML file in Figure 16.5 are automatically generated when we create the Web form, and drag and drop several controls to the Web form. Hence, Visual Studio allows us to create a Web page using Toolbox without knowing the syntax of HTML tags and manually entering HTML tags. However, we can manually add tags in the HTML file of the Web page. For example, we can make the text for the label "Select one of the following:" in bold by adding the bold tags as follows:

```
<asp:Label ID="Label1" runat="server" Text="<b>Select one of the
following:</b>"></asp:Label>
```

which has the bold tags embedded in the text tags. In Figure 16.7, the title of the Web page shows the default text "localhost" because there is no text in the title tags: <title> </title>. We can change the title of the Web page as follows:

```
<title>Healthcare Information System</title>
```

We can also add a hyperlink for another Web page as follows.

```
<a href="http://www.asu.edu">Arizona State University</a>
```

Figure 16.8 gives the new HTML file of Main.aspx after adding the above tags. Figure 16.9 shows the display of the Web page with "Select one of the following:" in bold, "Healthcare Information System" as the title of the Web page, and the hyperlink. When we click the hyperlink, we are directed to the website of Arizona State University.

16.3 Multiple Web Forms

After Example 16.1, we have one Web form, FormMain, in the Healthcare Information System-Web application. In Example 16.2, we add FormEmployers, FormPatients, FormProviders, FormCompanies, FormPhamarcies, and FormUpdates at the middle level of the hierarchy shown in Figure 11.13 and show how to move from one Web form to another Web form.

```
FormMain.aspx
<%@ Page Language="VB" AutoEventWireup="false" CodeFile="FormMain.aspx.vb" Inherits="Main" %>

<!DOCTYPE html>

<html xmlns="http://www.w3.org/1999/xhtml">
<head runat="server">
    <title>Healthcare Information System</title>
</head>
<body>
    <form id="form1" runat="server">
    <div>

        <asp:Label ID="Label1" runat="server" Text="<b>Select one of the following:</b>"></asp:Label>

    </div>
        <asp:RadioButtonList ID="RadioButtonList1" runat="server">
            <asp:ListItem>Employers</asp:ListItem>
            <asp:ListItem>Patients</asp:ListItem>
            <asp:ListItem>Healthcare Providers</asp:ListItem>
            <asp:ListItem>Insurance Companies</asp:ListItem>
            <asp:ListItem>Phamarcies</asp:ListItem>
            <asp:ListItem>Updates</asp:ListItem>
        </asp:RadioButtonList>
        <asp:Button ID="ButtonNext" runat="server" Text="Next" />
    </form>
    <div>
        <a href="http://www.asu.edu">Arizona State University</a>
    </div>
</body>
</html>
```

FIGURE 16.8
Updated HTML file for Main.aspx.

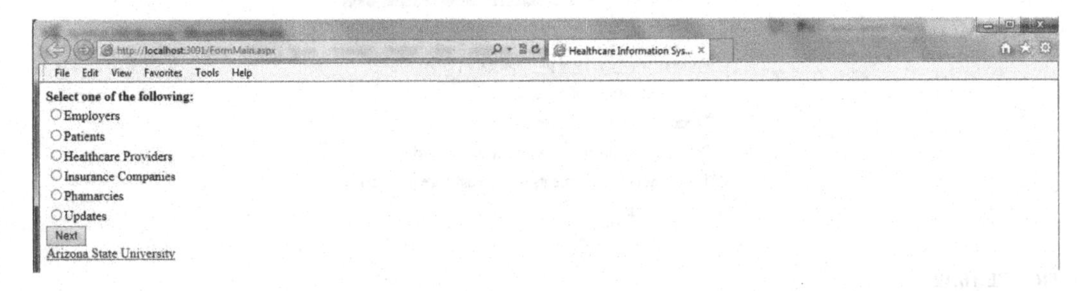

FIGURE 16.9
Display of the Web page in Figure 16.7 after manually adding new tags.

Example 16.2

Add `FormEmployers`, `FormPatients`, `FormProviders`, `FormCompanies`, `Form Phamarcies`, and `FormUpdates` in the Healthcare Information System–Web application and link these forms to `FormMain` created in Example 16.1.

Following the steps in Example 16.1, we create `FormEmployers`, `FormPatients`, `FormProviders`, `FormCompanies`, `FormPhamarcies`, and `FormUpdates`, which are shown in Figures 16.10 to 16.15, respectively. Figure 16.16 gives the Code View of `FormMain` that redirects from `FormMain` to `FormEmployers`, `FormPatients`, `FormProviders`, `FormCompanies`, `FormPhamarcies`, or `FormUpdates`, depending on which item of the `RadioButtonList` is selected. As shown in Figure 16.16, the code for redirecting from one page to another page is

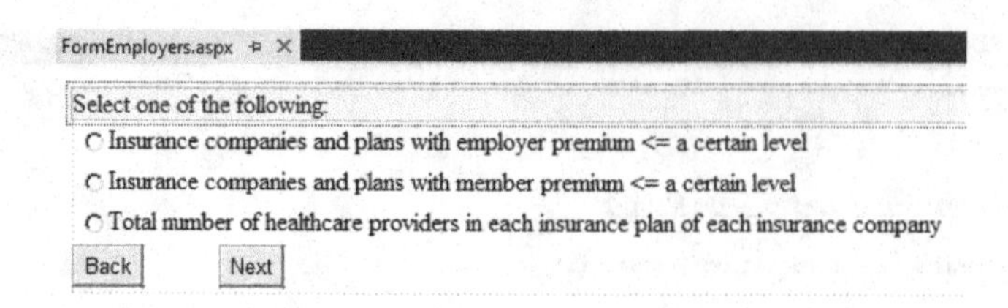

FIGURE 16.10
Design View of FormEmployers.aspx in the Healthcare Information System–Web application.

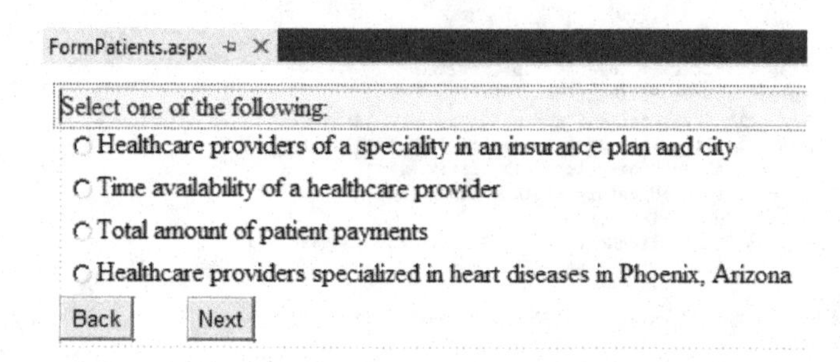

FIGURE 16.11
Design View of FormPatients.aspx in the Healthcare Information System–Web application.

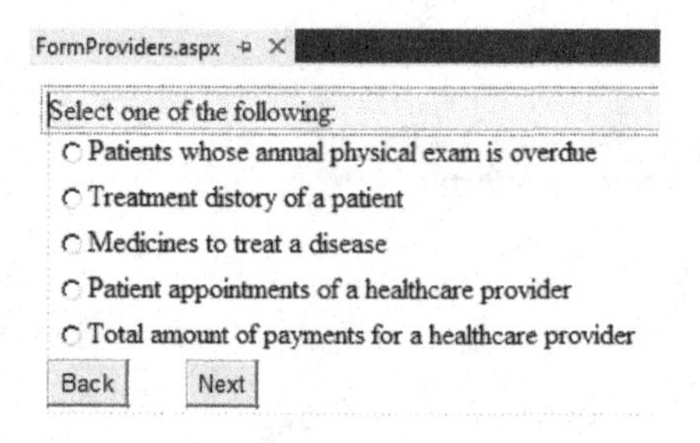

FIGURE 16.12
Design View of FormProviders.aspx in the Healthcare Information System–Web application.

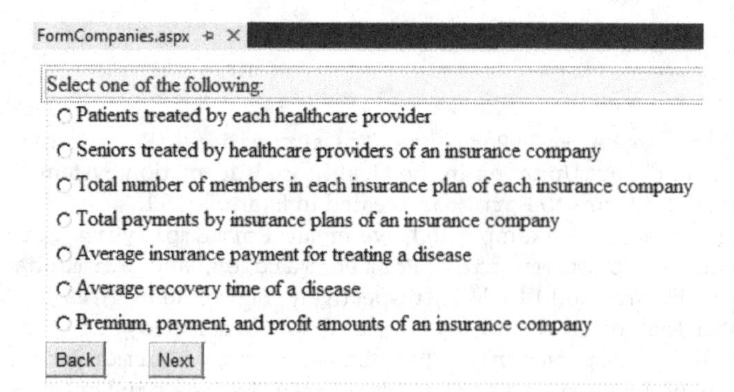

FIGURE 16.13
Design View of FormCompanies.aspx in the Healthcare Information System–Web application.

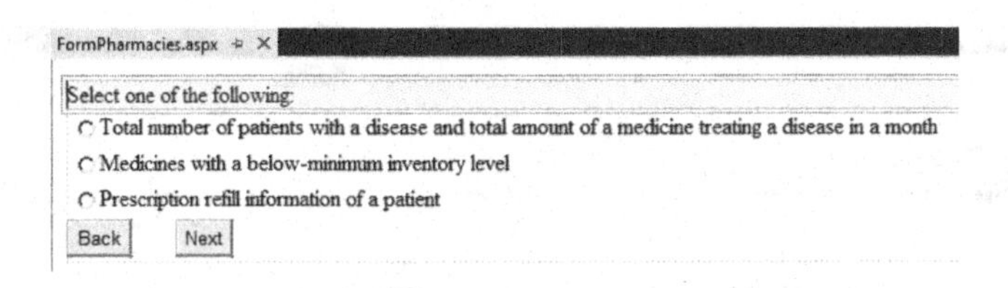

FIGURE 16.14
Design View of FormPharmacies.aspx in the Healthcare Information System–Web application.

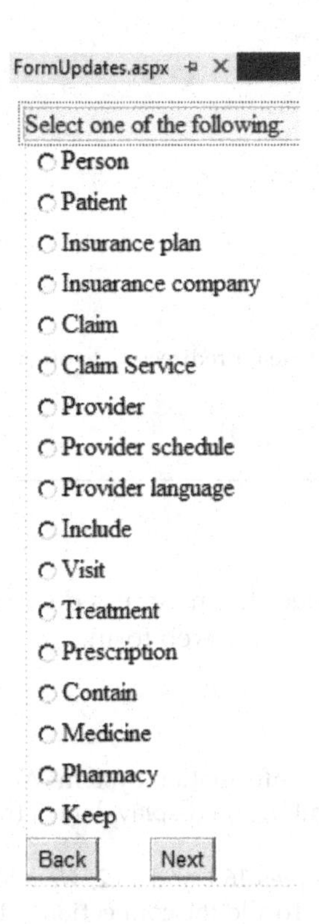

FIGURE 16.15
Design View of FormUpdates.aspx in the Healthcare Information System–Web application.

```
Response.Redirect("<a given Web form>")
```

For example, the code for redirecting from `FormMain` to `FormEmployers` is

```
Response.Redirect("FormEmployers.aspx")
```

The tags for a hyperlink can also be used to redirect from one Web page to another Web page. For example, we can add the following hyperlink tags in the HTML file of `FormMain.aspx` to redirect from the Web page for `FormMain.aspx` to that for `FormEmployers.aspx`.

```
<a href=FormEmployers.aspx>Employers</a>
```

```
FormMain.aspx.vb  FormMain.aspx
ButtonNext                                    Click

Partial Class Main
    Inherits System.Web.UI.Page

    Protected Sub Page_Load(sender As Object, e As EventArgs) Handles Me.Load

    End Sub

    Protected Sub ButtonNext_Click(sender As Object, e As EventArgs) Handles ButtonNext.Click
        Select Case RadioButtonListMain.SelectedIndex
            Case Is = 0
                Response.Redirect("FormEmployers.aspx")
            Case Is = 1
                Response.Redirect("FormPatients.aspx")
            Case Is = 2
                Response.Redirect("FormProviders.aspx")
            Case Is = 3
                Response.Redirect("FormCompanies.aspx")
            Case Is = 4
                Response.Redirect("FormPharmacies.aspx")
            Case Is = 5
                Response.Redirect("FormUpdates.aspx")
            Case Else
        End Select
    End Sub

    Protected Sub RadioButtonListMain_SelectedIndexChanged(sender As Object, e As EventArgs) Handles RadioButtonListMain.SelectedIndexChanged

    End Sub
End Class
```

FIGURE 16.16
Code View of FormMain, including the code for redirecting to other forms.

16.4 Database Connection

Example 16.3 shows how to connect to an Access database in a Web application and display data of a table in the database on a Web form.

Example 16.3

Add FormU1 to the Healthcare Information System–Web application, add code to redirect from FormUpdates to FormU1, and display data of the Person table in the healthcare Access database on FormU1.

Following the steps in Examples 16.1 and 16.2, we add FormU1 and code to redirect from FormUpdates to FormU1. To add the connection to the healthcare Access database, we bring up Server Explorer by selecting Server Explorer in the View menu. We right-click Data Connection and select Add Connection in the pop-up menu (see Figure 16.17). In the Add Connection window, we select the healthcare Access database file using the Browse button and click the OK button. Healthcare Access Database.accdb now appears in Server Explorer under Data Connection.

We double-click Healthcare Access Database.accdb in Server Explorer and double-click the Tables folder to expand and view the list of tables. We right-click the Person table and select Retrieve Data in the pop-up menu. We drag and drop the Person table in Server Explorer to the Design View of FormU1. This adds the GridView control and an AccessDataSource to FormU1. When we run the application and go to FormU1, we have the data of the Person table displayed. Figure 16.18a shows the Design View of FormU1, and Figure 16.8b gives the display of data in the Person table on FormU1.

We can select the GridView on FormU1, click the right arrow button at the top-right corner, select AutoFormat in the pop-up menu, and use the AutoFormat window to select a different display format for the GridView.

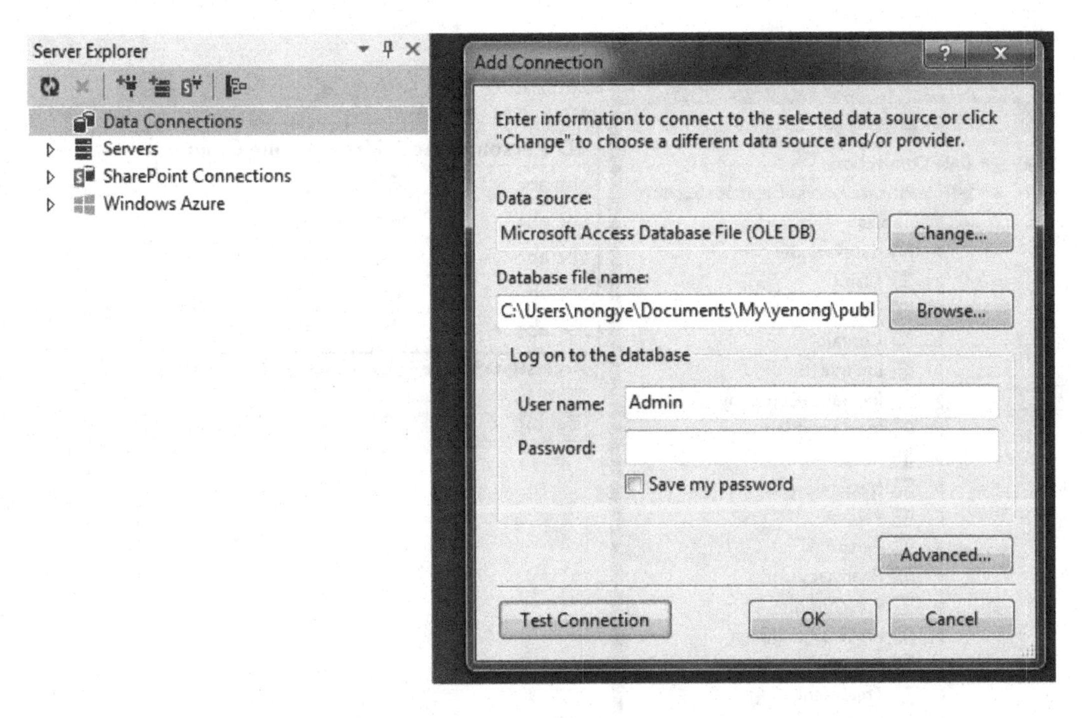

FIGURE 16.17
Server Explorer and Add Connection window for making connection to an Access database.

We click the right arrow for the GridView on FormU1 again; check Enable Paging, Enable Sorting, Enable Editing, Enable Deleting, and Enable Selection; and run the application. Figure 16.19 shows the display of FormU1 after checking Enable Paging, Enable Sorting, Enable Editing, Enable Deleting, and Enable Selection. We now have the data of the Person table displayed in two pages now with Enable Paging. When we click the heading of a given data field, data are sorted in ascending order of the values in the data field. Clicking the heading of the data field again changes the order of the values from ascending to descending. When we click Edit for a row, we can update the value of any data field in the row. When we click Delete for a row, we delete the row. Clicking Select for a row selects the row (Figure 16.19).

Data in each of the other tables in the Healthcare Access database can be displayed in the corresponding Web form in the Healthcare Information System–Web application by following the steps described in Example 16.3.

To display the query result of a query on an Access database, we need to design the query in an AccessDataSource. Example 16.4 shows how to design a query in an AccessDataSource.

Example 16.4

Add FormPA4 to the Healthcare Information System–Web application and display the query result of PAQ4 in the healthcare Access database on FormPA4. Figure 16.20 shows the query design of PAQ4 in Microsoft Access. This query presents a list of healthcare providers in Phoenix, Arizona, who are specializing in heart diseases.

Following the steps in Example 16.1, we add FormPA4 to the Healthcare Information System–Web application and add code that redirects from FormPatients to FormPA4 and from FormPA4 back to FormPatients. As shown in Figure 16.20, PAQ4 uses two

(a)

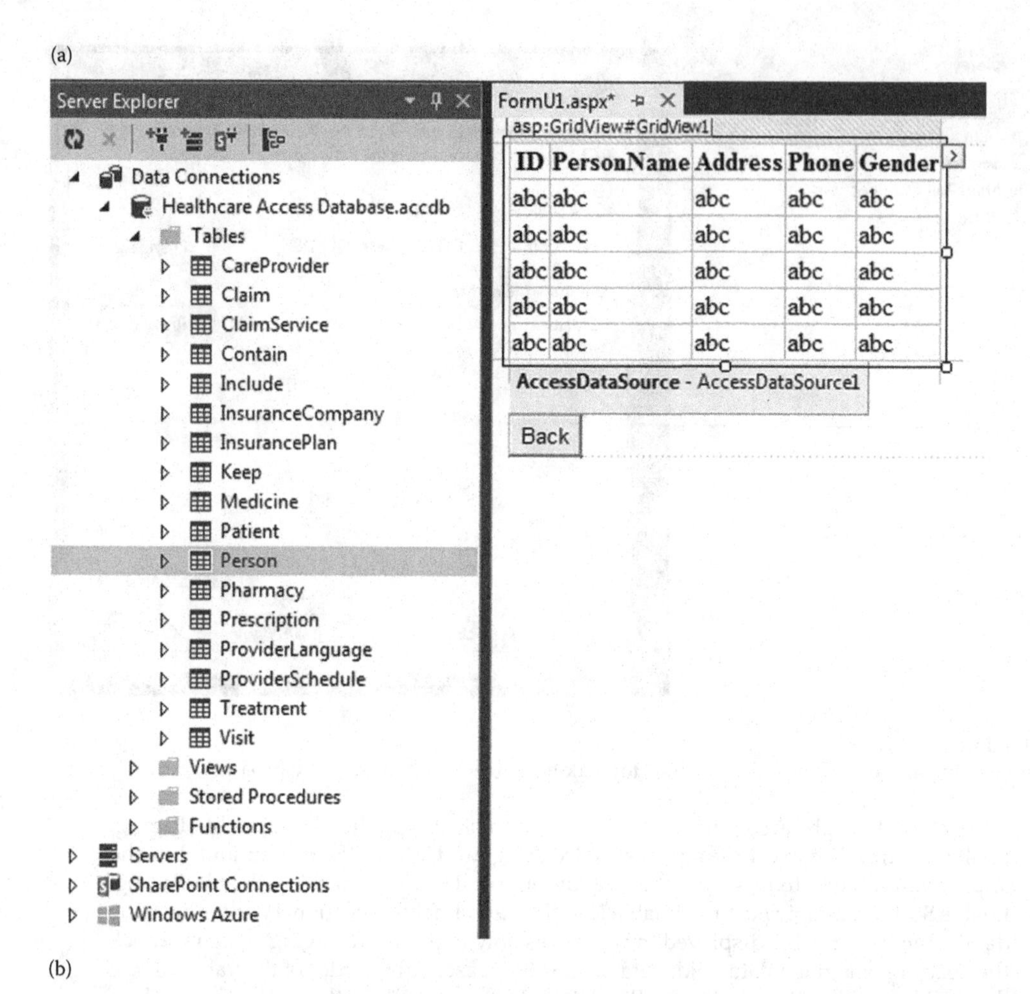

(b)

ID	PersonName	Address	Phone	Gender
PR1	John Smith	1 Main Street, Phoenix, Arizona	6021111111	
PR2	Mary Fowler	1 Mill Avenue, Tempe, Arizona	4802222222	
PA1	Mike Smith	1 University Street, Tempe, Arizona	4803333333	
PR3				
PR4				
PR5				
PR6				
PR7				
PR8				
PR9				
PR10				

Back

FIGURE 16.18
(a) Design View of FormU1 and (b) the display of FormU1 when the Healthcare Information System–Web application is running.

(a)

FormU1.aspx

asp:GridView#GridView1

	ID	PersonName	Address	Phone	Gender
Edit Delete Select	abc	abc	abc	abc	abc
Edit Delete Select	abc	abc	abc	abc	abc
Edit Delete Select	abc	abc	abc	abc	abc
Edit Delete Select	abc	abc	abc	abc	abc
Edit Delete Select	abc	abc	abc	abc	abc
Edit Delete Select	abc	abc	abc	abc	abc
Edit Delete Select	abc	abc	abc	abc	abc
Edit Delete Select	abc	abc	abc	abc	abc
Edit Delete Select	abc	abc	abc	abc	abc
Edit Delete Select	abc	abc	abc	abc	abc
1 2					

AccessDataSource - AccessDataSource1

Back

(b)

http://localhost:3091/FormU1.aspx Healthcare Information Sys...

File Edit View Favorites Tools Help

	ID	PersonName	Address	Phone	Gender
Edit Delete Select	PR1	John Smith	1 Main Street, Phoenix, Arizona	6021111111	
Edit Delete Select	PR2	Mary Fowler	1 Mill Avenue, Tempe, Arizona	4802222222	
Edit Delete Select	PA1	Mike Smith	1 University Street, Tempe, Arizona	4803333333	
Edit Delete Select	PR3				
Edit Delete Select	PR4				
Edit Delete Select	PR5				
Edit Delete Select	PR6				
Edit Delete Select	PR7				
Edit Delete Select	PR8				
Edit Delete Select	PR9				
1 2					

Back

FIGURE 16.19
(a) Design View of FormU1 and (b) display of FormU1 when the Healthcare Information System–Web application is running after checking Enable Sorting, Enable Sorting, Enable Editing, Enable Deleting, and Enable Selection.

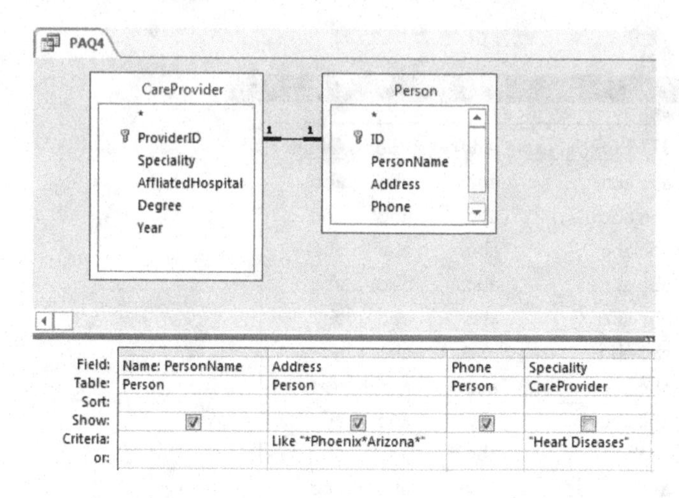

FIGURE 16.20
Query design of PAQ4 in the healthcare Access database.

tables: the CareProvider table and the Person table. We drag and drop one of the two tables, the CareProvider table, to FormPA4. This adds the GridView control and the AccessDataSource to FormPA4. We click the right arrow at the top-right corner of the GridView control and select Configure Data Source. Figure 16.21 shows the procedure of designing a query and its structured query language (SQL) using Configure Data Source. In the Choose a Database window, the healthcare Access database file is already selected, and we click the Next button. In the Configure the Select Statement window, we check Specify a Custom SQL Statement or Stored Procedure and click on the Next button. In the Define Custom SQL Statements or Stored Procedures window, we click on the Query Builder button. In the Query Builder window, we design the query as we did in Chapter 13. We right-click in the table pane area and select Add Table in the pop-up menu. In the Add Table window, we select the Person table, click the Add button, and click the Close button to go back to the Query Builder window, which now has the CareProvider and the Person tables in the table pane. Following the same steps described in Chapter 13, we enter the query design in Figure 16.20 in the Query Builder window, click the Execute Query button to see the query result, and click the OK button. The Define Custom SQL Statements or Stored Procedures window now has the completed SQL statement for the query, and we click on the Next button.

Note that, in Visual Studio 2013, we use

```
LIKE '%Phoenix%Arizona%'
```

in the selection criterion for the Address field, whereas in Access, we use

```
LIKE "*Phoenix*Arizona*"
```

for the same selection criterion. Hence, in Visual Studio 2013, % is the wild card to match any string, and a string is enclosed in `' '` rather than in `" "`. Visual Studio 2013 uses the operator "+" to put strings together into one string, whereas Access uses the operator "&" to do it. For example,

```
LIKE '%' + ? + '%'
```

matches any string that contains a string entered by a user, where "?" prompts the user to enter a string.

(a)

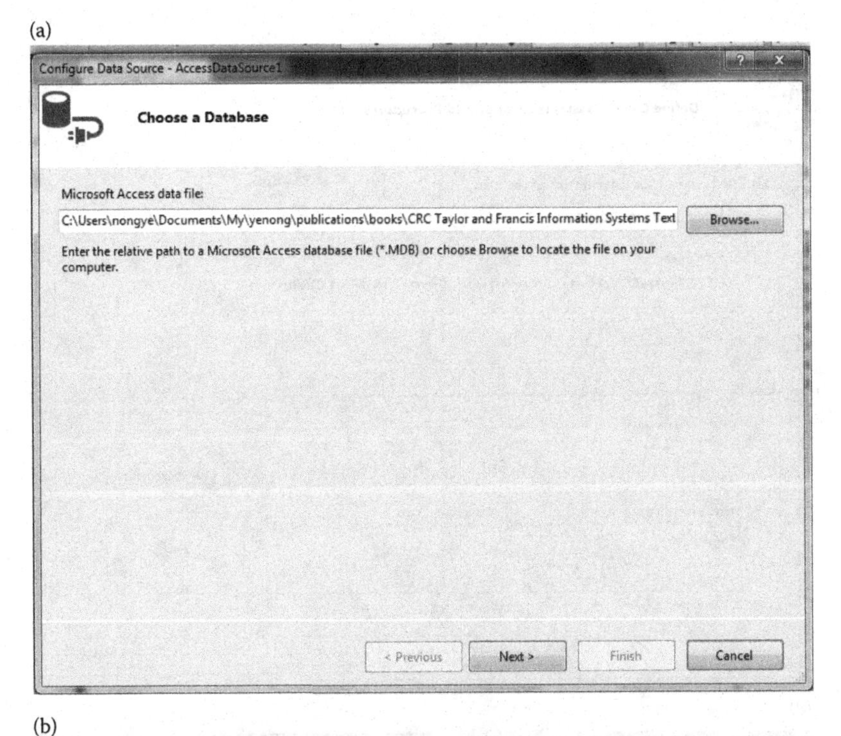

(b)

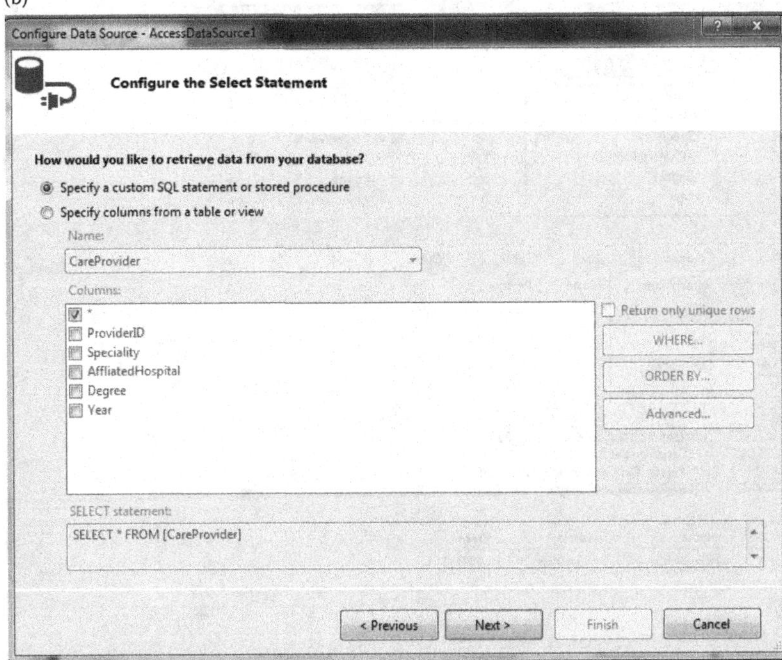

FIGURE 16.21
Procedure of designing a query and its SQL statement for a Web application. (a) Choose a Database window. (b) Configure the Select Statement window.

(c)

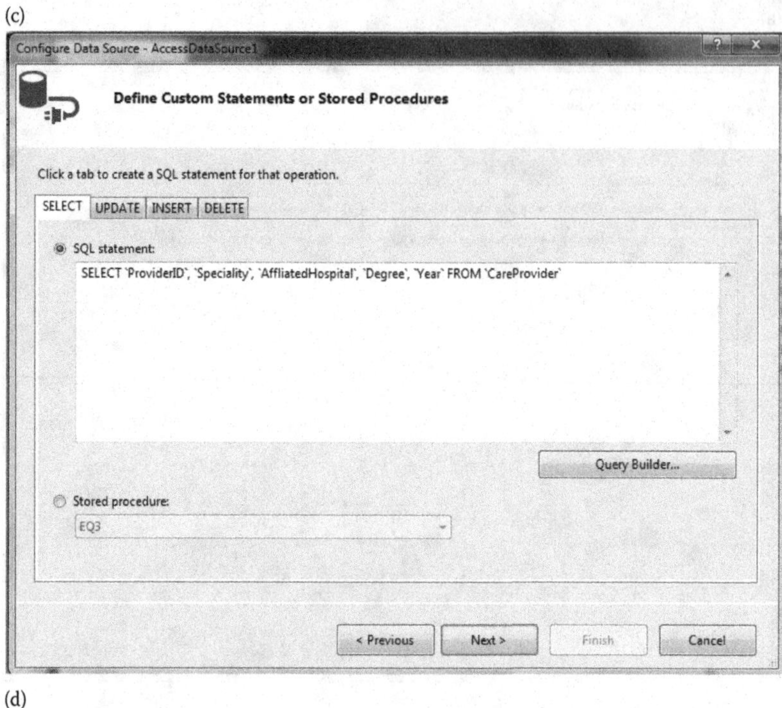

(d)

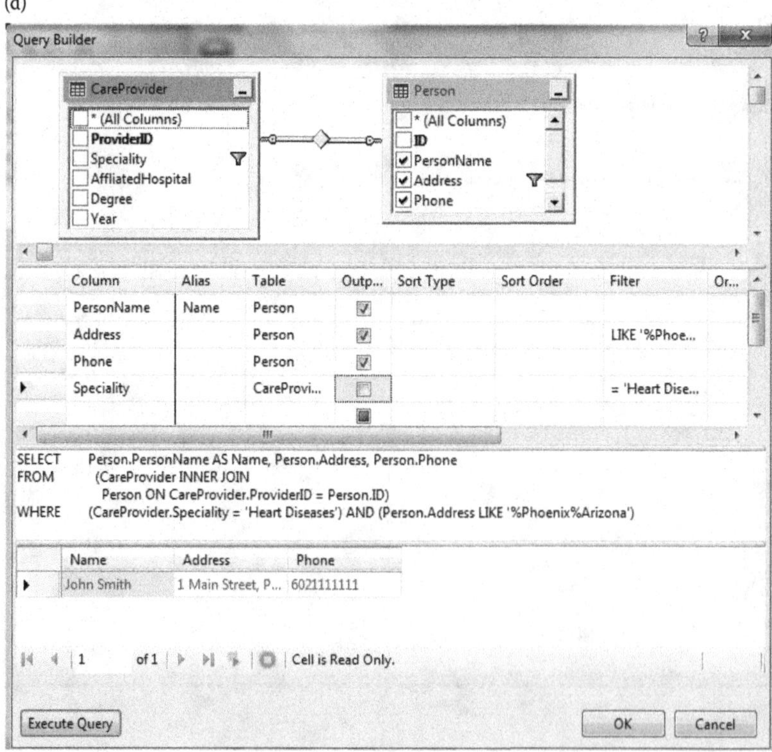

FIGURE 16.21 (Continued)
Procedure of designing a query and its SQL statement for a Web application. (c) Define Custom Statements or Stored Procedures window I. (d) Query Builder window.

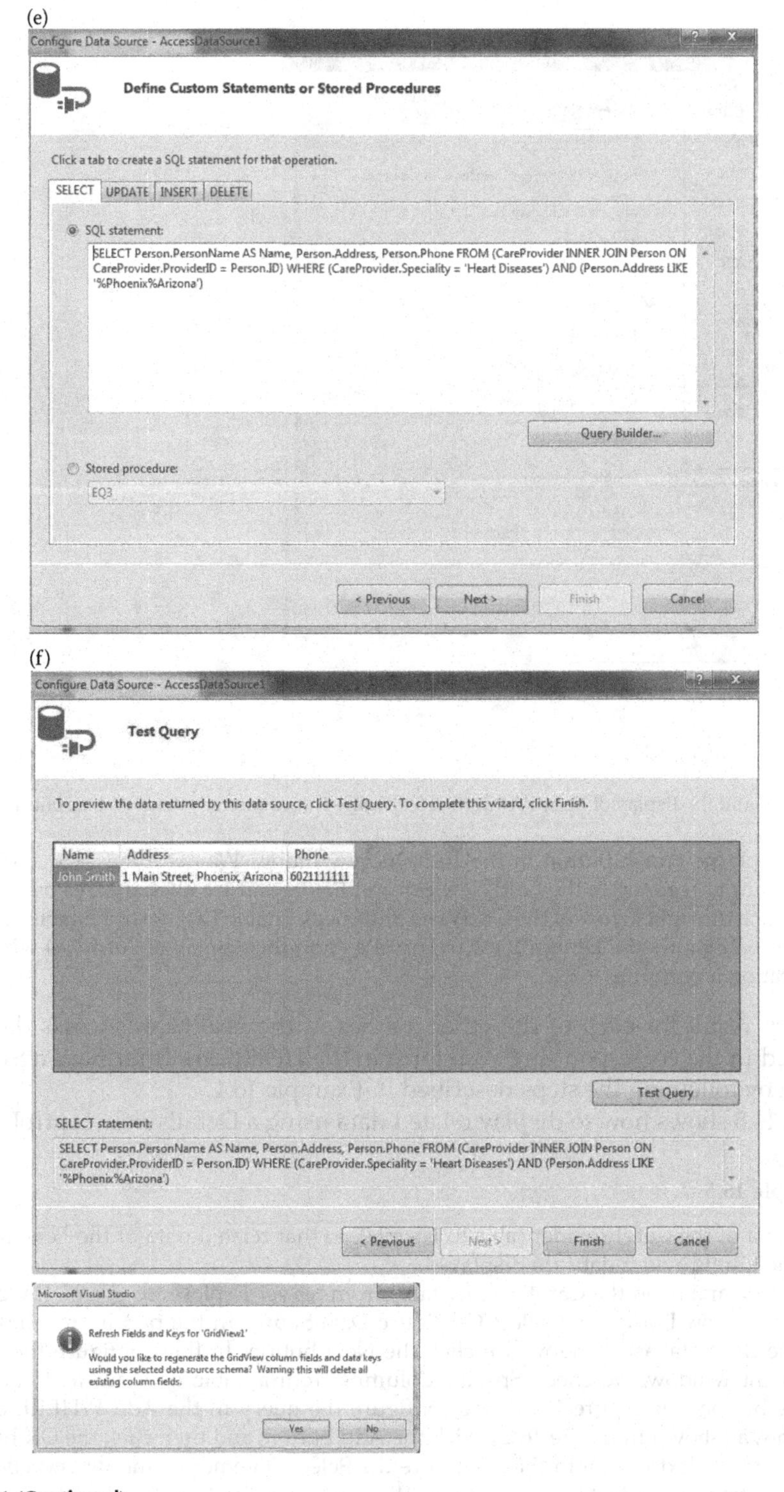

FIGURE 16.21 (Continued)
Procedure of designing a query and its SQL statement for a Web application. (e) Define Custom Statements or Stored Procedures window II. (f) Test Query window.

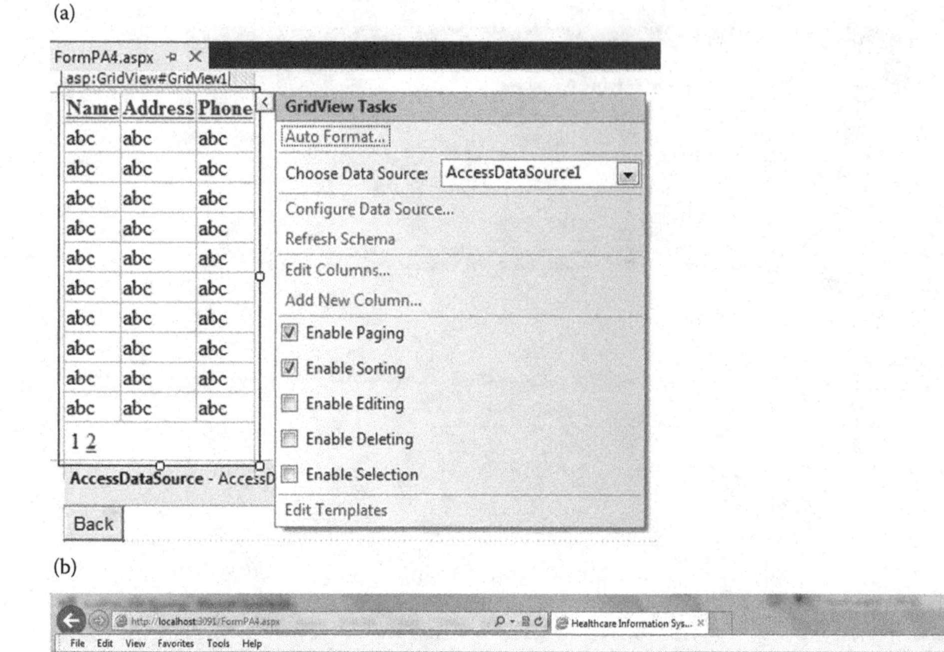

FIGURE 16.22
(a) Design View and (b) display of FormPA4 in the Healthcare Information System–Web application.

We click the Finish button in the Test Query window. We get a message asking us if we want to regenerate the GridView column fields, and we click the Yes button. We right-click the right arrow of the GridView and check Enable Paging and Enable Sorting. Figure 16.22 shows the Design View of FormPA4 and the display of FormPA4 when the application is running.

The query result for each of the other queries in the healthcare Access database can be displayed in the corresponding Web form in the Healthcare Information System–Web application by following the steps described in Example 16.4.

Example 16.5 shows how to display related data using a DetailsView control.

Example 16.5

Add data of the CareProvider table to FormU1 so that related data of the Person table and the CareProvider data are displayed.

We drag and drop the CareProvider table from Server Explorer to FormU1. We click the right arrow button and select Configure Data Source in the pop-up menu. In the Choose the Database window, we click the Next button. In the Configure the Select Statement window, we check Specify Columns from a Table or View, and click the Where button (see Figure 16.23). We configure the query in the Add WHERE clause window, as shown in Figure 16.23; click the Add button; and then click the OK button. We click the Next button in the Configure the Select Statement window and click the Finish button in the Test Query window. We run the application. On the Web page for FormU1, we click Select for the row with the ProviderID value of PR1, and we see the corresponding record in the CareProvider table displayed.

(a)

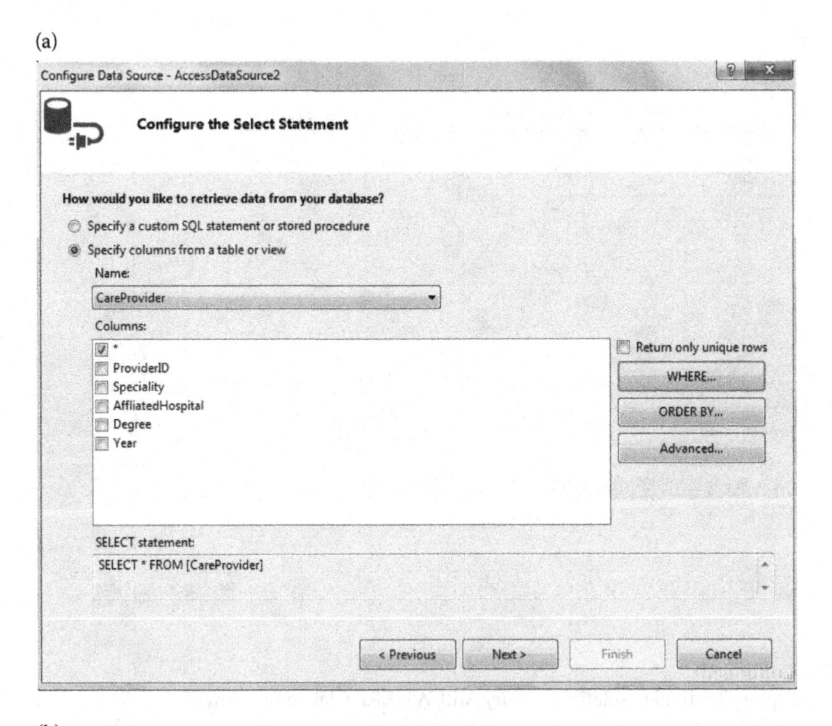

(b)

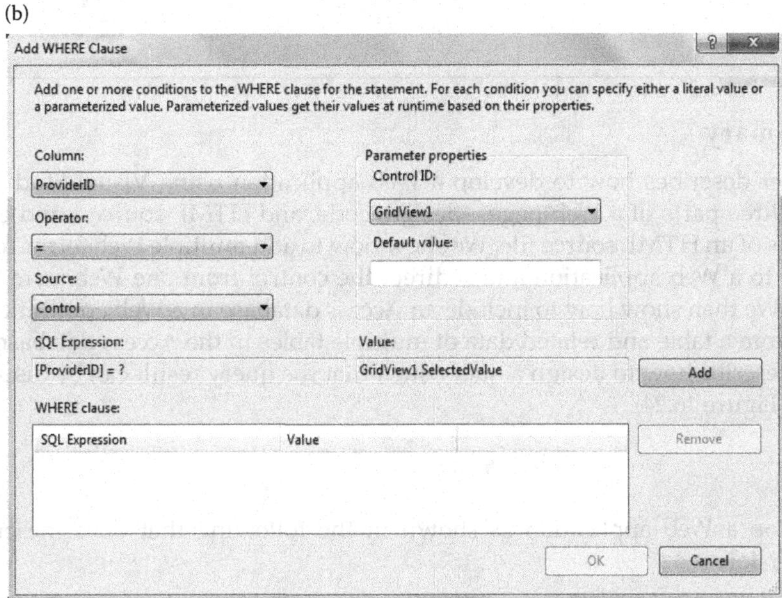

FIGURE 16.23
Configuring the query to display related data. (a) Configure the Select Statement window. (b) Add WHERE Clause window I.

(c)

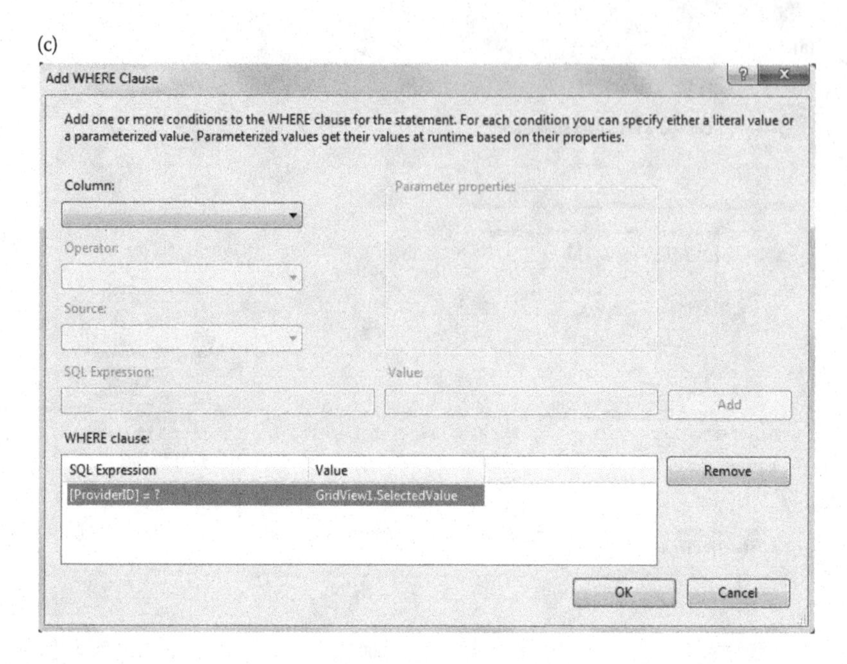

FIGURE 16.23 (Continued)
Configuring the query to display related data. (c) Add WHERE Clause window II.

16.5 Summary

This chapter describes how to develop a Web application using Visual Studio 2013. We introduce three parts of a Web page—design, code, and HTML source—and explain the components of an HTML source file. We show how to add multiple Web forms for multiple Web pages to a Web application and redirect the control from one Web form to another Web form. We then show how to include an Access database in a Web application and display data from a table and related data of multiple tables in the Access database on a Web form. We describe how to design a query such that the query result can be displayed on a Web form (Figure 16.24).

EXERCISES

1. Develop a Web application as shown in the following that does the currency conversion.

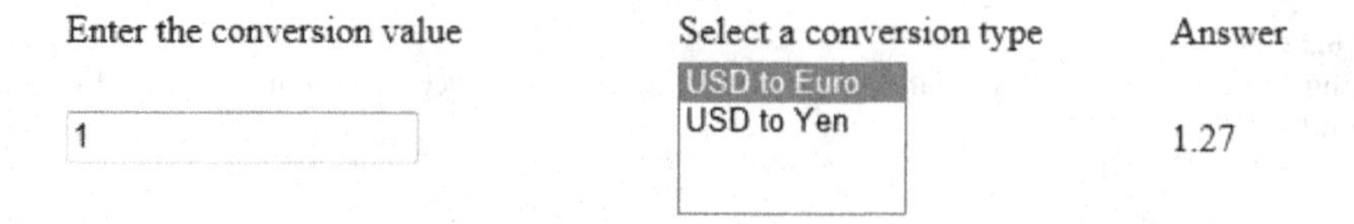

2. Develop a Web application for a simple calculator that can perform addition, subtraction, multiplication, and division.

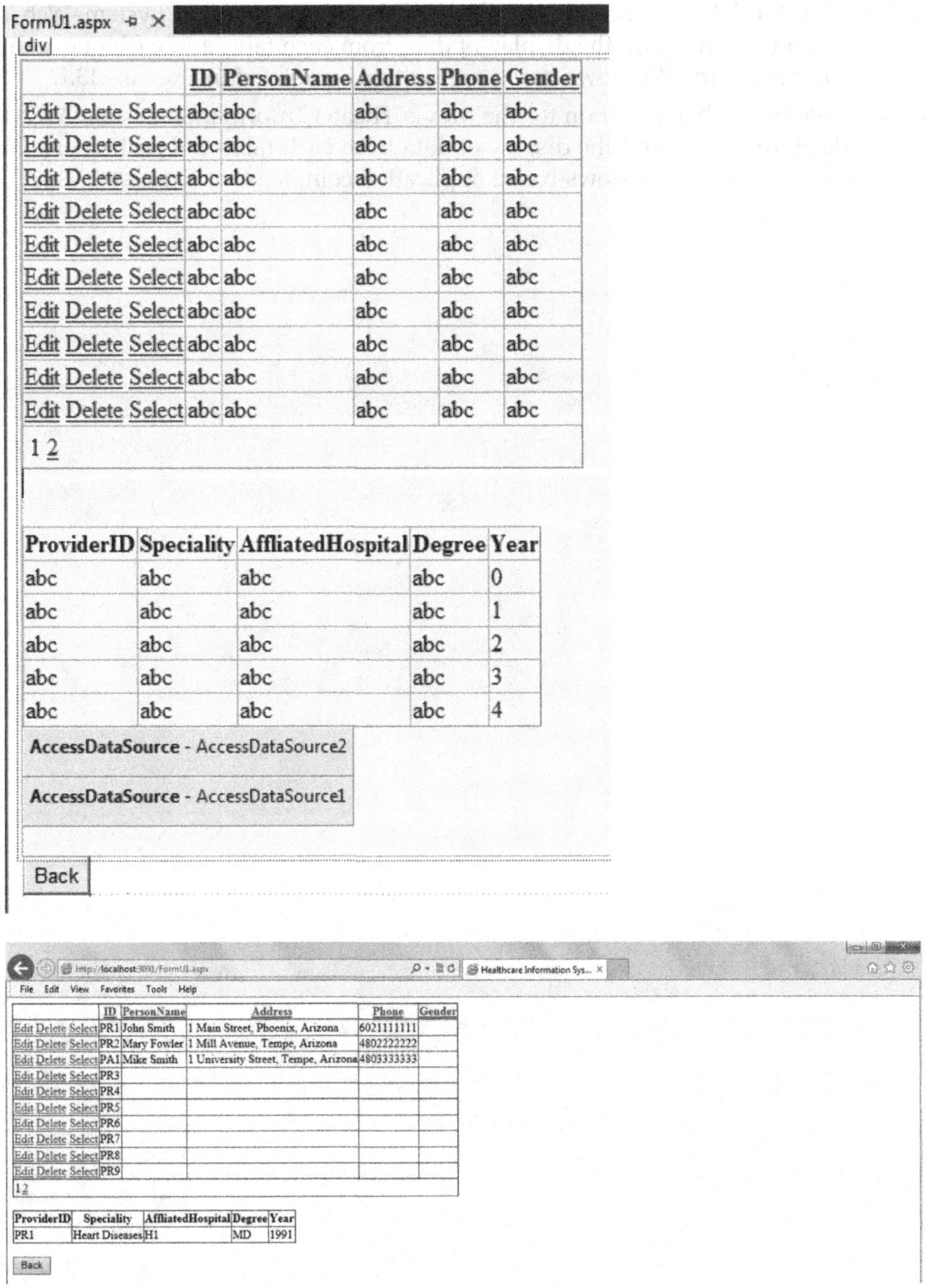

FormU1.aspx

div

	ID	PersonName	Address	Phone	Gender
Edit Delete Select	abc	abc	abc	abc	abc
Edit Delete Select	abc	abc	abc	abc	abc
Edit Delete Select	abc	abc	abc	abc	abc
Edit Delete Select	abc	abc	abc	abc	abc
Edit Delete Select	abc	abc	abc	abc	abc
Edit Delete Select	abc	abc	abc	abc	abc
Edit Delete Select	abc	abc	abc	abc	abc
Edit Delete Select	abc	abc	abc	abc	abc
Edit Delete Select	abc	abc	abc	abc	abc
Edit Delete Select	abc	abc	abc	abc	abc
1 2					

ProviderID	Speciality	AffliatedHospital	Degree	Year
abc	abc	abc	abc	0
abc	abc	abc	abc	1
abc	abc	abc	abc	2
abc	abc	abc	abc	3
abc	abc	abc	abc	4

AccessDataSource - AccessDataSource2

AccessDataSource - AccessDataSource1

Back

http://localhost:3091/FormU1.aspx

Healthcare Information Sys...

File Edit View Favorites Tools Help

	ID	PersonName	Address	Phone	Gender
Edit Delete Select	PR1	John Smith	1 Main Street, Phoenix, Arizona	6021111111	
Edit Delete Select	PR2	Mary Fowler	1 Mill Avenue, Tempe, Arizona	4802222222	
Edit Delete Select	PA1	Mike Smith	1 University Street, Tempe, Arizona	4803333333	
Edit Delete Select	PR3				
Edit Delete Select	PR4				
Edit Delete Select	PR5				
Edit Delete Select	PR6				
Edit Delete Select	PR7				
Edit Delete Select	PR8				
Edit Delete Select	PR9				
1 2					

ProviderID	Speciality	AffliatedHospital	Degree	Year
PR1	Heart Diseases	H1	MD	1991

Back

FIGURE 16.24
Design View and display of FormU1 showing related data.

3. Complete the Web application for the Healthcare Information System–Web to include all the forms and the display of data from each table and each query that are contained in the Windows-based application completed in Exercise 13.1.
4. Complete the Web application for the Movie Theater Information System–Web to include all the forms and the display of data from each table and each query that are contained in the Windows-based application completed in Exercise 13.2.

17

Working with XML (I)

Extensible markup language (XML) is a markup language that defines the structures of documents. Like hypertext markup language (HTML), XML originated from standard generalized markup language (SGML). This chapter first briefly reviews the concepts of markup languages, specifically, SGML and HTML. Then, the basic XML elements are discussed. Furthermore, two important concepts of XML, well-formed and validated XML, are explained in detail.

17.1 What Is Markup Language?

A markup language is designed to process, define, and present information in text format. In 1974, SGML was first created from an IBM document-sharing project, and it became an International Organization for Standardization (ISO) standard (ISO 8879) in 1986. As the Internet grew, especially the World Wide Web (WWW), two popular adaptations of SGML emerged: one is HTML, a standard to describe document layout and display pages, and the other one is XML, a standard to describe document semantics, thus enabling data exchange over the Internet.

17.1.1 Standard Generalized Markup Language

An SGML document is composed of several basic elements. An element is used to define the document structure and to present information contained within the document. Moreover, it should be noted that different SGML elements can be given different names. Since SGML was initially designed to manage and format large documents subject to frequent revisions, it is only concerned with annotating the syntax instead of the semantics of each element. Additionally, the most important rule in SGML is that all elements need to be explicitly marked. Specifically, the start-tag and end-tag pair is used to bracket off the element, as the way of using parentheses or quotation marks in conventional punctuation. An example SGML is shown in Table 17.1.

This example shows that the document type is a *Book*, and it consists of a number of book chapters. Each *Chapter* has a *title* and several sections, and each *section* has several *paragraphs*. This structure is represented by start-tag and end-tag pairs as follows:

- *<Book> </Book>*
- *<title> </title>*
- *<section> </section>*
- *<paragraph> </paragraph>*

TABLE 17.1

Example SGML Document (ch17_SGMLExample.sgml)

```
<Book>
  <Chapter ID="17">
    <title>
      Working with XML
    </title>
    <section>
      <paragraph>SGML stands for Standard Generalized Markup Language.</paragraph>
      <paragraph>XML stands for Extensible Markup Language.</paragraph>
    </section>
  </Chapter>
  </Chapter ID="18">
  ....
  </Chapter>
</Book>
```

From this example, it is learned that SGML rules emphasize the use of start and end tags to enclose each element. It does not, however, specify strict rules on how the tags should be placed. For example, whether a *<title>* can appear in places other than preceding the *<section>*, or whether a *<paragraph>* can appear outside a *<section>*, is not strictly required. In such cases, the beginning and the end of every element must be explicitly marked because there are no identifiable rules as to where an element can appear. As a result, SGML is not able to extract the semantic meaning of the elements.

17.1.2 Hypertext Markup Language

The first popular derivative of SGML is HTML, which was inspired by SGML's tagging concept. It is the main markup language for displaying Web pages as well as its basic building block. A typical Web browser does not display an HTML document's tags. Instead, it uses the tags to interpret and display the corresponding HTML Web page. Some standard HTML tags are headings (e.g., h1), formatting (e.g., b for bold), font (e.g., font size), and image (e.g., img src) to name a few. An HTML document example is shown in Table 17.2, with its corresponding display using a Web browser shown in Figure 17.1.

As shown in the example, HTML follows the tagged structure recommended by SGML. The tags <html>, <head>, <body>, and <h1> are standard and can be interpreted by any Web browser. On the other hand, HTML is less strict on the syntax of its code. For example, HTML is not case sensitive in the sense that <h1> and <H1> will be treated the same way. HTML does not require the start-tag and end-tag pairs to be closed: an HTML document having <h1> without being closed by </h1> is still accepted by the Web browser.

The two aforementioned examples illustrate that SGML is used to define the structure for document sharing without concerning the semantics of the elements, while HTML is

TABLE 17.2

Example HTML Document (ch17_HTMLExample.html)

```
<html>
  <head>
    <title>Hello HTML</title>
  </head>
    <h1>Hello World!</h1>
</html>
```

FIGURE 17.1
Display of the example HTML (ch17_HTMLExample.html).

used as a standard to define the layout for document presentation. However, there was no standardized way to describe and share data contained within the document. Motivated by the success of SGML and HTML, the WWW Consortium (W3C) initiated the process to develop a standard data sharing mechanism and came up with the first XML recommendation in 1998.

17.1.3 Extensible Markup Language

XML is developed to describe the content of a document in a standardized, text-based format that can easily be transported over the Internet. One notable advantage of XML is its extensibility. That is, the developer has the freedom to extend data elements provided by the basic XML standard as long as it follows the XML syntax and is validated to be correct. As such, the structure of an XML document would be usually complex for any human being to digest. However, there are a number of tools available (some are discussed in Chapter 18) to make XML validation and interpretation easy even in its most complex form. Compared to HTML, XML has much more rigid rules in its syntax. In addition, an XML document's structure and content need to be validated based on some validation standard, which includes data type definition (DTD) and XML schema. Although optional, data validation is becoming increasingly important, especially for new XML document developments.

17.2 XML Basics: Elements and Attributes

The W3C XML Recommendation is strict about the format requirements. This provides a major difference between an SGML and an XML document. These requirements are then used to define a well-formed XML document that is composed of elements, attributes, and text.

17.2.1 Elements

Elements are the basic building blocks of an XML document. An element in an XML starts with a "< >" tag and ends with a "</>" tag, and it looks like the following:

```
<name_of_the_element> text </name_of_the_element>
```

The names of the elements are developer defined and must adhere to the following basic rules:

- Element names must begin with a letter or an underscore (_) and cannot start with "xml" (reserved by XML).
- Element names cannot contain spaces. Underscores are usually used to replace space.
- Element names can contain any number of letters, digits, and underscores.
- A colon (:) is not recommended as it is generally reserved for namespaces (discussed later in the chapter).
- A dot (.) is also not recommended as it could be troublesome for object operation.
- Hyphen (-) is not recommended as it may confuse the program with the subtraction (–) operator.
- All names are case sensitive for start and end tags, which must match perfectly.

EXERCISE 17.1

Correct and incorrect examples of XML element names

<1stChapter> </1stChapter>	Incorrect since the element begins with invalid characters
<first Chapter> </first Chapter>	Incorrect since no space is allowed in the element name
<first_Chapter> </first_Chapter>	Correct
<_firstChapter> </_firstChapter>	Correct
<xmlbook> </xmlbook>	Will generate errors since "xml" is reserved by the namespace
<title> </Title>	Incorrect since XML is case sensitive
<!title> </!title>	Will generate errors since the declaration has an invalid name
<Chapter.title> </Chapter.title>	Will generate errors since the declaration has an invalid name
<Chapter:title> </Chapter:title>	Incorrect; reference to undeclared namespace prefix

17.2.2 Attributes

XML elements can have attributes that provide additional information about an element. It is presented as part of an element's starting tag as follows:

```
<name_of_the_element name_of_the_attribute="value">   </name_of_the_
   element>
```

The naming rules of attributes follow those for element names, with some additional requirements, including the following:

- Attribute values must be quoted. Either single or double quotes can be used. If the value itself contains one type of quote (e.g., double quote), the other type of quote (e.g., single quote) shall be used for the attribute value. For example,

```
<gangster name="George 'Shotgun' Ziergler"> </gangster>
```

- Attributes cannot contain multiple values.

There are no explicit rules on when to use attributes or when to use elements. In the context of documents, attributes are part of the markup, while elements are part of the basic document structure. Sometimes, ID references are assigned to elements to identify each element. These metadata (data about data) are recommended to be stored as an attribute, and the data itself should be stored as element. For example,

```
<Chapter ID="C1"> text </Chapter>
```

is usually preferred by the developer than

```
<Chapter>
      <ID> C1 </ID>
</Chapter>
```

The concepts of ID references will be discussed more in the DTD section later in this chapter.

17.2.3 Text

Text is located between the starting and closing tags of an element, and usually represents the actual data associated with the elements:

```
<element attribute="value"> text </element>
```

Note that the text is not constrained by the same syntax rules of elements and attributes, so any text can be used for the elements. The data type of the text (e.g., String, Time, Date) is usually specified in a validation standard via a DTD or an XML schema.

17.2.4 Empty Elements

Sometimes, elements with no attribute or text can also be represented as follows:

```
<element> </element> or
<element/> or
<element attribute="value"/>
```

This is usually used to accommodate a predefined data structure.

17.2.5 Well-Formed XML

Other than syntax rules on elements, attributes, and text, structural rules are defined to ensure a well-formed XML. Let us use the following example to explain the structure of a well-formed XML document (Table 17.3).

Example 17.1

Line 1: XML document declaration: Most XML documents start with a `<?xml?>` element, which refers to an XML document declaration. In the W3C XML 1.0 specification, this is not required for a well-formed XML. However, it is very useful to determine the version of XML and the encoding type of the source data. In this example, line 1 indicates that the version of the XML document is "1.0" and that the encoding type is "UTF-8."

TABLE 17.3

Example XML Document (ch17_SimpleXMLExample.xml)

```
1. <?xml version="1.0" encoding="UTF-8"?>
2. <Book>
3. <Part ID="P1">This is the first part
4.    <Chapter ID="C1">This is the first chapter.
5.    </Chapter>
6.    <Chapter ID="C2">This is the second chapter.
7.    </Chapter>
8. </Part>
9. <Part ID="P2">This is the second part
10.   <Chapter ID="C3">This is the third chapter.
11.   </Chapter>
12.   <Chapter ID="C4">This is the fourth chapter.
13.   </Chapter>
14. </Part>
15. </Book>
```

The XML version is used to identify version-specific syntax rules to parse the XML document. As of today, there are two versions of XML: version 1.0 and version 1.1. Version 1.1 improves upon XML 1.0 by fixing a few errors and by making the support for Unicode stronger (Table 17.4).

The encoding type refers to the character set (e.g., ASCII, Unicode). Since the XML Recommendation was developed by W3C, an international organization, Unicode was chosen as the standard text format to accommodate the world's language. Commonly used values for encoding attributes are "UTF-8" and "UTF-16," where "UTF" stands for "universal character transformation format," and the number "8" or "16" refers to the number of bits that the character is stored in. Using either UTF-8 or UTF-16, XML editors, generators, and parsers can identify and work with all major world languages and alphabets.

Line 2 and line 15: Root element: Every XML document starts with a root element, and in this example, we have: *<Book>...</Book>*. The start-tag *<Book>* is in line 2, and the end-tag *</Book>* is in line 15. Other elements and text values can be nested under the root element; however, the root element must be the first in the document and it must be unique.

Line 3 to line 8: Under the root element, the first nested sub-element is *<Part>*. This element has an attribute ID ("ID"), which has a value of "P1," while the text for this element is "This is the first part." Underneath the first sub-element *Part*, there are two next-level nested sub-elements of type: *Chapter* (*Chapter 1* and *Chapter 2*). The first sub-element

TABLE 17.4

XML Specifications and Timeline

Specification	Recommendation
XML 1.0	Feb. 10, 1998
XML 1.0 (2nd Edition)	Oct. 06, 2000
XML 1.0 (3rd Edition)	Feb. 04, 2004
XML 1.0 (5th Edition)	Nov. 26, 2008
XML 1.1	Feb. 04, 2004
XML 1.1 (2nd Edition)	Aug. 16, 2006

Source: w3c, 2014. http://www.w3.org.

<Chapter> has an attribute ID ("ID") having a value of "C1," and the text for this element is "This is the first Chapter." Moreover, the end tag has to be nested appropriately to close the element, specifically, we write *</Chapter>*. The second *<Chapter>* sub-element has an attribute ID ("ID") having a value of "C2," and the text is "This is the second Chapter." followed by the closing tag *</Chapter>*, which needs to be prior to the *</Part>* (line 8) sub-element.

Line 9 to line 14: Similar to lines 3 to 8, this is another instance of a nested sub-element *<Part>* under the root element. It has an attribute ID having a value of "P2," while its corresponding text is "This is the second part." In addition, it has two sub-elements *Chapter* (*Chapter 3* and *Chapter 4*).

As seen from the example, a well-formed XML needs to follow rigid syntax and structural rules. Specifically,

- A well-formed XML document must have a corresponding end tag for all its start tags.
- Proper nesting of elements within each other in an XML document is a must.
- In each element, each attribute needs to be distinctly specified.
- An XML document can contain only one root element.
- The root element shall appear only once in an XML document, and it cannot appear as a sub-element within any other element.

EXERCISE 17.2

Well-formed XML versus not well-formed XML

`<Book> </Book>`	Not well formed since XML is case sensitive; start and closing tags do not match
`<Part ID=1> </Section>`	Not well formed since the attribute value needs to be enclosed within quotation marks
`<Part ID="1">`	Not well formed since closing tag is missing
`<Part ID="P1" ID="P2"/>`	Not well formed since attribute name needs to be unique
`<Part id="P1" ID="P2"/>`	Well formed! XML is case sensitive (note that, this, however, is not a good idea)

`<Part ID="P1">`
`    <Chapter ID="C1"/>`
`</Part>`
	Well formed! Attribute name needs to be unique within one start tag or empty-element tag

`<Part>`
`    <Chapter>`
`    </Part>`
`</Chapter>`
	Not well formed since the elements are not properly nested

`<!---correct way of comments --->`	Well formed! This follows the SGML comment tag format

17.3 XML Namespaces

The namespace is a great concept implemented in XML to support extensibility. Since XML allows developers to choose their own element names, duplicated names may appear. A namespace is implemented to identify and differentiate identical names from different vocabularies in an XML document. This is done by prefixing a name to an element and an attribute so that elements with the same name but different prefixes will not be in conflict with each other. The namespace declaration is defined as follows:

```
<prefix:localname xmlns:prefix='namespace URI'/> or
<prefix:localname xmlns:prefix='namespace URI'></prefix:localname/> or
<prefix:localname xmlns:prefix='namespace URI'>children</
prefix:localname/>
```

Note that, in defining namespace declarations, a reserved **xmlns:** prefix is used. The attribute name after the **xmlns:** prefix identifies the name for the defined namespace. An example namespace declaration is as follows:

```
<eb:textbook xmlns:eb= http://www.crcpress.com/engineeringbooks>
```

The name of the namespace is *eb*, which has a uniform resource identifier (URI) http://www.crcpress.com/engineeringbooks. All elements from this namespace will have the prefix *eb*, for example, *eb*:`industrialengineering`, *eb*:`mechanicalengineering`.

Let us take a look at two XML examples shown in Table 17.5.

Apparently, if these two XML documents are put together, the XML parser will not be able to process this document since both have *<Book>* and *<Title>* elements. Using namespace, the problem is solved as shown in Table 17.6.

TABLE 17.5

Two XML Examples Without Using Namespace

Example I: CRC Press Publishes Engineering Textbooks

```
<?xml version="1.0" encoding="UTF-8"?>
<Book>
  <Title>Discrete Event Simulation
  </Title>
</Book>
```

Example II: CRC Press Publishes Computing, Information Technology Textbooks

```
<?xml version="1.0" encoding="UTF-8"?>
<Book>
  <Title>Developing Information Systems for Windows and Web Applications
  </Title>
</Book>
```

TABLE 17.6

Using Namespace

```
<?xml version="1.0" encoding="UTF-8"?>
<eb:Book xmlns:eb="http://www.crcpress.com/engineeringbooks">
  <eb:Title>Discrete Event Simulation
  </eb:Title>
</eb:Book>
<cit:Book xmlns:cit="http://www.crcpress.com/computingitbooks">
  <cit:Title>Developing Information Systems for Windows and
   Web Applications
  </cit:Title>
</cit:Book>
```

Although namespaces are optional components of basic XML documents, they are recommended if the XML documents have any current or future potential of being shared with other XML documents that may have the same element names. In addition, XML-based technologies such as XML schemas, simple object access protocol, and Web services description language make heavy use of namespaces to identify the encoding types and the important elements of the structure.

17.4 XML Validation

Given a well-formed XML, the document needs to be validated against a pre-defined structure. XML validation encompasses element data type verification, proper sub-element nesting, and correct element sequencing, just to name a few. This validation process is based on either a DTD or an XML schema.

17.4.1 DTD and XML Validation

DTD is the traditional way to (1) define the legal building blocks of an XML document and (2) define the specific structure and format of the document using the building blocks. Thus, DTD has been commonly used to validate an XML document.

Example 17.2

For the XML example in Table 17.3, create the DTD document to define its structure (Table 17.7).

Line 1: The *Book* element has one sub-element—*Part*. The "+" sign means that there may be more than one *Part* in the *Book*.

Line 2: The *Part* element has one sub-element—*Chapter*. The "+" sign means that there may be more than one *Chapter* within each *Part*.

Line 3: The element *Part* has an attribute "*ID*." Each instance of the *Part* element is required (#REQUIRED) to have this attribute and must have regular character data (CDATA).

Line 4: The element *Chapter* can contain any data as text.

Line 5: The element *Chapter* has an attribute "*ID*." Each instance of the *Chapter* element is required (#REQUIRED) to have this attribute and must have regular character data (CDATA).

This example illustrates the basic DTD document that can be used to validate an XML document. Next, how the detailed DTD structures are defined is explained.

TABLE 17.7

DTD for Example XML from Table 17.3 (ch17_DTDonSimpleXMLExample.dtd)

```
1. <!ELEMENT Book (Part+)>
2. <!ELEMENT Part (Chapter+)>
3. <!ATTLIST Part ID CDATA #REQUIRED>
4. <!ELEMENT Chapter ANY>
5. <!ATTLIST Chapter ID CDATA #REQUIRED>
```

17.4.1.1 DTD Structure

DTD element structure: There are a number of different syntaxes to define element structures in DTD documents. However, all of these structures use the `!ELEMENT` statement and are mandatory.

- `<!ELEMENT element-name>`: A statement that declares an element. As it is unspecified, the sub-elements can be in any order.

Example 17.3

DTD example: `<!Element Chapter>`
XML example: `<Chapter> text </Chapter>`

- `<!ELEMENT element-name EMPTY>`: Declares an empty element. It is an empty element having no sub-elements, but attributes may be defined for the element.

Example 17.4

DTD example: `<!ELEMENT Appendix EMPTY>`
XML example: `<Appendix/>`

Example 17.5

DTD example: `<!ELEMENT Appendix EMPTY>`

`<!ATTLIST Appendix Appendix_ID>`

XML example: `<Appendix Appendix_ID="Appendix for Chapter 1"/>`

Note: <!ATTLIST> is DTD attribute which will be discussed later in the chapter.

- `<!ELEMENT element-name (#PCDATA)>`: A statement that implies that an element that can contain only text data, but no sub-elements. Note that the non-markup text for the element is commonly taken as parsed character data (PCDATA) in DTD. The text defined as PCDATA will be processed and analyzed for the content by the XML parser.

Example 17.6

DTD example: `<!ELEMENT Section (#PCDATA)>`
XML example: `<Section> text </Section>`

- `<!ELEMENT element-name (child)>`: Means that the element has one sub-element.

Example 17.7

DTD example: `<!ELEMENT Book (Title)>`
XML example:

```
<Book>
    <Title> text </Title>
</Book>
```

- `<!ELEMENT element-name (child1, child2...)>`: A statement that declares an element that has more than one sub-element that follow the order specified.

Example 17.8

DTD example: <!ELEMENT Book (Title, Part)>
XML example:

```
<Book>
   <Title> text </Title>
   <Part> text </Part>
</Book>
```

DTD element cardinality: After the element structure is defined, the element cardinality is used to define how many times each element may appear within the current element. These include the following:

- x: x can appear once.

Example 17.9

DTD example: <!Element Book (Title)>. This DTD cardinality statement means that there is only one *Title* within a book element.
XML example:

```
<Book>
   <Title> text </Title>
</Book>
```

- x+: x can appear one or more times.

Example 17.10

DTD example: <!Element Part (Chapter+)>. This implies that there can be one or more *Chapters* within a *Part* element.
XML example:

```
<Book>
   <Chapter> chapter text </Chapter>
   <Chapter> more chapter text </Chapter>
</Part>
```

- x*: x can appear zero or more times.

Example 17.11

DTD example: <!Element Chapter (Section*)>. This means that there can be zero or more than one *Sections* within a *Chapter* element.
XML example:

```
<Chapter>
   <Section> section text </Section>
   <Section> more section text </Section>
</Chapter>
```

or

```
</Chapter>
   <Section>
</Chapter>
```

- x?: x can appear once or not at all.

Example 17.12

DTD example: <!Element Chapter (Appendix?)>. This statement means that some *Chapters* may have one *Appendix* or some *Chapters* may have no *Appendix*.

XML example:
```
<Chapter>
   <Appendix> appendix text </Appendix>
</Chapter>
</Chapter>
   <Appendix>
</Chapter>
```

- *x, y*: Means that *x* is followed by *y*.

Example 17.13

DTD example: `<!Element Book (Title, Part)>`. This means that the *Book* element has two sub-elements, with the first sub-element being the *Title* element followed by the *Part* element.
XML example:
```
<Book>
   <Title> title text </Title>
   <Part> title text </Part>
</Book>
```

Note that this is the same as Example 17.8.

- *x* | *y*: *x* or *y*, but not both.

Example 17.14

DTD example: `<!Element Books (EngineeringBook|ComputerITBook)>`. This implies that the *Book* element is either an *EngineeringBook* or a *ComputerITBook*, but not both.
XML example:
```
<Book>
   <EngineeringBook> text </EngineeringBook>
</Book>
```

or

```
<Book>
   <ComputerITBook> text </ComputerITBook>
</Book>
```

DTD attribute declaration: In a DTD, attributes are declared with an `ATTLIST` declaration along with their corresponding data type. `CDATA` (character data) is the most common data type to declare for an attribute. This data type is not parsed by the XML processor. Other data types include (1) `ID`: the value is a unique ID; (2) `IDREF`: the value is the ID of another element; (3) `IDREFS`: the value is a list of other IDs; (4) `NMTOKEN`: the value is a valid XML name; (5) `NMTOKENS`: the value is a list of valid XML names; and (6) `ENTITY`: the value is an entity. While there are a number of ways to declare an attribute, some commonly used ones are provided as follows:

- `<!ATTLIST element-name attribute-name CDATA>`: A declaration that implies that the attribute is non-parsed character data.

Example 17.15

DTD example: `<!ATTLIST Book Author>`
XML example: `<Book Author="William Johnson">`

- `<!ATTLIST element-name attribute-name CDATA "default value">`: Means that the attribute is non-parsed character data with the default-value given.

Example 17.16

DTD example: <!ATTLIST square size "10">
XML example: <square size="100"> (Note: If no size is specified, it has a default value of 10.)

- <!ATTLIST element-name attribute-name CDATA #REQUIRED>: Used in cases where the default value is not available and there is still a need to have values set for the attribute. The #REQUIRED keyword is usually used to handle this requirement.

Example 17.17

DTD example: <!ATTLIST Book Total_Pages #REQUIRED>
XML example: <Book Total_Pages="500">

- <!ATTLIST element-name attribute-name CDATA #IMPLIED>: Means that the attribute is optional.

Example 17.18

DTD example: <!ATTLIST Book Co-Authors #IMPLIED>
XML example: <Book/>

- <!ATTLIST element-name attribute-name CDATA #FIXED "value">: A declaration that the attribute is optional in the XML document but is specified in the DTD. The keyword #FIXED can be used to prevent XML developers to change its value.

Example 17.19

DTD example: <!ATTLIST Book Publisher #FIXED "CRC Press">
XML example: <Book Publisher="CRC Press"> or <Book/>

- <!ATTLIST element-name attribute-name ID #REQUIRED attribute-name IDREFS>: Implies that the ID attribute value of each element in an XML document must be distinct, so the ID attribute value is an object identifier. Furthermore, an attribute of type IDREF must contain the ID value of an element in the same document.

Example 17.20

DTD example: <!ATTLIST individual individual_id ID #REQUIRED parent_id IDREFS>
XML example:

```
<individuals>
      <individual individual_id="e10001" parent_id="e10002 e10003">
         <first_name>Bart</first_name>
         <last_name>Simpson</last_name>
      </individual>
      <individual individual_id="e10002">
         <first_name>Homer</first_name>
         <last_name>Simpson</last_name>
      </individual>
      <individual individual_id="e10003">
         <first_name>Marge</first_name>
         <last_name>Simpson</last_name>
      </individual>
</individuals>
```

In this example, the *individual* with ID (*individual_id*) "e10001" is the child of parents with ID e10002" and "e10003." This relationship is described using two attributes: ID and IDREF.

- <!ENTITY entity-name "entity-value">: Enables the reference to data that are outside of the current document. An ENTITY has three parts: an ampersand (&), an entity name, and a semicolon (;).

Example 17.21

DTD example: <!ENTITY copyright http://www.crcpress.com/entities.dtd>
XML example: <author> ©right; </author>

17.4.1.2 Internal versus External DTD

A DTD can be declared inline inside an XML document or as an external reference. To include a DTD inside an XML document, the <!DOCTYPE> element is declared. Taking the examples from Table 17.3 (XML) and Table 17.7 (DTD), an XML document with an internal DTD is shown in Table 17.8.

On the other hand, we can also create a DTD document externally and add a reference in the XML document. Public and private settings are two ways to refer to a DTD document externally.

Let us create a DTD document "Ch17_DTDonSimpleXMLExample.dtd" from Table 17.7. Now, we can specify that we are using an external private DTD by using the SYSTEM keyword in the <!DOCTYPE> element, such as this:

```
<!DOCTYPE Book SYSTEM "Ch17_DTDonSimpleXMLExample.dtd">
```

TABLE 17.8

Example XML Document with Internal DTD

```
<?xml version="1.0" encoding="UTF-8"?>
<!DOCTYPE Book [
<!ELEMENT Book (Part+)>
<!ELEMENT Part (Chapter+)>
<!ATTLIST Part ID CDATA #REQUIRED>
<!ELEMENT Chapter ANY>
<!ATTLIST Chapter ID CDATA #REQUIRED
]>
<Book>
     <Part ID="S1">This is the first part.
        <Chapter ID="C1">This is the first chapter.
        </Chapter>
        <Chapter ID="C2">This is the second chapter.
        </Chapter>
     </Part>
     <Part ID="S2">This is the second part.
        <Chapter ID="C3">This is the third chapter.
        </Chapter>
        <Chapter ID="C4">This is the fourth chapter.
        </Chapter>
     </Part>
</Book>
```

TABLE 17.9

Referring to an External DTD in Private

```
<?xml version="1.0" encoding="UTF-8">
<!DOCTYPE Book SYSTEM "Ch17_DTDonSimpleXMLExample.dtd">
<Book>
   <Part>
   ...
   </Part>
</Book>
```

The corresponding XML document is written as shown in Table 17.9.

In this example, it is assumed that the DTD is located in the same folder as the XML document. In case that the DTD is located in a different place, a uniform resource locator (URL) can be used to point to the DTD as:

```
<!DOCTYPE document SYSTEM "http://www.crcpress.com/dtds/Ch17_
  DTDonSimpleXMLExample.dtd">
```

To create and use a public external DTD, the keyword PUBLIC is used instead of SYSTEM in the `<!DOCTYPE>` DTD. In addition, a formal public identifier (FPI) needs to be created, which is a quoted string of text that is made up of four fields separated by "//." An example of an FPI is "-//W3C//DTD HTML 4.01//EN." The first field indicates whether the DTD is written for a formal standard. For DTDs that are created on our own, this field should be written as "-." If a non-official standard body has created the DTD, the "+" character is used. For formal standard bodies, this field is a reference to the standard itself (such as ISO/IEC 19775:2003). The second field holds the name of the group or person responsible for the DTD. Note that the name used should be unique (e.g., W3C just uses W3C). Furthermore, the third field specifies the type of the document that the DTD is used for and should be followed by a unique version number of some kind (e.g., XML 1.1). The fourth field specifies the language in which the DTD is written for (EN for English).

When a public external DTD is used, the `<!DOCTYPE>` element is defined as:

```
<!DOCTYPE rootname PUBLIC FPI URI>
```

Let us assume that we create a public DTD using a made-up FPI as follows: -//CRCPRESS// My DTD Version 1.0//EN. This document uses Ch17_DTDonSimpleXMLExample.dtd as the external DTD, as in the previous example, but as we can see, it treats the DTD as a public external DTD, complete with its own FPI (Table 17.10).

While DTD is still in use for XML validation, it has several major limitations. First, it is not written in XML syntax; as a result, the developers have to learn a new syntax to develop

TABLE 17.10

Referring to an External DTD in Public

```
<?xml version="1.0" encoding="UTF-8">
<!DOCTYPE Book PUBLIC "-//CRCPRESS//My DTD Version 1.0//EN"
   "http://www.crcpress.com/dtds/Ch17_DTDonSimpleXMLExample.dtd">
<Book>
   <Part>
   ...
   </Part>
</Book>
```

the DTD. Second, the DTD does not support namespaces, which limits its extensibility. In addition, there are no constraints imposed on the type of character data (e.g., Date, Time). To address these issues, an XML schema is recommended by W3C as an updated document format for XML validation.

17.4.2 XML Schema and XML Validation

XML schema is an XML-based alternative to DTD that is used to describe the structure of an XML document and, thus, is utilized to validate it. The XML schema language is also referred to as "XML schema definition" (XSD).

17.4.2.1 XML Schema Data Types

One drawback of DTDs is that it has limited support for pre-defined data types. An XML schema can support all DTD data types, including `CDATA` (replaced by primitive String data type), `ID`, `IDREF`, `IDREFS`, `ENTITY`, `ENTITIES`, `NMTOKEN`, `NMTOKENS`, and `NOTATION`. In addition, other data types are supported in a different format. Some of the primary types of data used in a schema are summarized in Table 17.11.

17.4.2.2 XML Schema Namespaces

One major advantage of XML schema over DTD is that XML schema supports namespaces as part of its document. For example, the `<schema>` element is the root element of every XML schema and is usually defined as shown in Table 17.12.

In this example,

Line 1: Indicate that the elements and data types used in the `schema` element (the root element) come from http://www.w3.org/2001/XMLSchema namespace, with the prefix being **`xs`**:, for example, **`xs`**:`schema`.

Line 2: Under the root element, the developer defines the elements coming from "http://www.crcpress.com" namespace.

TABLE 17.11

XML Schema Data Types

Name	Description
`String`	Any well-formed XML string
`Date`	Date value in the format YYYY-MM-DD
`Time`	Time value in the format HH:MM:SS
`dateTime`	Combined date and time value in the format YYYY-MM-DD HH:MM:SS
`Number`	Any numeric value up to 18 decimal places
`Decimal`	Any decimal value number
`Float`	Any 32-bit floating point–type real number
`Double`	Any 64-bit floating point–type real number
`Integer`	Any integer
`Byte`	Any signed 8-bit integer
`Int`	Any signed 32-bit integer
`Long`	Any signed 64-bit integer
`Boolean`	Standard binary logic, in the format of 1, 0, true, or false

TABLE 17.12

Schema Element in an XML Schema Document

```
1. <xs:schema xmlns:xs=http://www.w3.org/2001/XMLSchema
2.  targetNamespace=http://www.crcpress.com
3.  xmlns=http://www.w3schools.com
4.  elementFormDefault="qualified">
    ...
5. </xs:schema>
```

Line 3: Indicate that the default name space is "http://www.w3schools.com."

Line 4: Indicate that any element declared in this XML schema must be namespace qualified.

Line 5: The closing tag of the root element `schema` to ensure that it is a well-formed XML itself.

17.4.2.3 XML Schema Elements and Attributes

Within the schema root element, the details of every element contained in the XML document are defined. There are two types of elements: simple and complex. Each element type is defined as follows.

17.4.2.3.1 Simple Elements

A simple element is an XML element that contains only text and does not contain sub-elements or attributes. The text can be of any different types, including some listed in Table 17.11, or customized as defined by the developer. The syntax for defining a simple element is

```
<xs:element name="xxx" type="yyy"/>
```

An example XML schema is

```
<xs:element name="person" type="xs:string"/>
<xs:element name="age" type="xs:integer"/>
```

A simple element may have a default value of a specified fixed value. For example,

```
<xs:element name="size" type="xs:integer" default="10"/>
<xs:element name="publisher" type="xs:string" fixed="CRCPress"/>
```

17.4.2.3.2 Complex Elements

On the other hand, a complex element is an XML element that contains nested elements and/or attributes.

- The syntax for defining a complex element that only contains nested elements is

```
<xs:element name="xxx">
      <xs:complexType>
            <xs:sequence>
                  <xs:element name="xxx-xxx" type="xs:yyy"/>
                  <xs:element name="xxx-xxx-xxx" type="xs:yyy"/>
            </xs:sequence>
      </xs:complexType>
</xs:element>
```

- The syntax for defining a complex element that only contains attributes is

```
<xs:element name="xxx">
      <xs:complexType>
            <xs:attribute name="xxx-xxx" type="xs:yyy"/>
      </xs:complexType>
</xs:element>
```

- For a complex element containing simple contents (text and attribute) only, the `simpleContent` element can be used. `simpleContent` will use an extension or a restriction, and the syntax is defined as follows:

```
<xs:element name="xxx">
      <xs:complexType>
            <xs:simpleContent>
                  <xs:extension base="basetype">
                    ....
                    ....
                  </xs:extension>
            </xs:simpleContent>
      </xs:complexType>
</xs:element>
```

or

```
<xs:element name="xxx">
      <xs:complexType>
            <xs:simpleContent>
                  <xs:restriction base="basetype">
                    ....
                    ....
                  </xs:restriction>
            </xs:simpleContent>
      </xs:complexType>
</xs:element>
```

- The syntax for defining a complex element containing both sub-elements and attributes will be a combination of the two discussed above. In addition, the mixed attribute must be set to "true." As an example, the syntax is

```
<xs:element name="xxx">
      <xs:complexType>
            <xs:sequence>
                  <xs:element name="xxx-xxx" type="xs:yyy"/>
                  <xs:element name="xxx-xxx-xxx" type="xs:yyy"/>
            </xs:sequence>
                  <xs:attribute name="xxx-xxx-xxx" type="xs:yyy"/>
      </xs:complexType>
</xs:element>
```

17.4.2.3.3 XML Schema Extensibility

One additional function provided by XML schema is the use of the `<any>` and `<anyattribute>` elements. The `<any>` element enables developers to extend the XML document with elements not specified by the schema. To serve the same purpose, the `<anyAttribute>` element enables the developers to extend the XML document with attributes not specified by the schema as well.

17.4.2.3.4 XML Schema Indicator

Similar to DTD element cardinality, XML schema introduces Indicators to control how elements may appear in XML documents. The syntax is as follows:

```
<xs:element name="xxx">
      <xs:complexType>
            <xs:orderindicator>
                  <xs:element name="xxx-xxx" type="xs:yyy"/>
                  <xs:element name="xxx-xxx-xxx" type="xs:yyy"/>
            </xs:orderindicator>
      </xs:complexType>
</xs:element>
```

The **orderindicator** can be one of the three: `all`, `choice`, and `sequence`. The `all` indicator means that the sub-elements can appear in any order and that each sub-element must occur only once. On the other hand, the `choice` indicator means that only one of the sub-elements can appear, while the `sequence` indicator implies that the sub-elements must appear in the order specified. Often, other types of indicators can be added into the schema to define the number of times that an element can appear. These **occurindicators** include `minOccurs` and `maxOccurs`. The default value for both indicators is "1." The syntax that includes this type of indicator is as follows:

```
<xs:element name="xxx">
      <xs:complexType>
            <xs:indicatortype>
               <xs:element name="xxx-xxx" type="xs:yyy" occurindicators
                ="zzz"/>
               <xs:element name="xxx-xxx-xxx" type="xs:yyy"/>
            </xs: indicatortype>
      </xs:complexType>
</xs:element>
```

17.4.2.3.5 Other XML Schema Element and Data Type Restriction

Other than the elements listed above, there are several other types of elements that define constraints on other elements in the schema, as summarized in Table 17.13.

17.4.3 Referencing XML Schema for Validation

Let us use an example to explain how XML schema is referenced for validation.

Example 17.22

For the XML example from Table 17.3, create the XML schema.

The result is shown in Table 17.14.

Line 1: Different from DTD, an XML schema is based on XML syntax. So, it starts with an XML declaration.

Line 2: Declare the root element of the XML, in this example, it is a *Book*.

TABLE 17.13

XML Schema Element Restrictions

Restriction	Description
`fractionDigits`	Maximum number of decimal places for a value. Integers are 0
`length`	Number of characters or, for lists, number of list choices
`maxExclusive`	Maximum up to, but not including, the number specified
`maxInclusive`	Maximum including the number specified
`maxLength`	Maximum number of characters or, for lists, number of list choices
`minExclusive`	Minimum down to, but not including, the number specified
`minInclusive`	Minimum including the number specified
`minLength`	Minimum number of characters or, for lists, number of list choices
`pattern`	Defines a pattern and sequence of acceptable characters
`totalDigits`	Number of non-decimal, positive, non-zero digits
`whiteSpace`	How line feeds, tabs, spaces, and carriage returns are treated when the document is parsed

Line 3 to line 11: The complex element *Part* is defined. The *Part* element is defined as a mixed element since it includes several *Chapter* sub-elements and one attribute named *ID*. The number of *Chapters* is unbounded. In addition, the attribute *ID* is required.

Line 12 to line 21: The complex element *Chapter* is defined. The *Chapter* element has an attribute named *ID*, which is required. Moreover, the text of the *Chapter* element uses

TABLE 17.14

XML Schema for Example XML from Table 17.3

```
1. <?xml version="1.0" encoding="UTF-8"?>
2. <xs:schema xmlns:xs="http://www.w3.org/2001/XMLSchema">
3.     <xs:element name="Part">
4.         <xs:complexType mixed="true">
5.             <xs:sequence>
6.                 <xs:element name="Chapter" maxOccurs="unbounded"/>
7.             </xs:sequence>
8.             <xs:attribute name="ID" use="required">
9.             </xs:attribute>
10.        </xs:complexType>
11.    </xs:element>
12.    <xs:element name="Chapter">
13.        <xs:complexType>
14.            <xs:simpleContent>
15.                <xs:extension base="xs:string">
16.                    <xs:attribute name="ID" use="required">
17.                    </xs:attribute>
18.                </xs:extension>
19.            </xs:simpleContent>
20.        </xs:complexType>
21.    </xs:element>
22.    <xs:element name="Book">
23.        <xs:complexType>
24.            <xs:sequence>
25.                <xs:element name="Part" maxOccurs="unbounded"/>
26.            </xs:sequence>
27.        </xs:complexType>
28.    </xs:element>
29. </xs:schema>
```

a string data type. Note that appropriate closing tags are required to ensure that XML schema itself is a well-formed XML.

Line 22 to line 28: The complex element *Book* is defined. The *Book* element includes a number of *Part* sub-elements.

Line 29: This line is the closing tag for the root element `schema`.

To reference this XML schema in an XML document, we need to add the following at the top of the XML document right after the XML document declaration (note that, here, we use a make-up URI http://www.CRCPress.com):

```
<Book xmlns:xsi="http://www.w3.org/2001/XMLSchema-instance" xsi:noName
  spaceSchemaLocation="http://CRCPress.com/BookExampleXSD.xsd">
```

In this case, the namespace declaration referencing to http://www.w3.org/2001/XMLSchema-instance resolves to an actual document at that location, which is a brief description of the way that the W3C schema should be referenced, and to a link to the actual schema that describes schema data types, elements, and other schema descriptions based on the current W3C Recommendation. The noNamespaceSchemaLocation value tells us that there is no pre-defined namespace for the schema, but the location of the schema is http://www.CRCPress.com/BookExampleXSD.xsd.

17.5 XML Editor Tools

There are a number of XML editors available online for creating and validating XML. In this chapter, all the examples are created using XMLSpy (http://www.altova.com/). There are also a number of applications from database vendors that are providing interfaces to XML. Some of these will be discussed in the next chapter.

17.6 Summary

This chapter covers the basics of XML. Starting with SGML, the markup language principles are discussed. The major differences between SGML, HTML, and XML are then explained. Noting that XML is developed to exchange data over the Internet, which, thus, requires a strict rule to define the syntax, the basic building blocks of XML, that is, element, attribute, and text are explained. Next, the concept of a well-formed XML is introduced. Lastly, DTD and XML schema are introduced and illustrated for XML validation.

REVIEW EXERCISES

1. SGML elements are used to define the structure of an SGML document. True or false?
2. An HTML will not be displayed correctly when `<body>` is closed with `</Body>`. True or false?
3. HTML rules emphasize the use of start and end tags while SGML does not. True or false?
4. List three characters that are not included in naming XML elements and give corresponding examples.

5. An XML attribute can be placed in an XML element's end tag. True or false?
6. The following are ways to declare an empty element except:
   a. `<name> </name>`
   b. `</name>`
   c. `</name>`
   d. All of the above are valid ways to declare an empty element.
   e. None of the above are valid ways to declare an empty element.
7. Namespaces are optional when XML documents are shared with other XML documents of different element names. True or false?
8. XML DTDs are used for the following except:
   a. Defining blocks that provide a good structure for an XML document
   b. Formatting the document
   c. Defining which elements can have sub-elements
   d. Defining which element attributes are required
   e. All of the above are valid DTD purposes
9. One major advantage of XML schemas over DTDs is that XML schema supports namespaces. True or false?
10. Differentiate a simple element as compared to a complex element and provide examples.
11. Given the attribute declaration: `<!ATTLIST Prod_N CDATA #REQUIRED>`, the following XML document codes are valid except:
    a. `<Product Prod_N="121123">`
    b. `<Product Prod_N="12-1223">`
    c. `<Product> Prod_N="12-1223" </Product>`
    d. `<Product Prod_N="_121123_">`
    e. All of the above are valid XML codes for the attribute declaration.
12. Given the element declaration: `<!ELEMENT Food_Item(Expiration?)>`, the following XML document codes are valid except:
    a. `<Food_Item Expiration=""> </Food_Item>`
    b. `<Food_Item> </Food_Item Expiration="">`
    c. `<Food_Item><Expiration> </Food_Item>`
    d. `<Food_Expiration> </Food_Expiration>`
    e. None of the above is a valid XML code for the element declaration.
13. Which element is used to declare a DTD inside an XML document?
    a. `<!DOCTYPE>`
    b. `<!DTD>`
    c. `<!DTDTYPE>`
    d. `<!SYSTEM>`
    e. `<!XMLTYPE>`
14. Enumerate two limitations of DTDs that gave rise to XML schemas.

15. Give an example of an FPI with the following properties: (1) A Non-Standards Institution: the Institute of Operations Research and Management Science (INFORMS) generated the XML, (2) INFORMS is responsible for DTD and (3) is used for the INFORMS Journal of Computing new article format with current version 1.1 with (4) English as its language.
16. Given the simple XML schema definition: <xs:element name="Sales_Order" type="xs:string" default="Standard">, which of the following statements is true?
    a. We are declaring an attribute named "Sales_Order" with default value "Standard".
    b. We are declaring an element named "Sales_Order" with attribute Type composed of integers and default value "Standard".
    c. We are declaring a string element named Sales_Order with attribute Type and default value "Standard".
    d. We are declaring an attribute named Sales_Order with sub-attribute Type and default value "Standard".
    e. The XML schema will not compile since this is a complex XML schema definition.
17. Valid XML schema order indicators are:
    a. Sequence
    b. Series
    c. All
    d. All of the above
    e. Both a and c
18. Given the XML element declaration: <xs:element name="interest_rate" type="xs:float" default="2.21"/>, which of the following is a valid XML implementation of the schema?
    a. <interest_rate>2.21<interest_rate>
    b. <interest_rate>3.67%</interest_rate>
    c. <interest_rate rate="3.67"/>
    d. <interest_rate>3.67</interest_rate>
    e. Both b and d are valid implementations.
19. The <any> element extends the XML document with attributes not specified by the schema. True or false?
20. Given the following complex element declaration:

```
<xs:element name="sales_order">
  <xs:complexType>
    <xs:attribute name="sales_order_n" type="integer"/>
  </xs:complexType>
</xs:element>
```

   Which of the following are valid declarations of XML schema?
    a. <sales_order sales_order_n=1221232></sales_order>
    b. <sales_order sales_order_n=454><sales_item>Text</sales_item></sales_order>

c. `<sales _ order sales _ order _ n=112.94></sales _ order>`
d. All of the above
e. Both a and b

PRACTICAL EXERCISES

1. The following is an XML document that describes a single Purchase Order with multiple line items of fine jewelry.

```
 1. <?xml version="1.0" encoding="UTF-16"?>
 2. <Purchase_Order>
 3. <Billing_Info>
 4.        <Bill_Name>Irish Maravillas</Bill_Name>
 5.        <Bill_Street>1975 East University</Bill_Street>
 6.        <Bill_City>Tempe</Bill_City>
 7.        <Bill_State_Code>AZ</Bill_State_Code
 8.        <Bill_Zip_Code>85281</Bill_Zip_Code>
 9. </Billing_Info>
10. <Ship_Info>
11.        <Ship_Name>Jennifer Jareau</Ship_Name>
12.        <Ship_Street>3600 Wilshire Blvd.</ship_Street>
13.        <Ship_City>Los Angeles</Ship_City>
14.        <Ship_State_Code>CA</Ship_State_Code>
15.        <Ship_Zip_Code>90010</Ship_Zip_Code>
16. </Billing_Info>
17. <Purchase_Order>
18. </Purchase_Order>
19. <Items>
20.        <Item itemNum=N28812>
21.               <Product.Name>Diamond Necklace</Product.Name>
22.               <Quantity=3</Quantity>
23.               <_UnitPrice>99.85</_UnitPrice>
24.        </Item>
25.        <Item itemNum=R15632>
26.               <Product.Name>Star Trek Ring</Product.Name>
27.               <Quantity>1</Quantity>
28.               <_UnitPrice>1985.90</_UnitPrice>
29.        </Item>
30. </Items>
31. </Purchase_Order>
```

Identify which lines of code do not adhere to the guidelines set forth by WC3 in writing well-formed XML documents.

2. Consider the following e-mail message with `ID="823324"`:
   1. From: gilbert@hatanalytics.com
   2. To: catherine@hatanalytics.com, nick@hatanalytics.com
   3. cc: warrick@hatanalytics.com
   4. Subject: Installation of ERP Server
   5. Content: We will be performing maintenance in our servers and, as such, our access to the cloud-based information systems will be intermittent.

   Create an XML document for this e-mail message.

3. Consider the form. This form is a report of a comprehensive examination that is filled out after a student takes the comprehensive exam.

Arizona State University
Report of Comprehensive Exam Results

Name: Aaron Mendez
Student Number: 2002-36001
Program: MS Computer Science
Thesis Title: Addressing Issues on Large-Scale Pairwise Comparison Matrices
Committee Member Results:

ID	Committee Position	Name	Result
1	Co-Chair	Erin Strauss	High Pass
2	Co-Chair	Conrad Grissom	Pass
3	Member	William Li	Pass

Department Chair: Michelle Lazaro
College Dean: Haley Hotchner

Create a well-formed XML document for this form.

4. A student's transcript of records in XML is shown as follows:

```
1.  <TORForm>
2.  <Title Text_Align="Center">
3.          Transcript of Records
4.  </Title>
5.  <Stud_M Stud_N="1202703393">
6.          <First_M>Theodore</First_M>
7.          <Middle_M>Evelyn</Middle_M>
8.          <Last_M>Mosby</Last_M>
9.  </Stud_M>
10. <Deg_Program>BS Systems Engineering</Deg_Program>
11. <Semester Sem_ID="2314" Year="2012" Term="Fall">
12.         <Course Course_N="74336">
13.                 <Course_M>Information Systems</Course_M>
14.                 <Units>3</Units>
15.                 <Grade>A</Grade>
16.         </Course>
17.         <Course Course_N="73979">
18.                 <Course_M>Stochastic Optimization</Course_M>
19.                 <Units>3</Units>
20.                 <Grade>B+</Grade>
21.         </Course>
22. </Semester>
23. <Semester Sem_ID="2314" Year="2013" Term="Spring">
24.         <Course Course_N="78557">
25.                 <Course_M>Research Methods</Course_M>
26.                 <Units>1</Units>
27.                 <Grade>C</Grade>
28.         </Course>
29. </Semester>
30. </TORForm>
```

Create a document-like representation for the content of the transcript of records.

5. The following is an example of a DTD for a newly hired employee form.

```
 1. <!DOCTYPE EmpForm[
 2. <!ELEMENT Emp_M (First_M, Middle_M, Last_M)>
 3. <!ATTLIST Emp_M Emp_N CDATA #REQUIRED>
 4. <!ELEMENT First_M CDATA>
 5. <!ELEMENT Last_M CDATA>
 6. <!ELEMENT Address (Street_M, City_M, State_M)>
 7. <!ELEMENT Acad_Background (Degree+)>
 8. <!ATTLIST Degree Degree_N CDATA #REQUIRED>
 9. <!ELEMENT Degree (University, Course, GradYear)>
10. <!ELEMENT University CDATA>
11. <!ELEMENT Course CDATA>
12. ]>
```

Generate a well-formed XML document that is validated by this DTD. The example should have at least two Academic Degrees.

6. Consider the following XML schema:

```
 1. <?xml version="1.0" encoding="UTF-8"?>
 2. <xs:schema xmlns:xs="http://www.w3.org/2001/XMLSchema">
 3.        <xs:element name="Service" type="Service_Contract"/>
 4.        <xs:complexType name="Service_Contract">
 5.               <xs:sequence>
 6.                      <xs:element name="Service_Desc"
                          type="xs:string"/>
 7.                      <xs:element name="Service_Cost"
                          type="CostType"/>
 8.                      <xs:element name="Service_Pay_Terms"
                          type="xs:string" default="NET30"/>
 9.                      <xs:element name="Service_Item"
                          type="Service_Items"/>
10.               </xs:sequence>
11.               <xs:attribute name="Service_N" type="xs:integer"
                   use="required"/>
12.               <xs:attribute name="Service_Start_D" type="xs:date"/>
13.               <xs:attribute name="Service_End_D" type="xs:date"/>
14.        </xs:complexType>
15.        <xs:complexType name="Service_Items">
16.               <xs:sequence>
17.                      <xs:element name="Service_M"
                          type="xs:string" maxOccurs="unbounded"/>
18.                      <xs:element name="Service_Cost"
                          type="xs:float"/>
19.               </xs:sequence>
20.               <xs:attribute name="Service_Item_N"
                   type="xs:integer" use="required"/>
21.        </xs:complexType>
22.        <xs:simpleType name="CostType">
23.               <xs:restriction base="xs:float">
24.                      <xs:minInclusive value="100.00"/>
25.                      <xs:maxInclusive value="unbounded"/>
26.               </xs:restriction>
```

```
27.         </xs:simpleType>
28. </xs:schema>
```

Generate a well-formed XML document that is validated by this XML schema. The example should have at least three Service Items.

7. The following is an example of a dental patient invoice in DTD format:

```
 1. <!DOCTYPE Patient_Invoices[
 2. <!ELEMENT Patient_Invoices(Patient_Invoice+)>
 3. <!ELEMENT Patient_Invoice(Patient_Info, Insurance_Info?, Dental_
      Procedure+)>
 4. <!ATTLIST Patient_Invoice Invoice_N CDATA #REQUIRED>
 5. <!ELEMENT Patient_Info (Patient_M, Phone, Address)>
 6. <!ATTLIST Patient_Info Patient_SSN CDATA #REQUIRED >
 7. <!ELEMENT Patient_M(First_M, Middle_M, Last_M)>
 8. <!ELEMENT Phone EMPTY>
 9. <!ATTLIST Phone Phone_N CDATA Phone_Type CDATA>
10. <!ELEMENT Insurance_Info (Company_M, Plan_N, Phone, Address)>
11. <!ELEMENT Dental_Procedure(Procedure_M, Cost)>
12. <!ATTLIST Dental Procedure Procedure_N CDATA #REQUIRED>
13. ]>
```

Construct an XML schema from this DTD document.

8. Generate an XML schema for the following business rules:
   a. A bank is composed of many accounts.
   b. There are two types of accounts. Savings and Checking accounts. Each account must have a unique 10-digit account number and is linked to a single customer account. It is composed of the following elements: available balance, current balance, and account tier type. An account can have a maximum daily withdrawal limit and a minimum maintaining balance.
   c. Each account has many transactions. Each transaction has the following elements: posting date, transaction details, and amount (positive for deposits, negative for withdrawals). A transaction has a required attribute called "transaction number."

18

Working with XML (II)

Data independence, that is, the separation of content from its presentation, makes extensible markup language (XML) an ideal framework for data exchange. Since XML is designed to describe the data presented in the document, an application that sends and receives XML would need interfaces to interpret the data before it is integrated into the applications. This chapter first introduces two application programming interfaces (APIs) for parsing XML documents: document object model (DOM) and simple API for XML (SAX). The XML query language is then briefly reviewed. The interface between some commonly used database applications and XML is illustrated.

18.1 XML Parsers

An XML parser is a tool that manages the data content of an XML document. It identifies and converts XML elements into either nested nodes in a tree structure (e.g., DOM) or document events (e.g., SAX).

The W3C DOM is a standardized API used to define the logical structure of an XML document and the way the document is accessed and manipulated. Because an XML document is hierarchically structured, the DOM defines a tree structure to represent the entire XML document with the DOM root node matching the root element in the XML document. In addition, every XML element is matched to a DOM element node, an XML text is matched to a DOM text node, an XML attribute is matched to a DOM attribute node, and an XML comment is matched to a DOM comment node. These DOM nodes are properly located in the tree that correspond to their structural relationship defined in the XML document.

Let us revisit the *Book* example from Chapter 17 (shown in Table 18.1). The DOM tree is illustrated in Figure 18.1. As seen in the figure, the root element of the XML is <*Book*>, which has a sub-element named <*Part*>. Furthermore, the <*Part*> sub-element has a sub-element named <*Chapter*>. The <*Part*> and <*Chapter*> elements are connected to an attribute node and a text node, respectively. Working with the DOM tree, developers have the freedom to modify, delete, and add nodes to the tree whenever it is necessary. In this simple example, only the parent–child relationship is illustrated. There is another type of relationship called the "sibling," which defines a relationship between two element nodes having the same parent. For example, assume that the <*Book*> element has another sub-element called <*Title*> at the same level as <*Part*>. A corresponding <*Title*> element node will be added at the same level as <*Part*> in the DOM tree.

Any DOM-based parser will first convert an XML into DOM objects before the data are retrieved and manipulated. Many applications (such as Web browsers) have a built-in parser to make the conversion. To understand how a DOM is used as an XML parser, we first summarize some important DOM interfaces (Table 18.2), the methods for Documents

TABLE 18.1
Example XML Document

```
 1.<?xml version="1.0" encoding="UTF-8"?>
 2.<Book>
 3.  <Part ID="P1">This is the first part.
 4.         <Chapter ID="C1">This is the first chapter.
 5.         </Chapter>
 6.         <Chapter ID="C2">This is the second chapter.
 7.         </Chapter>
 8.  </Part>
 9.  <Part ID="P2">This is the second part.
10.         <Chapter ID="C3">This is the third chapter.
11.         </Chapter>
12.         <Chapter I D="C4">This is the fourth chapter.
13.         </Chapter>
14.  </Part>
15.</Book>
```

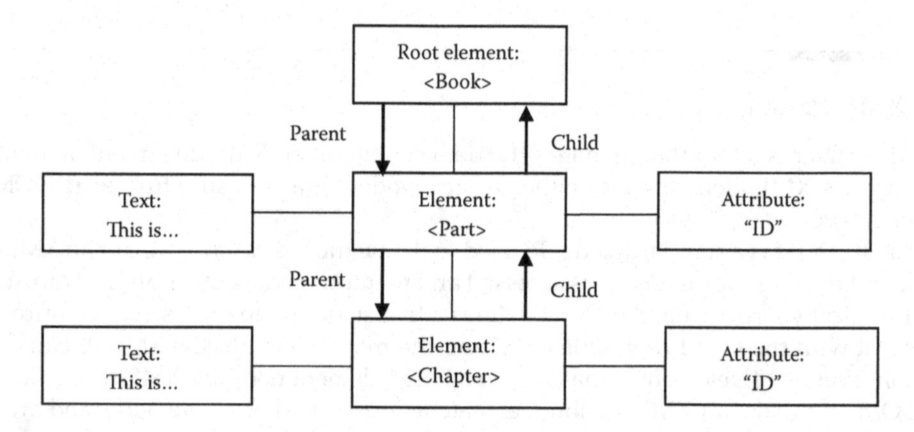

FIGURE 18.1
DOM tree representation of the XML document.

TABLE 18.2
Some DOM Interfaces

DOM Interfaces	Description
Document	Represents the XML documents' top-level node, which provides access to all the document nodes
Node	Represents an XML document node
NodeList	Represents a read-only list of Node objects
Element	Represents an element node; derives from Node
Attr	Represents an attribute node; derives from Node
Text	Represents a text node; derives from Node
Comment	Represents a comment node; derives from CharacterData
ProcessingInstruction	Represents a processing instruction node; derives from Node
DATASection	Represents a CDATA section; derives from Text

(Table 18.3), the methods for Nodes (Table 18.4), the methods for Elements (Table 18.5), and the several types of Nodes (Table 18.6).

The DOM-based parsers have been written in a variety of programming languages, including Java, Javascript, C#, and VB.NET. In this chapter, two illustrative examples are provided: the first one is written in Javascript, and the second one is written in VB.NET. These examples are used to illustrate the use of some of the interfaces, the methods of which are listed above.

TABLE 18.3

Some Document Methods

Document Methods	Description
`createElement`	Creates an element node
`createAttribute`	Creates an attribute node
`createTextNode`	Creates a comment node
`createProcessingInstruction`	Creates a processing instruction node
`createCDATASection`	Creates a CDATA section node
`getDocumentElement`	Returns the document's root element
`appendChild`	Appends a child node
`getChildNodes`	Returns the child nodes
`createXmlDocument`	Parses an XML document
`write`	Outputs the XML document

TABLE 18.4

Some Node Methods

Node Methods	Description
`appendChild`	Appends a child node
`cloneNode`	Duplicates the node
`getAttributes`	Returns the node's attributes
`getChildNodes`	Returns the node's child nodes
`getNodeName`	Returns the node's name
`getNodeType`	Returns the node's type (e.g., element, attribute, text, etc.)
`getNodeValue`	Returns the node's value
`getParentNode`	Returns the node's parent
`hasChildNodes`	Returns true if the node has child nodes
`removeChild`	Removes a child node from the node
`replaceChild`	Replaces a child node with another node
`setNodeValue`	Sets the node's value
`insertBefore`	Appends a child node in front of a child node

TABLE 18.5

Some Element Methods

Element Methods	Description
`getAttribute`	Returns an attribute's value
`getTagName`	Returns an element's name
`removeAttribute`	Removes an element's attribute
`setAttribute`	Sets an attribute's value

TABLE 18.6
Some Node Types

Node Types	Description
`Node.ELEMENT_NODE`	Represents an element node
`Node.ATTRIBUTE_NODE`	Represents an attribute node
`Node.TEXT_NODE`	Represents a text node
`Node.COMMENT_NODE`	Represents a comment node
`Node.PROCESSING_INSTRUCTION_NODE`	Represents a processing instruction node

Example 18.1

A DOM-based parser written in Javascript (Table 18.7).
In this JavaScript example,

Line 1: Import (i.e., specifies the location of) the classes needed by the program. In this example, some classes from the package *org.apache.xerces.parsers.DOMParser* are imported.
Line 5: Declare and create an instance of a DOM parser object.

TABLE 18.7
Example DOM (JavaScript)

```
 1.import org.apache.xerces.parsers.DOMParser;
 2.public class DOMWriter {
 3.public static void main( String[] args) {
 4.try {
 5.DOMParser parser=new DOMParser();
 6.parser.parse("Chapter 17.xml");
 7.   Document document=parser.getDocument();
 8.   processElement(document.
   getDocumentElement());
 9.} catch (SAXException e) {
10.e.printStacktrace();
11.} catch (IOException e) {
12.e.printStacktrace();
13.}
14.}
15.private static void processElement( Element
   element) {
16.NodeList nodes=element.getChildNodes();
17.System.out.println("<"+element.
   getNodeName()+">");
18.for (int i=0; i < nodes.getLength(); i++) {
19.Node node=nodes.item(i);
20.if (node.getNodeType()==Node.ELEMENT_NODE)
   {
21.processElement((Element)node);
22.} else {
23.System.out.println( node.getNodeValue());
24.}
25.}
26.System.out.println("</"+element.
   getNodeName()+">");
27.}
28.}
```

Line 6: The DOM parser is assigned to read the "Chapter 17.xml" document.
Line 7: Declare and create an instance of a Document object, called `document`. This object will start to parse the XML document via calling the `getDocument` method.
Line 8: First, the `document.getDocumentElement` method will get the root element of the XML document, in this example, the *<Book>* element. Next, a function `processElement` is called.
Line 16: For the root element, *<Book>*, its child elements are obtained by calling the method `element.getChildNodes` and are then assigned to an instance of the *NodeList*.
Line 17: For the root element, its name (in this example, *Book*) is obtained by calling the method `element.getNodeName`. The output is the root element name with "<" added at the left and ">" added at the right.
Line 18: The number of child elements for the root element is obtained by calling `nodes.getLength`.
Line 19: This is the start of a FOR loop for each child element (the index is used in `nodes.item(i)`).
Line 20: If the node is an element node, a recursive call of `processElement` is triggered. This will parse through the tree from the root element to the leaf elements.
Line 21: If the node is not an element node, but instead, it is an attribute node or a text node, the value of the node is obtained by calling the `node`.getNodeValue function.
Line 26: Finally, as the whole XML document is parsed through, it is closed with a closing tag, which is presented by the value of the root element adding "</" at the left and ">" at the right.

The output of this DOM is an XML document shown in Table 18.1 but without line 1, which is an XML declaration. This example only illustrates parsing through the XML document that mainly uses the `getNodeValue` function; other methods can be called to make changes (e.g., delete the Node) as the original XML document is parsed through.

Example 18.2

A DOM-based parser written in VB.NET.

The Microsoft .NET platform provides its own classes to support a DOM parser and, thus, defines Microsoft XML core services. Note that the functionalities of the basic interfaces and methods listed in Tables 18.2 to 18.6 are the same, but the names may vary. In this example, we illustrate a Windows application. Please refer to Chapters 11 and 12 for details on Windows forms, controls, and VB.NET coding. Here, we will briefly explain the interfaces and methods related to the parser (Table 18.8).

In this VB.NET example,

Lines 1 to 2: Import (i.e., specify the location of) the classes needed by the program. For example, some classes from package *System.xml* and *System.IO* are imported.
Lines 3 to 4: This is a VB.NET Windows application with a Button control on the Windows form.
Line 5: Declare and create an instance of a DOM document.
Line 6: Declare xmlnode as XmlNodeList object, a list of node objects (functions as NodeList from Table 18.2).
Line 10: Load the XML document.
Line 11: The root node of the XML is obtained and is assigned to the xmlnode instance.

TABLE 18.8

Example DOM (VB.NET)

```
1. Imports System.Xml
2. Imports System.IO
3. Public Class Form1
4. Private Sub Button1_Click(ByVal sender As System.Object, ByVal e As System.
   EventArgs) Handles Button1.Click
5.          Dim xmldoc As New XmlDataDocument()
6.          Dim xmlnode As XmlNodeList
7.          Dim i As Integer
8.          Dim str As String
9.          Dim fs As New FileStream("book.xml", FileMode.Open, FileAccess.Read)
10.                  xmldoc.Load(fs)
11.                  xmlnode=xmldoc.GetElementsByTagName("Part")
12.                  For i=0 To xmlnode.Count - 1
13.                           xmlnode(i).ChildNodes.Item(0).InnerText.Trim()
14.                           str=xmlnode(i).ChildNodes.Item(0).InnerText.Trim()
15.                           MsgBox(str)
16.                   Next
17. End Sub
18. End Class
```

Lines 12 to 16: The start of the FOR loop. Here, the element name is specified, that is, *<Part>*. As a result, for each *<Part>* element within the XML, its value and the information from its sub-elements are retrieved, and a message box is used to pop up the outputs.

In summary, both DOM parser examples indicate that the DOM is centered on a tree structure. DOM-based parsers will always scan through the entire XML document, no matter how much it is actually needed by the developer. By contrast, SAX is an event-based API that is known to be faster than DOM as it retrieves only related data from the document.

However, DOM will store the entire document's data in memory, and data can be quickly accessed since the SAX parser will pass the data to the application as it is found. There have always been discussions on DOM versus SAX. One recommendation is that SAX performs well for read-only access, while DOM works well if the developers need to alter, delete, and add data into the XML tree. Since SAX is not a W3C standard, we will not discuss this topic in detail here.

18.2 XML Query

XML parsers (e.g., DOM) are able to access and manipulate XML data using programming languages. XQuery is another W3C standard, developed to query XML documents. As a point of reference, XQuery is to XML as SQL is to databases. It is designed to replace proprietary middleware languages as well as Web application development languages. XQuery is built on XPath expressions. Indeed, XQuery 1.0 and XPath 2.0 share the same data model and support the same functions and operators. In this chapter, we list some

TABLE 18.9

XPath (XQuery) Location Operators

Operator	Description
.	The current node
..	The parent node
/	The root element
//	All descendants
@	Attribute identifier
*	All child nodes

important XPath operators (Table 18.9). For a comprehensive list of XPath functions, please visit the W3C website: http://www.w3.org/2005/xpath-functions (Table 18.10).

Please note that XQuery supports the use of namespaces and is case sensitive. In XQuery, all elements, attributes, and variables must be valid XML names. An XQuery variable is defined with a $ followed by a name (e.g., $x), and an XQuery comment is delimited by "(:" and ":)" (e.g., (: comments :)). The basic expression in XQuery is FLWOR where F refers to **for**, L refers to **let**, W refers to **where**, O refers to **order by**, and R refers to **return.**

Using an XML example (Table 18.11), some basics of XQuery are illustrated. This XML contains the information about best-selling single-volume books. For each book, the title, authors, language, first publication year, sales volume, price, and the types of the book are presented.

Example 18.3

An XQuery example on the XML from Table 18.11 is as follows:

```
for $x in doc ("books.xml")/Best-selling-single-volume-books//Book
where $x/Price >10.99
order by $x/title
let $x/Price: =10.99*0.80
return $x/Title, $x/Price
```

TABLE 18.10

List of Operators Used in XPath (XQuery)

Operator	Description
|	Computes two node sets, for example: `//Book	//Part` will return a node set with all Book and Part elements
+	Addition
-	Subtraction
*	Multiplication
div	Division
=	Equal
!=	Not equal
<	Less than
<=	Less than or equal to
>	Greater than
>=	Greater than or equal to
or	Or
and	And
mod	Modulus (division remainder)

TABLE 18.11

Example XML Document on Best-Selling Single-Volume Books (books.xml)

```
<?xml version="1.0" encoding="UTF-8"?>
<Best-selling-single-volume-books>
      <Book Type="Kindle">
            <Title>A Tale of Two Cities</Title>
            <Authors>
                  <Author>
                        <Lastname>Dickens</Lastname>
                        <Firstname>Charles</Firstname>
                  </Author>
            </Authors>
            <Language>English</Language>
            <FirstPublished>1859</FirstPublished>
            <Sales>Over 200 Million</Sales>
            <Price>3.50</Price>
      </Book>
      <Book Type="Kindle">
            <Title>The Lord of Rings</Title>
            <Authors>
                  <Author>
                        <Lastname>Tolkien</Lastname>
                        <Firstname>J.R.R.</Firstname>
                  </Author>
            </Authors>
            <Language>English</Language>
            <FirstPublished>1954-1955</FirstPublished>
            <Sales>150 million</Sales>
            <Price>3.10</Price>
      </Book>
      <Book Type="Kindle">
            <Title>The Little Prince</Title>
            <Authors>
                  <Author>
                        <Lastname>Saint-Exupéry</Lastname>
                        <Firstname>Antoine</Firstname>
                  </Author>
            </Authors>
            <Language>French</Language>
            <FirstPublished>1937</FirstPublished>
            <Sales>Over 100 million</Sales>
            <Price>8.74</Price>
      </Book>
      <Book Type="eBook">
            <Title>Dream of the Red Chamber</Title>
            <Authors>
                  <Author>
                        <Lastname>Cao</Lastname>
                        <Firstname>Xueqing</Firstname>
                  </Author>
            </Authors>
            <Language>Chinese</Language>
            <FirstPublished>1754-1791</FirstPublished>
            <Sales>100 million</Sales>
            <Price>18.59</Price>
      </Book>
</Best-selling-single-volume-books>
```

The **for** clause selects all *Book* elements within the *Books* element and assigns it to a variable **x** (**$x**). The **where** clause selects only the *Book* elements that has a *Price* sub-element, and the value of the *Price* sub-element is greater than 10.99. Additionally, the **order by** clause defines the sort order. By default, the ascending order is used. The **let** clause sets the discounted prices (80%) for the book, while the **return** clause returns the titles of the books satisfying the condition and the new prices. The result of the XQuery example is:

```
Dream of the Red Chamber, 14.82
```

The purpose of querying an XML document is to retrieve the related data within the document as well as to transmit it to other applications. Transforming the data into other formats (e.g., HTML, other XML formats) is often desired, which leads to another W3C standard, extensible stylesheet language (XSL).

18.3 XML Transformation

Cascading style sheets (CSS) is a commonly used tool for data transformation and presentation. However, similar to DTD, CSS does not support namespaces and suffers from extensibility, which, in turn, limits its application to XML. On the other hand, these limitations are addressed by XSL, which is a W3C standard. Together with XPath, XSL Transformation (XSLT) is able to create different presentations of XML content. This is done by the use of an XSL stylesheet, which explicitly defines how to transform the values of a queried XML element, attribute, and text into a desired output format. Note that an XSL stylesheet in itself is a well-formed XML and that it supports namespaces. Therefore, syntax rules on XML elements, attributes, and text discussed in Chapter 17 will also apply to the stylesheet. The vocabularies of XSLT though are mostly made up of XML elements. A partial list of XSLT elements are summarized in Table 18.12.

Example 18.4

Let us use an example on the best-selling book XML (Table 18.11) to illustrate the application of XSLT. In this example, the transformed output is in HTML format (Figure 18.2). To connect the XSLT file to an XML document, the `<?xml-stylesheet?>` processing function is used. To do so, the following element is added at the beginning of the XML document from Table 18.11:

```
<?xml-stylesheet type="text/xsl" href="BestSellingBooks.xsl"?>
```

Here, it is assumed that the XSL file is located in the same folder as the XML document. Considering the XSLT example (BestSellingBooks.xsl) (Table 18.13):

Line 1: Mandatory XML declaration.
Line 2: Declare the `xsl:stylesheet` element. In this document, there are four namespaces declared specifically: an XSL transform namespace with prefix `xsl`, an XSL format namespace with prefix `fo`, an XMLschema namespace with prefix `xs`, and an XPath functions namespace with prefix `fn`.
Line 3: Declare that the output method is hypertext markup language (HTML), that is, the source XML document is to be transformed to an HTML document.

TABLE 18.12
XSLT Elements

Elements	Description
`xsl:stylesheet`	Defines a root element of a stylesheet. This element can be used interchangeably with transform, but most stylesheets use stylesheet as a *de facto* standard
`xsl:template`	Applies rules in a match or select action
`xsl:value-of`	Retrieves a string value of a node and writes it to the output node tree
`xsl:for-each`	Iteratively processes each node in a node set defined by an XPath expression
`xsl:sort`	Defines a sort key used by apply-templates to a node set and by for-each to specify the order of iterative processing of a node set
`xsl:if`	Conditionally applies a template if the test attribute expression evaluates to true
`xsl:choose`	Makes a choice based on multiple options; used with elements when and otherwise
`xsl:apply-templates`	Applies templates to all children of the current node or a specified node set using the optional select attribute. Parameters can be passed using the with-param element
`xsl:element`	Adds an element to the output node tree. Names, namespaces, and attributes can be added with the names, namespaces, and use-attribute-sets attributes
`xsl:attribute`	Adds an attribute to the output node tree; must be a child of an element
`xsl:attribute-set`	Adds a list of attributes to the output node tree; must be a child of an element
`xsl:text`	Adds text to the output node tree
`xsl:comment`	Adds a comment to the output node tree
`processing-instruction`	Adds a processing instruction to the output node tree

Best Selling Book (Single Volume)

Title	Type	Author	Sales	Price
A Tale of Two Cities	Kindle	Dickens	Over 200 Million	3.50
The Lord of Rings	Kindle	Tolkien	150 million	3.10
The Little Prince	Kindle	Saint-Exupéry	over 100 million	8.74
Dream of the Red Chamber	eBook	Cao	100 million	18.59

FIGURE 18.2
Display of XML after applying XSLT.

As a result, the necessary HTML tags will be added to the formatting, including <html>, <body>, <h2>, <tr>, <th>, and <td>.

Line 4: Declare the `<xsl:template>` element. The stylesheet starts to search the XML document using the `template` element. The `match` attribute specifies the matching pattern. In this example, the matching pattern is "/" (root element, see Table 18.9: XPath location operator); the stylesheet will start from the root element.

Lines 5 to 15: Format the information for presentation using HTML tags.

Lines 16 to 34: Instruct the XSL processor to scan through the document. For each *<Book>* element under *<Best-selling-single-volume-books>*, the value of the *<Title>* element, the value of the attribute *<@Type>* (see Table 18.9), the value of the element *<Authors/Author/Lastname>*, the value of the *<Sales>* element, and the value of the *<Price>* element are retrieved and displayed in a table format.

TABLE 18.13

XSLT Example (BestSellingBooks.xsl)

```
1. <?xml version="1.0" encoding="UTF-8"?>
2. <xsl:stylesheet version="2.0" xmlns:xsl="http://www.w3.org/1999/XSL/Transform"
   xmlns:fo="http://www.w3.org/1999/XSL/Format" xmlns:xs="http://www.w3.org/2001/
   XMLSchema" xmlns:fn="http://www.w3.org/2005/xpath-functions">
3. <xsl:output method="html"/>
4. <xsl:template match="/">
5. <html>
6. <body>
7.      <h2>Best Selling Book (Single Volume)</h2>
8.      <table border="1">
9.      <tr>
10.          <th>Title</th>
11.          <th>Type</th>
12.          <th>Author</th>
13.          <th>Sales</th>
14.          <th>Prices</th>
15.     </tr>
16.     <xsl:for-each select="Best-selling-single-volume-books/Book">
17.     <tr>
18.          <td>
19.                    <xsl:value-of select="Title"/>
20.          </td>
21.          <td>
22.                    <xsl:value-of select="@Type"/>
23.          </td>
24.          <td>
25.                    <xsl:value-of select="Authors/Author/Lastname"/>
26.          </td>
27.          <td>
28.                    <xsl:value-of select="Sales"/>
29.          </td>
30.          <td>
31.                    <xsl:value-of select="Price"/>
32.          </td>
33.     </tr>
34.     </xsl:for-each>
35.     </table>
36. </body>
37. </html>
38. </xsl:template>
39. </xsl:stylesheet>
```

This example demonstrates the use of XSLT and XPath to transform an XML document to an HTML document. Indeed, an XML document can be transformed to a number of other document formats as well as generate other XML documents. For example, the developers can search for the *Books* having an "eBook" *Type* and then construct a new XML document that contains only eBook information. Some other common output file formats are text and portable document format (PDF). Unlike HTML, XML, and text format, converting an XML document to a PDF file will require a formatting objects processor (FOP). Currently, the most popular FOP engine is the Apache FOP processor.

18.4 Interfacing XML with Database Applications

As W3C continues to improve XML-related technologies and standards, most applications, specifically database systems, are providing corresponding interfaces with XML. Database systems are well known for consistent data storage, retrieval, and manipulation. At the same time, XML has been accepted as a standard for Web-based information systems and a medium for information exchange between organizations. Since a database is essentially the backbone for any information system, the integration between XML and a relational database system (RDBS) is a must.

18.4.1 Relational Database versus XML

To understand the relationship between XML and RDBS, we will first explain the differences between XML and RDBS in terms of (1) data modeling, (2) data schema, and (3) instance as follows.

The concept of data modeling refers to the definition of the structure of data. Since an RDBS is designed to store large volumes of data and provide efficient access with ensured integrity, an RDBS requires explicit definitions on the data types. In contrast, XML is designed to exchange data embedded in a structured document.

By definition, a data schema is the utilization of data modeling concepts to structure certain domain data. For example, an RDBS schema can describe the database structure as well as its relational table structure. It has to be defined before any data can be stored in the system. On the other hand, XML is more like a semi-structured document as it allows implicit, partial, or even an incomplete schema. This is mainly because an XML document may evolve over time. When the structure schema is well defined, developers usually use a DTD or an XML schema to specify the schema. XML documents are validated through a DTD or an XML schema, as discussed in Chapter 17. Additionally, although both DTD and XML schema support multiple data types, an XML schema is relatively limited and is not as strong as an RDBS. Furthermore, a DTD or an XML schema is not mandatory to be applied on an XML document.

At the instance level, XML is self-describing, which means that parts of the schema definition in terms of tags are replicated within each XML document, no matter if the schema is defined explicitly or not. For the example in Table 18.11, the *Author* tags are repeated for each *Book* element. This is in contrast to an RDBS where there is only one schema for the whole database. It is the way that the schema, along with the data, is stored that provides XML the flexibility in integrating heterogeneous data sources and allowing changes to its structure. However, the replicated schema information implies space cost for storage, time cost for retrieval, and the danger of inconsistencies in case of schema updates (Deutsch et al., 1999).

Example 18.5

Use Healthcaresystem as an example to compare relational database and XML.

Consider the relational database for Healthcaresystem, which has three tables: Drugs, Stores, and Sales. The Drugs table contains the information about the ID, name, price, and general description of the drug. The Stores table has the information on the Store_ID, the name of the store, and the contact phone number. The Sales table summarizes the quantity sold by the store for each specific drug.

From an RDBS perspective, first of all, the data types of each should be explicitly defined. For example, "Price" is a number while "Description" is a string. Next, the relational database schema needs to be defined. That is, there are four columns in the "Drugs" table, specifically, Drug_ID, Name, Price, and Description. Additionally, there are three columns in the "Stores" table, which includes Store_ID, Name, and Telephone. Lastly, there are three columns in the "Sales" table: Transaction_ID, Drug_ID, Quantity, and Store_ID. In addition, constraints should be specified when defining the schema, that is, the primary key in the "Drugs" (Drug_ID) table is a foreign key in the "Sales" (Drug_ID) table; the primary key in the "Stores" (Store_ID) table is a foreign key in "Sales" (Store_ID); and the primary key in the "Sales" table is a composite key, which is composed of Drug_ID and Store_ID. Let the tables be populated with some records as follows:

Drugs

Drug_ID	Name	Price	Description
T_001	Tylenol	$17.99	100 325mg Tablets Bottle
T_003	Tylenol	$8.99	
T_002	Tri_Vitamin	$14.99	1500-400-35 Solution 50mL Bottle
C_001	Claritin	$35.99	10 Tablets

Stores

Store_ID	Name	Phone_Number
S_001	Walgreen	1-800-746-7287
S_002	CVS	1-800-925-4733
S_003	Fry's Pharmacy	1-866-221-4141

Sales

Transaction_ID	Drug_ID	Quantity	Store_ID
ST_001	T_001	25	S_001
ST_002	T_001	300	S_002
ST_003	C_002	150	S_001

If this information is presented in an XML document, it is

```
<?xml version="1.0" encoding="UTF-8"?>
<Healthcaresystem>
<Drugs>
   <Drug Drug_ID="T_001">
   <Name>Tylenol</Name>
   <Price>$17.99</Price>
   <Description>100 325mg Tablets Bottle</Description>
   </Drug>

   <Drug Drug_ID="T_003">
   <Name>Tylenol</Name>
   <Price>$8.99</Price>
   <Description></Description>
   </Drug>

   <Drug Drug_ID="T_002">
   <Name>Tri_Vitamin</Name>
   <Price>$14.99</Price>
   <Description>1500-400-35 Solution 50ml Bottle</Description>
   </Drug>
   <Drug Drug_ID="C_001">
   <Name>Claritin</Name>
   <Price>$35.99</Price>
   <Description>10 Tablets</Description>
```

```
    </Drug>
  </Drugs>

  <Stores>
    <Store Store_ID="S_001">
    <Name>Walgreen</Name>
    <Telephone>1-800-746-7287</Telephone>
    </Store>

    <Store Store_ID="S_002">
    <Name>CVS</Name>
    <Telephone>1-800-925-4733</Telephone>
    </Store>

    <Store Store_ID="S_003">
    <Name>Fry's Pharmacy</Name>
    <Telephone>1-866-221-4141</Telephone>
    </Store>
  </Stores>

  <Sales>
    <Sale Drug_ID="T_001" Store_ID="S_001">
    <Quantity>25</Quantity>
    </Sale>

    <Sale Drug_ID="T_001" Store_ID="S_002">
    <Quantity>300</Quantity>
    </Sale>

    <Sale Drug_ID="C_002" Store_ID="S_001">
    <Quantity>501</Quantity>
    </Sale>
  </Sales>
  </Healthcaresystem>
```

An associated DTD (Table 18.14) or an XML schema (Table 18.15) for this XML document are as follows.

TABLE 18.14

DTD Document

```
<!ELEMENT Healthcaresystem (Drugs, Stores, Sales)*>
<!ELEMENT Drugs (Drug)*>
<!ELEMENT Drug (Name, Price, Description)*>
<!ATTLIST Drug Drug_ID ID #REQUIRED>
<!ELEMENT Name (#PCDATA)>
<!ELEMENT Price (#PCDATA)>
<!ELEMENT Description (#PCDATA)>
<!ELEMENT Stores (Store)*>
<!ELEMENT Store (Name | Telephone)*>
<!ATTLIST Store Store_ID ID #REQUIRED>
<!ELEMENT Name (#PCDATA)>
<!ELEMENT Telephone (#PCDATA)>
<!ELEMENT Sales (Sale)*>
<!ELEMENT Sale (Quantity)*>
<!ATTLIST Sale Transaction_ID ID #REQUIRED DrugID IDREF #REQUIRED Store_ID IDREF
 #REQUIRED>
<!ELEMENT Quantity (#PCDATA)>
```

TABLE 18.15

XML Schema Document

```
<?xml version="1.0" encoding="UTF-8"?>
<xsd:schema elementFormDefault="qualified" xmlns:xsd="http://www.w3.org/2001/XMLSchema">
 <xsd:element name="Healthcaresystem">
     <xsd:complexType>
         <xsd:sequence minOccurs="0" maxOccurs="unbounded">
                  <xsd:element ref="Drugs"/>
                  <xsd:element ref="Stores"/>
                  <xsd:element ref="Sales"/>
         </xsd:sequence>
     </xsd:complexType>
 </xsd:element>
 <xsd:element name="Drugs">
     <xsd:complexType>
         <xsd:sequence minOccurs="0" maxOccurs="unbounded">
                  <xsd:element ref="Drug"/>
         </xsd:sequence>
     </xsd:complexType>
 </xsd:element>
 <xsd:element name="Drug">
     <xsd:complexType mixed="true">
         <xsd:sequence minOccurs="0" maxOccurs="unbounded">
                  <xsd:element ref="Name"/>
                  <xsd:element ref="Price"/>
                  <xsd:element ref="Description"/>
         </xsd:sequence>
     <xsd:attribute name="Drug_ID" type="xsd:string" use="required"/>
         </xsd:complexType>
 </xsd:element>
 <xsd:element name="Name" type="xsd:string"/>
 <xsd:element name="Price" type="xsd:string"/>
 <xsd:element name="Description" type="xsd:string"/>
 <xsd:element name="Stores">
         <xsd:complexType>
         <xsd:sequence minOccurs="0" maxOccurs="unbounded">
                  <xsd:element ref="Store"/>
            </xsd:sequence>
         </xsd:complexType>
 </xsd:element>
 <xsd:element name="Store">
         <xsd:complexType mixed="true">
             <xsd:sequence minOccurs="0" maxOccurs="unbounded">
                  <xsd:element ref="Name"/>
                  <xsd:element ref="Telephone"/>
             </xsd: sequence>
 <xsd:attribute name="Store_ID" type="xsd:string" use="required"/>
         </xsd:complexType>
 </xsd:element>
 <xsd:element name="Name" type="xsd:string"/>
 <xsd:element name="Telephone" type="xsd:string"/>
 <xsd:element name="Sales">
         <xsd:complexType>
            <xsd:sequence minOccurs="0" maxOccurs="unbounded">
                          <xsd:element ref="Sale"/>
            </xsd:sequence>
         </xsd:complexType>
 </xsd:element>
```

(continued)

TABLE 18.15 (Continued)

XML Schema Document

```
<xsd:element name="Sale">
    <xsd:complexType mixed="true">
          <xsd:sequence minOccurs="0" maxOccurs="unbounded">
                       <xsd:element ref="Quantity"/>
          </xsd:sequence>
<xsd:attribute name="Drug_ID" type="xsd:string" use="required"/>
<xsd:attribute name="Store_ID" type="xsd:string" use="required"/>
        </xsd:complexType>
</xsd:element>
<xsd:element name="Quantity" type="xsd:string"/>
</xsd:schema>
```

From the example, it is concluded that both RDBS and XML are able to store data, but in different ways. Therefore, almost all relational database applications provide the interfaces with XML for importing and exporting.

18.4.2 XML and Microsoft Access

Since Office XP, Microsoft Access has enhanced its capability to export and import data by fully embracing the XML standard and embedding XML support in its application and programmability features. In this section, we will learn about how XML data can be imported and exported by using the Access graphical user interface (GUI). We will learn how XML schema can be used to ensure data integrity for imports and exports. We will also learn how to leverage XSL to convert semi-structured XML data properly for importing into a format that can be directly used by Access.

18.4.2.1 Exporting Access Database Tables to XML

Let us first create three tables in an Access database (ch18_AccesstoXML.mdb)—Drugs, Stores, and Sales—following Example 18.5. The relationships between the tables are shown in Figure 18.3.

The simplest way to export XML from this Access database is to right-click on a table. An export menu appears on the pop-up menu (see Figure 18.4).

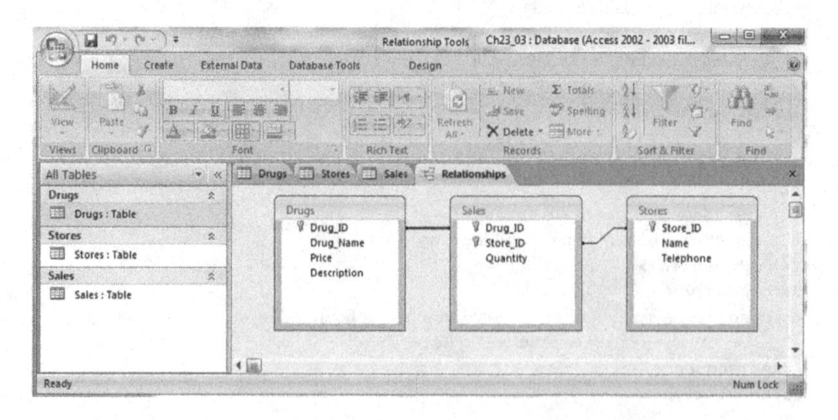

FIGURE 18.3

Access database example: three tables—Drugs, Sales, Stores and the relationship among the three.

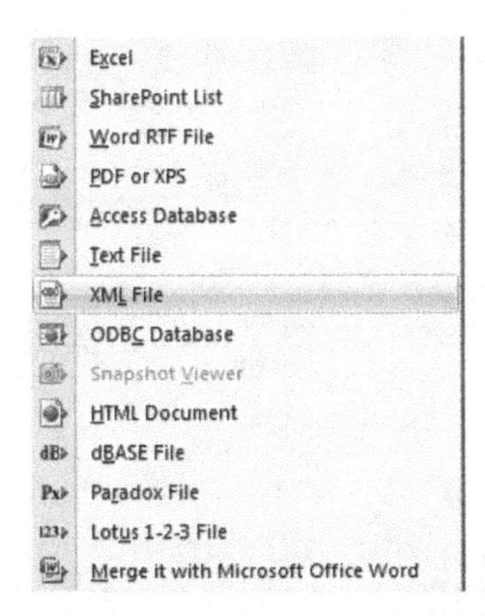

FIGURE 18.4
Export XML via user interface in Access.

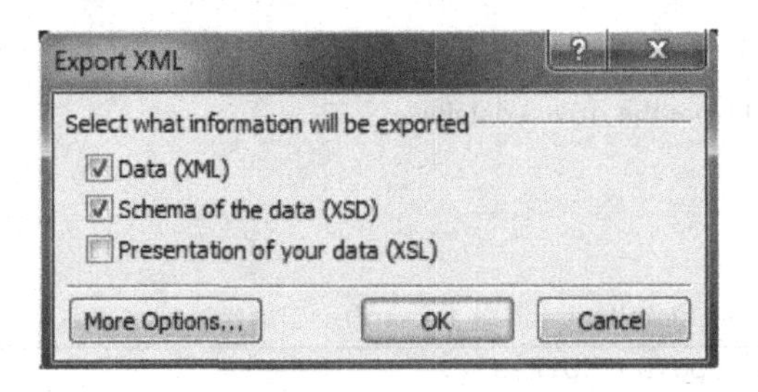

FIGURE 18.5
Export configuration settings.

Access will then prompt for a location where it can place the XML file containing the data from the table. In addition to the tables, we can also export the data from a query, as well as the datasheet, a form, or a report into an XML file.

Before the objects are to be exported, a prompt will be shown to ask for the configurations (see Figure 18.5). The resulting XML file contents will depend on the options that we select.

Let us look at the generated XML file. This file contains the data contents of the object exported from Access (Figure 18.6). Table 18.16 shows the final XML from the export. We can see how the fields from the Access table are related to the elements in the XML document.

In addition, Access has exported an XSD schema to preserve the definitions that describe how the fields relate one to the other. As seen, the XSD contains information such as whether there is a key constraint on the database table, whether the key enforces uniqueness, what the field names are, as well as their corresponding data types and lengths (see Table 18.17).

18.4.2.2 Importing XML to an Access Database

As is the case with XML exports, there is the ability to import an XML document into Access through its GUI. During the importing process, the most important step is to ensure that the structure of the data in the XML file conforms to the structure of the database table. If

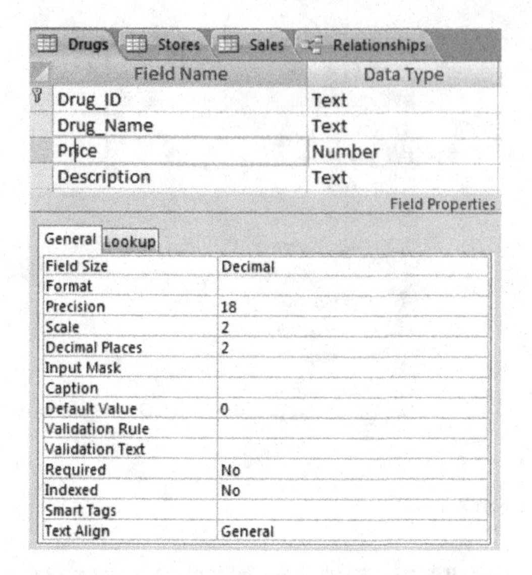

FIGURE 18.6
Design of the table exported from the Access database.

TABLE 18.16

XML Exported from Access Database (ch18_04.xml)

```
<?xml version="1.0" encoding="UTF-8"?>
<dataroot xmlns:od="Zurn:schemas-microsoft-com:officedata" xmlns:xsi="http://www.
 w3.org/2001/XMLSchema-instance" xsi:noNamespaceSchemaLocation="Drugs.xsd"
 generated="2011-07-03T13:15:14">
<Drugs>
        <Drug_ID>T_001</Drug_ID>
        <Name>Tylenol</Name>
        <Price>17.99</Price>
        <Description>100 325mg Tablets Bottle</Description>

        <Drug_ID>T_003</Drug_ID>
        <Name>Tylenol</Name>
        <Price>8.99</Price>
        <Description></Description>

        <Drug_ID>T_002</Drug_ID>
        <Name>Tri_Vitamin</Name>
        <Price>14.99</Price>
        <Description>1500-400-35 Solution 50ml Bottle</Description>

        <Drug_ID>C_001</Drug_ID>
        <Name>Claritin</Name>
        <Price>35.99</Price>
        <Description>10 Tablets</Description>
</Drugs>
</dataroot>
```

TABLE 18.17

XML Schema Exported from Access Database

```
<?xml version="1.0" encoding="UTF-8"?>
<xsd:schema xmlns:xsd="http://www.w3.org/2001/XMLSchema"
 xmlns:od="urn:schemas-microsoft-com:officedata">
<xsd:element name="dataroot"></xsd:element>
<od:index index-name="PrimaryKey" index-key="Drug_ID" primary="yes" unique="yes"
 clustered="no" order="asc"/>
<xsd:element name="Drugs">
     <xsd:complexType>
         <xsd:sequence>
             <xsd:element name="Drug_ID" minOccurs="0" od:jetType="text" od:sqlSType="nvarchar">
             </xsd:element>
             <xsd:element name="Name" minOccurs="0" od:jetType="text" od:sqlSType="nvarchar">
             </xsd:element>
             <xsd:element name="Price" minOccurs="0" od:jetType="decimal" od:sqlSType="decimal">
                <xsd:simpleType>
                 <xsd:restriction base="xsd:decimal">
                    <xsd:totalDigits value="18"/>
                    <xsd:fractionDigits value="2"/>
                 </xsd:restriction>
                </xsd:simpleType>
             </xsd:element>
           <xsd:element name="Description" minOccurs="0" od:jetType="text" od:sqlSType="nvarchar">
           </xsd:element>
        </xsd:sequence>
       </xsd:complexType>
</xsd:element>
</xsd:schema>
```

Access encounters a problem when inserting data into the database table, it will place the errors in a new table called "ImportErrors." When importing a new XML document and the table does not exist, Access will create it; otherwise, new errors will be appended to the Access table.

To begin an import attempt (an example import file is shown in Table 18.18), the user can choose the File → Get External Data → Import menu in Access, or choose a pop-up menu when right-clicking in the pane that lists tables, queries, or other objects in Access. Either way, a dialog box is popped up to navigate to the XML file that contains the data to be imported. After selecting the file and clicking the button to import the file, another dialog box prompts to configure the import options (Figure 18.7).

The import behavior can be different according to the setting that the users choose. For example:

- The import option is "Structure Only:" This imports the table structure only. If the table does not exist, it will be created. If the table already exists, a new one will be added with a number appended to differentiate it from the pre-existing table of the same name.
- The import option is "Structure and Data:" This imports the table and the data into the database. If the table exists, a new one will be added with a number appended. The difference between this and the previous option is that this one tells Access to import the data as well.

TABLE 18.18
XML Document for Importing

```
<Stores>
        <Store Store_ID="S_001">
        <Name>Walgreen</Name>
        <Telephone>480-111-2222</Telephone>
        </Store>

        <Store Store_ID="S_002">
        <Name>CVS</Name>
        <Telephone>480-333-4444</Telephone>
        </Store>
</Stores>
```

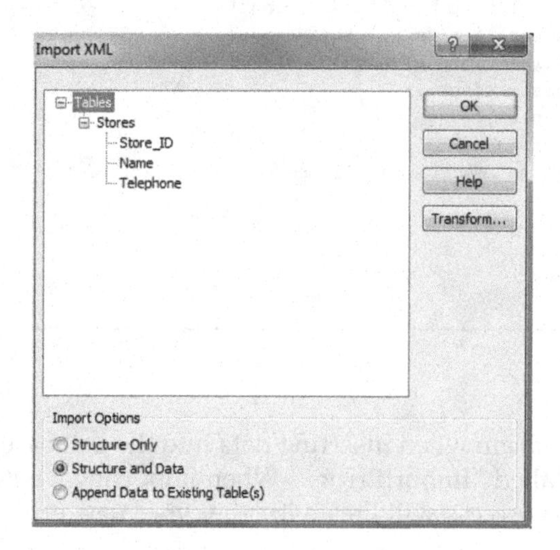

FIGURE 18.7
Import dialog box.

- The import option is "Append Data to Existing Table(s):" This option tells Access to merely append data to existing tables. If the tables do not exist, the import will fail.

Another aspect of the import is that Access can figure out if the data imported should be placed into a single table or multiple tables. For example, if an XML document contains a hierarchical structure, such as, for instance, customer and order information are nested within a single XML file, these will be split into two tables in Access. In addition, it is recommended to provide XML structure information, for example, XSD, to Access so that Access has an explicit instruction to digest the XML document for importing.

18.4.3 XML and MySQL

In MySQL, one powerful tool is Workbench, which is a client program providing a GUI for a number of management tasks. Some basic functionalities of Workbench are discussed in Chapter 9. In this chapter, the use of Workbench to export XML is briefly explained. However, Workbench does not provide the utility to import XML. The MySQL command-line client tool is used to demonstrate the basic command for importing XML.

18.4.3.1 Exporting MySQL to XML

To get ready, let us first create a table "Drugs" following the schema (see Figure 18.8). The data will be inserted in the table, as discussed in Chapter 8.

To export the MySQL table to XML, the first step is to create a query to retrieve the data, as shown in Figure 18.9.

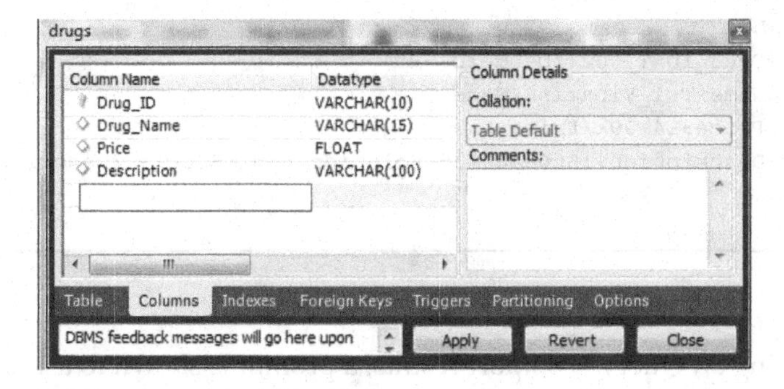

FIGURE 18.8
MySQL Workbench schema definition for the Drugs table.

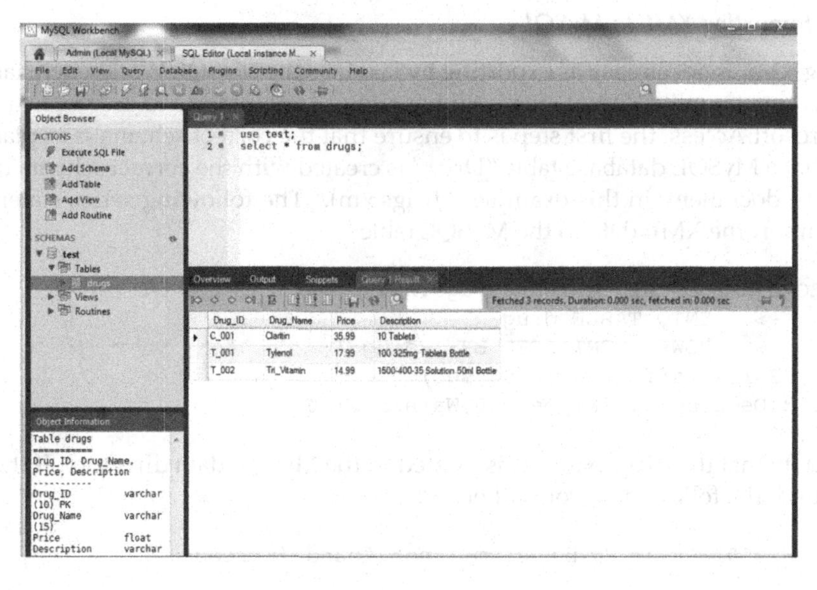

FIGURE 18.9
MySQL Workbench query.

TABLE 18.19
XML Document Exported from MySQL

```
<DATA>
    <ROW>
      <Drug_ID>C_001</Drug_ID>
      <Name>Claritin</Name>
      <Price>35.99</Price>
      <Description>10 Tablets</Description>
    </ROW>
    <ROW>
      <Drug_ID>T_001</Drug_ID>
      <Name>Tylenol</Name>
      <Price>17.99</Price>
      <Description>100 325mg Tablets Bottle</Description>
    </ROW>
    <ROW>
      <Drug_ID>T_002</Drug_ID>
      <Name>Tri_Vitamin</Name>
      <Price>14.99</Price>
      <Description>1500-400-35 Solution 50ml Bottle</Description>
    </ROW>
</DATA>
```

Next, by clicking on Query → Export results, a prompt is shown to ask the user to specify the file format as well as the destination of the file. The output XML document looks like Table 18.19.

18.4.3.2 Importing XML to MySQL

Importing XML is not as easy as exporting by far since the MySQL Workbench has not provided this functionality. Developers will have to use commands to accomplish this task. Like Microsoft Access, the first step is to ensure that the correct schema is in place. Let us assume that a MySQL database table "Drug" is created with the correct schema in relation to the XML document, in this example, "drugs.xml." The following series statements are used to import the XML data to the MySQL table:

```
mysql> LOAD XML LOCAL INFILE 'drugs.xml'
          ->    INTO TABLE drug
          ->    ROWS IDENTIFIED BY '<drug>';
Query OK, 3 rows affected (0.00 sec)
Records: 3 Deleted: 0 Skipped: 0 Warnings: 0
```

Please note that the drugs.xml file is located in the MySQL data directory. If the file cannot be found, the following error will occur:

```
ERROR 2 (HY000): File '/drugs.xml' not found (Errcode: 2)
```

To find the data directory, the following statement is used:

```
mysql> SHOW VARIABLES LIKE 'datadir';
```

Drug_ID	Drug_Name	Price	Description
T_001	Tylenol	17.99	100 325mg Tablets Bottle
T_002	Tri_Vitamin	14.99	1500-400-35 Solution 50ml Bottle
C_001	Claritin	35.99	10 Tablets

FIGURE 18.10
Excel loaded by an XML document.

The `ROWS IDENTIFIED BY '<drug>'` clause means that each *<drug>* element in the XML file is considered equivalent to a row in the table into which the data are to be imported. In this case, this is the Drugs table in the test database.

18.4.4 XML and Microsoft Excel

Even though Excel is not a relational database, as a component of Microsoft Office, it has great integration capabilities with XML. In addition, an important learning point of this book is to develop VBA in Excel. Thus, it is of our interest here to briefly introduce how Excel can consume and produce XML documents.

When Excel attempts to open an XML file, it has no idea about the source from which the XML came or what its ultimate purposes may be. In the end, Excel attempts to display the data into a two-dimensional spreadsheet format, which is the only format available to the user once the data is loaded. This is accomplished by an XML flattener, which interprets and alters the XML document. The flattener contains processing logic so that the data in an XML file can be converted to be displayed on the spreadsheet. An exception to this is when the file is already saved in the XML spreadsheet format. In this case, no flattening is required because the file is already in the Microsoft Excel native XML file format. This means that the XML stylesheet is being used to make sure that the XML is in a structure that Excel can understand.

There are several ways to get XML data into the Excel spreadsheet. The easiest way is to go to File → Open and navigate to an XML file. A loaded Excel spreadsheet will then be shown, as shown in Figure 18.10.

Exporting an XML from Excel is as easy as importing. The users can choose to save the file with .xml extension. However, the XML document rendered is very complex as it encodes all the formatting as well as the data element.

18.5 Summary

This chapter covers some important applications of XML. The first application is to parse an XML document using a programming language. DOM, an API for parsing XML, is introduced with two examples: one is written in JavaScript and the other is written in VB.NET. The second application is standard query and transformation of XML. Two W3C standards, XQuery/XPath for querying XML data and XSL for transforming XML into other documents, are discussed. The third application is to interface XML with a database. Given that XML is also known as a semi-structured database, a detailed comparison between XML and relational databases in terms of data modeling, data schema, and instance is provided. The interfaces between XML and Access database, MySQL database, and Excel for importing and exporting are demonstrated.

REVIEW EXERCISES

1. An XML parser identifies and converts the elements of an XML document into a tree structure. True or false?
2. The following XML components are matched into a DOM tree node except:
   a. Root element
   b. Comments
   c. Texts
   d. Attributes
   e. All of the above are matched into a DOM tree node.
3. A type of relationship in which two elements share the same parent node is called what?
4. The following are valid DOM interfaces, except:
   a. NodeList
   b. Root
   c. Element
   d. DATASection
   e. ProcessingInstruction
5. Which DOM method queries all child elements of a parent element?
   a. `element.getChilds`
   b. `document.getChildNodes`
   c. `element.getChildNodes`
   d. `document.getChilds`
   e. `element.getQueryChildNodes`
6. The `element.getNodeValue` function is used to obtain the value of the attribute within the element. True or false?
7. DOM is an event-driven API as compared to SAX. True or false?
8. In terms of usage, DOM is recommended for ____________ while SAX is recommended for ____________.
   a. reading data – altering data
   b. modifying data – reading and writing data
   c. adding, modifying, and deleting data – reading data
   d. altering data – adding data
   e. None of the above
9. The following are valid properties of the XQuery, except:
   a. XQuery language uses XPath expressions.
   b. XQuery supports the use of namespaces.
   c. XQuery variable is defined by the "$" character followed by a name.
   d. XQuery is case insensitive.
   e. All of the above are valid XQuery properties.

10. Using the SQL Query `SELECT Student _ M, Student _ N FROM Students`, which of the following is a valid corresponding FLWOR command?
    a. `for $x in doc("Students")//Student return $x/Student _ M, $x/Student _ N`
    b. `for $x in doc("Students")//Student let $x/Student _ M, $x/Student _ N`
    c. `when $x in doc("Students")//Student return $x/Student _ M, $x/Student _ N`
    d. `when $x in doc("Students")//Student order by $x/Student _ M, $x/Student _ N`
    e. `when $x in doc("Students")//Student let $x/Student _ M, $x/Student _ N`
11. An XSL stylesheet needs to be well formed. True or false?
12. The function `<?xml-stylesheet?>` is appended at the beginning of each XML document for it to be displayed accordingly. True or false?
13. The syntax rules that govern CSS also apply to XML stylesheets. True or false?
14. Which of the following statements is/are true with regard to RDBS and XML data modeling properties?
    a. RDBS data modeling refers to the use of data modeling principles to structure data.
    b. RDBS and XML both necessitate explicit definition of data types.
    c. An XML document requires referential integrity.
    d. All of the above are true.
    e. None of the above is true.
15. An XML schema needs to be defined before data can be stored in an XML document. True or false?
16. The following statements are true except:
    a. A replicated schema in an XML document implies storage, query time, and inconsistency risks.
    b. A single data schema exists for an entire RDBS and must be explicitly defined.
    c. An XML document is self-describing.
    d. RDBS has the flexibility to allow future changes to its structure.
    e. All of the above are true.
17. Which of the following is/are valid options to export information from an MS Access database file?
    a. DTD
    b. XSD
    c. XML
    d. All of the above
    e. Only b and c

18. When importing the data and structure of an XML document and a table of the same name already exists in the MS Access file, an error will be written into ImportErrors. True or false?
19. Which of the following is a valid option to import information from an XML document to an MS Access database file?
    a. Import Elements
    b. Import Attributes
    c. Import Root
    d. Append Data to Existing Table(s)
    e. Append Relationships to Existing Relationships
20. Exporting is easy as compared to importing XML documents in MySQL. True or false?
21. Any XML document is converted to a three-dimensional spreadsheet format when imported to MS Excel. True or false?
22. An XML flattener is used to:
    a. Convert XML into spreadsheet format
    b. Is not required when the XML file is in MS Excel XML format
    c. Convert XLS spreadsheets to XML
    d. Both a and b
    e. Both a and c

PRACTICAL EXERCISES

1. Consider the following sales order. Generate a corresponding DOM tree.

```
 1. <?xml version="1.0" encoding="UTF-16"?>
 2. <Sales_Order>
 3.     <Billing_Info>
 4.        <Bill_Name>Irish Maravillas</Bill_Name>
 5.        <Bill_Street>1975 East University Dr.</Bill_Street>
 6.        <Bill_City>Tempe</Bill_City>
 7.        <Bill_State_Code>AZ</Bill_State_Code>
 8.        <Bill_Zip_Code>85281</Bill_Zip_Code>
 9.     </Billing_Info>
10.     <Customer_Info>
11.        <Ship_Name>Jennifer Jareau</Ship_Name>
12.        <Ship_Street>3600 Wilshire Blvd.</Ship_Street>
13.        <Ship_City>Los Angeles</Ship_City>
14.        <Ship_State_Code>CA</Ship_State_Code>
15.        <Ship_Zip_Code>90010</Ship_Zip_Code>
16.     </Customer_Info>
17.     <Items>
18.        <Item item_N="N28812">
19.            <Product_M>Diamond Necklace</Product_M>
20.            <Quantity>3</Quantity>
21.            <UnitPrice>99.85</UnitPrice>
22.        </Item>
23.     </Items>
24. </Sales_Order>
```

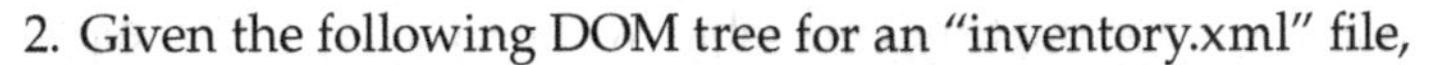

2. Given the following DOM tree for an "inventory.xml" file,

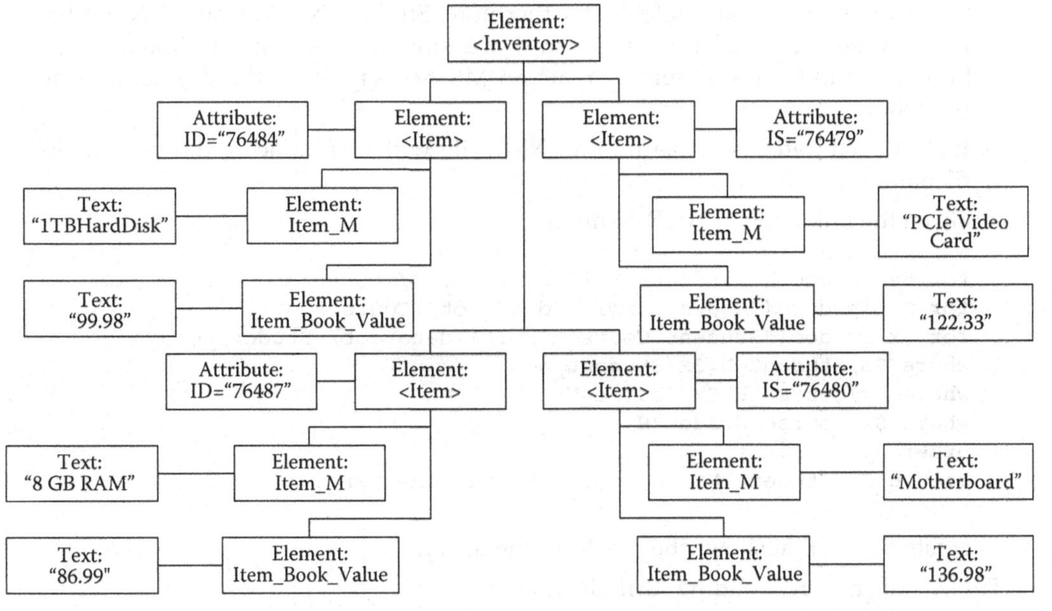

create a table that shows the query results of the FLWOR query as follows:

```
for $x in doc ("inventory.xml")/Inventory//Item
where $x/Item_Book_Value > 95.00
order by $x/ID
return $x/ID, $x/Item_M, $x/Item_Book_Value
```

For Exercises 3 to 6, consider the following dataset:

Students

Student_N	Student_M	Program_C
36001	Barns Stinson	MS IE
56987	Marsha Erikson	MS CS
36369	Lilia Aldrin	PhD IE

Courses

Course_N	Course_M	Credits
556987	Business Intelligence	3
554477	Data Mining	3

Student_Courses

Student_N	Course_N	Course_Grade	Semester_C
36001	556987	A+	Fall 2012
56987	556987	B	Fall 2012
36369	554477	B	Spring 2013
36001	554477	A	Spring 2013

Student_N is the primary key for the Students table, while Course_N is the primary key for the Courses table. Furthermore, Student_N and Course_N are the primary keys of Student_Courses and are foreign keys that are linked to the Students and Courses tables. Create an MS Access file for the dataset and do the following.

3. Export each database table into an XML file with the name of the table as its filename.
4. Given the following FLWOR command,

```
for $s in doc ("Students.xml")/dataroot//Students
for $c in doc ("Courses.xml")/dataroot//Courses
for $x in doc("Student_Courses.xml")/dataroot//Student_Courses
where $s/Student_N=$x/Student_N
where $c/Course_N=$x/Course_N
where $x/Course_Grade="B"
order by $s/Student_M
return $s/Student_M, $x/Course_M, $x/Semester_C,
```

   create a table that shows the results of the query.
5. Generate a corresponding XML document.
6. Import the XML document from Exercise 5 into an Excel spreadsheet.

19

Web Services

This chapter introduces Web services, extensible markup language (XML)-based information exchange systems that enable direct application-to-application interaction via hypertext transfer protocol (HTTP). The Web services architecture is first reviewed. The elements that support Web services, including HTTP; simple object access protocol (SOAP); Web services definition language (WSDL); and universal discovery, description, and integration (UDDI) language, are then discussed. At the end of the chapter, an example on how to create and use a Web service is illustrated.

19.1 Introduction to Web Services

A Web service is a method of making information available in a standardized way that could be accessed by any applications over the Internet. A Web service is not an application on its own; it is up to the developer to use and integrate them into applications to provide value adding services to the end user.

The information contained in a Web service is wrapped up as an XML document that handles cross-platform–specific concerns. Web services are not really anything new. With the emergence of distributed computing, a distributed structure usually contains a presentation layer, a middle tier layer, and a back-end data handling layer. A number of middleware models, such as Common Object Request Broker Architecture (CORBA), Distributed Component Object Model (DCOM), Remote Method Invocation (RMI), were introduced to implement the middleware functionalities. However, the middleware components had to be distributed by the developer, and they had to be downloaded and installed. These things alone meant that they tend to be tied to one platform, whereas a Web service is platform independent as it utilizes Web publishing standards. Everything that has to do with Web service is standardized: the method of transmission, the method used to wrap up the Web service, the way the Web service is defined—all have W3C standards associated with the technologies involved. Figure 19.1 outlines the architecture for a Web service. As seen, it consists of three components: a service broker, a service requester (client), and a service provider (server). A service broker acts as a look-up service between a service provider and a service requestor. It uses WSDL to describe the services provided by a specific provider. A service provider publishes it as a service to the service broker via SOAP and UDDI language. Also, the service provider uses WSDL to describe its services. A service requester will first query the service broker where to locate a suitable service via SOAP and UDDI language and then bind to the service provider that provides the specific service via SOAP. To accomplish this, a number of important protocols are in place to support the Web service architecture.

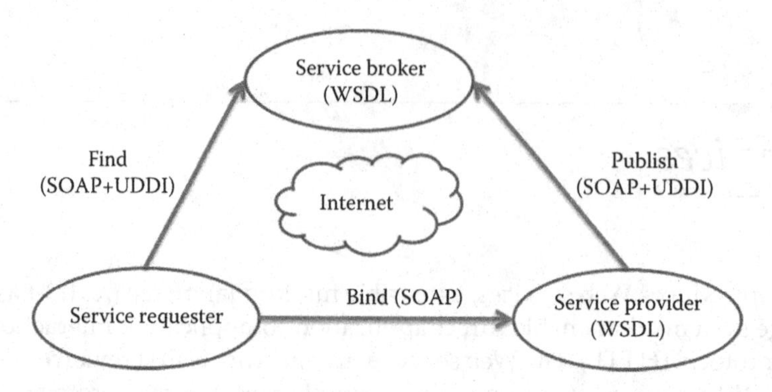

FIGURE 19.1
Web service architecture.

19.2 Web Service Protocols

Web services are built using a number of industry-standard protocols. Although covering these in detail is beyond the scope of this book, the following overview will help one understand the basics of Web services.

19.2.1 Hypertext Transfer Protocol

Although not highlighted in Figure 19.1, a Web service is centered on an Internet implementation, where HTTP is used to enable servers and client browsers to communicate. HTTP is primarily used to establish connections between servers and browsers and transmit data, such as hypertext markup language (HTML) pages or images, to the client browser.

The client sends an HTTP request to the server, which then processes the request. The server typically returns HTML pages to be rendered by the client browser. Additionally, HTTP supports a set of common methods. The typical methods are `GET` and `POST` (Web servers also support `PUT` for creating new resources, `DELETE` for deleting them, `HEAD` for getting the metadata for a resource, and others). In a typical HTML form page, `GET` is used to access the HTML that shows the form and `POST` is used to process it. These guidelines may not always be followed on Web pages, but Web service standards are written to expect closer adherence to these guidelines.

19.2.1.1 HTTP-GET

The `GET` method is used to retrieve whatever information is identified by the Request-universal resource locator (URL). In `GET` requests, the name/value pairs are appended directly to the URL. The data are encoded (which guarantees that only legal American Standard Code for Information Interchange [ASCII] characters are passed) and then appended to the URL, separated from the URL by a question mark. For example, consider the following URL:

http://www.google.com/search?q=webservices

The question mark indicates that this is an HTTP-GET request; the name of the variable passed to the Search method is "q" and the value is "webservices." This HTTP-GET will return the search results of webservices from the Google search engine.

GET requests are suitable when all the data that need to be passed can be handled by name/value pairs, there are few fields to pass, and the length of the fields is relatively short. GET requests are also suitable when security is not an issue. This last point arises because the URL is sent over the Web and is included in server logs as plain text. As such, this can be easily captured by a network sniffer or an unscrupulous person.

19.2.1.2 HTTP-POST

In POST requests, the name/value pairs are also encoded, but instead of being appended to the URL, they are sent as part of the request message. POST requests are suitable for a large number of fields or when lengthy parameters need to be passed. If security is an issue, a POST request is safer than a GET request since the HTTP request can be encrypted.

19.2.2 Simple Object Access Protocol

SOAP is an XML-based simple, lightweight protocol for exchanging Web service data. The typical components of a SOAP is the message itself, plus one or more header blocks, all wrapped within the SOAP envelope. Since SOAP envelope and messages all conform to XML syntax, which is plain text, it can easily pass through firewall, unlike many proprietary, binary formats.

As compared to both the HTTP-GET and the HTTP-POST methods, SOAP is not limited to name/value pairs. Instead, SOAP can be used to send more complex objects, including data sets, classes, and other objects. One drawback, however, is that SOAP messages tend to be verbose due to the nature of XML. Therefore, if bandwidth or transmission performance is an issue, it may be more appropriate to use the HTTP-GET and HTTP-POST methods.

19.2.3 Web Services Definition Language

WSDL describes the Web services using the XML syntax. It provides documentation for distributed systems and has the goal of enabling applications to communicate with each other in an automated way. While SOAP specifies the communication between a requester and a provider, WSDL describes the services offered by the provider (an "end point") and that might be used as a recipe to generate the proper SOAP messages to access the services. When any Web service is created, the service provider will first render a WSDL. Such information will be passed to the service broker when the Web service is published so that the service requesters can search for the appropriate services.

19.2.4 UDDI Language

The UDDI language standard is intended to provide a searchable directory of businesses and their Web services. Thus, it represents the service broker that enables service requesters to find a suitable service provider. In many ways, UDDI language is designed like a phone book. UDDI language was first written in 2000, at a time when the authors had a vision of a world in which consumers of Web services would be linked up with providers through a public or private dynamic brokerage system. Unfortunately, such a vision did not last long, and UDDI language has not been as widely adopted as its designers had hoped. The first leading UDDI language providers such as IBM, Microsoft, and SAP announced their closing UDDI language service in 2006. As of today, UDDI language has becoming more commonly used inside companies where they are used to dynamically bind client systems to implementations.

19.3 Web Service Example

In this section, an example on creating and then using the Web service is provided using Visual Studio 2010 (VS 2010). This example will reinforce the learning concepts of Web service discussed above (Table 19.1).

19.3.1 Creating Web Service

In VS 2010, developers may choose Visual Basic, Visual C#, Visual C++, or other languages to create a Web service. To better fit the scope of the book, this chapter chooses to use Visual Basic for the Web template. By choosing ASP.NET as the Web service application, a project window will then be launched. This looks similar to what was used in Chapters 11 and 12 except for the Code window and the Solution Explorer window, which list the code and the project structure for Web services, respectively (Figure 19.2).

In the Code window, a simple function to calculate the area of a circle given the radius under the WebMethod is created as follows:

First, the Web service–related namespaces need to be specified to ensure that appropriate classes and functions within the Web service can be used. By default, ASP.NET 4.0 adds three namespaces as in lines 1 to 3. Let this Web service have a class named "Service1." It functions just like a normal object except that it will be called upon over the Web. To signify that a particular method (or property) is called via the Web, we need to add a <WebMethod> declaration (line 6). Note that more than one Web method can be defined within a Web service project. In this example, our Web method is publically accessible. Otherwise, a protected keyword can be used to restrict the access. Lines 7 to 11 define the details of the Web function, which accepts the parameter as radius based on which the area of the circle is calculated and returned. Note that, in the Web function, a MATH.PI function is called, which returns the value of π (3.14).

After the Web service is created, a corresponding WSDL is generated to describe the details of services.

TABLE 19.1

Web Service Example

```
1.   Imports System.Web.Services
2.   Imports System.Web.Services.Protocols
3.   Imports System.ComponentModel
4.   Public Class Service1
5.   Inherits System.Web.Services.WebService
6.   <WebMethod()> _
7.   Public Function AreaCalculator(ByVal radius) As Decimal
8.   Dim area As Decimal
9.   area = 2 * Math.PI * radius
10.  Return area
11.  End Function
12.  End Class
```

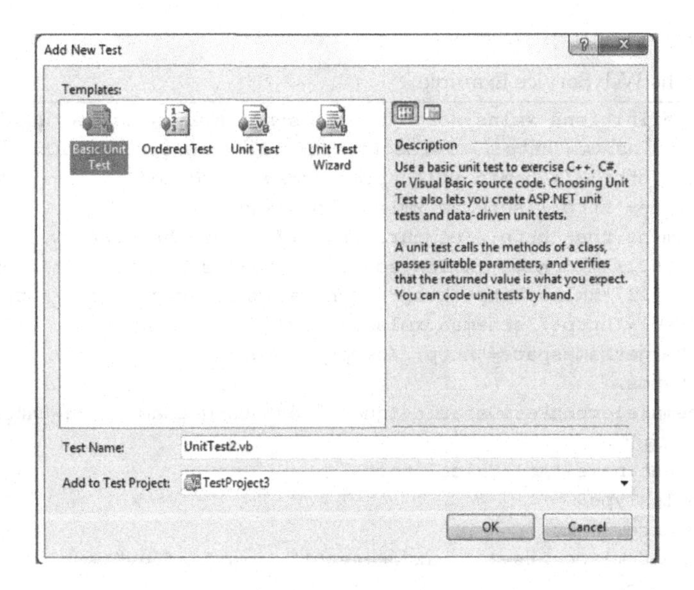

FIGURE 19.2
Creating a test project for the Web service.

19.3.1.1 Web Services Definition Language

As seen in Table 19.2, WSDL has the marks of an XML document.

In line 1, WSDL defines the namespaces used in the XML document. To better digest the document, it is important to know the following prefixes:

- wsdl is defined as in xmlns:wsdl="http://schemas.xmlsoap.org/wsdl/"
- soap is defined in xmlns:soap="http://schemas.xmlsoap.org/wsdl/soap/"
- tns is defined in xmlns:tns="http://tempuri.org/"
- s is defined in xmlns:s="http://www.w3.org/2001/XMLSchema"

In this example, http://temuri.org is used as the default namespace. This needs to be changed into a more permanent namespace before this Web service is made public.

Lines 2 to 19 explicitly define the Web service, which includes two elements: the name of the service, AreaCalculator, and the return of the Web service, AreaCalculatorResponse. The AreaCalculator contains one sub-element, radius, and the AreaCalculatorResponse contains one sub-element, AreaCalculatorResult, which is a decimal data type. In this section, the wsdl namespace is used to define the WSDL type, and the s namespace is used to define the standard data type in XML.

Next, in lines 20 to 22, a WSDL message named "AreaCalculatorSoapIn" is defined. The message contains an element, AreaCalculator, from the tns namespace. In lines 23 to 25, another WSDL message named "AreaCalculatorSoapOut" is defined, which contains an element, AreaCalculatorResponse, from the tns namespace. The communication port information is defined in lines 26 to 31. If a client wants to connect to the service, the binding information defined in lines 32 to 43 provides details such as the service name

TABLE 19.2

WSDL for the Web Service Example

```
1. <wsdl:definitions xmlns:soap="http://schemas.xmlsoap.org/wsdl/
   soap/" xmlns:tm="http://microsoft.com/wsdl/mime/textMatching/"xmlns:
   soapenc="http://schemas.xmlsoap.org/soap/encoding/"
   xmlns:mime="http://schemas.xmlsoap.org/wsdl/
   mime/"xmlns:tns="http://tempuri.org/" xmlns:s="http://www.
   w3.org/2001/XMLSchema" xmlns:soap12="http://schemas.xmlsoap.org/
   wsdl/soap12/"xmlns:http="http://schemas.xmlsoap.org/wsdl/http/"
   xmlns:wsdl="http://schemas.xmlsoap.org/
   wsdl/"targetNamespace="http://tempuri.org/">
2. <wsdl:types>
3. <s:schema elementFormDefault="qualified" targetNamespace="http://
  tempuri.org/">
4. <s:element name="AreaCalculator">
5. <s:complexType>
6. <s:sequence>
7. <s:element minOccurs="0" maxOccurs="1" name="radius"/>
8. </s:sequence>
9. </s:complexType>
10. </s:element>
11. <s:element name="AreaCalculatorResponse">
12. <s:complexType>
13. <s:sequence>
14. <s:element minOccurs="0" maxOccurs="1" name="AreaCalculatorResult"
  type="s:decimal"/>
15. </s:sequence>
16. </s:complexType>
17. </s:element>
18. </s:schema>
19. </wsdl:types>
20. <wsdl:message name="AreaCalculatorSoapIn">
21. <wsdl:part name="parameters" element="tns:AreaCalculator"/>
22. </wsdl:message>
23. <wsdl:message name="AreaCalculatorSoapOut">
24. <wsdl:part name="parameters" element="tns:AreaCalculator
  Response"/>
25. </wsdl:message>
26. <wsdl:portType name="Service1Soap">
27. <wsdl:operation name="AreaCalculator">
28. <wsdl:input message="tns:AreaCalculatorSoapIn"/>
29. <wsdl:output message="tns:AreaCalculatorSoapOut"/>
30. </wsdl:operation>
31. </wsdl:portType>
32. <wsdl:binding name="Service1Soap" type="tns:Service1Soap">
33. <soap:binding transport="http://schemas.xmlsoap.org/soap/http"/>
34. <wsdl:operation name="AreaCalculator">
35. <soap:operation soapAction="http://tempuri.org/AreaCalculator"
  style="document"/>
36. <wsdl:input>
37. <soap:body use="literal"/>
```

(continued)

TABLE 19.2 (Continued)

WSDL for the Web Service Example

```
38. </wsdl:input>
39. <wsdl:output>
40. <soap:body use="literal"/>
41. </wsdl:output>
42. </wsdl:operation>
43. </wsdl:binding>
44. <wsdl:service name="Service1">
45. <wsdl:port name="Service1Soap" binding="tns:Service1Soap">
46. <soap:address location="http://localhost:59483/Service1.asmx"/>
47. </wsdl:port>
48. </wsdl:service>
49. </wsdl:definitions>
```

(Service1Soap), the method name (AreaCalculator), the input, and the output. Next, given the binding and port information, the Web service is defined in lines 44 to 48 with added information of SOAP address, in this example, http://localhost:59483/Service1.asmx.

After the Web service is created, it can be made publicly available by publishing it to the public directory (e.g., UDDI language). The public Web service host will contain the WSDL for the client to search and match the service. SOAP is used for the client to bind with the service provider and get the response from the service provider.

19.3.1.2 SOAP Message

The following is the SOAP 1.1 request for the Web service created above. The placeholders shown need to be replaced with actual values.

This SOAP message is split into two sections. The first section (lines 1–5) is a set of HTTP headers that are used to communicate some details of the document. Specifically, line 1 indicates that information is sent via the HTTP-POST method. This might seem to contradict the fact that the SOAP is used here, but it is noted that the SOAP message has to be sent as or in an HTTP request to allow it to go through most Web servers. This section also indicates that the host is localhost (line 2) and that the service is located at http://tempuri.org/AreaCalculator (line 5).

The second section is an XML document, where line 6 declares the XML header and line 7 declares the namespaces (e.g., xsi, xsd, and soap). The root element is named "Envelope," which contains a SOAP header and a SOAP body. The SOAP header is optional and is not included in the code, but the SOAP body is a mandatory element. All documents sent by SOAP will follow the structure as follows:

```
<soap:Envelope>
<soap:Body>
...Web service content here...
<soap:Body>
<soap:Envelope>
```

Note that the SOAP here is the namespace declared in line 7. In this example, the Web service content is defined in lines 9 to 11. The AreaCalculator element is the Web service, and the xmlns attribute outlines the location of the Web service. Inside the AreaCalculator element is one empty sub-element radius. This is the parameter that the client user needs to supply to the Web service when it is invoked. The parameter is serialized into a SOAP document, which is parceled up in the HTTP data and transmitted to the Web service.

Next, let us take a look at the SOAP 1.1 response. Again, the placeholders shown need to be replaced with actual values. The structure of the SOAP response is very similar to that of the SOAP request except that the SOAP contains the information sent back from the server, so just the HTTP protocol needs to be specified in the first section (Tables 19.3 and 19.4). Second, the SOAP body element contains one AreaCalculatorResponse sub-element which, in turn, contains an AreaCalculatorResult sub-element. The value of the AreaCalculatorResult is in decimal data type.

With an understanding of HTTP, SOAP, and WSDL in a Web service created, it is time to learn how to use this area calculator Web service. Web service discovery is like the process of locating an item using a search engine. Most of the time, a URL can be used to

TABLE 19.3

SOAP Request

```
1.   POST/Service1.asmx HTTP/1.1
2.   Host: localhost
3.   Content-Type: text/xml; charset=utf-8
4.   Content-Length: length
5.   SOAPAction: "http://tempuri.org/AreaCalculator"
6.   <?xml version="1.0" encoding="utf-8"?>
7.   <soap:Envelope xmlns:xsi="http://www.w3.org/2001/XMLSchema-instance"
 xmlns:xsd="http://www.w3.org/2001/XMLSchema" xmlns:soap="http://schemas.xmlsoap.
 org/soap/envelope/">
8.   <soap:Body>
9.   <AreaCalculator xmlns="http://tempuri.org/">
10.  <radius/>
11.  </AreaCalculator>
12.  </soap:Body>
13.  </soap:Envelope>
```

TABLE 19.4

SOAP Response

```
1.   HTTP/1.1 200 OK
2.   Content-Type: text/xml; charset=utf-8
3.   Content-Length: length
4.   <?xml version="1.0" encoding="utf-8"?>
5.   <soap:Envelope xmlns:xsi="http://www.w3.org/2001/XMLSchema-instance"
 xmlns:xsd="http://www.w3.org/2001/XMLSchema" xmlns:soap="http://schemas.
 xmlsoap.org/soap/envelope/">
6.   <soap:Body>
7.   <AreaCalculatorResponse xmlns="http://tempuri.org/">
8.   <AreaCalculatorResult>decimal</AreaCalculatorResult>
9.   </AreaCalculatorResponse>
10.  </soap:Body>
11.  </soap:Envelope>
```

locate Web services. Since the example Web services created in this chapter are not publicly published, the next example demonstrates the use of VS 2010 to create an application that integrates the Web service locally.

19.3.2 Using the Web Service

To use the Web service hosted locally, a Web reference using the VS Developer needs to be added. (This is similar to the way of adding references into a VB.NET project or a VBA project.) We will illustrate how to use the Web service step by step as follows:

Step 1: Before the Web service is invoked, it needs to be compiled. We may choose Test, then New Test, and a pop-up window will show. Let us choose Basic Unit test, which will create a VB project.

Step 2: Under the VB test project in the Solution Explorer window, right-click on the project, and choose the Add Service Reference option. In the earlier VS version, the Web service is an option available under the same panel as adding reference (Figure 19.3). In VS 2010, we will take more steps to achieve the addition of a Web service, as outlined in the following.

Step 3: Add Service Reference will bring up the following window (Figure 19.4). To connect to a Web service, click on Advanced… (as shown in Figure 19.5).

Step 4: There are several options to locate the Web service, as shown in Figure 19.6. In this example, Web Services on the Local Machine is chosen.

Step 5: Once the Web service is located, the following window will show the details, including the URL, which, in this case, is http://localhost:59483/Service.asmx. In the Service1 window, if we click Service Description, the WSDL discussed above will be shown. Next, the Web reference name needs to be specified, in this example, let us use "localhost." Click on Add Reference, and this Web service is now integrated in the VB test project.

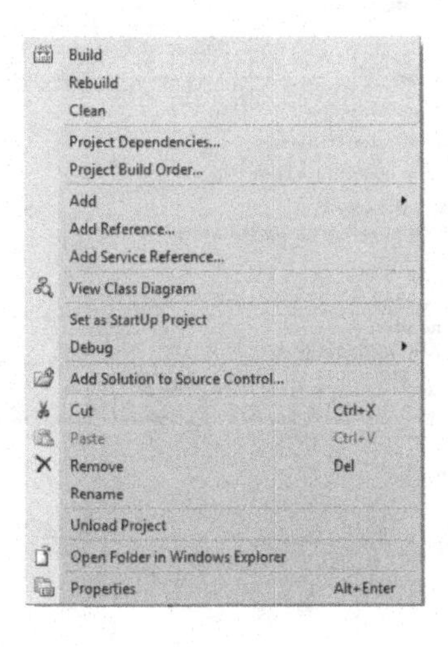

FIGURE 19.3
Adding service reference.

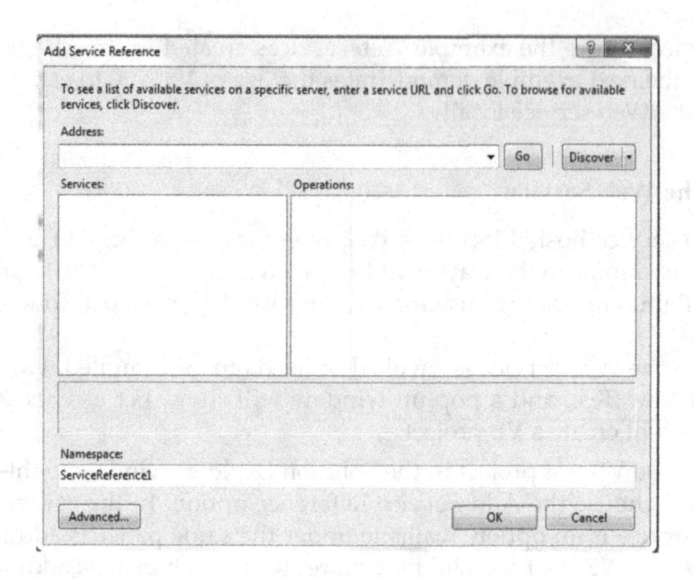

FIGURE 19.4
Add Service Reference window.

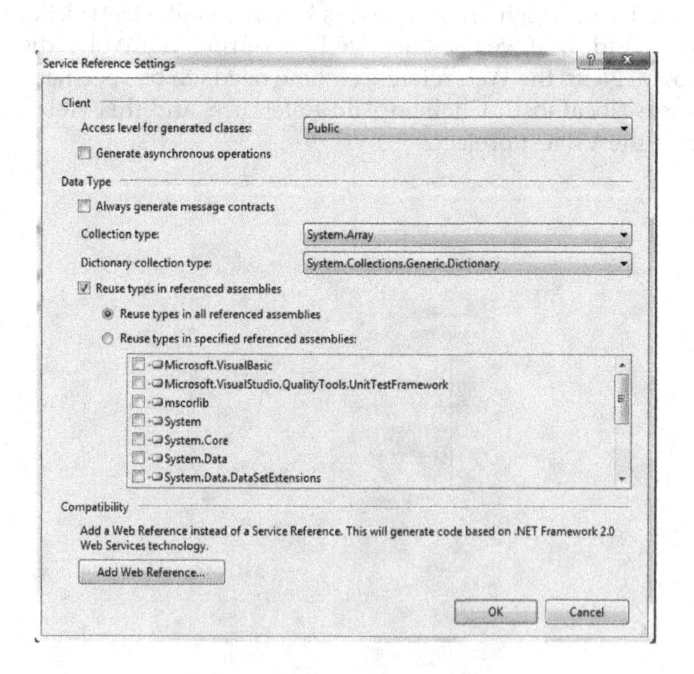

FIGURE 19.5
Adding Web reference.

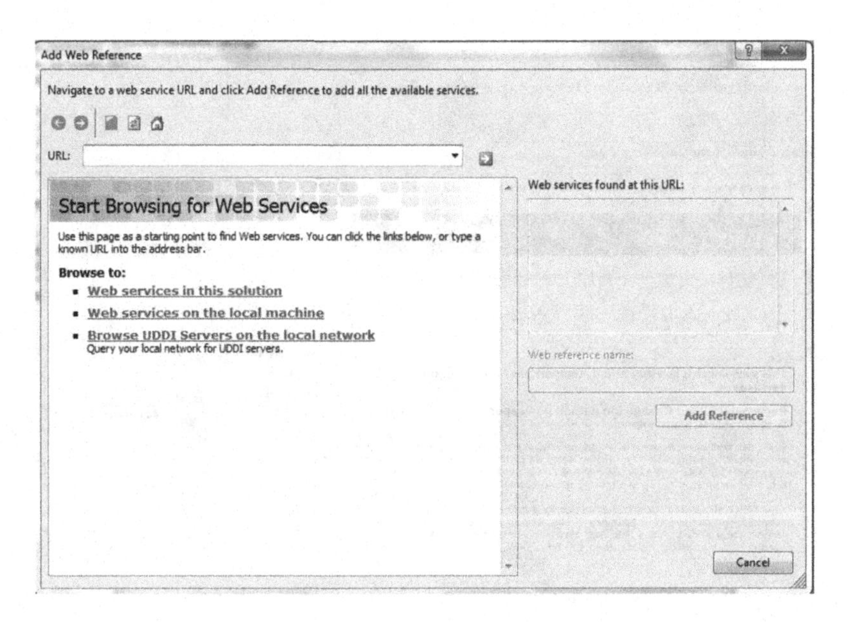

FIGURE 19.6
Discovering Web services.

Step 6: Now, in the Code window, which we should be very familiar with after Chapter 12, type the code as in Table 19.5. The use of a Web service is just like the use of a regular class in the program. First, a new object testWS as a localhost is declared. "localhost" is the Web reference name, and Service1 is the class created in the Web service. Next, the Web method AreaCalculator is invoked with the parameter (radius) set as 10 (Figure 19.7).

Step 7: Since we create a test project within the same solution as the Web service, to execute the test project, we need to do one more thing, that is, to choose the test project. Go to Project, then Set as Startup Project. Now, we are ready to test the Web service. Upon successful execution, a message box will pop up, showing the result, which is 62.8318530717959.

TABLE 19.5
Testing the Web Service

```
Imports System.Text
<TestClass()>
Public Class UnitTest1
  <TestMethod()>
  Public Sub TestMethod1()
     Dim testWS As New localhost.Service1
     Dim WSResult As Decimal
         WSResult = testWS.AreaCalculator(10)
            MsgBox(tt)
         End Sub
End Class
```

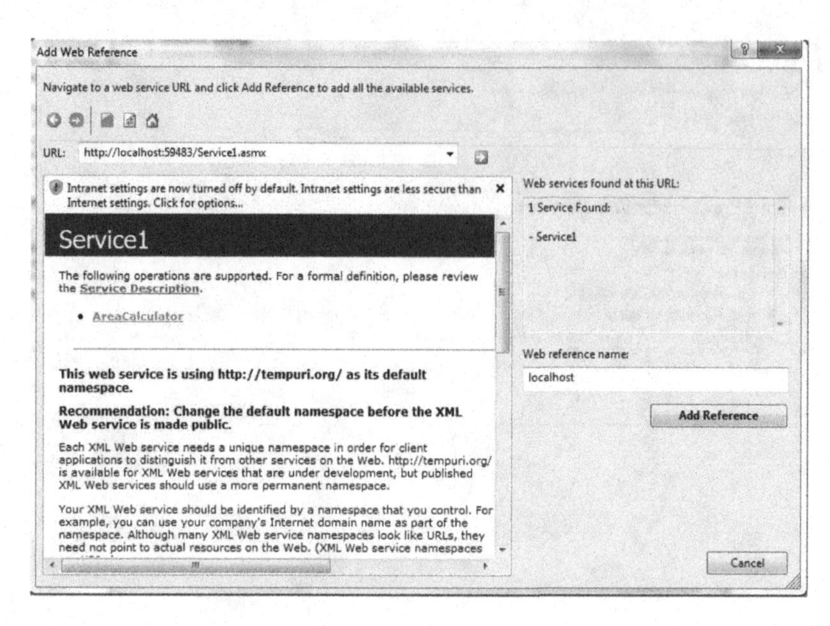

FIGURE 19.7
Adding Web services to the application.

This example demonstrates how to integrate and use the Web services deployed locally. If the Web service is published to a public directory (e.g., UDDI language), the developer just needs to locate it via the UDDI language server, and the remaining steps follow the same.

19.4 Summary

This chapter introduces the basic concepts and the architecture of Web services. Some important protocols to support Web services, including HTTP, WSDL, SOAP, and UDDI language, are reviewed. Using VS 2010, an example on creating Web service, and then integrating the Web service to a VB.NET Windows application is provided.

EXERCISES

1. A Web service provides custom-built services that can be accessed by other Web services over the Internet. True or false?
2. Web services are standardized in terms of the following areas except:
   a. Method of transmission
   b. Definition of Web services
   c. Methodology to publish a Web service
   d. Architecture
   e. All of the above are valid areas in which Web services are standardized

3. A Web service can be used to accommodate the __________ layer of distributed computing due to its platform agnostic property.
    a. Presentation
    b. Middle
    c. Back-end data
    d. Server
    e. Client
4. What are the three components of a Web service architecture?
5. Which of the following is true about the Web service architecture model?
    a. A service requester publishes a Web service to a service broker. The broker finds a Web service from multiple service providers.
    b. A service requester binds Web services to a service broker, which, in turn, finds a Web service from multiple service providers.
    c. The Internet is not used in the Web service architecture.
    d. A service requester finds Web services from a service broker, while a service provider publishes Web services to a service broker.
    e. None of the above are valid statements.
6. The Web URL http://queryauthors.aspx/query?name=Mickey is an example of a HTTP-POST method. True or false?
7. The HTTP-POST method is more secure as compared to the HTTP-GET method since it uses "https" in the URL instead of "http." True or false?
8. A SOAP message is sent using plain text and, as such, is optimal for limited bandwidth. True or false?
9. When sending an X-ray image of a patient to a physician using a local hospital Web service, which of the following protocols is the most efficient?
    a. HTTP
    b. HTTP-POST
    c. HTTP-GET
    d. SOAP
    e. None of the above
10. A WSDL is used to generate well-defined SOAP messages. True or false?
11. The following are valid properties of WSDL methods except:
    a. WSDL is written in XML.
    b. WSDL is used to describe the services shared publicly by the service provider.
    c. WSDL provides automated communication between a service requester and a service provider.
    d. All of the above are valid properties.
    e. None of the above is a valid property.

Section V

Design of Information Systems

In Sections II to IV, we described how to develop a relational database and build a Windows-based or Web-enabled application with the database embedded. This part introduces additional design issues of information systems concerning the following:

- Computing efficiency of algorithms (in Chapter 20)
- Interface design and usability (in Chapter 21)
- Computer and network security (in Chapter 22)
- Data mining (in Chapter 23) and expert systems (in Chapter 24), which help transform an information system to a decision support system by discovering useful information and knowledge from data in databases and utilizing information and knowledge to support decision making.

Many different forms of decision support systems are also introduced in Chapter 25. Hence, this part addresses many design aspects of information systems and their transformation to decision support systems for enhancing functionalities, usability, and performance of information systems and decision support systems that are built as Windows-based or Web-enabled applications.

20

Computing Efficiency of Algorithms

Windows and Web applications in the previous chapters consist of event-driven subroutines. Each subroutine implements an algorithm that takes an input and produces the correct output. The running time of an algorithm is the time from taking the input to producing the correct output. The running time of an algorithm determines the computing efficiency of the algorithm, which has a direct impact on the performance and the user satisfaction of an application. In this chapter, we introduce the basics of analyzing the computing efficiency of an algorithm by describing how to analyze the running time of a non-recursive algorithm. A non-recursive algorithm does not involve a function or a procedure calling the function or the procedure itself. We introduce asymptotic notations that specify the computing efficiency of an algorithm by defining the order of the running time when the size of inputs is large. With an understanding of the computing efficiency of an algorithm, we illustrate why the computing efficiency of an algorithm is important even when the processing speed of computers has become fast.

20.1 Analysis of the Running Time of a Non-Recursive Algorithm

The running time of the algorithm, T, typically grows with the size of an input. Hence, T is often a function of the input size. When analyzing the running time of an algorithm, we determine the time of running each line of the code once and the number of times to run each line of the code. We consider that it takes a constant amount of time to run a given line of the code each time. However, different lines of code may take different amounts of constant times due to the different operations involved in different lines of code. Hence, we denote the running time (the computing cost) of each line of the code by c_i for the ith line of code and the number of times to run each line of the code by t_i for the ith line of code. The running time of the algorithm, T(input size), is the sum of $c_i t_i$ for all i as follows:

$$T(\text{input size}) = \sum_i c_i t_i \tag{20.1}$$

We illustrate how to analyze the running time of an algorithm by analyzing the running time of the selection sort algorithm in Example 20.1.

Example 20.1

The algorithm of the selection sort written in Visual Basic is given as follows:

```
For j = 1 to n - 1
   smallest = j
   For i = j + 1 to n
```

```
        If A(i) < A(smallest) Then
            smallest = i
        End If
    Next
    value = A(smallest)
    A(smallest) = A(j)
    A(j) = value
Next
```

The algorithm takes an array of n values as the input and produces an array of sorted values in a non-decreasing order of these values as the output. The algorithm takes the first value ($j = 1$ in the first For loop) in the array, compares it with each of the other values in the array ($i = j + 1, \ldots, n$ in the second For loop) to determine the smallest value, and switches the first value with the smallest value in the array. The algorithm repeats this for the second value ($j = 2$), the third value, …, and the second last value ($j = n - 1$) in the array.

Table 20.1 lists the lines of the code in the algorithm in the first column and gives the computing cost of running each line of code in the second column and the number of times to run each line of code in the third column. For the second column, we put in the constant computing cost of c_i for the ith line of the code.

The first line of the code:

```
For j = 1 to n - 1
```

is executed n times, including $n - 1$ times for $j = 1, \ldots, n - 1$, to get into the For loop and one more time for $j = n$ to examine that $j = n$ does not satisfy the condition of getting into the For loop and getting out of the For loop.

TABLE 20.1

Selection Sort Algorithm and Analysis of the Running Time

Selection Sort Algorithm	Cost	Times
`For j = 1 to n - 1`	c_1	n
`smallest = j`	c_2	$n - 1$
`For i = j + 1 to n`	c_3	$\sum_{j=1}^{n-1}(n-j-1+1+1) = \sum_{j=1}^{n-1}(n-j+1)$
`If A(i) < A(smallest) Then`	c_4	$\sum_{j=1}^{n-1}(n-j-1+1) = \sum_{j=1}^{n-1}(n-j)$
`smallest = i`	c_5	$\sum_{j=1}^{n-1}\sum_{i=j+1}^{n} t_i$
`End If`	c_6	$\sum_{j=1}^{n-1}(n-j-1+1) = \sum_{j=1}^{n-1}(n-j)$
`Next`	c_7	$\sum_{j=1}^{n-1}(n-j-1+1) = \sum_{j=1}^{n-1}(n-j)$
`value = A(smallest)`	c_8	$n - 1$
`A(smallest) = A(j)`	c_9	$n - 1$
`A(j) = value`	c_{10}	$n - 1$
`Next`	c_{11}	$n - 1$

The second line of the code:

```
smallest = j
```

is executed $n - 1$ times inside the For loop for $j = 1, \ldots, n - 1$.

The third line of the code:

```
For i = j + 1 to n
```

is inside the first For loop and executed each time that the control gets into the first For loop. In each time when this line of the code is executed inside the first For loop, this line of the code is executed $n - (j + 1) + 1 = n - j$ times to get into the second For loop plus one more time to check $i = n + 1$ and to exit the second For loop. Hence, the total number of times to execute the third line of the code is $\sum_{j=1}^{n-1}(n-j+1)$.

The fourth line of the code:

```
If A(i) < A(smallest) Then
```

is inside the second For loop and executed one time less than the third line for each time of getting inside the first For loop. Hence, the total number of times to execute the fourth line of the code is $\sum_{j=1}^{n-1}(n-j)$.

The fifth line of the code:

```
smallest = i
```

is inside the If–Then statement. Hence, the fifth line of the code may or may not be executed, depending on whether the condition of the If–Then statement is satisfied. How many times the condition in the If–Then statement is satisfied varies with different arrays and depends on the input array. Hence, we cannot give a known number of times running the fifth line of the code, and we denote it using a variable, t_j, with the subscript i indicating that t_i varies for i when $A(i)$ is compared with A(smallest). $A(i)$ takes the value of 0 or 1.

The sixth line of the code:

```
End If
```

is a part of the If–Then statement in the fourth line and executed the same number of times as the fourth line.

The seventh line of the code:

```
Next
```

is a part of the second For loop statement, performs $i = i + 1$, and is executed $n - j$ times each time inside the first For loop. Hence, the total number of times to run this line of the code is $\sum_{j=1}^{n-1}(n-j)$.

The eighth line of the code:

```
value = A(smallest)
```

is inside the first For loop and executed $n - 1$ times. Similarly, the ninth and the tenth lines of the code is executed $n - 1$ times.

The eleventh line of the code:

```
Next
```

is a part of the first For loop statement, performs $j = j + 1$, and is executed $n - 1$ times.

The number of times running each line of the code depends on n, which measures the size of the input array. Hence, the total running time of the algorithm is a function of n:

$$T(n) = c_1 n + c_2(n-1) + c_3 \sum_{j=1}^{n-1}(n-j+1) + c_4 \sum_{j=1}^{n-1}(n-j) + c_5 \sum_{j=1}^{n-1}\sum_{i=j+1}^{n} t_i + c_6 \sum_{j=1}^{n-1}(n-j)$$
$$+ c_7 \sum_{j=1}^{n-1}(n-j) + c_8(n-1) + c_9(n-1) + c_{10}(n-1) + c_{11}(n-1) \quad (20.2)$$

where

$$\sum_{j=1}^{n-1}(n-j) = \sum_{j=1}^{n-1} n - \sum_{j=1}^{n-1} j = n(n-1) - \frac{(1+n-1)(n-1)}{2} = n^2 - n - \frac{1}{2}n^2 - \frac{1}{2}n$$
$$= \frac{1}{2}n^2 - \frac{3}{2}n \quad (20.3)$$

$$\sum_{j=1}^{n-1}(n-j+1) = \frac{1}{2}n^2 - \frac{3}{2}n + \sum_{j=1}^{n-1} 1 = \frac{1}{2}n^2 - \frac{3}{2}n + n - 1 = \frac{1}{2}n^2 - \frac{1}{2}n - 1 \quad (20.4)$$

$$T(n) = c_1 n + c_2(n-1) + c_3\left(\frac{1}{2}n^2 - \frac{1}{2}n - 1\right) + c_4\left(\frac{1}{2}n^2 - \frac{3}{2}n\right) + c_5 \sum_{j=1}^{n-1}\sum_{i=j+1}^{n} t_i + c_6\left(\frac{1}{2}n^2 - \frac{3}{2}n\right)$$
$$+ c_7\left(\frac{1}{2}n^2 - \frac{3}{2}n\right) + c_8(n-1) + c_9(n-1) + c_{10}(n-1) + c_{11}(n-1)$$
$$= \frac{1}{2}(c_3 + c_4 + c_6 + c_7)n^2$$
$$+ \left(c_1 + c_2 - \frac{1}{2}c_3 - \frac{3}{2}c_4 - \frac{3}{2}c_6 - \frac{3}{2}c_7 + c_8 + c_9 + c_{10} + c_{11}\right)n$$
$$- (c_2 + c_3 + c_8 + c_9 + c_{10} + c_{11}) + c_5 \sum_{j=1}^{n-1}\sum_{i=j+1}^{n} t_i \quad (20.5)$$

In Equation 20.5, t_i depends on the input array. If the input array has values already in a non-decreasing order, the condition of the If–Then statement is never satisfied, and thus, $t_i = 0$. If the input array has values in a decreasing order without any two values being equal, the condition of the If–Then statement is always satisfied, and we have

$$t_i = 1 \quad (20.6)$$

and the number of times to run the fifth line of the code is the same as the number of times to run the fourth line of the code as follows:

$$\sum_{j=1}^{n-1}(n-j) = \frac{1}{2}n^2 - \frac{3}{2}n \quad (20.7)$$

Hence, in the best case, the running time of the selection sort algorithm is

$$T(n) = c_1 n + c_2(n-1) + c_3\left(\frac{1}{2}n^2 - \frac{1}{2}n - 1\right) + c_4\left(\frac{1}{2}n^2 - \frac{3}{2}n\right) + c_5\sum_{j=1}^{n-1}\sum_{i=j+1}^{n} t_i + c_6\left(\frac{1}{2}n^2 - \frac{3}{2}n\right)$$

$$+ c_7\left(\frac{1}{2}n^2 - \frac{3}{2}n\right) + c_8(n-1) + c_9(n-1) + c_{10}(n-1) + c_{11}(n-1) = \frac{1}{2}(c_3 + c_4 + c_6 + c_7)n^2$$

$$+\left(c_1 + c_2 - \frac{1}{2}c_3 - \frac{3}{2}c_4 - \frac{3}{2}c_6 - \frac{3}{2}c_7 + c_8 + c_9 + c_{10} + c_{11}\right)n - (c_2 + c_3 + c_8 + c_9 + c_{10} + c_{11})$$

$$+ c_5\sum_{j=1}^{n-1} 0 = \frac{1}{2}(c_3 + c_4 + c_6 + c_7)n^2 + \left(c_1 + c_2 - \frac{1}{2}c_3 - \frac{3}{2}c_4 - \frac{3}{2}c_6 - \frac{3}{2}c_7 + c_8 + c_9 + c_{10} + c_{11}\right)n$$

$$-(c_2 + c_3 + c_8 + c_9 + c_{10} + c_{11}) \tag{20.8}$$

In the worst case, the running time of the selection sort algorithm is

$$T(n) = c_1 n + c_2(n-1) + c_3\left(\frac{1}{2}n^2 - \frac{1}{2}n - 1\right) + c_4\left(\frac{1}{2}n^2 - \frac{3}{2}n\right) + c_5\sum_{j=1}^{n-1}\sum_{i=j+1}^{n} t_i + c_6\left(\frac{1}{2}n^2 - \frac{3}{2}n\right)$$

$$+ c_7\left(\frac{1}{2}n^2 - \frac{3}{2}n\right) + c_8(n-1) + c_9(n-1) + c_{10}(n-1) + c_{11}(n-1) = \frac{1}{2}(c_3 + c_4 + c_6 + c_7)n^2$$

$$+\left(c_1 + c_2 - \frac{1}{2}c_3 - \frac{3}{2}c_4 - \frac{3}{2}c_6 - \frac{3}{2}c_7 + c_8 + c_9 + c_{10} + c_{11}\right)n - (c_2 + c_3 + c_8 + c_9 + c_{10} + c_{11})$$

$$+ c_5\left(\frac{1}{2}n^2 - \frac{3}{2}n\right) = \frac{1}{2}(c_3 + c_4 + c_5 + c_6 + c_7)n^2 + \left(c_1 + c_2 - \frac{1}{2}c_3 - \frac{3}{2}c_4 - \frac{3}{2}c_5\right.$$

$$\left. - \frac{3}{2}c_6 - \frac{3}{2}c_7 + c_8 + c_9 + c_{10} + c_{11}\right)n - (c_2 + c_3 + c_8 + c_9 + c_{10} + c_{11}) \tag{20.9}$$

Both the best-case and the worst-case running times of the selection sort algorithm are in the form of $an^2 + bn + c$.

20.2 Asymptotic Notations for the Computing Efficiency of Algorithms

The running time of the algorithm grows with the input size. For the running time of the selection sort algorithm in the form of $an^2 + bn + c$, the order of the input size in the highest-order term, n^2, is most critical to the computing efficiency of the algorithm and determines the order of growth. Table 20.2 shows the order of growth with the input size that is determined by different terms in $an^2 + bn + c$. As shown in Table 20.2, n^2 and $10n^2$ produce much larger values for large n values (i.e., $n > 1000$, such as $n = 10{,}000$; 1,000,000; and 1,000,000,000) than $10{,}000n$, although the coefficient 10,000 in $10{,}000n$ is much larger than the coefficients in n^2 and $10n^2$. Hence, the highest-order term of the input size is more critical than any lower-order term when we consider the running time of an algorithm for

TABLE 20.2

Order of Growth with the Input Size n Determined by Different Terms in $an^2 + bn + c$

Input Size n	n^2	$10n^2$	$10{,}000n$
10	100	1000	100,000
1000	1,000,000	10,000,000	10,000,000
10,000	100,000,000	1,000,000,000	100,000,000
1,000,000	1,000,000,000,000	10,000,000,000,000	10,000,000,000
1,000,000,000	1,000,000,000,000,000,000	10,000,000,000,000,000,000	10,000,000,000,000

a large input size. The coefficient in the highest-order term is less significant than the order of the input size in the highest-order term because the order determines the rate of growth with the input size.

We emphasize the running time of an algorithm for a large input size because the small running time of the algorithm for a small input size can be easily handled by high-speed computers nowadays, but the large running time of an algorithm for a large input size can still present a challenge for super-fast computers to produce the output quickly.

When considering the computing efficiency of an algorithm, we are mostly concerned with the worst-case running time than the best-case running time because the order of growth in the worst-case running time presents more challenge for the algorithm to produce the output quickly. Hence, if the worst-case running time of one algorithm has a smaller order of growth, the algorithm is more computing efficient when the algorithm is compared with another algorithm.

The computing efficiency of an algorithm is more important than the speed of a computer with regard to the time that the computer takes to produce the output for a given input. Suppose that Computer A is 1000 times as fast as Computer B, Computer A processes 1 billion of instructions per second, and Computer B processes 1 million of instructions per second. If we run the selection sort algorithm with the order of growth at n^2 and another sorting algorithm called the "merge sort algorithm" with the order of growth at $n\log_2 n$ (Cormen et al., 2009), the merge sort algorithm will be more efficient than the selection sort algorithm. We run both algorithms to sort 1 million of values in the input array with $n = 1{,}000{,}000$ on Computer A and Computer B. It takes Computer A $(1{,}000{,}000)^2/1{,}000{,}000{,}000 = 1000$ s or 2.78 h to run the selection sort algorithm and $(1{,}000{,}000)(\log_2 1{,}000{,}000)1{,}000{,}000{,}000 = 0.0199$ s to run the merge sort algorithm. It takes Computer B $(1{,}000{,}000)^2/1{,}000{,}000 = 1{,}000{,}000$ s or 166 days to run the selection sort algorithm and $(1{,}000{,}000)\log_2 1{,}000{,}000/1{,}000{,}000 = 19.93$ s to run the merge sort algorithm. That is, the more efficient algorithm (the merge sort algorithm) takes less time (19.93 s each) to produce the output for a large input on the slower computer (computer B) than the time (2.78 h) that the less efficient algorithm (the selection sort algorithm) takes on the faster computer (computer A). Hence, the computing efficiency of an algorithm is more important than the speed of a computer with regard to the time of producing the output.

Asymptotic notations of the computing efficiency, O, Ω, Θ, o, and ω, define the order of growth for large inputs. The definition of the O notation is as follows (Cormen et al., 2009):

$$O(g(n)) = \{f(n) \mid 0 \leq f(n) \leq cg(n) \text{ for all } n \geq n_0 \text{ and some positive constants } c \text{ and } n_0\} \quad (20.10)$$

where n is a measure of the input size such as the number of values in an array for a sorting algorithm. For example, $O(n^2)$ includes $10n^2$, $50n$, $1000n^2 + 3n + 16$, $n\log_2 n$, and so on.

Example 20.2

Use the definition of the O notation to show $10n^2 \in O(n^2)$ and $50n^2 \in O(n^2)$.
We have $f(n) = 10n^2$ satisfy Equation 20.10 because

$$0 \le 10n^2 \le cn^2 \text{ holds for } n \ge n_0, n_0 = 1, \text{ and}$$

$$c = 10, 11, \text{ or any value greater than or equal to } 10.$$

We have $f(n) = 50n$ satisfy Equation 20.10 because

$$0 \le 50n \le cn^2 \text{ holds for } n \ge n_0, n_0 = 10, \text{ and } c = 5.$$

Example 20.3

Use the definition of the O notation to show that $50n^3$ is not in $O(n^2)$.
To have $f(n) = 50n^3$ satisfy Equation 20.10 or

$$0 \le 50n^3 \le cn^2 \text{ holds for } n \ge n_0 \text{ and some positive constants } c \text{ and } n_0,$$

we need $50n \le c$ or $n < \frac{c}{50}$. However, for any positive constant c, there are always large n values to make $n \ge \frac{c}{50}$. Hence, Equation 20.10 does not hold for $f(n) = 50n^3$, and $50n^3$ is not in $O(n^2)$.

When we determine a function, $f(n)$, for the worst-case running time of an algorithm, we can take the highest-order term without the coefficient in $f(n)$ and put that in the O notation because the function for the running time of the algorithm for any other input is in the same or a lower order of growth in comparison with $f(n)$ and is in the collection of functions represented by the O notation. For example, with the function in Equation 20.9 in the form of $an^2 + bn + c$ as the worst-case running time of the selection sort algorithm, we have $O(n^2)$ for the selection sort algorithm.

The definition of the Ω notation is as follows (Cormen et al., 2009):

$$\Omega(g(n)) = \{f(n) \mid 0 \le cg(n) \le f(n) \text{ for all } n \ge n_0 \text{ and some positive constants } c \text{ and } n_0\} \quad (20.11)$$

For example, $\Omega(n^2)$ includes $10n^2$, $50n^3$, $1000n^2 + 3n + 16$, $n^2\log_2 n$, and so on.

Example 20.4

Use the definition of the Ω notation to show $10n^2 \in \Omega(n^2)$ and $50n^2 \in \Omega(n^2)$.
We have $f(n) = 10n^2$ satisfy Equation 20.11 because

$$0 \le cn^2 \le 10n^2 \text{ holds for } n \ge n_0, n_0 = 1, \text{ and } c = 9, 8, \text{ or any value less than or equal to } 10.$$

We have $f(n) = 50n^3$ satisfy Equation 20.11 because

$$0 \le cn^2 \le 50n^2 \text{ holds for } n \ge n_0, n_0 = 1, \text{ and } c = 50 \text{ or any value less than } 50.$$

Example 20.5

Use the definition of the Ω notation to show that $50n$ is not in $\in \Omega(n^2)$.
To have $f(n) = 50n$ satisfy Equation 20.11 or

$$0 \le cn^2 \le 50n \text{ holds for } n \ge n_0 \text{ and some positive constants } n_0 \text{ and } c,$$

we need $n \le \frac{50}{c}$. However, for the given positive constants c and n_0, we always have some $n \ge n_0$ to make $n > \frac{50}{c}$. Hence, Equation 20.11 does not hold for $f(n) = 50n$, and $50n$ is not in $\in \Omega(n^2)$.

When we determine a function, $f(n)$, for the best-case running time of an algorithm, we can take the highest-order term without the coefficient in $f(n)$ and put that in the Ω notation because the function for the running time of the algorithm for any other input is in the same or a higher order of growth in comparison with $f(n)$ and is in the collection of functions represented by the Ω notation. For example, with the function in Equation 20.8 in the form of $an^2 + bn + c$ as the best-case running time of the selection sort algorithm, we have $\Omega(n^2)$ for the selection sort algorithm.

The definition of the Θ notation is as follows (Cormen et al., 2009):

$$\begin{aligned}&\Theta(g(n)) = \\ &\{f(n) \mid 0 \le c_1 g(n) \le f(n) \le c_2 g(n) \text{ for all } n \ge n_0 \\ &\text{and some positive constants } c_1, c_2, \text{ and } n_0\}\end{aligned} \tag{20.12}$$

$f(n) \in \Theta(g(n))$ if and only if $f(n) \in O(g(n))$ and $f(n) \in \Omega(g(n))$. For example, $\Omega(n^2)$ includes $10n^2$, $1000n^2 - 3n + 16$, and so on. Because we have both $O(n^2)$ and $\Omega(n^2)$ for the selection sort algorithm, the computing efficiency of the selection sort algorithm can be defined by $\Theta(n^2)$.

Example 20.6

Use the definition of the Θ notation to show $10n^2 \in \Theta(n^2)$.

We have $f(n) = 10n^2$ satisfy Equation 20.12 because

$$0 \le c_1 n^2 \le 10n^2 \le c_2 n^2 \text{ holds for } n \ge n_0,\ n_0 = 1,\ c_1 = 1, \text{ and } c_2 = 11.$$

The definition of the o notation is as follows (Cormen et al., 2009):

$$\begin{aligned}&o(g(n)) = \\ &\{f(n) \mid 0 \le f(n) < cg(n) \text{ for all } n \ge n_0, \\ &\text{all positive constants } c, \text{ and some positive constant } n_0\}\end{aligned} \tag{20.13}$$

For example, $o(n^2)$ includes $50n$, $n\log_2 n$, and so on.

Example 20.7

Use the definition of the o notation to show $50n \in o(n^2)$.

We have $f(n) = 50n$ satisfy Equation 20.13 because

$$0 \le 50n < cn^2 \text{ holds for } n \ge n_0, \text{ all positive constant } c, \text{ and } n_0 = \frac{50}{c} + 1.$$

Example 20.8

Use the definition of the o notation to show that n^2 is not in $o(n^2)$.

To have $f(n) = n^2$ satisfy Equation 20.13 or have

$0 \le n^2 < cn^2$ holds for $n \ge n_0$, all positive constant c, and some positive constant n_0,

we need $c > 1$. For c in (0, 1), $n^2 < cn^2$ does not hold. Hence, $o(n^2)$ does not include n^2.

The definition of the ω notation is as follows (Cormen et al., 2009):

$$\begin{aligned} &\omega(g(n)) = \\ &\{f(n) \mid 0 \le cg(n) < f(n) \text{ for all } n \ge n_0, \\ &\text{all positive constant } c, \text{ and some positive constant } n_0\} \end{aligned} \tag{20.14}$$

For example, $\omega(n^2)$ includes $50n^3$, $n^2\log_2 n$, and so on.

Example 20.9

Use the definition of the ω notation to show $50n^3 \in \omega(n^2)$.

We have $f(n) = 50n^3$ satisfy Equation 20.14 because

$0 \le cn^2 < 50n^3$ holds for $n \ge n_0$, all positive constant c, and $n_0 = \frac{c}{50} + 1$.

Example 20.10

Use the definition of the o notation to show that n^2 is not in $\omega(n^2)$.

To have $f(n) = n^2$ satisfy Equation 20.14 or have

$0 \le cn^2 < n^2$ holds for $n \ge n_0$, all positive constant c, and some positive constant n_0,

we need $c < 1$. For c in $(1, \infty)$, $nc^2 < n^2$ does not hold. Hence, $\omega(n^2)$ does not include n^2.

The running times of many algorithms are often in a polynomial function of the input size (e.g., n):

$$f(n) = \sum_{i=0}^{k} a_i n^i, \tag{20.15}$$

in an exponential function of the input size:

$$f(n) = a^n \tag{20.16}$$

or in a polylogarithmic function of the input size:

$$f(n) = (\log_a n)^k \tag{20.17}$$

A polylogarithmic function is more efficient than a polynomial function, which is more efficient than an exponential function. The following equations indicate the computing efficiency of a factorial function (Cormen et al., 2009):

$$n! = o(n^n) \tag{20.18}$$

$$n! = \omega(2^n) \tag{20.19}$$

$$\log_a n! = \Theta(n \log_a n). \tag{20.20}$$

20.3 Summary

In this chapter, we demonstrate how to determine the computing efficiency of an algorithm by describing how to analyze the running time of a non-recursive algorithm and how to use asymptotic notations for defining the computing efficiency of an algorithm. Cormen et al. (2009) describe the methods of analyzing the running time and the computing efficiency of a recursive algorithm.

EXERCISES

1. The following table gives the insertion sort algorithm written in Visual Basic to sort values in an array in non-decreasing order. The running time of the algorithm is a function of n, which is the number of values in the input array. Use the table to determine the running time of the algorithm. Give an array that produces the best-case running time of the algorithm and define the best-case running time in an asymptotic notation. Give an array that produces the worst-case running time of the algorithm and define the worst-case running time in an asymptotic notation.

Insertion Sort Algorithm	Cost	Times
`For j = 2 to n`		
`key = A(j)`		
`i = j - 1`		
`Do While i > 0 AND A(i) > key`		
`A(i + 1) = A(i)`		
`i = i - 1`		
`Loop`		
`A(i + 1) = key`		
`Next`		

2. The following table gives the merge algorithm written in Visual Basic, which merges two sorted arrays: L with the k number of values and R with the n-k number of values into one large sorted array with the n number of values, A. Use the table to determine the running time of the algorithm. Give L and R arrays that produce the best-case running time of the algorithm and define the best-case

running time in an asymptotic notation. Give L and R arrays that produce the worst-case running time of the algorithm and define the worst-case running time in an asymptotic notation.

Merge Algorithm **Cost** **Times**

```
L(k) = ∞
R(n-k) = ∞
i = 0
j = 0
For k = 0 to n - 1
        If L(i) <= R(j) Then
                A(k) = L(i)
                i = i + 1
        Else
                A(k) = R(j)
                i = j + 1
        End If
Next
```

3. Use the definition of the O notation to show $1000n^2 + 3n + 16 \in O(n^2)$.
4. Use the definition of the Ω notation to show $1000n^2 + 3n + 16 \in \Omega(n^2)$.
5. Use the definition of the o notation to show $3n + 16 \in o(n^2)$.
6. Use the definition of the ω notation to show $n^2\log_2 n \in \omega(n^2)$.

21

User Interface Design and Usability

The user interface design of a software system plays a critical role in the usability and marketability of the software system. This chapter describes the usability criteria that set the goals for the user interface design. We then give the principles and guidelines of user interface design for achieving usability goals.

21.1 Usability Criteria

The user interface of a software system should be designed to make the system easy to learn and use, cause few errors, and allow error recovery. Wickens et al. (1998) give the following usability criteria:

- *Learnability*: Measures how much time it takes for novice users to learn how to use the software system.
- *Efficiency*: Measures how well the software system supports user performance after users have become familiar with the system.
- *Memorability*: Measures the use of the working memory and the long-term memory required by the software system. The working memory of a user typically holds 7 ± 2 information items (Wickens et al., 1998) for the user to access on a short term. Overloading the working memory can cause disorientation and other problems due to loss of information. A software system should also support intermittent users to carry out tasks without recalling information from the long-term memory. Intermittent users use the software system intermittently with relatively long periods of time between uses.
- *Low error rate*: Measures how frequently users make errors. For example, users have done something that they do not mean to do. This criterion is also related to how well the software system supports error recovery.
- *User satisfaction*: A subjective measure of many factors that determine how well users like the software system.

These usability criteria involve different types of users:

- Novice users who have little knowledge of the software system versus expert users who are knowledgeable of and familiar with the software system
- Users who use the software system intermittently versus users who use the software system frequently

The learnability criterion is especially important for novice users. The memorability criterion is especially important for intermittent and novice users. The efficiency criterion is especially important for expert frequent users.

21.2 Principles and Guidelines of User Interface Design

Table 21.1 gives some principles and guidelines for designing the user interface of a software system to address the usability criteria. Some examples are included in Table 21.1 to illustrate these principles and guidelines.

The learnability of the user interface can be enhanced through many interface design features. Using a familiar metaphor from a non-computer domain helps users to immediately transfer knowledge of how the metaphor works in the non-computer domain to knowledge of how a software system works. For example, the Windows operating

TABLE 21.1

Some Principles and Guidelines of User Interface Design

Usability Criteria	Principles and Guidelines of User Interface Design	Examples
Learnability	Use a familiar metaphor for knowledge transfer.	Use the desktop metaphor.
	Be consistent.	Keep the same information or function in the same way at the same location throughout the interface (e.g., different screens and different versions of software).
	Present associated items.	Provide intelligent sensing.
	Make the invisible visible.	Show that a user input has been received and is being processed.
		Provide status feedback (e.g., loading, searching).
	Provide online documentation.	Provide online user manual.
Efficiency	Use shortcuts.	Press Ctrl and F5 at the same time for debugging and executing a Windows application.
Memorability	Provide an exhaustive list of all available.	Have menus include all available, with menu items organized based on the frequency of use or the logical sequence of operations, or in alphabetical order.
	WYSIWYG	Use icons, GUI controls, and direct manipulation.
	Provide guided procedures.	Have wizards in Microsoft Visual Studio 2013 set up connection to an Access database.
Low error rate	Make the invisible visible.	Provide status feedback (e.g., loading, searching).
		Show that a user input has been received and is being processed.
	Separate items with severe consequence from frequently used items.	Separate Delete from Rename in a menu.
	Provide error recovery.	Provide undo.
User satisfaction	Increase interestingness.	Use color to attract attention.

system uses the desktop metaphor to organize and represent application icons, folders, files, etc., on the computer screen. The trash basket icon directly corresponds to a trash basket in the real world to hold unwanted papers and files. Moving a file to the trash basket throws away or deletes it. Opening the trash basket allows users to see its contents. Using a familiar metaphor can facilitate users' development of a correct mental model about how a software system works. A mental model is a user's internal conceptualization and understanding of the software system works. A mental model can be used by the user to understand what is being presented on the software system, what the user is required to do for a task, what would happen if the user chooses a particular option, what the system is currently doing, and so on.

Being consistent helps users find the same information or function at the same location in the same way throughout the interface. As shown in Chapters 13 and 16, connecting to an Access database for a Windows-based application is carried out differently from connecting to an Access database for a Web-enabled application using the same software of Microsoft Visual Studio. Such an inconsistency requires users to take additional amounts of time to learn how to make a database connection for a Web-enabled application even when they already know how to make a database connection for a Windows-based application using the same software. The interface inconsistency often exists between different operating systems, which forces users to take time and effort in learning how to use a computer with a new operating system. In the Windows operating system, every window has the same layout, with the same set of icons at the top-right corner for closing the window, switching between the full screen view and the partial screen view of the window, etc. On an Apple computer, users find the similar functions being represented differently (e.g., the red dot for closing a window). Not only are different software systems by different software vendor inconsistent, but it is also a common practice for the same software vendor to create many changes and thus inconsistencies in the interface design between different versions of the same software system, possibly because it takes less development effort to change the user interface than the core set of system functions while giving users a feel of something new.

Presenting associated items to what users are working on can reduce demands for users' learning and can facilitate users' task performance. For example, many software systems now use intelligent sensing to help users with what they are doing. When we enter a code in the Code View of a Windows form using Microsoft Visual Studio 2013, intelligent sensing presents what is available to select based on what we enter so far. This helps eliminate the need for users to learn the syntax of code and helps reduce syntax errors in the program code. Microsoft Excel shows us choices of mathematical functions, the purpose of each available function, and the syntax of a mathematical function when we select the function. For example, when we enter the mathematical function, power, we get the purpose of the power function, "Return the result of a number raised to a power," and the syntax of the power function, power(number, power). When we enter a text message on iPhone, we are presented with a suggested word to complete what we are entering. However, intelligent sensing should not automatically force what is suggested as what is wanted by users. Intelligent sensing for entering code in Microsoft Visual Studio 2013 does not automatically put in the suggested code unless users specifically select the suggested code. However, a suggested word is automatically entered into a text on iPhone if users do not specifically cancel the suggested word. For example, "Jay" can be automatically replaced by "Kay" while a user enters the remaining text and then presses the Send button, causing the embarrassment of addressing the text to a wrong person with a different gender. Keeping cancelling suggested words or keeping going back to correct unwanted words while entering a text hinders user performance. Hence, it is recommended to present items associated with what users are doing but keep suggested items as just suggested items rather than as forced-upon items. Many new users do not know that right-clicking can bring up a menu of available functions in some operating systems. If the systems presents a message to users indicating that a menu is available through right-clicking when users move the cursor to a screen area, it helps reduce demands for users' learning of this interface feature.

A software system is often in a state of internal processing (e.g., loading files and searching) without the need of user input. It is recommended to make such an invisible, internal processing visible to users. For example, when a user presses the Submit button for an online order, a status message can be shown to let the user know that the user's input has been received and that the order is being processed so that the user will not press the Submit button again.

The online documentation of a software system can be an important information source for users to learn the functions and the features of the system. The online documentation should be comprehensive. However, many software systems do not provide a comprehensive online documentation anymore, which forces users to search on the Internet for answers to many questions about the system without guarantee to find correct answers.

Expert and frequent users of a software system usually prefer to use shortcuts (e.g., shortcut icons or combination of keys) for executing functions efficiently without using menus to locate and execute functions. For example, a user may press the Ctrl key and the F5 key together to debug and execute a Windows application in Microsoft Visual Studio rather than go to the Debug menu and select Start Debugging. However, shortcuts should be designed in a way that they do not get easily mixed up with what users normally do. For example, on the keyboard, the Shift key and the Ctrl key are next to each other. If a shortcut uses the Ctrl key and a letter key together, the shortcut function can be executed by a user by simply entering a text and wanting to press the Shift key and a letter key to enter an uppercase letter but accidently pressing the Ctrl key and a letter key instead.

Users should not be required to memorize all system functions. Menus are an effective method of presenting all available system functions to users. When menus are used to present all available system functions, they should be visible and easily accessible. Right-clicking to bring up a menu is not an obvious or intuitive way of presenting the menu to users. Because many software systems have a large set of functions, menus need to cover many items. A menu hierarchy is often used to organize many items. An important design decision for the menu hierarchy is about depth versus width. The depth of the menu hierarchy measures how many levels of menus the menu hierarchy has. The width of the menu hierarchy measures how many menu items each menu contains. If the menu hierarchy is too deep with many levels of menus, users may have to go through these many levels of menus to search and find an item. Going through many levels of menus may result in users' disorientation and, thus, is not recommended. If we do not use many levels of menus in the menu hierarchy, we need to include many items in a menu. Organizing items in a menu in a meaningful way, for example, based on the frequency of use, the logical sequence of operations, or in alphabetical order, helps users search and locate a specific menu item.

An important interface design principle for the graphical user interface (GUI) is What You See Is What You Get (WYSIWYG), which helps eliminate the need for users to memorize system functions. Icons and other GUI controls can be used to present system functions directly to users and allow users to directly manipulate GUI controls. Right-clicking to bring up a menu does not comply to the WYSIWYG principle because the menu is not visible unless a user right-clicks a screen area.

Guided procedures, for example, through the wizard for making connection to an Access database in Microsoft Visual Studio, also help eliminate the need for users to memorize procedural details of carrying out a task.

Making the invisible visible for the learnability criterion also helps the criterion of low error rate. For example, showing that the system has received and is processing a user input prevents a user from pressing a Submit Order button more than once. It is also recommended to separate items with severe consequences from frequently used items. For example, having Delete and Rename put together in a menu may cause a file to be accidently deleted when a user wants to rename a file. Hence, it is important to provide error recovery such as Undo. However, error recovery is usually not available for every action. For example, a user may accidently select and move a folder into another folder while moving the cursor around, which may make the user feel like having lost the folder or not being able to locate the folder again. Undoing such an action may not be available.

Increasing the interestingness of a software system by using colors or animated features can draw users' attention and can help users like the system. User satisfaction can be enhanced by improving other usability criteria (learnability, efficiency, memorability, and low error rate) and many other factors.

In addition to some examples of principles and guidelines for the user interface design in Table 21.1, there are principles and guidelines for presenting information on a single screen. The screen design usually involves the screen layout, the presentation of text, the presentation of numbers, coding techniques, and the use of colors. We give some examples of principles and guidelines for the screen design (Wickens et al., 1998) in the following.

- Screen layout
  - Show only what users need to know
  - Use grouping, ordering, and highlighting through font type and size, spacing, border, grid, and so on to present structured information and avoid clutter
  - Provide symmetry and balance
- Text
  - Avoid the heavy use of all-uppercase letters
  - Use simple, specific text
  - Express a message in a simple, specific, affirmative, and constructive way
- Numbers
  - Break up long numbers into groups of three or four digits
- Coding techniques
  - Use attention-getting techniques (e.g., blinking, bold, underline) sparingly
  - Less is better for size coding (e.g., font sizes) and color coding
- Colors
  - Use colors to show qualitative differences, attract attention, or highlight information
  - Use color as a redundant cue because at least 9% of the population, mostly male, is color deficient

21.3 Summary

This chapter addresses the user interface design for a software system. We introduce several usability criteria and give some examples of principles and guidelines to achieve various usability goals. Although the user interface design plays a critical role in the marketability and user satisfaction of a software system, the user interface design often depends on the creativity of system developers for designing user interface features to support various usability criteria and creating an attractive, user-friendly interface.

EXERCISES

1. Give interface design examples on existing software products running on electronic devices (e.g., computers, cellphones, control panels in cars) that positively or negatively affect each usability criterion in Table 21.1.
2. Give interface design examples on existing software products running on electronic devices (e.g., computers, cellphones, control panels in cars) that support or violate each principle and guideline in Table 21.1.
3. Add your own principles and guidelines of user interface design based on your experience of using the interface of software products, and give examples from the interface design of existing software products that support or violate each of your principles and guidelines.

22

Computer and Network Security

Computer and network attacks, called "cyber attacks," compromise security attributes (e.g., availability, confidentiality, and integrity) of computer and network assets (e.g., computer and network systems, databases, and applications running on systems). When developing databases, Windows application and Web applications described in Sections II to IV of this book, we need to be aware of security threats from computer and network attacks and learn how to protect computer and network assets from security threats. This chapter gives an overview of security risks by introducing three risk elements: assets, vulnerabilities, and threats. Various means of threats are illustrated through various types of cyber attacks. This chapter then describes security protection methods. Materials in this chapter are reprinted from Chapters 1 and 2 of the book by Ye (2008) with editing and copyright permission.

22.1 Security Risks Involving Assets, Vulnerabilities, and Threats

A risk such as a security risk exists when there is a possibility of a threat to exploit a vulnerability of a valuable asset (Ye, 2008). That is, the three elements of a risk are asset, vulnerability, and threat. A valuable asset makes it become a target for an attacker. A vulnerability of an asset presents an opportunity of a possible asset damage or loss. A threat is a potential attack that can exploit a vulnerability to attack an asset.

For example, a network interface is a network asset on a computer and network system. The network interface has an inherent vulnerability due to its limited bandwidth capacity. In a threat of a distributed denial-of-service (DDoS) attack, an attacker can first compromise a number of computers on the Internet and then instruct these victim computers to send large amounts of network traffic data to the target computer all at once and, thus, flood the network interface of the target computer with the attacker's traffic data. The constant arrival of large amounts of traffic data launched by the attack at the target computer leaves no bandwidth capacity of the target computer available to handle legitimate users' traffic data, thus denying network services to legitimate users. In this attack, the vulnerability of the limited bandwidth capacity is exploited by the attacker who uses up all the available bandwidth capacity with the attacker's traffic data.

An asset value can be assigned to measure the relative importance of an asset. For example, both a password file and a Microsoft Word help file are information storage assets on a computer and network system. The password file typically has a higher asset value than the help file due to the importance of passwords. A vulnerability value can be assigned to indicate the severity of a vulnerability, which is related to the severity of the asset damage or loss due to the vulnerability. For example, a system administrator account with a default password on a computer is a vulnerability whose exploitation could produce more severe damage or loss of any asset on the computer than a vulnerability of a regular user account with an easy-to-guess password. A threat value determines the likelihood of a threat, which depends on many factors, such as purpose (e.g., malicious versus non-malicious), means (e.g., gaining access versus denial of service), and so on. For example, one means of threat may be easier to execute and, thus, is more likely to occur than another means of threat. The risk value for a possibility of a given threat to exploit a given vulnerability of a given asset can be determined as follows:

$$\text{Risk value} = \text{Asset value} \times \text{Vulnerability value} \times \text{Threat value} \quad (22.1)$$

A higher asset value, a higher vulnerability value, and/or a higher threat value lead to a higher risk value.

Computer and network systems have many security vulnerabilities. Each computer or network asset has a limited service capacity, which is an inherent vulnerability exposing the asset to DoS attacks through flooding. Errors can also be made by system administrators when they configure software. For example, a system administrator may forget to remove the default system administrator account that comes with new computers. The default system administrator account with the commonly known username and password allows any user to gain access to these computers.

Most of system software and application software, which enable users to operate computers and networks, are large in size and complex in nature. Large-scale, complex software presents considerable challenges in specification, design, implementation, testing, configuration, and operation management. As a result, system and application software is often released without being fully tested and evaluated for being free from errors, due to the complexity of large-scale software, and contains many vulnerabilities. Suppose that a programmer writes the program code for a Web server application. The Web server application takes the uniform resource locator (URL) and sends it to a subroutine, with URL as a parameter for the subroutine. The subroutine places the URL in a variable. Suppose that the programmer makes a program coding error by not including the program code to check if the size of the variable is large enough to hold the URL before placing it in the variable. A malicious user can launch a buffer overflow attack by entering a large string with a well-crafted attack code as the URL. When such a user input is placed in the variable, the attack code is stored in the computer memory, and the buffer holding the variable in the computer memory is overflowed to the next memory location which holds the memory address of the program code to be executed after the execution of the subroutine ends. The buffer overflow causes this memory address to be replaced by the memory address where the attack code is stored and, thus, gets the attack code to be executed after the execution of the subroutine ends. Buffer overflow attacks are often used by attackers to gain access to a computer and network system. Buffer overflow attacks are attributed to coding faults because a program does not have a code for checking and limiting the length of a user input within the maximum length, which is used to allocate the memory space.

Cyber security threats can be characterized by many factors, such as motive, objective, origin, means, skills and resources, and so on. For example, there may be a political motive for a massive destruction of computer and network assets at a national level, a financial motive for gathering and stealing information at a corporate level, and a personal motive for a technical challenge to gain access to a computer and network system. Objectives can vary from gathering or stealing information to gaining access, disrupting or denying service, and modifying or deleting data. In general, a threat can come internally or externally. An internal threat or insider threat comes from an attacker who has access rights but abuses them. An external threat comes from an attacker who is not authorized to access a computer and network system. An attacker can have no sophisticated skills and little resources but can simply execute a downloaded attack script. Nation- or organization-sponsored attacks can use sophisticated skills and knowledge about computers and networks with unlimited resources. Section 22.2 describes various types of cyber attacks to illustrate various means of producing security threats.

A security threat exploiting a vulnerability of a computer and network system produces impacts on computer and network assets. There are three types of assets on a computer and network system: resources, processes, and users. A user calls for a process, which requests and receives service from a resource. There are processing resources (e.g., CPU, processes, and threads), storage resources (e.g., memory, hard drive, and files), and communication resources (e.g., network interface and ports) on a computer and network system at the hardware level and the software level.

A resource has a certain state at a given time. For computer and network security, we are concerned mainly with the availability, confidentiality, and integrity aspects of a resource state. The availability state of a resource indicates how much of the resource is available to serve a process. For example, 30% of a memory section may be used, making 70% available for storing additional information. The confidentiality state of a resource measures how well the resource keeps information that is stored, processed, or transmitted by the resource from an unauthorized leak. For example, the confidentiality state of an unencrypted e-mail message, which is an asset being transmitted over a network, is low. The integrity state of a resource indicates how well the resource executes its service correctly. For example, if the routing table of a router is corrupted, the integrity state of the routing table as an asset is low because it contains erroneous routing information, which leads to the incorrect routing of network data.

The performance of a process depends on the state of the resource serving the process. Three primitive aspects of the process performance are timeliness, accuracy, and precision. Timeliness measures the amount of time to produce the output of a process. Accuracy measures the correctness and, thus, the quality of the output. Precision measures the amount and, thus, the quantity of the output. The three primitive aspects of performance can be measured individually or in combination. For example, the response time, which is the elapsed time from the time when the input of a process is entered to the time when the output of the process is received, is a measure of timeliness. The data transmission rate (e.g., bandwidth) measures the time taken to transmit a given amount of data, a metric reflecting both timeliness and precision. Different computer and network applications usually have different performance requirements. For example, some applications, such as e-mail, come with no hard timeliness requirements. Others, such as audio broadcasting, video streaming, and internet protocol (IP) telephony, are time sensitive and place strict timeliness requirements.

A cyber attack can change the security state of a resource and degrade the performance of a process on a computer and network system. For example, a DoS attack of flooding the network bandwidth changes the availability state of the network interface. A buffer overflow attack compromises the integrity state of the memory and the executing process.

22.2 Various Types of Cyber Attacks

This section describes various types of cyber attacks to illustrate various means of producing cyber security threats. Table 22.1 gives a list of threat means and cyber attacks.

TABLE 22.1

Examples of Threat Means with Known Attacks Using Those Threat Means*

Means of Threats	Known Attacks
1. Brute force attack	1.1 Remote dictionary attack
2. Bypassing	2.1 Bypassing service access
	2.1.1 Buffer overflow, e.g., WarFTP, RootKit, botnets, Slammer worm
	2.1.2 Backdoor, e.g., RootKit
	2.1.3 Trojan program, e.g., Netbus Trojan
	2.1.4 Malformed message command attack, e.g., EZPublish and SQL query injection
	2.2 Bypassing information access
	2.2.1 Covert channel exploitation, e.g., steganography
3. Code attachment	3.1 Virus
	3.2 Adware and spyware
	3.3 Embedded objects in files, e.g., macros in Microsoft Word and Excel
4. Mobile code	4.1 Worm
5. DoS	5.1 Flooding, e.g., fork bomb attack, Trinoo network traffic DoS, UDP storm, TCP SYN flood
	5.2 Malformed message, Apache Web server attack, LDAP
	5.3 Destruction
6. Tampering	6.1 Network tampering, e.g., Ettercap ARP poison, DNS poison
	6.2 File and process trace hiding, e.g., RootKit
7. Man in the middle	7.1 Eavesdropping, e.g., Ettercap sniffing
	7.2 Software and hardware keylogger
8. Probing and scanning	8.1 NMAP, Nessus, traceroute
9. Spoofing	9.1 Masquerading and misdirecting, e.g., e-mail scams through phishing and spam, ARP poison attack, DNS poison attack
10. Adding	10.1 Adding new device, user, etc., e.g., Yaga
11. Insider threat	11.1 User error
	11.2 Abuse/misuse, e.g., security spill, data exfiltration, coerced actions, privilege elevation, etc.

*Reprinted with permission from Table 1.4 in Ye, N. 2008. *Secure Computer and Network Systems: Modeling, Analysis, and Design*. London, United Kingdom: John Wiley & Sons, p. 16.

22.2.1 Brute Force Attack

A brute force attack involves many repetitions of the same action. A known example of a brute force attack is a remote dictionary attack, for example, using TScrack 2.1 (http://www.web.archive.org/web/20021014015012/), which attempts to uncover the administrator's password on a computer with a Windows operating system and terminal services or a remote desktop enabled. The attack is scripted to try words from a dictionary one by one as a password for a user account until a login is successful. Most user accounts will be locked out after about three incorrect login attempts. However, the administrator's account is not locked out.

22.2.2 Bypassing Attack

A bypassing attack avoids a regular way of accessing an asset or elevating access privileges but, instead, uses an unauthorized or covert way. For example, a WarFTP attack using WarFTPd (http://www.metasploit.com/projects/Framework/exploits.html#warftpd_165_use/)

exploits a buffer overflow vulnerability to load an attack code through an input to a running process and to execute the attack code with the same privileges of the running process, thus bypassing the regular procedure and method of loading a program code and starting the corresponding process. The attack code installed through WarFTPd opens a shell environment for an attacker to remotely control the victim computer.

In addition to exploiting a buffer overflow vulnerability, RootKit (http://www.iamaphex.cjb.net/) installs a backdoor, which is a program running at an uncommonly used network port to avoid notice. The program listens to the port and accepts an attacker's request to access a computer, thus allowing the attacker to bypass regular network ports (e.g., e-mail and Web) and corresponding service processes of accessing a computer. RootKit typically alters its trace on the operating system to hide itself.

Bots (short for "robots") (http://www.honeynet.org/papers/bots/) are programs that are covertly installed on a user's computer to allow an unauthorized user to control the computer remotely. In a botnet, bots or zombies are controlled by their masters. There have been botnets established through the internet relay chat (IRC) communication protocol or a control protocol.

Slammer worm (http://www.cert.org/advisories/CA-2003-04.html) spreads from an infected host by sending out user datagram protocol (UDP) packets to port 1434 for Microsoft SQL Server 2000 at random IP addresses. Each packet contains a buffer overflow attack and a complete copy of the worm. When the packet hits a vulnerable computer, a buffer overflow occurs, allowing the worm to execute on the new victim computer. Once executing on the new victim computer, the worm installs itself and then begins sending out packets to try and locate more computers to infect.

In a Netbus Trojan attack (http://www.securityfocus.com/archive/), an attacker tricks a user to install a game file in an e-mail attachment containing a copy of the Netbus server program or to click a Web link. When the user installs the game, the Netbus server also gets installed. The attacker can then use the Netbus server as a backdoor to gain access to the computer with the same privileges as the user who installs it. Hence, the Netbus Trojan server is installed without the user noticing it, thus bypassing the regular procedure and method of loading a program code and starting the corresponding process.

EZPublish is a Web application for content management. In an EZPublish attack (http://www.[target]/settings/site.ini), a remote user sends a specially crafted URL, which gives the user the site.ini file in the settings directory, which would have not been accessible to a non-administrative user. The file contains the username, password, and other system information.

A covert channel is used to pass information between two parties without others noticing it. What makes the channel covert is that information is not expected to flow over the channel. For example, a digital image is expected to convey the image only. However, steganography hides secret information in a digital image by changing a small number of binary digits in the digital image. As a result, the change of the image is hardly noticeable. For example, the following digital image:

00101001
00101001
00101010
00101100
00101001
01110010

can be used to hide a message, 010001, by embedding the digits of the message at the last column of the digits in the image as follows and, thus, changing three digits in the original image:

00101000
00101001
00101010
00101100
00101000
01110011

22.2.3 Code Attachment

Many forms of virus, adware, spyware, and other forms of malware are installed on a computer through a file in an e-mail attachment or an embedded object such as a macro in a file. When a user clicks and executes the file in the e-mail attachment, the malware is installed on the computer.

22.2.4 Mobile Code

Mobile code is a software program sent from a remote computer, transferred across a network, and downloaded and executed on a local computer without the explicit installation or execution by a user. For example, unlike virus code, which must attach itself to another executable code such as a boot sector program or an application program, a worm propagates from one computer to another without the assistance of a user.

22.2.5 Denial of Service

A DoS attack can be accomplished by consuming all the available capacity of a resource or destroying the resource. Generating a flood of service requests is a common way of consuming all the available capacity of a resource. Some examples of DoS attacks through flooding are fork bomb attack, Trinoo network traffic DoS, UDP storm, and transmission control protocol synchronize (TCPSYN) flood.

A fork bomb attack, for example, Winfb.pl (http://www.iamaphex.cjb.net/), floods the process table by creating a fork bomb in which a process falls into a loop of iterations. In each iteration, a new process is spawned. These new processes clog the process table with many new entries.

Trinoo (http://www.packetstormsecurity.org/distributed/trinoo.tgz/) produces a DDoS attack. The Trinoo master controls an army of Trinoo zombies that send massive amounts of network traffic to a victim computer and, thus, flood the network bandwidth of the victim computer.

A UDP storm attack (Skoudis, 2002) creates a never-ending stream of data packets between the UDP echo ports of two victim computers by sending a single spoofed data packet. First, an attacker forges and sends a single data packet, which is spoofed to appear as if coming from the echo port of the first victim computer to the echo port of the second victim computer. The echo service of the second victim computer blindly responds to the request by echoing the data of the request back to the echo port of the first victim computer, which appears to send the echo request. The loop of echo traffic thus starts and continues endlessly.

A TCP SYN flood attack (Skoudis, 2002) exploits a design fault of a network protocol, TCP, which requires a three-way hand shaking to establish a connection session between two computers (Stevens, 1994). The three-way hand shaking, which is also described in Chapter 3 starts with an SYN data packet from one computer to another, which registers a half-open connection into a queue. Once the three-way hand shaking is completed with the connection being established, its corresponding half-open connection entry in the queue is removed. In a TCP SYN flood attack, an attacker sends a large number of TCP SYN packets using a spoofed source IP address to a victim computer, making the victim computer busy in responding to these connection requests, which fill up the half-connection queue and make the victim computer incapable of responding to other legitimate connection requests.

A malformed message is also used by some attacks to create an overwhelming amount of service requests for DoS. In an Apache Web server attack (http://www.apache.org/), a malformed Web request with a large header is sent to an Apache Web server, which is fooled into allocating more and more memory to satisfy the request. This results in either a crash or a significant performance degradation of the Web server. A lightweight directory across protocol (LDAP) attack (http://www.microsoft.com/technet/security/bulletin/ms04-011.mspx) exploits a vulnerability on a Windows 2000 operating system, which allows an attacker to send a specially crafted LDAP message to a Windows 2000 domain controller, causing the service responsible for authenticating users in an Active Directory domain to stop responding.

22.2.6 Tampering

Tampering has been used to corrupt network assets, such as the address resolution protocol (ARP) table and the domain name system (DNS) table, and host assets, such as process and file logs. In an Ettercap ARP poison attack (http://www.ettercap.sourceforge.net), an attacker sends out an ARP request to every IP address on a local network for the corresponding media access control (MAC) address. The attacker then sends spoofed ARP replies, which contain the mapping of the MAC address of the attacker's computer to the IP addresses of other computers on the network. Other computers on the network take the false information in the ARP replies and update their ARP tables accordingly. Consequently, network traffic data sent by all computers on the network are directed to the attacker's computer, which can then direct network traffic to their intended destinations, modify traffic data, or drop traffic data. Ettercap automatically pulls out usernames and passwords if they are present in the network traffic data. It also has the ability to filter and inject network traffic. In a DNS poison attack (Skoudis, 2002), the DNS table, which is used to convert a user-readable IP address in a text format into a computer-readable IP address in a numeric format, is corrupted. RootKit (http://www.iamaphex.cjb.net/) hides its trace on a computer by altering file and process logs.

22.2.7 Man in the Middle

Threats through the means of man in the middle have an attacker positioned in the middle of two parties to intercept or redirect information between the two parties. Eavesdropping through a network sniffer such as Ettercap (http://www.ettercap.sourceforge.net) passively intercepts network data traveling through one point (e.g., a router) on a network, without significantly disturbing the data stream. Ettercap is also capable of performing decryption and traffic analysis, which collects measures to give an indication of actions taking place, their location, source, etc.

A hardware keylogger, such as KeyKatcher 64K Mini (http://www.keykatcher.com/), plugs in between the back of the computer and the keyboard and intercepts keystrokes. A software keylogger, such as Windows Keylogger 5.0 (http://www.littlesister.de/), intercepts system calls related to keyboard events and records every keystroke to a file. System calls are used by a user program to have the operating system perform an act on behalf of the user program.

22.2.8 Probing and Scanning

Probing accesses an asset to determine its characteristics. Scanning checks a set of assets sequentially to look for a specific characteristic of these assets. Network mapped (NMAP) (http://www.insecure.org/nmap/) and Nessus (http://www.nessus.org/) are common network scanning and probing tools to find open ports on a range of computers as well as the operating system and the network applications running on those ports and to test for numerous vulnerabilities applicable to identified operating systems and network applications.

A traceroute attack (Skoudis, 2002) exploits a network mechanism, which uses the time-to-live (TTL) field of a packet header to prevent endless traveling of a data packet on a network. When a router receives a data packet, the router decreases the TTL value of the data packet by 1. If the TTL value becomes 0, the router sends an internet control message protocol (ICMP) Time Exceeded message containing the router's IP address to the source of the data packet. In the attack, a series of data packets with incrementally increasing TTL values in their packet headers is sent out to a network destination. As a result, the attacker at the source receives a number of ICMP Time Exceeded messages that reveal the IP addresses of consecutive routers on the path from the source to the destination.

22.2.9 Spoofing

Spoofing usually involves one subject masquerading as another subject to the victim and, consequently, misguiding the victim. In e-mail scams through phishing and spam, attackers send out bogus e-mails to trick and misdirect users to fake websites that resemble legitimate ones to obtain personal or confidential information of users. In an ARP poison attack (http://www.ettercap.sourceforge.net), a spoofed MAC address is used to redirect network traffic.

22.2.10 Adding

Adding a user account, a device, or another kind of computer and network asset can also occur in an attack. For example, Yaga is a user-to-root attack on a Windows NT computer (http://www.cert.org/advisories/). An attacker puts a program file on a victim computer and edits the victim's registry entry for: **HKEY_LOCAL_MACHINE_SOFTWARE\Microsoft\WindowsNT\CurrentVersion\AeDebug** through a telnet session. The attacker then remotely crashes a service on the victim computer. When the service crashes, the attacker's program, instead of the standard debugger, is invoked. The attacker's program runs with administrative privileges and adds a new user to the Domain Admins group. Once the attacker gains administrative access, the attacker executes a cleanup script, which deletes the registry entry and removes the attacker's program file for covering up the attack activities.

22.2.11 Insider Threat

Insider threats represent any attack means that can be employed by those who have access to computers and networks and, thus, pose threats from within. For example, attacks, such as Yaga (http://www.cert.org/advisories/) involving the privilege elevation of a non-privileged user, can be considered as insider threats.

In general, insider threats fall into two categories: user error and abuse/misuse. For example, a user error occurs when a user unintentionally deletes a file, modifies data, or introduces other kinds of asset damage. Abuse/misuse involves an insider's inappropriate use of access rights and privileges.

Abuse/misuse includes, for example, elevating privileges (e.g., in Yaga [http://www.cert.org/advisories/]), exceeding permissions, providing unapproved access, circumventing security controls, damaging resources, accessing or disclosing information without authorization or in an inappropriate manner (i.e., security spill and data exfiltration), and conducting other kinds of malicious or inappropriate activities. Security spill borrows a concept from the discipline of toxic waste management to indicate a release or a disclosure of information of a higher sensitivity level to a system of a lower sensitivity level or to a user not cleared to see information of a higher sensitivity level. Data exfiltration indicates a situation in which data go to where it is not supposed to be. When an insider is captured by the enemy, coerced actions of the insider produce a misuse situation. Google AdSense abuse and online poll abuse are also examples of insider abuse/misuse.

22.3 Cyber Security Protection

Protecting the security of computer and network systems against cyber threats requires three areas of work: prevention, detection, and response. Prevention aims at strengthening a computer and network system to make it difficult to launch a threat against the system and, thus, reduce the likelihood of a threat. However, determined, organized, and skilled attackers can overcome attack difficulty created by prevention mechanisms to break into a computer and network system by exploiting known and unknown system vulnerabilities. Hence, detection is required to detect an attack acting on a computer and network system, identify the nature of the attack, and assess the impacts (e.g., the origin, path, and damage) of the attack. A detection of an attack calls for an appropriate response to stop the attack, recover the system, and correct the exploited vulnerability, all based on diagnostic information from the attack assessment part of the attack detection. The following sections go over each area in more details.

22.3.1 Cyber Attack Prevention

This section describes three commonly used prevention mechanisms: data encryption, firewalls, and authentication/authorization.

22.3.1.1 Data Encryption

Data transmitted over computer networks and stored on computers can be encrypted so that messages carried by data cannot be read directly from data. There are many data encryption methods. We introduce some substitution-based methods and a public key cryptographic algorithm.

The Caesar cipher is a classical substitution method for text encryption. The Caesar cipher encrypts a text by substituting each letter with one that shifts a number of places to the right (e.g., three spaces to the right, as shown in Table 22.2) and decrypts a cipher text by substituting each letter with one that shifts the same number of spaces to the left. A variation of the Caesar cipher is to use a random sequence of letters for replacement, as shown in Table 22.3.

TABLE 22.2
Caesar Cipher

Letter	a	b	c	d	e	f	g	h	i	j	k	l	m	n	o	p	q	r	s	t	u	v	w	x	y	z
Substitution letter	d	e	f	g	h	i	j	k	l	m	n	o	p	q	r	s	t	u	v	w	x	y	z	a	b	c

TABLE 22.3
Variation of the Caesar Cipher

Letter	a	b	c	d	e	f	g	h	i	j	k	l	m	n	o	p	q	r	s	t	u	v	w	x	y	z
Substitution letter	k	l	q	w	z	x	m	z	s	h	j	e	r	t	y	u	i	o	p	c	v	b	n	d	f	g

TABLE 22.4
Relative Use Frequencies of Alphabet Letters

Letter	Relative Use Frequency
e	0.125
t	0.090
r	0.083
n	0.075
i	0.074
o	0.073
a	0.071
s	0.058
d	0.040
l	0.036
h	0.034
c	0.033
f	0.029
u	0.028
m	0.026
p	0.025
y	0.022
g	0.020
w	0.015
v	0.014
b	0.012
k	0.005
x	0.004
q	0.004
j	0.002
z	0.002

Example 22.1

Use the Caesar cipher to encrypt the following text: Caesar cipher.

The cipher text is given as follows:

Text: Caesarcipher
Cipher text: fdwhwduflskhu

One problem with Caesar ciphers and variations of Caesar cyphers is that letter substitutions do not change the relative use frequencies of letters or letter groups in texts. Table 22.4 gives relative use frequencies of alphabet letters (Pfleeger, 1997) and shows that the letter "e" is used most frequently. If we use the Caesar cipher in Table 22.2 to replace the letter "e" with the letter "h" in a text, the letter "h" becomes the most frequently used letter in the cipher text, and we can then guess that the original letter for "h" is the letter "e." That is, computing the relative use frequencies of letters in a cipher text and comparing the use frequencies with those in Table 22.4, we can guess some letters or letter groups and then use guessed letters or letter groups to figure out the remaining letters. To overcome the problem of the cipher text being analyzed based on the relative use frequencies of letters, there are many substitution-based methods that generate different substitution letters for the same letter at different places in the plain text. For example, the book cipher (Pfleeger, 1997), which is a variation of the Vigenère cipher, uses the Vigenère table (see Table 22.5) and a running text from a given book as the key for encryption. The running text is selected to have the same length as that of the plain text. A letter in the key is used as the row index and a letter in the plain text is used as the column index to look up the Vigenère table and find the substitution letter for the letter in the plain text. Example 22.2 shows how the book cipher works.

TABLE 22.5

Vigenère Table

	a	**b**	**c**	**d**	**e**	**f**	**g**	**h**	**i**	**j**	**k**	**l**	**m**	**n**	**o**	**p**	**q**	**r**	**s**	**t**	**u**	**v**	**w**	**x**	**y**	**z**
a	a	b	c	d	e	f	g	h	i	j	k	l	m	n	o	p	q	r	s	t	u	v	w	x	y	z
b	b	c	d	e	f	g	h	i	j	k	l	m	n	o	p	q	r	s	t	u	v	w	x	y	z	a
c	c	d	e	f	g	h	i	j	k	l	m	n	o	p	q	r	s	t	u	v	w	x	y	z	a	b
d	d	e	f	g	h	i	j	k	l	m	n	o	p	q	r	s	t	u	v	w	x	y	z	a	b	c
e	e	f	g	h	i	j	k	l	m	n	o	p	q	r	s	t	u	v	w	x	y	z	a	b	c	d
f	f	g	h	i	j	k	l	m	n	o	p	q	r	s	t	u	v	w	x	y	z	a	b	c	d	e
g	g	h	i	j	k	l	m	n	o	p	q	r	s	t	u	v	w	x	y	z	a	b	c	d	e	f
h	h	i	j	k	l	m	n	o	p	q	r	s	t	u	v	w	x	y	z	a	b	c	d	e	f	g
i	i	j	k	l	m	n	o	p	q	r	s	t	u	v	w	x	y	z	a	b	c	d	e	f	g	h
j	j	k	l	m	n	o	p	q	r	s	t	u	v	w	x	y	z	a	b	c	d	e	f	g	h	i
k	k	l	m	n	o	p	q	r	s	t	u	v	w	x	y	z	a	b	c	d	e	f	g	h	i	j
l	l	m	n	o	p	q	r	s	t	u	v	w	x	y	z	a	b	c	d	e	f	g	h	i	j	k
m	m	n	o	p	q	r	s	t	u	v	w	x	y	z	a	b	c	d	e	f	g	h	i	j	k	l
n	n	o	p	q	r	s	t	u	v	w	x	y	z	a	b	c	d	e	f	g	h	i	j	k	l	m
o	o	p	q	r	s	t	u	v	w	x	y	z	a	b	c	d	e	f	g	h	i	j	k	l	m	n
p	p	q	r	s	t	u	v	w	x	y	z	a	b	c	d	e	f	g	h	i	j	k	l	m	n	o
q	q	r	s	t	u	v	w	x	y	z	a	b	c	d	e	f	g	h	i	j	k	l	m	n	o	p
r	r	s	t	u	v	w	x	y	z	a	b	c	d	e	f	g	h	i	j	k	l	m	n	o	p	q
s	s	t	u	v	w	x	y	z	a	b	c	d	e	f	g	h	i	j	k	l	m	n	o	p	q	r
t	t	u	v	w	x	y	z	a	b	c	d	e	f	g	h	i	j	k	l	m	n	o	p	q	r	s
u	u	v	w	x	y	z	a	b	c	d	e	f	g	h	i	j	k	l	m	n	o	p	q	r	s	t
v	v	w	x	y	z	a	b	c	d	e	f	g	h	i	j	k	l	m	n	o	p	q	r	s	t	u
w	w	x	y	z	a	b	c	d	e	f	g	h	i	j	k	l	m	n	o	p	q	r	s	t	u	v
x	x	y	z	a	b	c	d	e	f	g	h	i	j	k	l	m	n	o	p	q	r	s	t	u	v	w
y	y	z	a	b	c	d	e	f	g	h	i	j	k	l	m	n	o	p	q	r	s	t	u	v	w	x
z	z	a	b	c	d	e	f	g	h	i	j	k	l	m	n	o	p	q	r	s	t	u	v	w	x	y

Example 22.2

Use the book cipher and the passage from this book chapter starting with "computer and network attacks…" as the key to encrypt the text: Caesar cipher.

The following table shows the key, the text, and the cipher text.

Key	c	o	m	p	u	t	e	r	a	n	d	n
Text	c	a	e	s	a	r	c	i	p	h	e	r
Cipher text	e	o	q	h	u	k	g	z	p	u	h	e

For the first letter, we use "c" in the key as the row index and "c" in the text as the column index to look up the Vigenère table and get the substitution letter "e." For the second letter, we use "o" in the key as the row index and "a" in the text as the column index to look up the Vigenère table and get the substitution letter "o." This process continues for the other letters in the text. Note that the letters in the cipher text for the first "a" and the second "a" in the plain text are "o" and "u," respectively, and the letters in the cipher text for the first "e" and the second "e" in the plain text are "q" and "h," respectively.

To decrypt the cipher text, we use the key and the cipher text to trace back the plain text. For example, for the first letter "e" in the cipher text, the first letter in the key is "c." We examine the "c" row of the Vigenère table and find the letter "e" in the "c" column. Hence, the letter in the plain text is "c."

The Rivest-Shamir-Adelman (RSA) algorithm is a public key cryptographic algorithm that assigns a public key and a private key to a user. The public key of the user is publicly known. Only the user knows his/her private key. We first describe how to generate a pair of public key and private key and then introduce how to use the pair of public key and private key to perform data encryption and produce a digital signature.

We take the following steps to generate a pair of public key and private key.

1. Select a pair of large prime numbers: p and q.
2. Determine the public key, which contains two numbers, n and e, where

$$n = p \times q \tag{22.2}$$

and e is chosen randomly and should be relatively prime to $(p - 1)(q - 1)$, that is, e has no factors in common with $(p - 1)(q - 1)$.

3. Choose the private key, d, to satisfy the following equation:

$$e \times d = 1 \bmod (p - 1)(q - 1) \quad \text{or} \quad e \times d = a(p - 1)(q - 1) + 1 \tag{22.3}$$

After the public key $\{e, n\}$ and the private key $\{d\}$ are generated, we can discard p and q, but still, we must keep them confidential. Note that e and d are interchangeable. That is, the public key $\{e, n\}$ and the private key $\{d\}$ are equivalent to the public key $\{d, n\}$ and the private key $\{e\}$.

Example 22.3

Use the RSA algorithm to generate a pair of public key and private key.

We select $p = 17$ and $q = 23$, which are two small prime numbers, for ease of computation in this example. We use p and q to determine the public key and the private key as follows:

$$n = p \times q = 17 \times 23 = 391$$

$$e = 7, \text{ which is relatively prime to } (p-1)(q-1) = 16 \times 22 = 352$$

$d = 151$, which satisfies:

$$e \times d = a(p-1)(q-1) + 1 \quad \text{or} \quad 7 \times 151 = 3 \times (17-1)(23-1) + 1$$

Hence, the public key is (7, 391), and the private key is 151.

A user encrypts a message, m, using his/her public key. At first, the message is broken into blocks, $m_1, \ldots, m_k$. The number of characters in each block is smaller than the number of characters in n. Each block, m_i, is encrypted into a cipher block, c_i, as follows:

$$c_i = m_i^e \bmod n \tag{22.4}$$

The cipher message is decrypted using the private key, d, as follows:

$$m_i = c_i^d \bmod n \tag{22.5}$$

Because the decryption requires the private key of the user and only the user knows the private key, no one else can decrypt the cipher message. It is difficult to decrypt a cipher text generated using the RSA algorithm without the private key because, currently, we cannot solve the factorization problem in the polynomial time.

Example 22.4

As described in Chapter 2, a text is represented on a computer using binary numbers that can be converted into decimal numbers. In this example, a text message is represented by decimal numbers, $m = 34511228919$. Encrypt this message using the public key from Example 22.3, and then decrypt the cipher message using the private key from Example 22.3.

We have $n = 391$ in the public key (7, 391) from Example 22.3. There are three characters in n. Hence, we break the message m into blocks of two characters: $m_1 = 34$, $m_2 = 51$, $m_3 = 12$, $m_4 = 28$, $m_5 = 91$, and $m_6 = 9$. The encryption of each message block is performed as follows:

$m_1 = 34, c_1 = 34^7 \bmod 391 = 306$
$m_2 = 51, c_2 = 51^7 \bmod 391 = 017$
$m_3 = 12, c_3 = 12^7 \bmod 391 = 177$
$m_4 = 28, c_4 = 28^7 \bmod 391 = 224$
$m_5 = 91, c_5 = 91^7 \bmod 391 = 139$
$m_6 = 9, c_6 = 9^7 \bmod 391 = 257$

Hence, the cipher text is 306017177224139257.

The decryption of the cipher message is performed as follows:

$c_1 = 306, m_1 = 306^{151} \bmod 391 = 34$
$c_2 = 017, m_2 = 17^{151} \bmod 391 = 51$
$c_3 = 177, m_3 = 177^{151} \bmod 391 = 12$
$c_4 = 224, m_4 = 224^{151} \bmod 391 = 28$
$c_5 = 139, m_5 = 139^{151} \bmod 391 = 91$
$c_6 = 257, m_6 = 257^{151} \bmod 391 = 9$

Hence, we recover the message $m = 34511228919$.

The RSA algorithm can also be used to generate a digital signature. If a user wants to sign a digital document, the user encrypts the document using his/her private key. The cipher document can be decrypted using only this user's public key. Hence, if others can decrypt the cipher document using the user's public key, it proves that the cipher document is encrypted by the user using the private key known by the user only, and it thus proves that it is the user who signs the document digitally.

22.3.1.2 Two Types of Firewalls: Screening Routers and Application Gateways

A firewall is usually installed on a router or an application gateway that controls incoming and outgoing traffic of a protected computer and network system. A firewall on a router, called a "screening router," filters traffic data between a protected system and its outside world by defining rules that are applicable to most header fields of data packets at the TCP and IP layers of the TCP/IP. The data portion of a network packet may not be readable due to the encrypted application data and, therefore, is not usually used to define filtering rules in the firewall. As described in Chapter 3, TCP/IP header fields include the following:

- Time
- Source IP address
- Source port
- Destination IP address
- Destination port
- Flags, a combination of TCP control bits: S (SYN), F (FIN), P(PUSH), and R(RST)
- Sequence number
- Acknowledge number
- Length of data payload
- Window size, indicating the buffer space available for receiving data to help the flow control between two host computers
- Urgent, indicating that there are "urgent" data
- Options, indicating TCP options, if there are any

A filtering rule can look for specific types of values in one or more header fields, and allow or deny data packets based on these values. Using the header information, a screening router can deny data packets from a specific source IP address; block data packets targeting specific network ports running vulnerable network services; prevent certain types of data packets, such as those containing ICMP Echo Reply messages from going out; and so on.

A firewall can also be installed on a computer running proxy network applications called a "proxy gateway" or an "application gateway." A proxy gateway transforms network data packets into application data by performing pseudo-application operations. Using information available at the application layer, a proxy gateway can block access to specific services of an application (e.g., certain file transfer protocol [FTP] commands) and block certain kinds of data to an application (e.g., a file attachment with a detected virus). For example, a proxy gateway running a pseudo–FTP application can screen FTP commands to allow only acceptable FTP commands. Sophisticated proxy gateways are called "guards," which carry out sophisticated computation tasks for filtering. For example, a guard may run an e-mail application, perform virus scanning on file attachments to e-mails, and determine whether to drop or allow file attachments.

22.3.1.3 Authentication and Authorization

Authentication and authorization work together to control a user's access to computer and network assets. Through authentication, a user is verified to be truly who the user claims to be. Through authorization, a user is granted access rights to computer and network assets based on the user's authenticated identity. In addition to access rights to individual computer and network assets, flow control policies can also be specified to control information flow between computer and network assets.

Username and password are commonly used for user authentication. In addition to information keys such as passwords, there are also physical keys such as magnetic cards and security calculators, and biometric keys such as voice print, finger print, and retinal print.

Just like a credit card, a magnetic card contains the identity of the card holder. In addition to a magnetic card, a user may have to enter a personal identification number. A disadvantage of using a magnetic card as a key is that a card reader needs to be attached to each computer. A security calculator is not uncommon in practice. To use a security calculator, a user first presents a username to a computer and network system. The system responds with a challenge value. The user enters a personal identification number along with the challenge value into the security calculator. The security calculator computes a response value. The user presents the response value as the key to the system. If the response value matches the expected response value computed by the system, the user is successfully authenticated. Hence, the system must store the personal identification number for each username. An advantage of using a security calculator is that the response value as the key to pass the authentication process changes with the challenge value. Since different challenge values are usually generated at different times, response values as keys change at different times when the system is used. This makes it difficult to guess the key each time when the computer is used. Even if a break-in is successful in using a key, the key cannot be used the next time for break-in. Moreover, a user must have the right personalized security calculator and the correct personal identification number to compute the correct response value as the key.

A voice print; a finger print; a retinal print, which is a blood vessel pattern on the back of an eye; and a hand geometry are examples of biometric keys. Like a magnetic card, those biometric keys need a special device to attach to a computer and network system for reading and recognizing these biometric keys.

A digital signature has become an increasingly popular method of authenticating the sender of a digital document. For example, using a public key cryptographic algorithm such as the RSA algorithm (Pfleeger, 1997), the sender of a digital document has a pair of a private key and a public key that is known to others. The sender first uses the private key to encrypt the document as a way of signing the document and then sends the encrypted document to a receiver. If the receiver of the document can use the sender's public key to decrypt the document, this proves that the document is truly signed by the sender since only the sender knows the private key, which matches with the public key.

A public key cryptographic algorithm can also be used to encrypt data in transmission or data in storage for protecting the confidentiality or the integrity state of those computer and network assets because encrypted data cannot be easily read or modified by others. Take protecting data in transmission over a network as an example. The sender of data can encrypt the data using the receiver's public key. The encrypted data can be decrypted by using only the private key, which is paired with the public key and is known by the receiver only. Hence, only the receiver can use the private key to decrypt the data.

User authentication is a part of the authorization process, which determines what access rights are granted for what computer and network assets to an authenticated user. Hence, authorization controls a user's access to computer and network assets, and can also limit information flow between computer and network assets. Authentication/authorization aims at access and flow control by limiting each user to his/her own workspace on a computer and network system.

22.3.2 Cyber Attack Detection

As long as a computer and network system allows access to the system even in limited ways, determined and organized attackers with sophisticated skills and plentiful resources (e.g., organization-sponsored attackers) can break into the system through limited access due to many known and unknown system vulnerabilities. In reality, a computer and network system usually includes software packages that are released by commercial software vendors without being fully tested and evaluated for being free from security holes. Patches of software vulnerabilities are usually developed and delivered only after security incidents occur and make the exploited vulnerabilities known.

Detection provides another layer of protection against security threats by monitoring system data, detecting security-related events, and analyzing security incidents to trace their origin and path, assess their impact, and predict their development. Data, events, and incidents, as well as detection methodologies are described in the following.

There are two kinds of data to capture activities, state changes, and performance changes on computers and networks: network data and host computer data. Network data come from either raw data packets or tools that provide network traffic statistics, network performance information, etc. Host data reflect activities, state changes, and performance changes on a host computer. There are facilities and tools to collect data from various computer and network platforms, such as Windows, Linux, and Unix-based operating systems. Different auditing/logging facilities and tools provide different kinds of system data. For example, system log data from Windows capture auditable events generated by given system programs (e.g., login, logout, and privileged programs for network services). Information recorded for each auditable event may reveal, for example, the following:

- Time of the event
- Type of the event
- User generating the event
- Process requesting the event
- Object accessed in the event
- Return status of the event

Windows performance objects collect activity, state, and performance data related to many computer objects, such as Cache, Memory, Network Interface, System, etc. An example of activity variables is Network Interface\Packets/sec, which records the number of packets sent and received through the network interface card. An example of state variables is Memory\Available Bytes, which measures the amount of memory space available. An example of performance variables is Process (_Total)\Page Faults/sec. A page fault occurs when a thread refers to a virtual memory page that is not in its working set in the main memory. Certain applications, for example, the Web application, come with their own logging facilities. Log data provided by a Web application may record information such as the source IP address of the user accessing a website, user ID, session ID, time of the Web request, Web file requested, number of bytes returned for the request, etc.

Security events, which are detected while monitoring computer and network data, are associated with special phenomena produced in a security incident of a threat attacking system assets by exploiting system vulnerabilities. There exist three methodologies of attack event detection: signature recognition, anomaly detection, and attack norm separation (Ye, 2008). Signature recognition uses signature patterns of attack data (e.g., three consecutive login failures) to look for matches in observed computer and network data. A match to an attack signature results in the detection of an attack event. Hence, signature recognition relies on the model of attack data to perform attack detection. Signature patterns of attack data are either manually captured by human analysts or automatically discovered by mining attack and norm data in contrast. Most of the existing commercial intrusion detection systems employ the methodology of signature recognition. Anomaly detection first defines a profile of normal use behavior (norm profile) for a computer or network subject of interest and then raises a suspicion of an ongoing attack when detecting a large deviation of the observed data from the norm profile. Hence, anomaly detection relies on the model of normal use data to perform attack detection. Unlike signature recognition and anomaly detection, attack norm separation (Ye, 2008) relies on both an attack model and a normal use data model to detect and identify an attack that often occurs at the same time when there are also ongoing normal use activities. The occurrence of an attack during ongoing normal use activities produces the observed data that contain the mixed data effects of the attack and normal use activities. Considering that the observed computer and network data are the mixed attack and norm data, attack norm separation first uses the normal use data model to filter out the effect of normal use activities from the mixed attack and norm data and then uses the attack data model to detect and identify the presence of the attack in the residual data after filtering out the effect of the normal use data.

Since a security incident has a series of events along its cause–effect chain, analyzing security incidents involves linking and correlating detected events of a security incident, producing an accurate picture of the incident's cause–effect chain with the origin, path, and impact information, and predicting the incident's development. That is, a security incident is defined to be a cause–effect chain of events produced by a threat attacking certain system assets through exploiting certain system vulnerabilities.

Attack assessment is analyzing a security incident by linking and correlating detected events of a security incident in the cause–effect chain of the security incident to reveal the origin, path, impact, and future development of the security incident. Existing solutions of attack assessment (Ye, 2008) rely mainly on prior knowledge of known threats. An event may manifest in several data features and, thus, may produce several detection outcomes from different techniques monitoring different features of the same data stream. An event may be involved in more than one attack. Hence, event optimization is desired to determine the optimized set of events that gives the smallest number of events with the largest coverage of various attacks.

22.3.3 Cyber Attack Response

Diagnostic information from attack assessment is the key input to planning the course of actions for attack response, which includes stopping an attack, recovering an affected system, and correcting the exploited vulnerabilities. In practice, attack response has been mostly planned and performed by system administrators or security analysts manually. Actions of stopping attacks often involve disconnecting a user; terminating a connection, process, or service; disabling a user account, etc. (Ye, 2008). Recovering an affected system often requires reinstalling programs and using backup data to bring the system to a pre-attack state. Actions of correcting vulnerabilities must specifically address the exploited vulnerabilities, which can be diagnosed during the attack assessment. It usually takes time for software or security product vendors (e.g., Microsoft) to identify the vulnerabilities exploited by previously unknown attacks and develop patches for them. Attack response in a quick, automotive manner still remains as a challenge.

22.4 Summary

In this chapter, we introduce the risks of computer and network attacks by describing three security risk elements—assets, vulnerabilities, and threats—and giving examples of computer and network vulnerabilities and cyber attacks. We then describe three areas of work for protecting computer and network systems: prevention, detection, and response. For cyber attack prevention, we describe methods of data encryption, firewalls, and authentication/authorization. For cyber attack detection, we consider data, events, and incidents and introduce three detection methodologies (signature recognition, anomaly detection, and attack norm separation) and security incident assessment. We specify tasks for cyber attack response.

EXERCISES

1. Use the following substitution:
   ABCDEFGHIJKLMNOPQRSTUVWXYZ
   SECURITYABDFGHJKLMNOPQVWXZ
   to encrypt the following text:

   "If a student has a reasonable excuse…"

2. Use the book cipher and the following passage "Development of Windows-based and Web-enabled applications for decision support…" to encrypt the same text in Exercise 1.
3. Decrypt the cipher text from Exercise 1 based on the relative frequency of alphabet letters in Table 22.4, and show your analysis and procedure of decryption.
4. Use the RSA algorithm to choose a pair of public key and private key different from that in Example 22.3, and use the pair of public key and private key to encrypt m = 91982211543 and then decrypt the cipher data.
5. Add Windows controls and code to the Windows-based application for the health-care information system in Exercise 1 in Chapter 13 so that the system keeps login names and encrypted passwords of some users and allows only these users to update all the database tables. When a user selects the radio button for Updates,

the user is asked to enter a login name and a password. The system encrypts the entered password and checks whether the login name and the encrypted password match those stored in the database. If a match is found, the system takes the user to the Windows form listing all the database tables for updates. If a match is not found, the system presents a message to the user stating that access to the Updates function is denied.

23

Data Mining

Data mining aims at discovering data patterns from massive amounts of data. Data patterns can reveal useful information and knowledge that can be used to support decision making. Information systems described in Sections II to IV use databases to store data and queries to retrieve data from databases. Although groupings of data records and aggregate functions (e.g., count, sum, average, and variance) available in database tools allow us to perform data analysis and produce statistics, data mining allows us to discover more sophisticated data patterns for useful information and knowledge from data; helps us enrich information systems with information, knowledge, and decision support functions; and thus, enables us to turn information systems to decision support systems.

As described in Ye (2013), there are many data mining algorithms to discover various types of data patterns. In this chapter, we describe one of the data mining algorithms, decision trees, to illustrate how data mining discovers data patterns from which useful information and knowledge are obtained. In the next chapter, we describe expert systems to illustrate how we include knowledge in an information system and use knowledge to support decision making. The materials in this chapter are reprinted from Chapters 1 and 4 of the book by Ye (2013) with editing and copyright permission.

Decision trees are used to learn classification patterns from data and express the relationship of attribute variable x with a target variable y, $y = F(x)$, in the form of a decision tree. A decision tree classifies the categorical target value of a data record using its attribute values. In this chapter, we first define a binary decision tree and give the algorithm to learn a binary decision tree from a data set with categorical attribute variables and a categorical target variable. Then, the method of learning a non-binary decision tree is described. Additional concepts are introduced to handle numeric attribute variables and missing values of attribute variables. A list of data mining software packages that support the learning of decision trees is provided.

23.1 Learning a Binary Decision Tree and Classifying Data Using a Decision Tree

In this section, we describe a data set that is used to illustrate the algorithm of a decision tree. We introduce elements of a decision tree. The rationale of seeking a decision tree with the minimum description length is provided and followed by the split selection methods. Finally, the top-down construction of a decision tree is illustrated.

23.1.1 Description of a Data Set

Table 23.1 gives the data set for fault detection of a manufacturing system (Ye et al., 1993). The manufacturing system consists of nine machines, M1, M2, …, M9, which process parts. Figure 23.1 shows the production flows of parts to go through the nine machines. There are

TABLE 23.1
Data Set for System Fault Detection

Instance (Faulty Machine)	Attribute Variables: Quality of Parts x_1	x_2	x_3	x_4	x_5	x_6	x_7	x_8	x_9	Target Variable: System Fault y
1 (M1)	1	0	0	0	1	0	1	0	1	1
2 (M2)	0	1	0	1	0	0	0	1	0	1
3 (M3)	0	0	1	1	0	1	1	1	0	1
4 (M4)	0	0	0	1	0	0	0	1	0	1
5 (M5)	0	0	0	0	1	0	1	0	1	1
6 (M6)	0	0	0	0	0	1	1	0	0	1
7 (M7)	0	0	0	0	0	0	1	0	0	1
8 (M8)	0	0	0	0	0	0	0	1	0	1
9 (M9)	0	0	0	0	0	0	0	0	1	1
10 (none)	0	0	0	0	0	0	0	0	0	0

Source: Reprinted with permission from Table 4.1 in Ye, N. 2013. *Data Mining: Theories, Algorithms, and Examples*, p. 38, Boca Raton, Florida: CRC Press.

some parts that go through M1 first, M5 second, and M9 last; some parts that go through M1 first, M5 second, and M7 last, and so on. There are nine variables, x_i, $i = 1, 2, \ldots, 9$, representing the quality of parts after they go through the nine machines. If parts after machine i pass the quality inspection, x_i takes the value of 0; otherwise, x_i takes the value of 1. There is a variable, y, representing whether the system has a fault. The system has a fault if any of the nine machines is faulty. If the system does not have a fault, y takes the value of 0; otherwise, y takes the value of 1. The fault detection problem is to determine whether the system has a fault based on the quality information. The fault detection problem involves the nine quality variables, x_i, $i = 1, 2, \ldots, 9$, and the system fault variable, y. There may be one or more machines that have a fault at the same time, or no faulty machine. For example, in instance 1, with M1 being faulty (y taking the value of 1), the quality of parts after M1, M5, M7, and M9 fails the inspection, with x_1, x_5, x_7, and x_9 taking the value of 1 and other quality variables, x_2, x_3, x_4, x_6, and x_8, taking the value of 0.

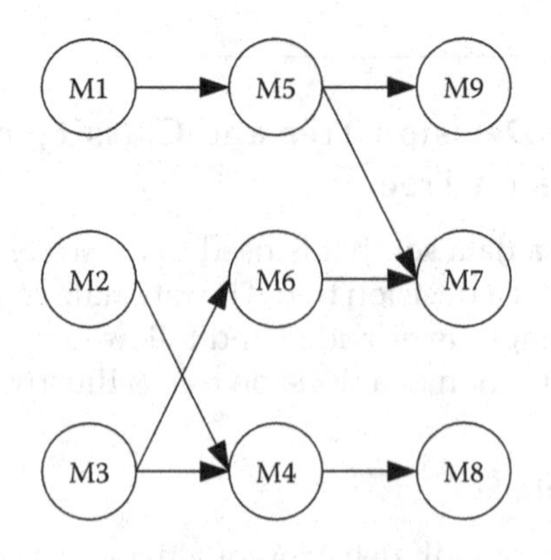

FIGURE 23.1
Manufacturing system. (Reprinted with permission from Figure 1.1 in Ye, N. 2013. *Data Mining: Theories, Algorithms, and Examples*, p. 8, Boca Raton, Florida: CRC Press.)

23.1.2 Elements of a Decision Tree

The data set in Table 23.1 is used as the training data set to learn a binary decision tree for classifying whether the system is faulty using values of the nine quality variables. Figure 23.2 shows the resulting binary decision tree to illustrate the elements of the decision tree. How this decision tree is learned from data is explained later.

As shown in Figure 23.2, a binary decision tree is a graph with nodes. The root node at the top of the tree consists of all data records in the training data set. For the data set of system fault detection, the root node contains a set with all the 10 data records in the training data set, {1, 2, ..., 10}. Note that the numbers in the data set are the instance numbers. The root node is split into two subsets, {2, 4, 8, 9, 10} and {1, 2, 5, 6, 7}, using the attribute variable, x_7, and its two categorical values, $x_7 = 0$ and $x_7 = 1$. All the instances in the subset, {2, 4, 8, 9, 10}, have $x_7 = 0$. All the instances in the subset, {1, 2, 5, 6, 7}, have $x_7 = 1$. Each subset is represented as a node in the decision tree. A Boolean expression is used in the decision tree to express that $x_7 = 0$ by $x_7 = 0$ is TRUE and that $x_7 = 1$ by $x_7 = 0$ is FALSE. $x_7 = 0$ is called a "split condition" or a "split criterion," and its TRUE and FALSE values allow a binary split of the set at the root node into two branches with a node at the end of each branch. Each of the two new nodes can be further divided using one of the remaining attribute variables in the split criterion. A node cannot be further divided if the data records in the data set at this node have the same value of the target variable. Such a node becomes a leaf node in the decision tree. Except for the root node and the leaf nodes, all other nodes in the decision tree are called "internal nodes."

The decision tree can classify a data record by passing the data record through the decision tree using the attribute values in the data record. For example, the data record for Instance 10 is first checked with the first split condition at the root node. With $x_7 = 0$, the data record is passed down to the left branch. With $x_8 = 0$ and then $x_9 = 0$, the data record is passed down to the left-most leaf node. The data record takes the target value for this leaf node, $y = 0$, which classifies the data record as not faulty.

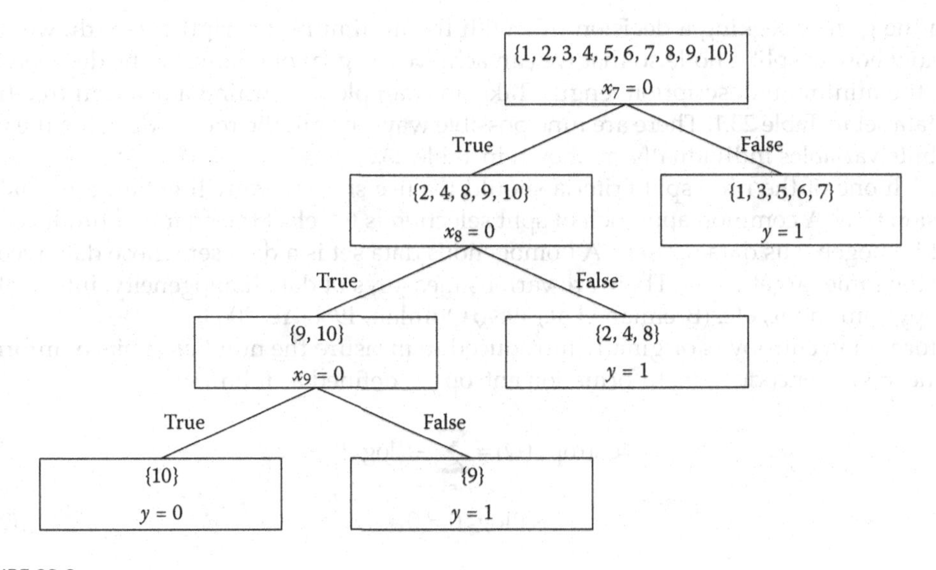

FIGURE 23.2
Decision tree for system fault detection. (Reprinted with permission from Figure 4.1 in Ye, N. 2013. *Data Mining: Theories, Algorithms, and Examples*, p. 38, Boca Raton, Florida: CRC Press.)

23.1.3 Decision Tree with the Minimum Description Length

Starting with the root node with all the data records in the training data set, there are nine possible ways to split the root node using the nine attribute variables individually in the split condition. For each node at the end of a branch from the split of the root node, there are eight possible ways to split the node using each of the remaining eight attribute variables individually. This process continues and can result in many possible decision trees. All the possible decision trees differ in their size and complexity. A decision tree can be large to have as many leaf nodes as data records in the training data set, with each leaf node containing each data record. Which one of all the possible decision trees should be used to represent *F*, the relationship of the attribute variables with the target variable? A decision tree algorithm aims at obtaining the smallest decision tree that can capture *F*, that is, the decision tree that requires the minimum description length. Given both the smallest decision tree and a larger decision tree, both of which classify all the data records in the training data set correctly, it is expected that the smallest decision tree generalizes classification patterns better than the larger decision tree, and that better-generalized classification patterns allow the better classification of more data points, including those not in the training data set. Consider a large decision tree that has as many leaf nodes as data records in the training data set, with each leaf node containing each data record. Although this large decision tree classifies all the training data records correctly, it may perform poorly in classifying new data records not in the training data set. Those new data records have sets of attribute values different from those of data records in the training data set and, thus, do not follow the same paths of the data records to leaf nodes in the decision tree. We need a decision tree that captures generalized classification patterns for the *F* relationship. The more generalized the *F* relationship is, the smaller description length it has because it eliminates specific differences among individual data records. Hence, the smaller a decision tree is, the more generalization capacity the decision tree is expected to have.

23.1.4 Split Selection Methods

With the goal of seeking a decision tree with the minimum description length, we need to know how to split a node so that we can achieve the goal of obtaining the decision tree with the minimum description length. Take an example of learning a decision tree from the data set in Table 23.1. There are nine possible ways to split the root node using the nine attribute variables individually, as shown in Table 23.2.

Which one of the nine split criteria should we use so that we will obtain the smallest decision tree? A common approach of split selection is to select the split that produces the most homogeneous data subsets. A homogenous data set is a data set whose data records have the same target value. There are various measures of data homogeneity: information entropy, gini-index, etc. (Breiman et al., 1984; Quinlan, 1986; Ye, 2003).

Information entropy is originally introduced to measure the number of bits of information needed to encode data. Information entropy is defined as follows:

$$\text{entropy}(D) = \sum_{i=1}^{c} -P_i \log_2 P_i \tag{23.1}$$

$$-0\log_2 0 = 0 \tag{23.2}$$

$$\sum_{i=1}^{c} P_i = 1 \tag{23.3}$$

TABLE 23.2

Binary Split of the Root Node and Calculation of Information Entropy for the Data Set of System Fault Detection

Split Criterion	Resulting Subsets and Average Information Entropy of the Split
$x_1 = 0$: TRUE or FALSE	{2, 3, 4, 5, 6, 7, 8, 9, 10}, {1} $\text{entropy}(S) = \frac{9}{10}\text{entropy}(D_{\text{true}}) + \frac{1}{10}\text{entropy}(D_{\text{false}})$ $= \frac{9}{10} \times \left(-\frac{8}{9}\log_2\frac{8}{9} - \frac{1}{9}\log_2\frac{1}{9}\right) + \frac{1}{10} \times 0 = 0.45$
$x_2 = 0$: TRUE or FALSE	{1, 3, 4, 5, 6, 7, 8, 9, 10}, {2} $\text{entropy}(S) = \frac{9}{10}\text{entropy}(D_{\text{true}}) + \frac{1}{10}\text{entropy}(D_{\text{false}})$ $= \frac{9}{10} \times \left(-\frac{8}{9}\log_2\frac{8}{9} - \frac{1}{9}\log_2\frac{1}{9}\right) + \frac{1}{10} \times 0 = 0.45$
$x_3 = 0$: TRUE or FALSE	{1, 2, 4, 5, 6, 7, 8, 9, 10}, {3} $\text{entropy}(S) = \frac{9}{10}\text{entropy}(D_{\text{true}}) + \frac{1}{10}\text{entropy}(D_{\text{false}})$ $= \frac{9}{10} \times \left(-\frac{8}{9}\log_2\frac{8}{9} - \frac{1}{9}\log_2\frac{1}{9}\right) + \frac{1}{10} \times 0 = 0.45$
$x_4 = 0$: TRUE or FALSE	{1, 5, 6, 7, 8, 9, 10}, {2, 3, 4} $\text{entropy}(S) = \frac{7}{10}\text{entropy}(D_{\text{true}}) + \frac{3}{10}\text{entropy}(D_{\text{false}})$ $= \frac{7}{10} \times \left(-\frac{6}{7}\log_2\frac{6}{7} - \frac{1}{7}\log_2\frac{1}{7}\right) + \frac{3}{10} \times 0 = 0.41$
$x_5 = 0$: TRUE or FALSE	{2, 3, 4, 6, 7, 8, 9, 10}, {1, 5} $\text{entropy}(S) = \frac{8}{10}\text{entropy}(D_{\text{true}}) + \frac{2}{10}\text{entropy}(D_{\text{false}})$ $= \frac{8}{10} \times \left(-\frac{7}{8}\log_2\frac{7}{8} - \frac{1}{8}\log_2\frac{1}{8}\right) + \frac{2}{10} \times 0 = 0.43$
$x_6 = 0$: TRUE or FALSE	{1, 2, 4, 5, 7, 8, 9, 10}, {3, 6} $\text{entropy}(S) = \frac{8}{10}\text{entropy}(D_{\text{true}}) + \frac{2}{10}\text{entropy}(D_{\text{false}})$ $= \frac{8}{10} \times \left(-\frac{7}{8}\log_2\frac{7}{8} - \frac{1}{8}\log_2\frac{1}{8}\right) + \frac{2}{10} \times 0 = 0.43$
$x_7 = 0$: TRUE or FALSE	{2, 4, 8, 9, 10}, {1, 3, 5, 6, 7} $\text{entropy}(S) = \frac{5}{10}\text{entropy}(D_{\text{true}}) + \frac{5}{10}\text{entropy}(D_{\text{false}})$ $= \frac{5}{10} \times \left(-\frac{4}{5}\log_2\frac{4}{5} - \frac{1}{5}\log_2\frac{1}{5}\right) + \frac{5}{10} \times 0 = 0.36$

(continued)

TABLE 23.2 (Continued)

Binary Split of the Root Node and Calculation of Information Entropy for the Data Set of System Fault Detection

Split Criterion	Resulting Subsets and Average Information Entropy of the Split
$x_8 = 0$: TRUE or FALSE	{1, 5, 6, 7, 9, 10}, {2, 3, 4, 8}
	$\text{entropy}(S) = \frac{6}{10}\text{entropy}(D_{\text{true}}) + \frac{4}{10}\text{entropy}(D_{\text{false}})$
	$= \frac{6}{10} \times \left(-\frac{5}{6}\log_2\frac{5}{6} - \frac{1}{6}\log_2\frac{1}{6}\right) + \frac{4}{10} \times 0 = 0.39$
$x_9 = 0$: TRUE or FALSE	{2, 3, 4, 6, 7, 8, 10}, {1, 5, 9}
	$\text{entropy}(S) = \frac{7}{10}\text{entropy}(D_{\text{true}}) + \frac{3}{10}\text{entropy}(D_{\text{false}})$
	$= \frac{7}{10} \times \left(-\frac{6}{7}\log_2\frac{6}{7} - \frac{1}{7}\log_2\frac{1}{7}\right) + \frac{3}{10} \times 0 = 0.41$

Source: Reprinted with permission from Table 4.2 in Ye, N. 2013. *Data Mining: Theories, Algorithms, and Examples*, pp. 41–42, Boca Raton, Florida: CRC Press.

where D denotes a given data set, c denotes the number of different target values, and P_i denotes the probability that a data record in the data set takes the ith target value. An entropy value falls in the range, $(0, \log_2 c)$. For example, given the data set in Table 23.1, we have $c = 2$ (for two target values, $y = 0$ and $y = 1$), $P_1 = 9/10$ (nine of the ten records with $y = 1) = 0.9$, $P_2 = 1/10$ (one of the ten records with $y = 0) = 0.1$, and

$$\text{entropy}(D) = \sum_{i=1}^{2} -P_i \log_2 P_i = -0.9\log_2 0.9 - 0.1\log_2 0.1 = 0.47.$$

Figure 23.3 shows how the entropy value changes with $P_1 (P_2 = 1 - P_1)$ when $c = 2$. Especially, we have the following:

- $P_1 = 0.5$, $P_2 = 0.5$, entropy$(D) = 1$
- $P_1 = 0$, $P_2 = 1$, entropy$(D) = 0$
- $P_1 = 1$, $P_2 = 0$, entropy$(D) = 0$

If all the data records in a data set take one target value, we have $P_1 = 0$, $P_2 = 1$ or $P_1 = 1$, $P_2 = 0$, and the value of the information entropy is 0, that is, we need 0 bit of information because we already know the target value that all the data records take. Hence, the entropy value of 0 indicates that the data set is homogenous with regard to the target value. If one-half set of the data records in a data set takes one target value, and the other half set takes another target value, we have $P_1 = 0.5$, $P_2 = 0.5$, and the value of the information entropy is 1, which means that we need 1 bit of information to convey what target value is. Hence, the entropy value of 1 indicates that the data set is inhomogeneous. When we use the information entropy to measure data homogeneity, the lower the entropy value is, the more homogenous the data set is with regard to the target value.

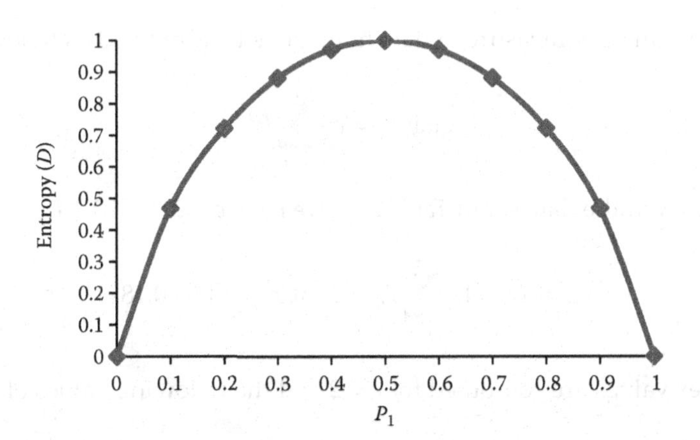

FIGURE 23.3
Information entropy. The decision tree for system fault detection. (Reprinted with permission from Figure 4.2 in Ye, N. 2013. *Data Mining: Theories, Algorithms, and Examples*, p. 42, Boca Raton, Florida: CRC Press.)

After a split of a data set into several subsets, the following formula is used to compute the average information entropy of subsets:

$$\text{entropy}(S) = \sum_{v \in Values(S)} \frac{|D_v|}{|D|} \text{entropy}(D_v) \tag{23.4}$$

where S denotes the split, *Values*(S) denotes a set of values that are used in the split, v denotes a value in *Values*(S), D denotes the data set being split, $|D|$ denotes the number of data records in the data set D, D_v denotes the subset resulting from the split using the split value v, and $|D_v|$ denotes the number of data records in the data set D_v.

For example, the root node of a decision tree for the data set in Table 23.1 has the data set, $D = \{1, 2, \ldots, 10\}$, whose entropy value is 0.47, as shown previously. Using the split criterion, $x_1 = 0$: TRUE or FALSE, the root node is split into two subsets: $D_{\text{false}} = \{1\}$, which is homogenous, and $D_{\text{true}} = \{2, 3, 4, 5, 6, 7, 8, 9, 10\}$, which is inhomogeneous, with eight data records taking the target value of 1 and one data record taking the target value of 0. The average entropy of the two subsets after the split is

$$\text{entropy}(S) = \frac{9}{10}\text{entropy}(D_{\text{true}}) + \frac{1}{10}\text{entropy}(D_{\text{false}})$$

$$= \frac{9}{10} \times \left(-\frac{8}{9}\log_2\frac{8}{9} - \frac{1}{9}\log_2\frac{1}{9}\right) + \frac{1}{10} \times 0 = 0.45.$$

Since this average entropy of subsets after the split is better than entropy(D) = 0.47, the split improves data homogeneity. Table 23.2 gives the average entropy of subsets after each of the other eight splits of the root node. Among the nine possible splits, the split using the criterion of $x_7 = 0$: TRUE or FALSE produces the smallest average information entropy, which indicates the most homogeneous subsets. Hence, the split criterion of $x_7 = 0$: TRUE or FALSE is selected to split the root node, resulting in two internal nodes, as shown in Figure 23.2. The internal node with the subset, {2, 4, 8, 9, 10}, is not homogenous. Hence, the decision tree is further expanded with more splits until all leaf nodes are homogenous.

The gini-index, another measure of data homogeneity, is defined as follows:

$$\text{gini}(D) = 1 - \sum_{i=1}^{c} P_i^2 \tag{23.5}$$

For example, given the data set in Table 23.1, we have $c = 2$, $P_1 = 0.9$, $P_2 = 0.1$, and

$$\text{gini}(D) = 1 - \sum_{i=1}^{c} P_i^2 = 1 - 0.9^2 - 0.1^2 = 0.18.$$

The gini-index values are computed for $c = 2$ and the following values of P_i:

- $P_1 = 0.5$, $P_2 = 0.5$, $\text{gini}(D) = 1 - 0.5^2 - 0.5^2 = 0.5$
- $P_1 = 0$, $P_2 = 1$, $\text{gini}(D) = 1 - 0^2 - 1^2 = 0$
- $P_1 = 1$, $P_2 = 0$, $\text{gini}(D) = 1 - 1^2 - 0^2 = 0$

Hence, the smaller the gini-index value is, the more homogeneous the data set is. The average gini-index value of data subsets after a split is calculated as follows:

$$\text{gini}(S) = \sum_{v \in \text{Values}(S)} \frac{|D_v|}{|D|} \text{gini}(D_v) \tag{23.6}$$

Table 23.3 gives the average gini-index value of subsets after each of the nine splits of the root node for the training data set of system fault detection. Among the nine possible splits, the split criterion of $x_7 = 0$: TRUE or FALSE produces the smallest average gini-index value, which indicates the most homogeneous subsets. The split criterion of $x_7 = 0$: TRUE or FALSE is selected to split the root node. Hence, using the gini-index produces the same split as using the information entropy.

TABLE 23.3

Binary Split of the Root Node and Calculation of the Gini-Index for the Data Set of System Fault Detection

Split Criterion	Resulting Subsets and Average Gini-Index Value of the Split
$x_1 = 0$: TRUE or FALSE	{2, 3, 4, 5, 6, 7, 8, 9, 10}, {1} $\text{gini}(S) = \frac{9}{10}\text{gini}(D_{\text{true}}) + \frac{1}{10}\text{gini}(D_{\text{false}})$ $= \frac{9}{10} \times \left(1 - \left(\frac{1}{9}\right)^2 - \left(\frac{8}{9}\right)^2\right) + \frac{1}{10} \times 0 = 0.18$
$x_2 = 0$: TRUE or FALSE	{1, 3, 4, 5, 6, 7, 8, 9, 10}, {2} $\text{gini}(S) = \frac{9}{10}\text{gini}(D_{\text{true}}) + \frac{1}{10}\text{gini}(D_{\text{false}})$ $= \frac{9}{10} \times \left(1 - \left(\frac{1}{9}\right)^2 - \left(\frac{8}{9}\right)^2\right) + \frac{1}{10} \times 0 = 0.18$

(*continued*)

TABLE 23.3 (Continued)

Binary Split of the Root Node and Calculation of the Gini-Index for the Data Set of System Fault Detection

Split Criterion	Resulting Subsets and Average Gini-Index Value of the Split
$x_3 = 0$: TRUE or FALSE	{1, 2, 4, 5, 6, 7, 8, 9, 10}, {3} $\text{gini}(S) = \frac{9}{10}\text{gini}(D_{\text{true}}) + \frac{1}{10}\text{gini}(D_{\text{false}})$ $= \frac{9}{10} \times \left(1 - \left(\frac{1}{9}\right)^2 - \left(\frac{8}{9}\right)^2\right) + \frac{1}{10} \times 0 = 0.18$
$x_4 = 0$: TRUE or FALSE	{1, 5, 6, 7, 8, 9, 10}, {2, 3, 4} $\text{gini}(S) = \frac{7}{10}\text{gini}(D_{\text{true}}) + \frac{3}{10}\text{gini}(D_{\text{false}})$ $= \frac{7}{10} \times \left(1 - \left(\frac{6}{7}\right)^2 - \left(\frac{1}{7}\right)^2\right) + \frac{3}{10} \times 0 = 0.17$
$x_5 = 0$: TRUE or FALSE	{2, 3, 4, 6, 7, 8, 9, 10}, {1, 5} $\text{gini}(S) = \frac{8}{10}\text{gini}(D_{\text{true}}) + \frac{2}{10}\text{gini}(D_{\text{false}})$ $= \frac{8}{10} \times \left(1 - \left(\frac{7}{8}\right)^2 - \left(\frac{1}{8}\right)^2\right) + \frac{2}{10} \times 0 = 0.175$
$x_6 = 0$: TRUE or FALSE	{1, 2, 4, 5, 7, 8, 9, 10}, {3, 6} $\text{gini}(S) = \frac{8}{10}\text{gini}(D_{\text{true}}) + \frac{2}{10}\text{gini}(D_{\text{false}})$ $= \frac{8}{10} \times \left(1 - \left(\frac{7}{8}\right)^2 - \left(\frac{1}{8}\right)^2\right) + \frac{2}{10} \times 0 = 0.175$
$x_7 = 0$: TRUE or FALSE	{2, 4, 8, 9, 10}, {1, 3, 5, 6, 7} $\text{gini}(S) = \frac{5}{10}\text{gini}(D_{\text{true}}) + \frac{5}{10}\text{gini}(D_{\text{false}})$ $= \frac{5}{10} \times \left(1 - \left(\frac{4}{5}\right)^2 - \left(\frac{1}{5}\right)^2\right) + \frac{5}{10} \times 0 = 0.16$
$x_8 = 0$: TRUE or FALSE	{1, 5, 6, 7, 9, 10}, {2, 3, 4, 8} $\text{gini}(S) = \frac{6}{10}\text{gini}(D_{\text{true}}) + \frac{4}{10}\text{gini}(D_{\text{false}})$ $= \frac{6}{10} \times \left(1 - \left(\frac{5}{6}\right)^2 - \left(\frac{1}{6}\right)^2\right) + \frac{4}{10} \times 0 = 0.167$
$x_9 = 0$: TRUE or FALSE	{2, 3, 4, 6, 7, 8, 10}, {1, 5, 9} $\text{gini}(S) = \frac{7}{10}\text{gini}(D_{\text{true}}) + \frac{3}{10}\text{gini}(D_{\text{false}})$ $= \frac{7}{10} \times \left(1 - \left(\frac{6}{7}\right)^2 - \left(\frac{1}{7}\right)^2\right) + \frac{3}{10} \times 0 = 0.17$

Source: Reprinted with permission from Table 4.3 in Ye, N. 2013. *Data Mining: Theories, Algorithms, and Examples*, pp. 45–46, Boca Raton, Florida: CRC Press.

23.1.5 Algorithm for the Top-Down Construction of a Binary Decision Tree

This section describes and illustrates the algorithm of constructing a complete decision tree. The algorithm for the top-down construction of a binary decision tree has the following steps:

1. Start with the root node, which includes all the data records in the training data set, and select this node to split.
2. Apply a split selection method to the selected node to determine the best split along with the split criterion, and partition the set of the training data records at the selected node into two nodes with two subsets of data records.
3. Check if the stopping criterion is satisfied. If so, the tree construction is completed; otherwise, go back to Step 2 to continue by selecting a node to split.

The stopping criterion based on data homogeneity is to stop when each leaf node has homogeneous data, that is, a set of data records with the same target value. Many large sets of real-world data are noisy, making it difficult to obtain homogeneous data sets at leaf nodes. Hence, the stopping criterion is often set to have the measure of data homogeneity to be smaller than a threshold value, for example, entropy(D) < 0.1.

We show the construction of the complete, binary decision tree for the data set of system fault detection in the following.

Example 23.1: Construct a Binary Decision Tree for the Data Set of System Fault Detection in Table 23.1.

We first use the information entropy as the measure of data homogeneity. As shown in Figure 23.2, the data set at the root node is partitioned into two subsets, {2, 4, 8, 9, 10} and {1, 3, 5, 6, 7}, which are already homogeneous with the target value, $y = 1$, and do not need a split. For the subset, $D = \{2, 4, 8, 9, 10\}$,

$$\text{entropy}(D) = \sum_{i=1}^{2} -P_i \log_2 P_i = -\frac{1}{5}\log_2\frac{1}{5} - \frac{4}{5}\log_2\frac{4}{5} = 0.72.$$

Except for x_7, which has been used to split the root node, the other eight attribute variables, x_1, x_2, x_3, x_4, x_5, x_6, x_8, and x_9, can be used to split D. The split criteria using $x_1 = 0$, $x_3 = 0$, $x_5 = 0$, and $x_6 = 0$ do not produce a split of D. Table 23.4 gives the calculation of information entropy for the splits using x_2, x_4, x_7, x_8, and x_9. Since the split criterion, $x_8 = 0$: TRUE or FALSE, produces the smallest average entropy of the split, this split criterion is selected to split $D = \{2, 4, 8, 9, 10\}$ into {9, 10} and {2, 4, 8}, which are already homogeneous with the target value, $y = 1$, and do not need a split. Figure 23.2 shows this split.

For the subset, $D = \{9, 10\}$,

$$\text{entropy}(D) = \sum_{i=1}^{2} -P_i \log_2 P_i = -\frac{1}{2}\log_2\frac{1}{2} - \frac{1}{2}\log_2\frac{1}{2} = 1.$$

Except for x_7 and x_8, which have been used, the other seven attribute variables, x_1, x_2, x_3, x_4, x_5, x_6, and x_9, can be used to split D. The split criteria using $x_1 = 0$, $x_2 = 0$, $x_3 = 0$, $x_4 = 0$, $x_5 = 0$, and $x_6 = 0$ do not produce a split of D. The split criterion, $x_9 = 0$: TRUE or FALSE, produces two subsets, {9} with the target value of $y = 1$ and {10} with the target

TABLE 23.4

Binary Split of an Internal Node with D = {2, 4, 5, 9, 10} and Calculation of Information Entropy for the Data Set of System Fault Detection

Split Criterion	Resulting Subsets and Average Information Entropy of the Split
$x_2 = 0$: TRUE or FALSE	{4, 8, 9, 10}, {2}
	$\text{entropy}(S) = \frac{4}{5}\text{entropy}(D_{\text{true}}) + \frac{1}{5}\text{entropy}(D_{\text{false}})$
	$= \frac{4}{5} \times \left(-\frac{3}{4}\log_2\frac{8}{9} - \frac{1}{4}\log_2\frac{1}{4}\right) + \frac{1}{5} \times 0 = 0.64$
$x_4 = 0$: TRUE or FALSE	{8, 9, 10}, {2, 4}
	$\text{entropy}(S) = \frac{3}{5}\text{entropy}(D_{\text{true}}) + \frac{2}{5}\text{entropy}(D_{\text{false}})$
	$= \frac{3}{5} \times \left(-\frac{2}{3}\log_2\frac{2}{3} - \frac{1}{3}\log_2\frac{1}{3}\right) + \frac{2}{5} \times 0 = 0.55$
$x_8 = 0$: TRUE or FALSE	{9, 10}, {2, 4, 8}
	$\text{entropy}(S) = \frac{2}{5}\text{entropy}(D_{\text{true}}) + \frac{3}{5}\text{entropy}(D_{\text{false}})$
	$= \frac{2}{5} \times \left(-\frac{1}{2}\log_2\frac{1}{2} - \frac{1}{2}\log_2\frac{1}{2}\right) + \frac{3}{5} \times 0 = 0.4$
$x_9 = 0$: TRUE or FALSE	{2, 4, 8, 10}, {9}
	$\text{entropy}(S) = \frac{4}{5}\text{entropy}(D_{\text{true}}) + \frac{1}{5}\text{entropy}(D_{\text{false}})$
	$= \frac{4}{5} \times \left(-\frac{3}{4}\log_2\frac{3}{4} - \frac{1}{4}\log_2\frac{1}{4}\right) + \frac{1}{5} \times 0 = 0.64$

Source: Reprinted with permission from Table 4.4 in Ye, N. 2013. *Data Mining: Theories, Algorithms, and Examples*, p. 47, Boca Raton, Florida: CRC Press.

value of $y = 0$, which are homogeneous and do not need a split. Figure 23.2 shows this split. Since all leaf nodes of the decision tree are homogeneous, the construction of the decision tree is stopped with the complete decision tree shown in Figure 23.2.

We now show the construction of the decision tree using the gini-index as the measure of data homogeneity. As described previously, the data set at the root node is partitioned into two subsets, {2, 4, 8, 9, 10} and {1, 3, 5, 6, 7}, which are already homogeneous with the target value, $y = 1$, and do not need a split. For the subset, D = {2, 4, 8, 9, 10},

$$\text{gini}(D) = 1 - \sum_{i=1}^{c} P_i^2 = 1 - \left(\frac{4}{5}\right)^2 - \left(\frac{1}{5}\right)^2 = 0.32.$$

The split criteria using $x_1 = 0$, $x_3 = 0$, $x_5 = 0$, and $x_6 = 0$ do not produce a split of D. Table 23.5 gives the calculation of the gini-index values for the splits using x_2, x_4, x_7, x_8, and x_9. Since the split criterion, $x_8 = 0$: TRUE or FALSE, produces the smallest average gini-index value of the split, this split criterion is selected to split D = {2, 4, 8, 9, 10} into {9, 10} and {2, 4, 8}, which are already homogeneous with the target value, $y = 1$, and do not need a split.

TABLE 23.5

Binary Split of an Internal Node with $D = \{2, 4, 5, 9, 10\}$ and Calculation of the Gini-Index Values for the Data Set of System Fault Detection

Split Criterion	Resulting Subsets and Average Gini-Index Value of the Split
$x_2 = 0$: TRUE or FALSE	$\{4, 8, 9, 10\}, \{2\}$ $\text{gini}(S) = \frac{4}{5}\text{gini}(D_{\text{true}}) + \frac{1}{5}\text{gini}(D_{\text{false}})$ $= \frac{4}{5} \times \left(1 - \left(\frac{3}{4}\right)^2 - \left(\frac{1}{4}\right)^2\right) + \frac{1}{5} \times 0 = 0.3$
$x_4 = 0$: TRUE or FALSE	$\{8, 9, 10\}, \{2, 4\}$ $\text{gini}(S) = \frac{3}{5}\text{gini}(D_{\text{true}}) + \frac{2}{5}\text{gini}(D_{\text{false}})$ $= \frac{3}{5} \times \left(1 - \left(\frac{3}{4}\right)^2 - \left(\frac{1}{4}\right)^2\right) + \frac{2}{5} \times 0 = 0.27$
$x_8 = 0$: TRUE or FALSE	$\{9, 10\}, \{2, 4, 8\}$ $\text{gini}(S) = \frac{2}{5}\text{gini}(D_{\text{true}}) + \frac{3}{5}\text{gini}(D_{\text{false}})$ $= \frac{2}{5} \times \left(1 - \left(\frac{1}{2}\right)^2 - \left(\frac{1}{2}\right)^2\right) + \frac{3}{5} \times 0 = 0.2$
$x_9 = 0$: TRUE or FALSE	$\{2, 4, 8, 10\}, \{9\}$ $\text{gini}(S) = \frac{4}{5}\text{gini}(D_{\text{true}}) + \frac{1}{5}\text{gini}(D_{\text{false}})$ $= \frac{4}{5} \times \left(1 - \left(\frac{3}{4}\right)^2 - \left(\frac{1}{4}\right)^2\right) + \frac{1}{5} \times 0 = 0.3$

Source: Reprinted with permission from Table 4.5 in Ye, N. 2013. *Data Mining: Theories, Algorithms, and Examples*, p. 48, Boca Raton, Florida: CRC Press.

For the subset, $D = \{9, 10\}$,

$$\text{gini}(D) = 1 - \sum_{i=1}^{c} P_i^2 = 1 - \left(\frac{1}{2}\right)^2 - \left(\frac{1}{2}\right)^2 = 0.5$$

Except for x_7 and x_8, which have been used, the other seven attribute variables, x_1, x_2, x_3, x_4, x_5, x_6, and x_9, can be used to split D. The split criteria using $x_1 = 0$, $x_2 = 0$, $x_3 = 0$, $x_4 = 0$, $x_5 = 0$, and $x_6 = 0$ do not produce a split of D. The split criterion, $x_9 = 0$: TRUE or FALSE, produces two subsets, $\{9\}$ with the target value of $y = 1$ and $\{10\}$ with the target value of $y = 0$, which are homogeneous and do not need a split. Since all leaf nodes of the decision tree are homogeneous, the construction of the decision tree is stopped with the complete decision tree that is the same as that from using the information entropy as the measure of data homogeneity.

23.1.6 Classifying Data Using a Decision Tree

A decision tree is used to classify a data record by passing the data record into a leaf node of the decision tree using the values of the attribute variables and assigning the target value of the leaf node to the data record. Figure 23.4 highlights in bold the path of passing the training data record, for instance, 10 in Table 23.1, from the root node to a leaf node with the target value, $y = 0$. Hence, the data record is classified to have no system fault. For the data records in the testing data set of system fault detection in Table 23.6, their target values are obtained using the decision tree in Figure 23.2 and are shown in Table 23.6.

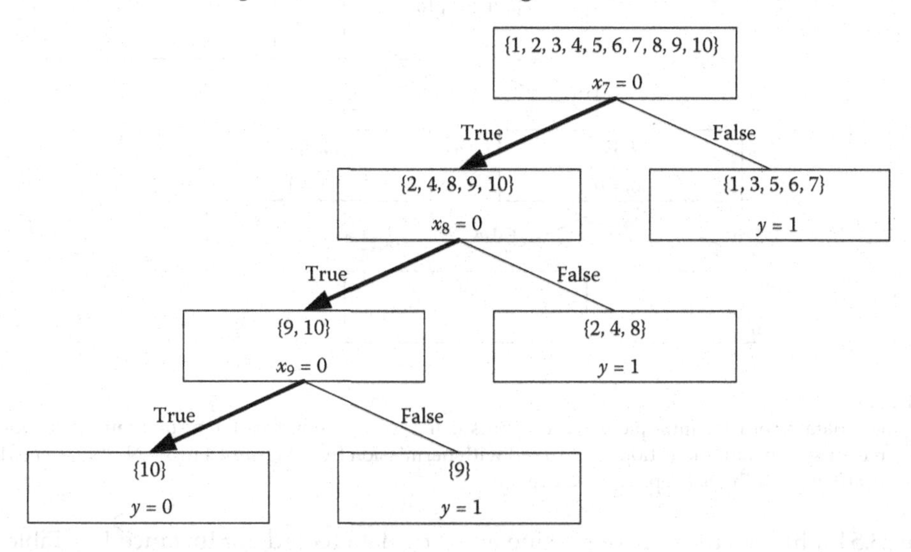

FIGURE 23.4
Classifying a data record for no system fault using the decision tree for system fault detection. The decision tree for system fault detection. (Reprinted with permission from Figure 4.3 in Ye, N. 2013. *Data Mining: Theories, Algorithms, and Examples*, p. 49, Boca Raton, Florida: CRC Press.)

TABLE 23.6
Classification of Data Records in the Testing Data Set for System Fault Detection

Instance (Faulty Machines)	Attribute Variables (Quality of Parts)									Target Variable, y (System Faults)	
	x_1	x_2	x_3	x_4	x_5	x_6	x_7	x_8	x_9	True Value	Classified Value
1 (M1, M2)	1	1	0	1	1	0	1	1	1	1	1
2 (M2, M3)	0	1	1	1	0	1	1	1	0	1	1
3 (M1, M3)	1	0	1	1	1	1	1	1	1	1	1
4 (M1, M4)	1	0	0	1	1	0	1	1	1	1	1
5 (M1, M6)	1	0	0	0	1	1	1	0	1	1	1
6 (M2, M6)	0	1	0	1	0	1	1	1	0	1	1
7 (M2, M5)	0	1	0	1	1	0	1	1	0	1	1
8 (M3, M5)	0	0	1	1	1	1	1	1	1	1	1
9 (M4, M7)	0	0	0	1	0	0	1	1	0	1	1
10 (M5, M8)	0	0	0	0	1	0	1	1	0	1	1
11 (M3, M9)	0	0	1	1	0	1	1	1	1	1	1
12 (M1, M8)	1	0	0	0	1	0	1	1	1	1	1
13 (M1, M2, M3)	1	1	1	1	1	1	1	1	1	1	1
14 (M2, M3, M5)	0	1	1	1	1	1	1	1	1	1	1
15 (M2, M3, M9)	0	1	1	1	0	1	1	1	1	1	1
16 (M1, M6, M8)	1	0	0	0	1	1	1	1	1	1	1

Source: Reprinted with permission from Table 4.6 in Ye, N. 2013. *Data Mining: Theories, Algorithms, and Examples*, p. 50, CRC Press.

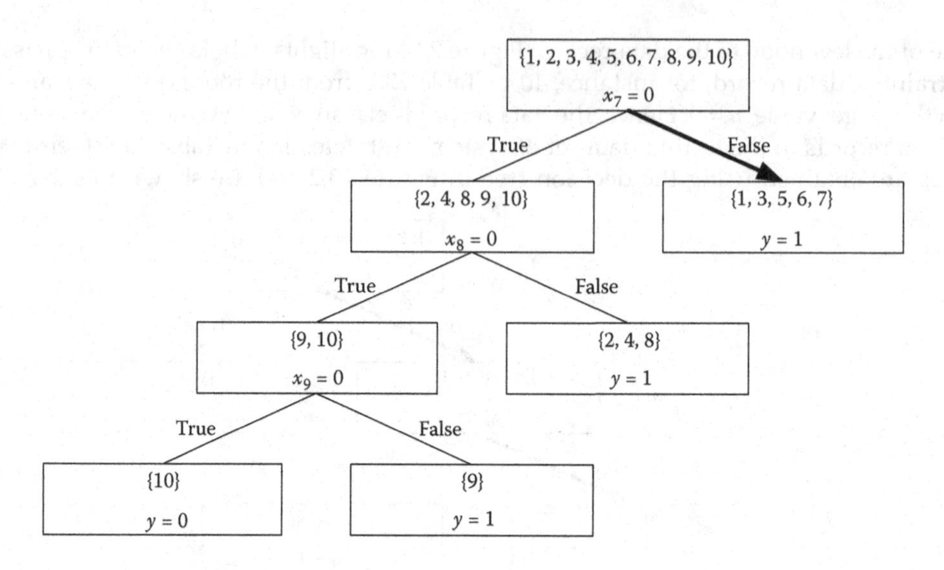

FIGURE 23.5
Classifying a data record for multiple machine faults using the decision tree for system fault detection. The decision tree for system fault detection. (Reprinted with permission from Figure 4.4 in Ye, N. 2013. *Data Mining: Theories, Algorithms, and Examples*, p. 50, CRC Press.)

Figure 23.5 highlights the path of passing a testing data record, for Instance 1 in Table 23.6, from the root node to a leaf node with the target value, $y = 1$. Hence, the data record is classified to have a system fault.

23.2 Learning a Non-Binary Decision Tree

In the lenses data set in Table 23.7 from the UCI Machine Learning Repository (Frank and Asuncion, 2010), the attribute variable, Age, has three categorical values: Young, Pre-Presbyopic, and Presbyopic. If we want to construct a binary decision tree for this data set, we need to convert the three categorical values of the attribute variable, Age, into two categorical values if Age is used to split the root node. We may have Young and Pre-Presbyopic together as one category, Presbyopic as another category, that is using Age = Presbyopic: TRUE or FALSE as the split criterion. We may also have Young as one category, Pre-Presbyopic and Presbyopic together as another category, that is using Age = Young: TRUE or FALSE as the split criterion. We may have Age = Pre-Presbyopic: TRUE or FALSE as another split criterion. However, we can construct a non-binary decision tree to allow partitioning a data set at a node into more than two subsets by using each of the multiple categorical values for each branch of the split. Example 23.2 shows the construction of a non-binary decision tree for the lenses data set.

Example 23.2: Construct a Non-Binary Decision Tree for the Lenses Data Set in Table 23.7.

If the attribute variable, Age, is used to split the root node for the lenses data set, all three categorical values of Age can be used to partition the set of 24 data records at the root node using the split criterion, Age = Young, Pre-Presbyopic, or Presbyopic, as

TABLE 23.7
Lenses Data Set

	Attributes				Target
Instance	Age	Spectacle Prescription	Astigmatic	Tear Production Rate	Lenses
1	Young	Myope	No	Reduced	Non-Contact
2	Young	Myope	No	Normal	Soft Contact
3	Young	Myope	Yes	Reduced	Non-Contact
4	Young	Myope	Yes	Normal	Hard Contact
5	Young	Hypermetrope	No	Reduced	Non-Contact
6	Young	Hypermetrope	No	Normal	Soft Contact
7	Young	Hypermetrope	Yes	Reduced	Non-Contact
8	Young	Hypermetrope	Yes	Normal	Hard Contact
9	Pre-Presbyopic	Myope	No	Reduced	Non-Contact
10	Pre-Presbyopic	Myope	No	Normal	Soft Contact
11	Pre-Presbyopic	Myope	Yes	Reduced	Non-Contact
12	Pre-Presbyopic	Myope	Yes	Normal	Hard Contact
13	Pre-Presbyopic	Hypermetrope	No	Reduced	Non-Contact
14	Pre-Presbyopic	Hypermetrope	No	Normal	Soft Contact
15	Pre-Presbyopic	Hypermetrope	Yes	Reduced	Non-Contact
16	Pre-Presbyopic	Hypermetrope	Yes	Normal	Non-Contact
17	Presbyopic	Myope	No	Reduced	Non-Contact
18	Presbyopic	Myope	No	Normal	Non-Contact
19	Presbyopic	Myope	Yes	Reduced	Non-Contact
20	Presbyopic	Myope	Yes	Normal	Hard Contact
21	Presbyopic	Hypermetrope	No	Reduced	Non-Contact
22	Presbyopic	Hypermetrope	No	Normal	Soft Contact
23	Presbyopic	Hypermetrope	Yes	Reduced	Non-Contact
24	Presbyopic	Hypermetrope	Yes	Normal	Non-Contact

Source: Reprinted with permission from Table 1.3 in Ye, N. 2013. *Data Mining: Theories, Algorithms, and Examples*, p. 6, Boca Raton, Florida: CRC Press.

shown in Figure 23.6. We use the data set of 24 data records in Table 23.7 as the training data set, *D*, at the root node of the non-binary decision tree. In the lenses data set, the target variable has three categorical values: Non-Contact in 15 data records, Soft Contact in 5 data records, and Hard Contact in 4 data records. Using the information entropy as the measure of data homogeneity, we have

$$\text{entropy}(D) = \sum_{i=1}^{3} -P_i \log_2 P_i = -\frac{15}{24}\log_2\frac{15}{24} - \frac{5}{24}\log_2\frac{5}{24} - \frac{4}{24}\log_2\frac{4}{24} = 1.3261.$$

Table 23.8 shows the calculation of information entropy to split the root node using the split criterion, Tear Production Rate = Reduced or Normal, which produces a homogenous subset of {1, 3, 5, 7, 9, 11, 13, 15, 17, 19, 21, 23} and an inhomogeneous subset of {2, 4, 6, 8, 10, 12, 14, 16, 18, 20, 22, 24}. Table 23.9 shows the calculation of information entropy to split the node with the data set of {2, 4, 6, 8, 10, 12, 14, 16, 18, 20, 22, 24} using the split criterion, Astigmatic = No or Yes, which produces two subsets of {2, 6, 10, 14, 18, 22} and {4, 8, 12, 16, 20, 24}. Table 23.10 shows the calculation of information entropy to split the node with the data set of {2, 6, 10, 14, 18, 22} using the split criterion,

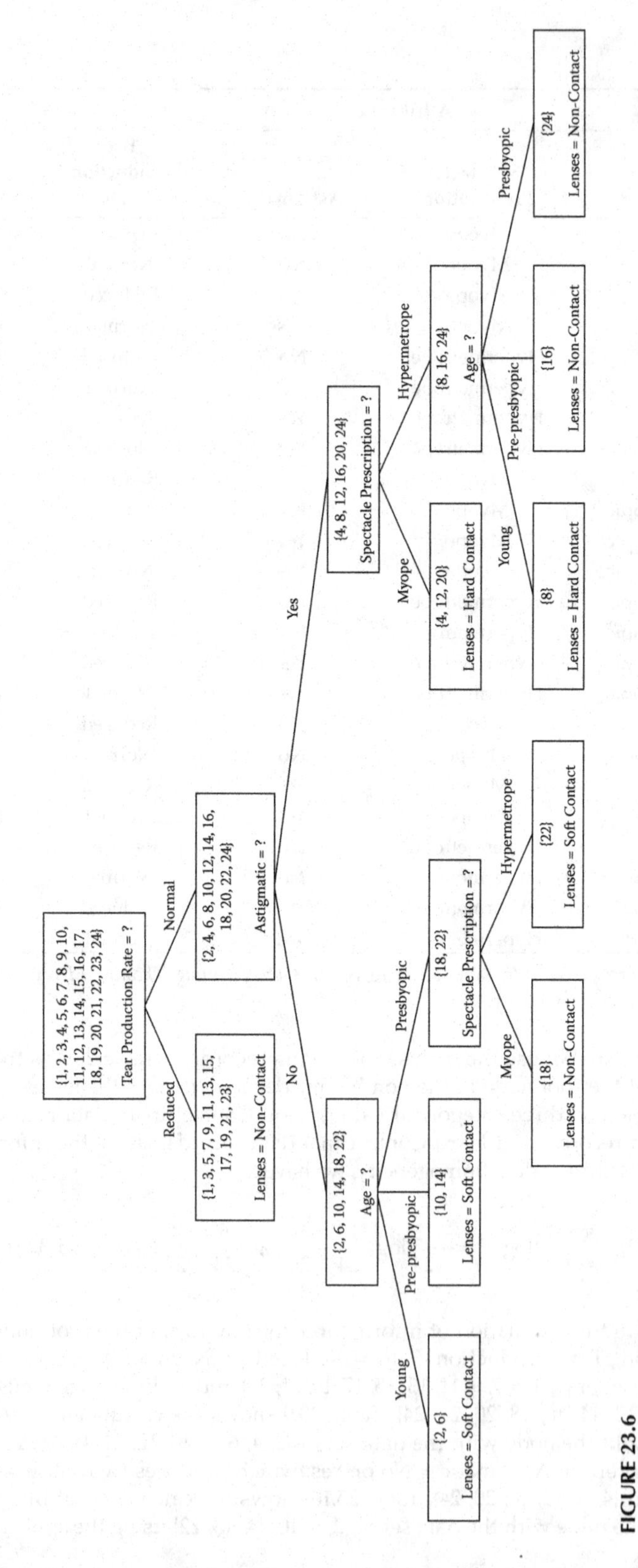

FIGURE 23.6
Decision tree for the lenses data set. The decision tree for system fault detection. (Reprinted with permission from Figure 4.5 in Ye, N. 2013. *Data Mining: Theories, Algorithms, and Examples*, p. 52, Boca Raton, Florida: CRC Press.)

TABLE 23.8

Non-Binary Split of the Root Node and Calculation of Information Entropy for the Lenses Data Set

Split Criterion	Resulting Subsets and Average Information Entropy of the Split
Age = Young, Pre-Presbyopic, or Presbyopic	{1, 2, 3, 4, 5, 6, 7, 8}, {9, 10, 11, 12, 13, 14, 15, 16}, {17, 18, 19, 20, 21, 22, 23, 24}
	$\text{entropy}(S) = \frac{8}{24}\text{entropy}(D_{\text{Young}}) + \frac{8}{24}\text{entropy}(D_{\text{Pre-presbyopic}}) + \frac{8}{24}\text{entropy}(D_{\text{Presbyopic}})$
	$= \frac{8}{24} \times \left(-\frac{4}{8}\log_2\frac{4}{8} - \frac{2}{8}\log_2\frac{2}{8} - \frac{2}{8}\log_2\frac{2}{8}\right) + \frac{8}{24} \times \left(-\frac{5}{8}\log_2\frac{5}{8} - \frac{2}{8}\log_2\frac{2}{8} - \frac{1}{8}\log_2\frac{1}{8}\right) + \frac{8}{24} \times \left(-\frac{6}{8}\log_2\frac{6}{8} - \frac{1}{8}\log_2\frac{1}{8} - \frac{1}{8}\log_2\frac{1}{8}\right) = 1.2867$
Spectacle prescription = Myope or Hypermetrope	{1, 2, 3, 4, 9, 10, 11, 12, 17, 18, 19, 20}, {5, 6, 7, 8, 13, 14, 15, 16, 21, 22, 23, 24}
	$\text{entropy}(S) = \frac{12}{24}\text{entropy}(D_{\text{Myope}}) + \frac{12}{24}\text{entropy}(D_{\text{Hypermetrope}})$
	$= \frac{12}{24} \times \left(-\frac{7}{12}\log_2\frac{7}{12} - \frac{2}{12}\log_2\frac{2}{12} - \frac{3}{12}\log_2\frac{3}{12}\right) + \frac{12}{24} \times \left(-\frac{8}{12}\log_2\frac{8}{12} - \frac{3}{12}\log_2\frac{3}{12} - \frac{1}{12}\log_2\frac{1}{12}\right) = 1.2866$
Astigmatic = Yes or No	{1, 2, 5, 6, 9, 10, 13, 14, 17, 18, 21, 22}, {3, 4, 7, 8, 11, 12, 15, 16, 19, 20, 23, 24}
	$\text{entropy}(S) = \frac{12}{24}\text{entropy}(D_{\text{No}}) + \frac{12}{24}\text{entropy}(D_{\text{Yes}})$
	$= \frac{12}{24} \times \left(-\frac{7}{12}\log_2\frac{7}{12} - \frac{5}{12}\log_2\frac{5}{12} - \frac{0}{12}\log_2\frac{0}{12}\right) + \frac{12}{24} \times \left(-\frac{8}{12}\log_2\frac{8}{12} - \frac{4}{12}\log_2\frac{4}{12} - \frac{0}{12}\log_2\frac{0}{12}\right) = 0.9491$
Tear production rate = Reduced or Normal	{1, 3, 5, 7, 9, 11, 13, 15, 17, 19, 21, 23}, {2, 4, 6, 8, 10, 12, 14, 16, 18, 20, 22, 24}
	$\text{entropy}(S) = \frac{12}{24}\text{entropy}(D_{\text{Reduced}}) + \frac{12}{24}\text{entropy}(D_{\text{Normal}})$
	$= \frac{12}{24} \times \left(-\frac{12}{12}\log_2\frac{12}{12} - \frac{0}{12}\log_2\frac{0}{12} - \frac{0}{12}\log_2\frac{0}{12}\right) + \frac{12}{24} \times \left(-\frac{3}{12}\log_2\frac{3}{12} - \frac{5}{12}\log_2\frac{5}{12} - \frac{4}{12}\log_2\frac{4}{12}\right) = 0.7773$

Source: Reprinted with permission from Table 4.7 in Ye, N. 2013. *Data Mining: Theories, Algorithms, and Examples*, p. 53, Boca Raton, Florida: CRC Press.

TABLE 23.9

Non-Binary Split of an Internal Node, {2, 4, 6, 8, 10, 12, 14, 16, 18, 20, 22, 24}, and Calculation of Information Entropy for the Lenses Data Set

Split Criterion	Resulting Subsets and Average Information Entropy of the Split
Age = Young, Pre-Presbyopic, or Presbyopic	{2, 4, 6, 8}, {10, 12, 14, 16}, {18, 20, 22, 24}
	$\text{entropy}(S) = \frac{4}{12}\text{entropy}(D_{\text{Young}}) + \frac{4}{12}\text{entropy}(D_{\text{Pre-presbyopic}}) + \frac{4}{12}\text{entropy}(D_{\text{Presbyopic}})$
	$= \frac{4}{12} \times \left(-\frac{0}{4}\log_2\frac{0}{4} - \frac{2}{4}\log_2\frac{2}{4} - \frac{2}{4}\log_2\frac{2}{4}\right) + \frac{4}{12} \times \left(-\frac{1}{4}\log_2\frac{1}{4} - \frac{2}{4}\log_2\frac{2}{4} - \frac{1}{4}\log_2\frac{1}{4}\right) + \frac{4}{12} \times \left(-\frac{2}{4}\log_2\frac{2}{4} - \frac{1}{4}\log_2\frac{1}{4} - \frac{1}{4}\log_2\frac{1}{4}\right) = 1.3333$
Spectacle prescription = Myope or Hypermetrope	{2, 4, 10, 12, 18, 20}, {6, 7, 14, 16, 22, 24}
	$\text{entropy}(S) = \frac{6}{12}\text{entropy}(D_{\text{Myope}}) + \frac{6}{12}\text{entropy}(D_{\text{Hypermetrope}})$
	$= \frac{6}{12} \times \left(-\frac{1}{6}\log_2\frac{1}{6} - \frac{2}{6}\log_2\frac{2}{6} - \frac{3}{6}\log_2\frac{3}{6}\right) + \frac{6}{12} \times \left(-\frac{2}{6}\log_2\frac{2}{6} - \frac{3}{6}\log_2\frac{3}{6} - \frac{1}{6}\log_2\frac{1}{6}\right) = 1.4591$
Astigmatic = Yes or No	{2, 6, 10, 14, 18, 22}, {4, 8, 12, 16, 20, 24}
	$\text{entropy}(S) = \frac{6}{12}\text{entropy}(D_{\text{No}}) + \frac{6}{12}\text{entropy}(D_{\text{Yes}})$
	$= \frac{6}{12} \times \left(-\frac{1}{6}\log_2\frac{1}{6} - \frac{5}{6}\log_2\frac{5}{6} - \frac{0}{6}\log_2\frac{0}{6}\right) + \frac{6}{12} \times \left(-\frac{2}{6}\log_2\frac{2}{6} - \frac{0}{6}\log_2\frac{0}{6} - \frac{4}{6}\log_2\frac{4}{6}\right) = 0.7842$

Source: Reprinted with permission from Table 4.8 in Ye, N. 2013. *Data Mining: Theories, Algorithms, and Examples*, p. 54, Boca Raton, Florida: CRC Press.

TABLE 23.10

Non-Binary Split of an Internal Node, {2, 6, 10, 14, 18, 22}, and Calculation of Information Entropy for the Lenses Data Set

Split Criterion	Resulting Subsets and Average Information Entropy of the Split
Age = Young, Pre-Presbyopic, or Presbyopic	{2, 6}, {10, 14}, (18, 22}
	$\text{entropy}(S) = \frac{2}{6}\text{entropy}(D_{\text{Young}}) + \frac{2}{6}\text{entropy}(D_{\text{Pre-presbyopic}}) + \frac{2}{6}\text{entropy}(D_{\text{Presbyopic}})$
	$= \frac{2}{6} \times \left(-\frac{0}{2}\log_2\frac{0}{2} - \frac{2}{2}\log_2\frac{2}{2} - \frac{0}{2}\log_2\frac{0}{2}\right) + \frac{2}{6} \times \left(-\frac{0}{2}\log_2\frac{0}{2} - \frac{2}{2}\log_2\frac{2}{2} - \frac{0}{2}\log_2\frac{0}{2}\right) + \frac{2}{6} \times \left(-\frac{1}{2}\log_2\frac{1}{2} - \frac{1}{2}\log_2\frac{1}{2} - \frac{0}{2}\log_2\frac{0}{2}\right) = 0.3333$
Spectacle prescription = Myope or Hypermetrope	{2, 10, 18}, {6, 14, 22}
	$\text{entropy}(S) = \frac{3}{6}\text{entropy}(D_{\text{Myope}}) + \frac{3}{6}\text{entropy}(D_{\text{Hypermetrope}})$
	$= \frac{3}{6} \times \left(-\frac{1}{3}\log_2\frac{1}{3} - \frac{2}{3}\log_2\frac{2}{3} - \frac{0}{3}\log_2\frac{0}{3}\right) + \frac{3}{6} \times \left(-\frac{0}{3}\log_2\frac{0}{3} - \frac{3}{3}\log_2\frac{3}{3} - \frac{0}{3}\log_2\frac{0}{3}\right) = 0.4591$

Source: Reprinted with permission from Table 4.9 in Ye, N. 2013. *Data Mining: Theories, Algorithms, and Examples*, p. 55, Boca Raton, Florida: CRC Press.

TABLE 23.11

Non-Binary Split of an Internal Node, {4, 8, 12, 16, 20, 24}, and Calculation of Information Entropy for the Lenses Data Set

Split Criterion	Resulting Subsets and Average Information Entropy of the Split
Age = Young, Pre-Presbyopic, or Presbyopic	{4, 8}, {12, 16}, (20, 24} $\text{entropy}(S) = \frac{2}{6}\text{entropy}(D_{\text{Young}}) + \frac{2}{6}\text{entropy}(D_{\text{Pre-presbyopic}}) + \frac{2}{6}\text{entropy}(D_{\text{Presbyopic}})$ $= \frac{2}{6}\times\left(-\frac{0}{2}\log_2\frac{0}{2} - \frac{0}{2}\log_2\frac{0}{2} - \frac{2}{2}\log_2\frac{2}{2}\right) + \frac{2}{6}\times\left(-\frac{1}{2}\log_2\frac{1}{2} - \frac{0}{2}\log_2\frac{0}{2} - \frac{1}{2}\log_2\frac{1}{2}\right) + \frac{2}{6}\times\left(-\frac{1}{2}\log_2\frac{1}{2} - \frac{0}{2}\log_2\frac{0}{2} - \frac{1}{2}\log_2\frac{1}{2}\right) = 0.6667$
Spectacle Prescription = Myope or Hypermetrope	{4, 12, 20}, {8, 16, 24} $\text{entropy}(S) = \frac{3}{6}\text{entropy}(D_{\text{Myope}}) + \frac{3}{6}\text{entropy}(D_{\text{Hypermetrope}})$ $= \frac{3}{6}\times\left(-\frac{0}{3}\log_2\frac{0}{3} - \frac{0}{3}\log_2\frac{0}{3} - \frac{3}{3}\log_2\frac{3}{3}\right) + \frac{3}{6}\times\left(-\frac{2}{3}\log_2\frac{2}{3} - \frac{0}{3}\log_2\frac{0}{3} - \frac{1}{3}\log_2\frac{1}{3}\right) = 0.4591$

Source: Reprinted with permission from Table 4.10 in Ye, N. 2013. *Data Mining: Theories, Algorithms, and Examples*, p. 56, Boca Raton, Florida: CRC Press.

Age = Young, Pre-Presbyopic, or Presbyopic, which produces three subsets of {2, 6}, {10, 14}, and {18, 22}. These subsets are further partitioned using the split criterion, Spectacle Prescription = Myope or Hypermetrope, to produce leaf nodes with homogeneous data sets. Table 23.11 shows the calculation of information entropy to split the node with the data set of {4, 8, 12, 16, 20, 24} using the split criterion, Spectacle Prescription = Myope or Hypermetrope, which produces two subsets of {4, 12, 20} and {8, 16, 24}. These subsets are further partitioned using the split criterion, Age = Young, Pre-Presbyopic, or Presbyopic, to produce leaf nodes with homogeneous data sets. Figure 23.6 shows the complete non-binary decision tree for the lenses data set.

23.3 Handling Numeric and Missing Values of Attribute Variables

If a data set has a numeric attribute variable, the variable needs to be transformed into a categorical variable before being used to construct a decision tree. We present a common method to perform the transformation. Suppose that a numeric attribute variable, x, has the following numeric values in the training data set: $a_1, a_2, \ldots, a_k$, which are sorted in an increasing order of values. The middle point of two adjacent numeric values, a_i and a_j, is computed as follows:

$$c_i = \frac{a_i + a_j}{2} \tag{23.6}$$

Using c_i for $i = 1, \ldots, k-1$, we can create the following $k+1$ categorical values of x:

Category 1: $x \le c_1$

Category 2: $c_1 < x \le c_2$

$\vdots$

Category k: $c_{k-1} < x \le c_k$

Category $k+1$: $c_k < x$

A numeric value of x is transformed into a categorical value according to the above definition of the categorical values. For example, if $c_1 < x \le c_2$, the categorical value of x is Category 2.

In many real-world data sets, we may find an attribute variable that does not have a value in a data record. For example, if there are attribute variables of name, address, and e-mail address for customers in a database for a store, we may not have the e-mail address for a particular customer. That is, we may have missing e-mail addresses for some customers. One way to treat a data record with a missing value is to discard the data record. However, when the training data set is small, we may need all the data records in the training data set to construct a decision tree. To use a data record with a missing value, we may estimate the missing value and use the estimated value to fill in the missing value. For a categorical attribute variable, its missing value can be estimated to be the value that is taken by the majority of data records in the training data set that have the same target value as that of the data record with a missing value of the attribute variable. For a numeric attribute variable, its missing value can be estimated to be the average of values that are taken by data records in the training data set that have the same target value as that of the data record with a missing value of the attribute variable. Other methods of estimating a missing value are given in Ye (2003).

23.4 Advantages and Shortcomings of the Decision Tree Algorithm

An advantage of using the decision tree algorithm to learn classification patterns is the explicit expression of classification patterns in the decision tree. The decision tree in Figure 23.2 uncovers the following three patterns of part quality leading to three leaf nodes with the classification of system fault:

- $x_7 = 1$
- $x_7 = 0$ and $x_8 = 1$
- $x_7 = 0$ and $x_8 = 0$ and $x_9 = 1$

and the following pattern of part quality to one leaf node with the classification of no system fault:

- $x_7 = 0$ and $x_8 = 0$ and $x_9 = 0$

The above four data patterns can be expressed in the form of decision rules as follows:

- If $x_7 = 1$ Then $y = 1$
- If $x_7 = 0$ and $x_8 = 1$ Then $y = 1$
- If $x_7 = 0$ and $x_8 = 0$ and $x_9 = 1$ Then $y = 1$
- If $x_7 = 0$ and $x_8 = 0$ and $x_9 = 0$ Then $y = 0$

The above explicit classification patterns and rules reveal the following key knowledge for detecting the fault of this manufacturing system:

- Among the nine quality variables, only three quality variables, x_7, x_8, and x_9, matter for system fault detection. This knowledge allows us to reduce the cost of part quality inspection by inspecting the part quality after M7, M8, and M9 only rather than all the nine machines.
- If one of these three variables, x_7, x_8, and x_9, shows quality failure, the system has a fault; otherwise, the system has no fault.

A decision tree has its shortcoming in expressing classification patterns because it uses only one attribute variable in a split criterion. This may result in a large decision tree. From a large decision tree, it is difficult to see clear patterns for classification. However, considering all the attribute variables and their combinations of conditions for each split would correspond to an exhaustive search of all combination values of all the attribute variables. This is computationally costly or, sometimes, impossible for a large data set with a large number of attribute variables.

23.5 Software

The website, http://www.knuggets.com, has information about various data mining tools. The following software packages support the learning of decision trees:

- Weka (http://www.cs.waikato.ac.nz/ml/weka/)
- SPSS AnswerTree (http://www.spss.com/answertree/)
- SAS Enterprise Miner (http://www.sas.com/products/miner/)
- IBM Intelligent Miner (http://www.ibm.com/software/data/iminer/)
- CART (http://www.salford-systems.com/)
- C4.5 (http://www.cse.unsw.edu.au/quinlan/)

23.6 Summary

This chapter describes one of the data mining algorithms, decision trees, to illustrate turning data into useful information and knowledge, which can then be used to support decision making. The structure of a decision tree is defined. The algorithm of constructing a binary decision tree is provided along with split selection methods based on information entropy and gini-index. The algorithm of constructing a binary decision tree is extended to construct a non-binary decision tree. The use of a decision tree to perform classification and extract useful information and knowledge is illustrated. Ye (2013) describes many other data mining algorithms for discovering useful information and knowledge from data.

EXERCISES

1. Construct a binary decision tree for the lenses data set in Table 23.7 using the information entropy as the measure of data homogeneity.
2. Construct a binary decision tree for the data set in Table 23.6 along with data of Instance 10 in Table 23.1 using the information entropy as the measure of the data homogeneity.

24

Expert Systems

Expert systems are software systems that imitate the decision-making ability of human experts. It is observed that a main distinction of experts and novices in a specialty field is experts' possession of vast amounts of heuristic knowledge acquired and accumulated over many years of experience in the specialty field. For example, a car driver may not know why the engine of a car does not start, whereas an auto repair technician can readily check a few things (e.g., battery), figure out the cause, and fix the problem because the technician has seen and fixed all causes of the same problem over many years and has put years of experience into heuristic knowledge such as rules of thumb. Hence, two main components of an expert system are the knowledge base and the inference engine, which performs knowledge-based reasoning to make decisions. During knowledge-based reasoning, the expert system uses a working memory to keep given or inferred facts. Knowledge in the knowledge base can be directly acquired from human experts or extracted through mining data, as described in Chapter 23. Expert systems can play an important role in developing decision support systems, which store and transform data into information and knowledge and then utilize knowledge to support decision making. In this chapter, we first describe the structure of the knowledge base in an expert system. Then, we present the inference engine and the working memory of an expert system and explain how they work to conduct knowledge-based reasoning and make a decision.

24.1 Structure of the Knowledge Base

The knowledge base of an expert system consists of rules in the following form:

IF condition THEN consequence

Hence, the knowledge base is also called the "rule base." Table 24.1 gives a set of rules that describes consumers' liking of nine products: P1, P2, …, P9. Take an example of Rule 1: IF P1 = Yes THEN P5 = Yes. This rule indicates that, if consumers like P1, then consumers also like P5. The rule base in Table 24.1 can be used to make the product recommendation for consumers. For example, if a consumer visits an e-commerce website and adds P1 to the shopping cart, the expert system can search the rule base to find associated products such as P5 and recommend P5 to the consumer.

The set of rules in Table 24.1 can be obtained through mining consumers' purchasing data to discover what products are purchased together by many consumers. The mining of such frequently associated products and the discovery of association rules can be carried out using the data mining method called "association rules" (see Chapter 12 in Ye, 2013).

TABLE 24.1

Rule Base that Describes Consumers' Liking of Products

Rule Number	Rule
1	IF P1 = Yes THEN P5 = Yes
2	IF P2 = Yes OR P3 = Yes THEN P4 = Yes
3	IF P3 = Yes THEN P6 = Yes
4	IF P4 = Yes THEN P8 = Yes
5	IF P5 = Yes OR P6 = Yes THEN P7 = Yes
6	IF P5 = Yes THEN P9 = Yes

Rule bases can be developed to capture knowledge in many fields. For example, in the rule base for selecting an appropriate car for a family based on the family size, budget, and workload, we may have rules such as:

IF family_size = large AND budget = moderate AND workload = high THEN car = minivan

In the rule base for identifying an animal, we may have rules such as:

IF *x* is a carnivore AND *x* has tawny color AND *x* has dark spots THEN *x* is a cheetah.

24.2 Forward Chaining and Backward Chaining by the Inference Engine

The inference engine of an expert system applies knowledge in the rule base to facts in the working memory and makes inferences for the goal of making a decision. There are two inference mechanisms: forward chaining and backward chaining.

Forward chaining, also called "data-driven reasoning," starts with given fact(s) kept in the working memory of the expert system and carries out the following steps:

1. Search in the rule base for a rule whose condition is satisfied by the fact(s) in the working memory.
2. Fire the rule from Step 1 by executing its consequence, which results in an addition of new facts to the working memory and, thus, an update of the working memory.
3. Repeat the cycle consisting of Steps 1 and 2 until the fact(s) concerning the goal is obtained or there are no new rules to fire.

Example 24.1 illustrates forward chaining.

Example 24.1

Table 24.1 gives the rule base of the expert system for making product recommendations to consumers. Use forward chaining to infer what other products a consumer likes when the consumer adds P1 to the shopping cart.

Table 24.2 shows the process of forward chaining, which starts with P1 = Yes in the working memory. In Cycle 1, the condition of Rule 1 is satisfied by the fact, P1 = Yes, in

TABLE 24.2

Illustration of Forward Chaining

Cycle	Rule to Fire	Working Memory
0		P1 = Yes
1	Rule 1	P1 = Yes P5 = Yes
2	Rule 5	P1 = Yes P5 = Yes P7 = Yes
3	Rule 6	P1 = Yes P5 = Yes P7 = Yes P9 = Yes

the working memory. Rule 1 is fired by executing the consequence to add P5 = Yes to the working memory. In Cycle 2, both the conditions of Rule 5 and Rule 6 are satisfied by the fact, P5 = Yes, in the working memory. When multiple rules can be fired, a conflict resolution method should be used to select one of the multiple rules to fire. Section 24.3 presents several conflict resolution methods. In this example, we simply select Rule 5 to fire first. Rule 5 is fired by executing the consequence to add P7 = Yes to the working memory. In Cycle 3, the condition of Rule 6 is satisfied by the fact, P5 = Yes, in the working memory. Rule 6 is fired by executing the consequence to add P9 = Yes to the working memory. After Cycle 3, no new rules are satisfied by the facts in the working memory to add new facts to the working memory. The process of forward chaining ends. All the facts in the working memory after the process of forward chaining show that the consumer also likes P5, P7, and P9, with P1 in the shopping cart.

Backward chaining, also called "goal-directed reasoning," starts with a given goal and searches for the rule base and the given facts in the working memory to verify if the goal can be satisfied by given facts. Backward chaining starts with a given goal at the top of the goal stack and carries out the following steps:

1. Examine the goal stack. If the goal stack is empty, stop with the initial goal satisfied; otherwise, go to Step 2.
2. Search in the rule base for a rule whose consequence gives the goal.
   a. If such a rule is found:
      i. If the condition of the rule is satisfied by fact(s) in the working memory, fire the rule by executing the consequence of the rule, which results in an update of the working memory, remove the goal from the goal stack, and go to Step 1 for another cycle.
      ii. Otherwise, fire the rule by placing the condition of the rule as the goal at the top of the goal stack, and go to Step 1 for another cycle.
   b. Otherwise, stop with the initial goal not satisfied.

Examples 24.2 and 24.3 illustrate the process of backward chaining.

Example 24.2

Use the expert system whose rule base is given in Table 24.1 to verify whether a consumer likes P7 when the consumer adds P3 to the shopping cart.

TABLE 24.3

Illustration of Backward Chaining to Verify the Goal P7 = Yes

Cycle	Rule to Fire	Goal Stack	Working Memory
0		P7 = Yes	P3 = Yes
1	Rule 5	P5 = Yes OR P6 = Yes P7 = Yes	P3 = Yes
2	Rule 3	P7 = Yes	P3 = Yes P6 = Yes
3	Rule 5		P3 = Yes P6 = Yes P7 = Yes

Table 24.3 shows the process of backward chaining, which starts with P3 = Yes in the working memory and the goal P7 = Yes at the top of the goal stack. In Cycle 1, the goal stack is not empty. Rule 5 has the consequence that gives the goal at the top of the goal stack, P7 = Yes. The condition of Rule 5 is not satisfied by the fact, P3 = Yes, in the working memory. Hence, the condition of Rule 5, P5 = Yes OR P6 = Yes, is pushed into the goal stack. In Cycle 2, Rule 1 and Rule 3 give the goal at the top of the goal stack, P5 = Yes OR P6 = Yes. However, only the condition of Rule 3 is satisfied by the fact, P3 = Yes, in the working memory. Hence, Rule 3 is fired by executing the consequence to add P6 = Yes to the working memory. The goal, P5 = Yes OR P6 = Yes, is removed from the goal stack. In Cycle 3, Rule 5 has the consequence that gives the goal at the top of the goal stack, P7 = Yes, and the condition of Rule 5 is satisfied by P6 = Yes in the working memory. Hence, Rule 5 is fired by executing the consequence to add P7 = Yes to the working memory, and the goal, P7 = Yes, is removed from the goal stack. In Cycle 4, the goal stack is empty, which ends the process of backward chaining with the goal satisfied.

Example 24.3

Use the expert system whose rule base is given in Table 24.1 to verify whether a consumer likes P9 when the consumer adds P3 to the shopping cart.

Table 24.4 shows the process of backward chaining, which starts with P3 = Yes in the working memory and the goal P9 = Yes at the top of the goal stack. In Cycle 1, the goal stack is not empty. Rule 6 has the consequence that gives the goal at the top of the goal stack, P9 = Yes. The condition of Rule 6 is not satisfied by the fact, P3 = Yes, in the working memory. Hence, the condition of Rule 6, P5 = Yes, is pushed to the goal stack. In Cycle 2, Rule 1 has the consequence that gives the goal at the top of the goal stack, P5 = Yes. The condition of Rule 1 is not satisfied by the fact, P3 = Yes, in the working memory.

TABLE 24.4

Illustration of Backward Chaining to Verify the Goal P9 = Yes

Cycle	Rule to Fire	Goal Stack	Working Memory
0		P9 = Yes	P3 = Yes
1	Rule 6	P5 = Yes P9 = Yes	P3 = Yes
2	Rule 1	P1 = Yes P5 = Yes P9 = Yes	P3 = Yes
3	No rule		

Hence, the condition of Rule 6, P1 = Yes, is pushed to the goal stack. In Cycle 3, no rule has the consequence to give the goal at the top of the goal stack, P1 = Yes. Hence, the process of backward chaining ends with the initial goal, P9 = Yes, not satisfied.

Forward chaining and backward chaining differ in many aspects. Figure 24.1a represents the conditions and the consequences of rules in Table 24.1 in a tree. Figure 24.1b and 24.1c illustrate the differences between backward chaining and forward chaining, both of which start with the same set of given facts—P1, P2, and P3. Backward chaining in Figure 24.1b verifies the goal, P9 = Yes, by performing a top-down, depth-first search for only rules and facts that are related to the goal. Forward chaining in Figure 24.1c infers all facts from given facts by performing a bottom-up, breadth-first search of all facts supported by given facts.

Backward chaining is typically used when we are interested in verifying a particular goal but do not know what facts to collect. Backward chaining guides the collection of facts relating to a goal. Moreover, if the knowledge tree of the rule base (e.g., the one shown in Figure 24.1) has a fan-out structure with more nodes at the top of the knowledge tree and fewer nodes at the bottom of the knowledge tree, a set of given facts leads to a large number of inferred facts while there are few ways to reach a goal. Hence, backward chaining is more efficient than forward chaining for a rule base that has a fan-out structure of the knowledge tree.

Forward chaining is typically used when we have many given facts but do not know what conclusions to draw. Moreover, if the knowledge tree of the rule base has a fan-in structure with fewer nodes at the top of the knowledge tree and more nodes at the bottom of the knowledge tree, a set of given facts leads to a small number of inferred facts while there are many ways to reach a goal. Hence, forward chaining is more efficient than backward chaining for a rule base that has a fan-in structure of the knowledge tree.

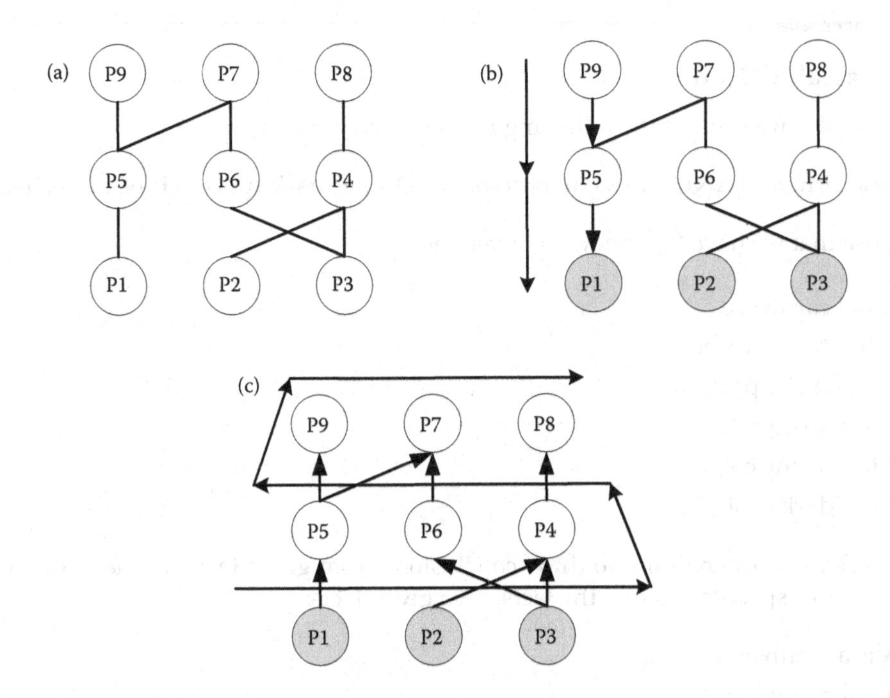

FIGURE 24.1
Differences between forward chaining and backward chaining. (a) The tree of conditions and consequences in the rule base for product recommendations. (b) Top-down, depth-first search in backward chaining. (c) Bottom-up, breadth-first search in forward chaining.

24.3 Conflict Resolution

In Example 24.1, there are multiple rules, Rule 5 and Rule 6, whose condition is satisfied. Both rules can be fired in Cycle 2. When there are multiple rules to fire, we need a conflict resolution method to select one of the multiple rules to fire. A list of commonly used conflict resolution methods is given as follows:

- *Rule priority ordering*: If all rules are arranged in a prioritized list, we fire the rule that has the highest priority.
- *Fact priority ordering*: If all facts are arranged in a prioritized list, we fire the rule that matches the fact with the higher priority.
- *Specific ordering*: If the condition of one rule is a superset of the condition of another rule, we fire the rule with the superset because the rule has the more specific condition.
- *Size ordering*: We fire the rule with the longest condition because such a condition is more specific and more difficult to match.

In Example 24.1, both Rule 5 and Rule 6 can be fired in Cycle 2. We fire Rule 5 if we use the size ordering method or the specific ordering method of conflict resolution because the condition of Rule 5 is longer and more specific than the condition of Rule 6.

24.4 Variable Binding

In Section 24.1, we present the following rules that contain a variable x:

IF x is a carnivore AND x has tawny color AND x has dark spots THEN x is a cheetah.

Suppose that we have the following given facts:

- A is a carnivore.
- A has tawny color.
- A has dark spots.
- B is a carnivore.
- B has tawny color.
- B has dark spots.

If we use forward chaining to draw conclusions from given facts, we need to bind the variable, x, to a specific value in the facts. Two given facts:

- A is a carnivore.
- B is a carnivore.

match the first condition of the rule, x is a carnivore. This creates two variable bindings: x = A and x = B. Each variable binding is used to match two other conditions of the rule.

Both variable bindings satisfy the condition of the rule. The rule is fired twice for both variable bindings to add two new facts to the working memory:

- A is a cheetah.
- B is a cheetah.

Hence, if there are multiple variable bindings and each variable binding satisfies the condition of a rule, the rule is fired multiple times, each time for one variable binding during forward chaining or backward chaining.

24.5 Certainty Factor

In an expert system called "MYCIN," which is developed to identify bacteria causing severe infections (Buchanan and Shortliffe, 1984; Heckerman and Shortliffe, 1992), the certainty factor (CF) is used to express the belief in a rule. The belief in a rule measures how well the evidence (E) in the condition of the rule supports the hypothesis (H) in the consequence of the rule, and is a belief that the rule holds true. A CF of a rule takes a value in (0, 1). For example, we may assign the CF of 0.8 to Rule 1 in Table 24.1:

IF P1 = Yes THEN P5 = Yes (CF = 0.8),

based on the percentage of consumers who purchase both P1 and P5 from consumer purchasing data. MYCIN uses the following formulas to make inference with CF:

$$\text{IF E THEN H: } CF(H) = CF(H \mid E) \times CF(E) \tag{24.1}$$

$$\text{IF E1 AND E2 THEN H: } CF(H) = CF(H \mid \text{E1 AND E2}) \times \min\{CF(E1), CF(E2)\} \tag{24.2}$$

$$\text{IF E1 OR E2 THEN H: } CF(H) = CF(H \mid \text{E1 OR E2}) \times \max\{CF(E1), CF(E2)\} \tag{24.3}$$

$$\text{IF NOT E THEN H: } CF(H) = CF(H \mid \text{NOT E}) \times \{-CF(E)\} \tag{24.4}$$

If two rules conclude the same hypothesis but with different CFs, the following formula is used to determine the combined CF for the hypothesis:

$$CFcombine(CF1, CF2) = \begin{cases} CF1 + CF2 \times (1 - CF1) & \text{if both CF1 and CF2} >= 0 \\ (CF1 + CF2)/(1 - \min\{|CF1|, |CF2|\}) & \text{if either CF1 or CF2} < 0 \\ CF1 + CF2 \times (1 + CF1) & \text{if both CF1 and CF2} < 0 \end{cases} \tag{24.5}$$

To perform forward chaining and backward chaining with CF, we first carry out each step of forward chaining and backward chaining and then compute the CF for any new fact added to the working memory. Example 24.4 illustrates forward chaining with CF.

TABLE 24.5
Illustration of Forward Chaining with CF

Cycle	Rule to Fire	Working Memory	CF
0		P1 = Yes	CF(P1) = 1
1	Rule 1	P1 = Yes	CF(P1) = 1
		P5 = Yes	CF(P5) = CF(Rule 1) × CF(P1) = 0.8 × 1 = 0.8
2	Rule 5	P1 = Yes	CF(P1) = 1
		P5 = Yes	CF(P5) = 0.8
		P7 = Yes	CF(P7) = CF(Rule 5) × max{CF(P5), CF(P6)} = 0.9 × max{0.8, 0.11} = 0.9 × 0.8 = 0.72
3	Rule 6	P1 = Yes	CF(P1) = 1
		P5 = Yes	CF(P5) = 0.8
		P7 = Yes	CF(P7) = 0.72
		P9 = Yes	CF(P9) = CF(Rule 6) × CF(P5) = 0.7 × 0.8 = 0.56

Example 24.4

Perform the forward chaining in Example 24.1 with the following CFs:

CF(P1) = 1, CF(P2) = 0.11, CF(P3) = 0.11, …, CF(P9) = 0.11
CF(Rule 1) = 0.8
CF(Rule 5) = 0.9
CF(Rule 6) = 0.7

The CF of P1 is 1 because the consumer adds P1 to the shopping cart. In this example, we assign 0.11 or 1/9 as the CF of P2, …, P9 based on the equal likelihood of a consumer purchasing one out of nine products. Instead of assigning the same value of CF for P2, …, P9 based on the assumed equal likelihood of purchasing each product, the value of CF can also be assigned based on the ratio of purchases for each product to purchases of all the nine products.

Table 24.5 gives the steps of forward chaining and the computation of CF for each step.

24.6 Software and Advantages of Expert Systems to Algorithms of Computer Programs

The commonly known software for developing expert systems is CLIPS (http://www.clipsrules.sourceforge.net/). Although knowledge and knowledge-based reasoning can be coded into algorithms of computer programs, the structure of expert systems with the separation of the rule base, the inference engine, and the working memory has many advantages. Algorithms of computer programs mix data, knowledge, and inference altogether in a fixed order of examining data and rules. Forward chaining or backward chaining by the inference engine of an expert system does not follow a fixed order of examining data and rules and, thus, offers flexibility in knowledge-based reasoning. For experts in a specialty field, rules in the rule base are much easier to construct, update, and understand than algorithms of computer programs. The process of forward chaining or backward chaining along with the facts and the rules used can be presented to the user of an expert system to explain how the conclusion is drawn, making the expert system user-friendly. With the explanation capability of an expert system, the expert system can be used not only as a decision support system, but also as a training system for knowledge and skill training.

24.7 Summary

By describing expert systems in this chapter, we introduce one way of turning information systems to decision support systems by incorporating knowledge into information systems and utilizing knowledge to support decision making. We describe the structure of rules in the knowledge base of an expert system and explain how forward chaining and backward chaining are conducted to make inferences and decisions using rules and given facts.

EXERCISES

1. Use the expert system whose rule base is given in Table 24.1 to perform forward chaining for inferring what other products a consumer likes when the consumer adds P3 and P5 to the shopping cart. Use the rule priority ordering method of conflict resolution if needed. The priority of rules follows the same order of rules in Table 24.1, with rule 1 having the highest priority.
2. Use the expert system whose rule base is given in Table 24.1 to perform backward chaining for verifying whether a consumer likes P7 when the consumer adds P1 and P3 to the shopping cart.
3. Use the expert system whose rule base is given in Table 24.1 to perform backward chaining for verifying whether a consumer likes P6 when the consumer adds P2 to the shopping cart.
4. Perform the forward chaining in Exercise 1 with the following CFs:

   CF(P3) = 1, CF(P5) = 1, CF(P1) = 0.11, CF(P2) = 0.11, CF(P4) = 0.11, CF(P6) = 0.11, CF(P7) = 0.11, CF(P8) = 0.11, CF(P9) = 0.11

   CF(Rule 1) = 0.8

   CF(Rule 2) = 0.6

   CF(Rule 3) = 0.5

   CF(Rule 4) = 0.6

   CF(Rule 5) = 0.9

   CF(Rule 6) = 0.7.

25

Decision Support Systems

This chapter studies the importance and benefits of computer-based decision support systems. First, the concepts of decision and decision making, its components, and its different classifications are discussed. A typical decision-making process performed by a decision maker is reviewed. Next, the concepts of decision support systems are introduced, including its design architecture: the different sources of data, the development of a data warehouse, the various mathematical decision-making models, as well as the graphical user interfaces (GUIs) and dashboards used by the decision maker to make a decision. Finally, a common development methodology for designing and implementing decision support systems is discussed.

25.1 What Is a Decision?

A decision is a problem faced by a decision maker in which they must choose from a set of alternatives given a set of criteria. Every human being has been faced with making decisions the entire life, which could be as simple as choosing which restaurant to eat out at later or as complex as buying a house to start a family. In this section, we describe the components of a decision as well as the different types or classifications of decisions through Example 25.1.

Example 25.1: Problem Statement

Consider that you are interested in buying a brand-new car based on the total cost of ownership. You have narrowed your alternatives into three cars: a 2013 Toyota Prius, a 2012 Honda Civic, and a 2013 Audi A3. However, since you are a fresh engineering graduate, you are faced with a budget constraint for this investment, and hence, the initial outright down-payment cost should be minimized. Furthermore, you are aware of the increasing cost of gasoline, and thus, a fuel-efficient car is a must. Additionally, you are concerned about the reliability of the car since you do not know anything about fixing and repairing a car.

Using this car procurement example, we list down the components of a decision as follows.

25.1.1 Components of a Decision

25.1.1.1 Decision Problem

Every decision starts with the realization of a problem by the decision maker. The decision problem is just like any problem that needs to be solved by choosing the best solution. In Example 25.1, the problem is to choose the best car given the constraints or criteria at hand. A clear and concise definition of the problem is a must for us to effectively solve the problem. Furthermore, by defining the problem, we can clearly glean the alternatives and the decision criteria considered by the decision maker.

25.1.1.2 Decision Alternatives

In the car selection example, we have identified three cars up for purchase: a 2013 Toyota Prius, a 2012 Honda Civic, and a 2013 Audi A3. These are considered as alternatives, which represent the possible solutions or actions to the identified decision problem. The number of alternatives is decision dependent and could be as few as two options or as many as an infinite number of alternatives.

25.1.1.3 Decision Constraints

Constraints limit the number of feasible decision alternatives. In Example 25.1, we have narrowed the alternatives to three cars. However, we are aware that there exist a lot of alternative cars available in today's car market. The main constraint here is attributed to cost, and as such, expensive luxury cars are not considered. The existence of constraints substantially reduced the number of alternatives, which could be interpreted as good or bad. When reducing the number of alternatives, constraints help make the decision maker focus on the feasible few; however, better alternatives are excluded from the analysis due to resource constraints.

25.1.1.4 Decision Criteria

By definition, a decision criterion is an objective measurement of the performance of each alternative with respect to a given decision problem. In our car selection example, we have identified the total cost of ownership as the main decision criterion. Furthermore, this criterion can be broken down into components, specifically the down payment, fuel, and maintenance costs. These criteria are used to quantitatively measure each alternative to select the best rational action. However, decision criteria may not only come in a quantitative sense, but, more often than that, it may also come in a subjective sense when making a decision. If we go back to our car selection example, we could include car color or car style as additional subjective criteria for choosing the best car.

25.1.2 Types of Decisions

We typically classify decisions based on their nature or scope. By effectively classifying decisions, an appropriate decision support system could be implemented to solve decisions with the same characteristics efficiently and accurately. We first explain the types of decisions according to their nature then according to their scope.

25.1.2.1 Decisions According to Their Nature

25.1.2.1.1 Structured Decisions

A structured decision is a decision that is well defined in terms of the decision components and occurs routinely or frequently within a specific time frame. Analytical methodologies are developed to handle this type of decision and are best suited to be automated. Example 25.2 is presented to illustrate a typical structured decision.

Example 25.2

A staff nurse is scheduling room cleaners for the week. This schedule is based on the current census of hospital rooms within the staff nurse's area. Furthermore, the number of cleaners to be requested is primarily based on the number of estimated patient discharges within the week. The primary concern is that the staff nurse could not request too many cleaners as they are expensive. On the other hand, a few cleaners will not suffice since cleaning and disinfecting rooms takes time and significant delays would be experienced by the next admitted patient. Hence, a decision model that can balance the costs of waiting and the costs of scheduling cleaners is of great value to the hospital. This decision can be solved using mathematical programming methodologies.

25.1.2.1.2 Unstructured Decisions

An unstructured decision is usually one of a kind or infrequent, in which the components of the decision are uncertain, imprecise, or cannot be described explicitly. This type of decision cannot be solved using mechanical methodologies and, hence, is done on a case-to-case basis and is not easily reproducible. Example 25.3 is presented to illustrate a typical unstructured decision.

Example 25.3

Consider the case when a hospital is considering implementing a third-party enterprise resource planning (ERP) system to streamline its financial and controlling business processes. However, the selection of an ERP software and its corresponding vendor is not an easy task. Several economic, political, as well as socioeconomic considerations need to be considered when making this one-of-a-kind decision.

25.1.2.2 Decisions According to Their Scope

25.1.2.2.1 Operational Decisions

Operational decisions are decisions that refer to day-to-day low-impact activities that do not significantly affect the entire organization. These decisions are typically made by low-level managers within their department and are typically accomplished in a short-range horizon of weeks to, at most, a month. We present Example 25.4 as an example of a typical operational decision.

Example 25.4

A purchasing supervisor of a real-estate company needs to supply a purchase requisition of construction materials, specifically aggregates, since the current contracted supplier would not be able to supply the entire order for next week due to labor strikes. Hence, the selection of another supplier must be performed immediately to fulfill the gap in supply for next week.

25.1.2.2.2 Tactical Decisions

Tactical decisions are mid-range decisions that encompass a 1- to 12-month time period. These decisions affect an entire organization's department and are made by mid-level managers. Consider Example 25.5 as an illustration of a tactical decision.

Example 25.5

The purchasing department director notices that the current scheduling agreement supplier of aggregates does not perform according to the specifications of the purchasing contract. He decides to let go of the current supplier and chooses to select a new one.

Hence, a new supplier that can fulfill a 1-year scheduling agreement of aggregates must be evaluated and selected to support future real-estate development projects.

25.1.2.2.3 Strategic Decisions

Strategic decisions are long-term decisions (greater than a year) that affect multiple or all business departments. Due to their high-impact and high-stakes effect, these decisions are made only by the top management. We present an example of a strategic decision in Example 25.6.

Example 25.6

The vice president is considering outsourcing the procurement of non-essential or non-critical construction materials to a third-party purchasing company. This would help relieve the current purchasing staff of value adding time to focus on critical construction materials.

25.2 Decision-Making Process

After identifying the components as well as the different classifications of decisions, we then illustrate a typical decision-making process using the flowchart presented in Figure 25.1.

The first two steps presented in Figure 25.1 can be considered as the exclusion phase. Every decision starts with the identification of the problem situation. Furthermore, after a careful examination and analysis of the problem, potential alternatives can be revealed and new alternatives are crafted. Constraints are also identified and are then used to exclude potential alternatives.

The next two steps can be considered as the evaluation phase. The decision criteria are identified to objectively measure each feasible alternative. Next, there exist several methodologies that support the decision maker in making and choosing the best alternative. We will discuss more of this in Section 25.3.4.

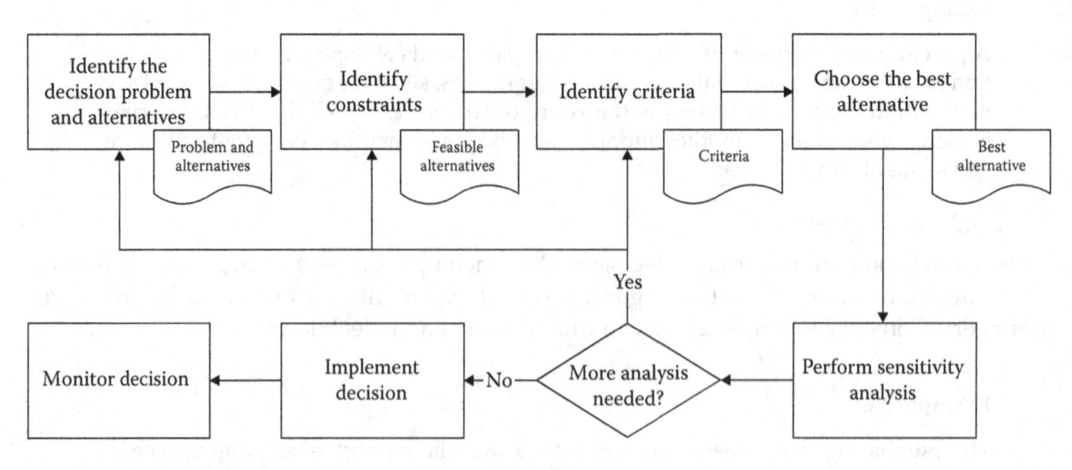

FIGURE 25.1
Decision-making process.

The last four steps are considered as the implementation and monitoring phase, in which a sensitivity analysis process is performed to determine the robustness of the optimal decision. By changing the scores of each alternative with respect to each criterion, the decision maker can determine the range of scores with which the best solution is still the best solution. If the decision maker is not satisfied, then a redefinition of the problem, alternatives, and/or criteria is necessitated. Otherwise, the decision maker can then implement the best solution. The last step is done to monitor the performance of the implemented best alternative over time to measure whether the best alternative performs according to the desired specifications.

25.3 Decision Support Systems

A decision support system is an integrated computer-based system. It helps a decision maker make a decision by providing useful insights based on processed data. It takes in data as input from multiple sources, utilizes mathematical models to crunch the data to transform it to useful information, and provides decision outputs using a GUI. We present a typical implementation of a decision support system in Figure 25.2.

We note that there is no one optimal design of a decision support system. Much of the design of the system can be considered as an art as well as a science and is implemented on a case-to-case basis. Hence, taking Figure 25.2 in mind, we discuss the following common components of a decision support system.

25.3.1 Data Sources

A typical decision support system is driven by data. Hence, the system usually takes in input data from multiple sources. This is due to the fact that, in a given business, it is unlikely that all of its entire data are stored in a single database; otherwise, it might collapse due to its weight and sheer dimension. Typical operational data sources include operational database systems that the company uses. These systems could be for point of sale, production management, inventory management, finance and HR, etc. Furthermore, external data sources such as the business environment, specifically competitors or

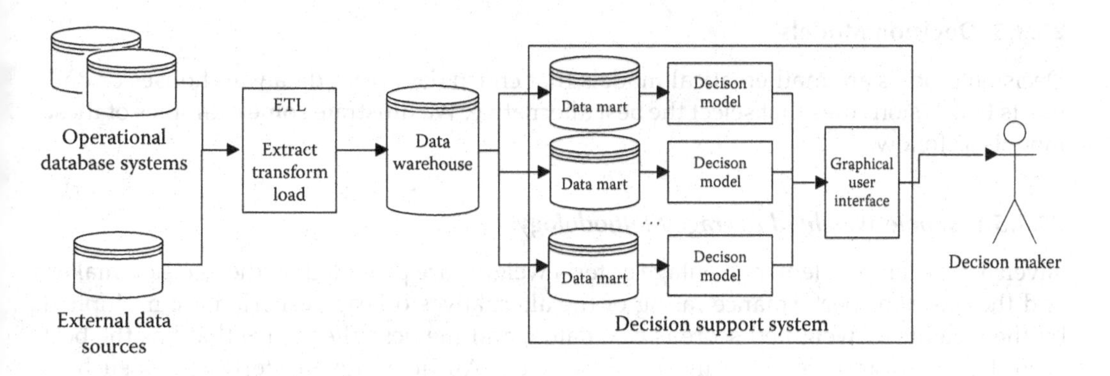

FIGURE 25.2
Typical decision support system.

business partners, or even government-acquired data that are out of the company's control, can also be considered as a valid source of data for building a decision support system.

25.3.2 Extraction, Transformation, and Loading

Data extraction, transformation, and loading (ETL) is a process that connects to all relevant data sources, queries the corresponding data from each, transforms the data to a specific format, and loads them into a data warehouse for storage. The connections are realized using standard database connectivity drivers while the queries are done using standard Structured Query Language (SQL) SELECT statements. Furthermore, the transformation process is included due to the fact that the different data stores may have different data formats such as "Y" and "N" to denote a Yes or a No answer, respectively, or "1" or "0." Additionally, the transformation process fixes missing data such as replacing them with default values or zeros. Lastly, the loading process loads the transformed data into the data warehouse using standard SQL INSERT statements.

25.3.3 Data Warehouse

A data warehouse is a data repository designed for quick data queries. One advantage of implementing data warehouses is that the processing cost of querying from multiple databases is greater than that of querying from a single data source. Hence, a data warehouse acts as an intermediary between various sources of data and decision models. A typical data warehouse has a "denormalized" data due to multiple SELECT statements collated from different normalized data stores. Data stored in the warehouse are designed to be easily accessible and are optimally designed for analytical processing or for further specialized storage in data marts.

25.3.4 Data Marts

A data mart is generally a smaller data warehouse that queries a subset of data from a data warehouse. The purpose of a data mart is to supply the data needed for a specific decision model for consumption of a decision maker. The inclusion of data marts within a decision support system is optional since decision models could connect directly to data warehouses. However, data marts make querying faster since specialized data sets are queried.

25.3.5 Decision Models

Decision models are mathematical models that churn data from data warehouses or data marts to decision rules that select the best alternative. We illustrate some examples of these models as follows.

25.3.5.1 Simple Weighted Average Methodology

Given a decision problem, percentage criteria weights are provided by the decision maker, and the scores or performance rating of the alternatives on each criterion are multiplied by the weights. A weighted score is calculated, and the best alternative that has the best score based on an objective function is selected. All qualitative criteria scores such as "Favorable" or "Good" are converted to a single quantitative scale. Example 25.7 illustrates a simple weighted average decision.

TABLE 25.1

Data for the Car Selection Example

Alternative	Criteria (weight)			
	Down Payment (30%)	**Annual Gasoline Cost (45%)**	**Maintenance and Reliability Cost (25%)**	**Weighted Average**
2013 Toyota Prius	$12,000	$1300	$500	$4310
2013 Honda Civic	$9000	$1700	$800	$3665
2013 Audi A3	$14,000	$1700	$700	$5140

Example 25.7

Suppose that data are collected for our car selection problem in Example 25.1, as summarized in Table 25.1. We further assume that you consider gasoline cost as the most important, and hence, you gave it a weight of 45%, followed by the initial down payment with a weight of 30%, and lastly, maintenance and reliability cost with a weight of 25%.

Based on Table 25.1, it is evident that the 2013 Honda Civic has the least weighted cost of ownership. Hence, it is then selected as the best alternative.

25.3.5.2 Compromise Programming

In this methodology, alternatives are chosen based on the ideal alternative's score. Hence, any alternative that is closest or least different from the ideal alternative is chosen. We seek to minimize the weighted deviation of an alternative from the ideal by using the following equation:

$$\min_j D_j = \sqrt{\sum_{i=1}^{M} \left(\frac{w_i (T_i - r_{ij})}{T_i} \right)^2} \tag{25.1}$$

where w_i is the weight of criteria i, T_i is the ideal or the best rating for criterion i, and r_{ij} is the score of alternative j on criterion i. Consider the data in Table 25.1 and assume that the best rating score for each criterion is the lowest one in each column. We then illustrate a compromise programming example as follows.

Example 25.8

The computed distances of each alternative to the best criterion are calculated using the normalizing equation $\left(\frac{w_i (T_i - r_{ij})}{T_i} \right)^2$, and the weighted average is calculated using Equation 25.1. The normalized distances and the weighted average are tabulated in Table 25.2.

It is observed that the best alternative now is the 2013 Toyota Prius.

TABLE 25.2

Normalized Distances

Alternative	Criteria (weight)			
	Down Payment (30%)	**Annual Gasoline Cost (45%)**	**Maintenance and Reliability Cost (25%)**	**Weighted Average**
2013 Toyota Prius	0.06	0.00	0.00	0.019
2013 Honda Civic	0.00	0.06	0.14	0.060
2013 Audi A3	0.13	0.06	0.08	0.084

TABLE 25.3

Simple Encoding of Decision Rules

	Bit		
Position	**1**	**2**	**3**
Criterion	**Down Payment (d)**	**Annual Gas Cost (g)**	**Annual Maintenance Cost (m)**
Value	$= \begin{cases} 1, & d < \$10{,}000 \\ 0, & d \geq \$10{,}000 \end{cases}$	$= \begin{cases} 1, & g < \$1800 \\ 0, & g \geq \$1800 \end{cases}$	$= \begin{cases} 1, & m < 850 \\ 0, & m \geq 850 \end{cases}$

25.3.5.3 Genetic Algorithms

A genetic algorithm is a computational decision-making algorithm derived from natural evolution. It is a stochastic search algorithm that searches for the optimal solution from a population of initial solutions. These initial solutions are coded as "chromosomes," and the algorithm chooses the best chromosome parents to do a crossover or a mutation for a child chromosome to provide an improved solution pool. The algorithm repeats itself until an optimal solution is obtained or chromosomes do not provide a better solution. We illustrate a genetic algorithm for the decision-making example as follows.

Example 25.9

Consider the same data from Table 25.1. Suppose that you have a decision rule that you can only afford a $10,000 down payment, a $1800 annual gasoline budget, and any car with, at most, $850 annual maintenance cost. It is acceptable that we want to choose the best alternative using genetic algorithm. First, we encode the rules into a chromosome. Table 25.3 shows a sample encoding of the 3-bit decision rules.

Therefore, an alternative with chromosome 101 states a car with less than $10,000 down payment, greater than $1800 fuel cost, and less than $850 annual maintenance cost. When encoding our three car alternatives, we have 2013 Toyota Prius = 011, 2013 Honda Civic = 111, and 2013 Audi A3 = 011. We can then code our genetic algorithm to choose the best chromosome based on the given decision rules or mutate to find other car alternatives that could satisfy our decision rules.

25.3.5.4 Analytic Hierarchy Process

The analytic hierarchy process (AHP) addresses the subjective limitation of simple weighted average methodologies in estimating criteria or alternative weights. The core principle of AHP is that it uses pairwise comparisons tabulated into a reciprocal matrix. To populate the matrix, each pairwise comparison is a numeric answer between (1/9, 9) provided by the decision maker based on the question "How important is Element A as compared to Element B?" The weights of the elements are approximated using an eigenvector methodology. Example 25.10 illustrates the AHP methodology using a pairwise comparison matrix.

Example 25.10

Consider the same data from Table 25.1. Suppose that we would want to estimate the criteria weights when choosing a car. Suppose that the annual gasoline cost criterion is twice as important as the annual maintenance cost and 1.5 times more important than the initial down payment. Furthermore, suppose that the importance of the annual

TABLE 25.4

Pairwise Comparison Matrix for the Car Selection Criteria

	Down Payment	Annual Gas Cost	Annual Maintenance Cost
Down payment	1	2/3	1
Annual gas cost	3/2	1	2
Annual maintenance cost	1	1/2	1

TABLE 25.5

Normalized Pairwise Comparison Matrix for the Car Selection Criteria

	Down Payment	Annual Gas Cost	Annual Maintenance Cost	Row Average (%)
Down payment	0.29	0.31	0.25	28.11
Annual gas cost	0.43	0.46	0.50	46.34
Annual maintenance cost	0.29	0.23	0.25	25.55

maintenance cost is approximately equal to that of the initial down payment. We tabulate these rules in a 3 × 3 pairwise comparison matrix (Table 25.4).

Next, we normalize each column such that the sum of each column would be one. Hence, Table 25.5 shows us the normalized pairwise comparison matrix. Furthermore, in Table 25.5, we average the normalized values across rows, and hence, we obtain the row sum.

These row average priorities are the estimated priority or weights of the corresponding criteria to be used for decision making.

25.3.6 Interfaces

GUI is the decision support system component that directly interacts with end users. A GUI is usually desktop based, but recent developments led to browser and Web interfaces as well as mobile applications. A GUI is composed of several panels that summarize the results provided by various decision models either in spreadsheet or graphical format such as a dashboard. Figure 25.3 shows a typical executive dashboard example that shows us information to be consumed by a decision maker. Furthermore, Figure 25.4 presents a housekeeping scheduling model using Excel that determines the number of hospital cleaners to be requested per time interval from Example 25.2.

25.4 Implementing Decision Support Systems

The basic principle of implementing decision support systems is that these systems are designed with the end goal in mind. For example, suppose that we need to develop a decision support system to aid bank customers in determining how much they would have to pay monthly when they are going to apply for a car loan. This system would be implemented online and installed together with the bank's online banking system. Hence, we need to backtrack these data requirements from the start. Specifically, we need to determine where to access relevant data, preprocess them, and transform them to useful information for consumption of the bank's customers. We present Figure 25.5 as a high-level flowchart to implement decision support systems.

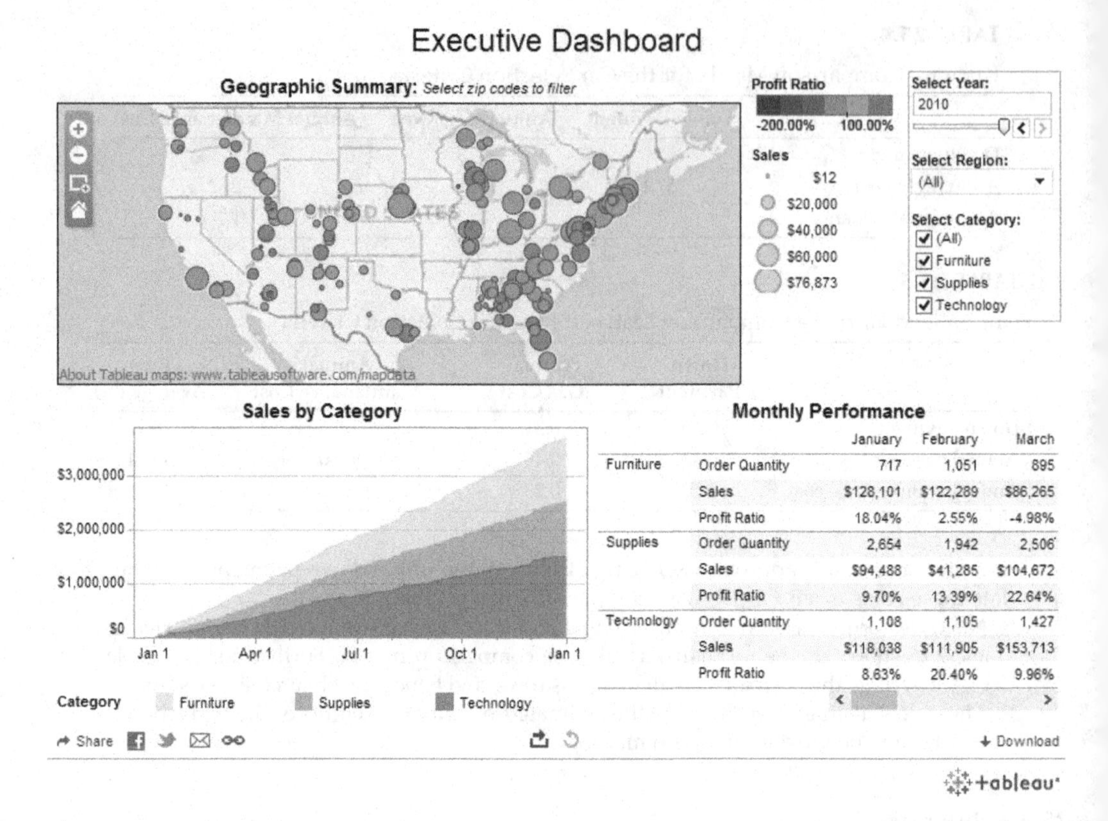

FIGURE 25.3
Dashboard GUI from Tableau. (From http://www.tableausoftware.com/.)

Housekeeping staffing model

Data table

Interval	# of Discharges	# of Cleaners
6–7 A.M.	0.23	1
7–8 A.M.	0.45	1
8–9 A.M.	0.55	1
9–10 A.M.	1.52	5
10–11 A.M.	3.06	5
11–12 N.N.	6.19	7
12–1 P.M.	7.13	11
1–2 P.M.	10.00	11
2–3 P.M.	8.00	11
3–4 P.M.	8.26	10
4–5 P.M.	7.68	10
5–6 P.M.	10.00	10
6–7 P.M.	5.71	6
7–8 P.M.	5.87	6
8–9 P.M.	3.42	4
9–10 P.M.	1.65	2
10–11 P.M.	1.19	2
11–12 M.N.	1.13	2
12–1 A.M.	0.77	2
1–2 A.M.	0.77	2
2–3 A.M.	0.97	2
3–4 A.M.	0.61	2
4–5 A.M.	0.32	2
5–6 A.M.	0.07	2

Graph

# of Cleaners
0 1 2 3 4 5 6 7 8 9 10 11 12
6–7 A.M. 7–8 A.M. 8–9 A.M. 9–10 A.M. 10–11 A.M. 11–12 N.N. 12–1 P.M. 1–2 P.M. 2–3 P.M. 3–4 P.M. 4–5 P.M. 5–6 P.M. 6–7 P.M. 7–8 P.M. 8–9 P.M. 9–10 P.M. 10–11 P.M. 11–12 M.N. 12–1 A.M. 1–2 A.M. 2–3 A.M. 3–4 A.M. 4–5 A.M. 5–6 A.M.
Time interval
Discharges Cleaners

FIGURE 25.4
Housekeeping staffing model dashboard GUI using Excel.

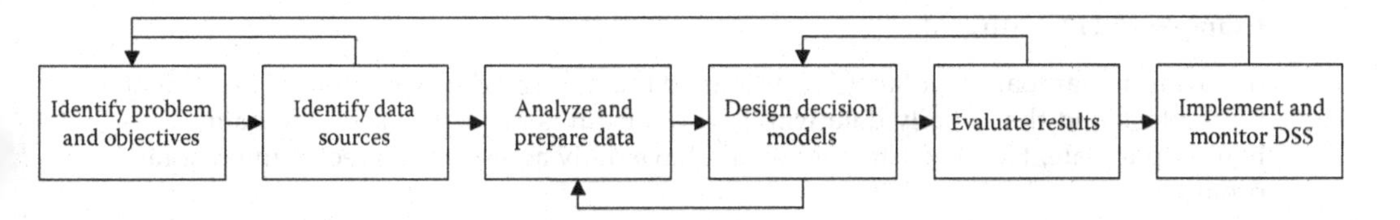

FIGURE 25.5
Flowchart for implementing decision support systems.

25.4.1 Identifying Problem and Objectives

The hardest part of implementing a decision support system is the identification of the problem and its objectives. Identifying the business scenario, the business case, the success criteria, and a potential project plan are some of the required outputs of this phase. Consider the car loan calculator application as an example for this phase. Substantial time and effort must be invested in this phase since the success of the next implementation phases hinges on proper problem definition.

Example 25.11

Based on this phase, we identified that the problem is to develop an accurate and useful decision support system that can help a customer decide on whether to apply for a car loan. Furthermore, we have identified potential success criteria such as accurate and useful.

25.4.2 Identifying Data Sources

After defining and understanding the problem, the search for sources of data that could support the decision-making process is performed. Sources of data are, but are not limited to, operational databases, user inputs, flat spreadsheet files, query reports, surveys, and even external data sources. We observe in Figure 25.5 that a feedback mechanism is in place to reevaluate our objectives if given the state of the feasible data sources.

Example 25.11 Continued

To help decide on the car loan, the decision support system must connect to the bank's chart of accounts database to access the client's credit history. This would affect the interest rate of the loan. Furthermore, the system may also utilize promotional packages from the bank's promotion database to complement the car loan application. Additionally, the current economic environment is also taken into consideration as well as industry standards in pricing the interest rate.

25.4.3 Data Analysis and Preparation

In this phase, an ETL system and the data warehouse are set up such that the data queried from the data sources are cleaned and preprocessed for the implementation of the decision model. Additionally, the frequency of data warehouse and data mart updates must also be agreed upon and configured.

Example 25.11 Continued

In the online car loan calculator, it was agreed that a single data warehouse is sufficient for the task and that a daily data warehouse refresh is needed to keep the data warehouse up to date. Hence, customers would have daily access on the recent interest rate promos.

25.4.4 Designing Decision Models

After designing the ETL and data warehouses, the appropriate decision model must be designed to transform the raw data queried from the data warehouses or data marts to useful information. This information can then be used to support the decision-making process.

Example 25.11 Continued

The model used here would just be a simple weighted average of the client's credit history, his or her current annual income, the number of years to pay, and the current industry and economic environment interest rates. A weighted score is obtained and a minimum passing score is used as the decisions rule whether a loan is offered to the customer.

25.4.5 Evaluation of Results

The performance of the decision models needs to be evaluated such that accurate and credible information can be presented to the end user. The results of the models can be compared to historical data to determine its effectiveness. If the results are not viable for operational go-live, then a reevaluation of the designed decision models is done by tweaking decision model parameters.

Example 25.11 Continued

Past loan applications with decisions are also evaluated in the model to determine its effectiveness. In a simple sense, past loan applications with an approved decision must also be approved when run using the decision support system. The same goes for rejected loan applications. The dashboard is evaluated for correct display of a possible decision on the car loan, monthly fees based on different payment terms, and various other graphics.

25.4.6 Implementing and Monitoring the Decision Support System

After an acceptable performance is obtained, the system is then implemented for go-live using dashboards and various other GUIs. Furthermore, the decision support system is also monitored for shifts in decisions, changes in policies, and changes in the decision environment that could affect the performance of the system.

Example 25.11 Continued

The car loan application system was then installed and implemented within 4 weeks of conceptualization. A simple dashboard that contains a spreadsheet of the computations and a bar graph containing different loan terms are provided. An example screenshot is shown in Figure 25.6.

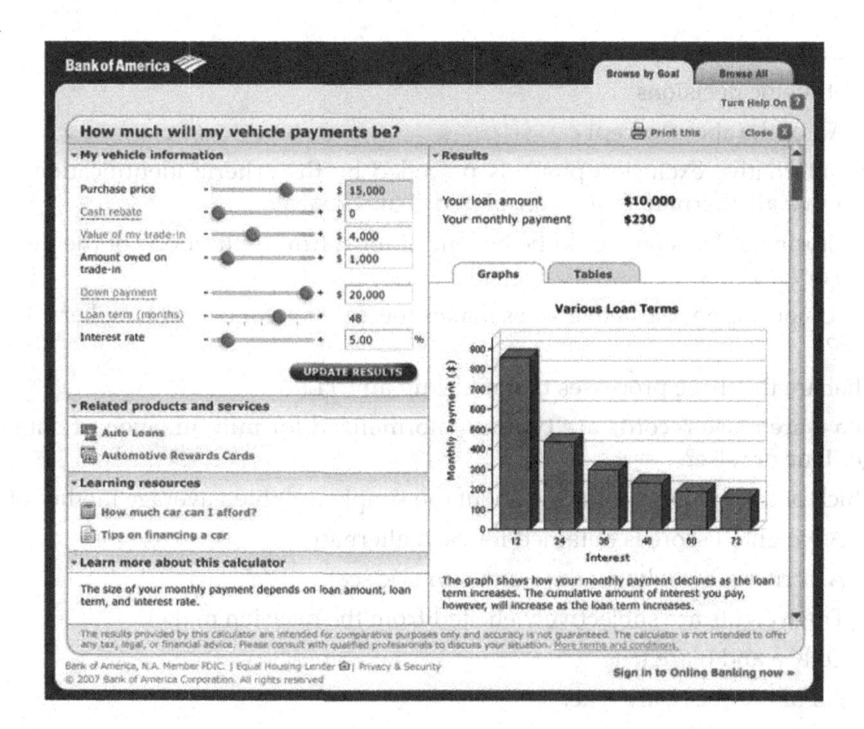

FIGURE 25.6
Online car loan application dashboard. (From http://www.behance.net/gallery/Bank-of-America-Dashboard/4451499.)

25.5 Summary

In this chapter, we have discussed some important applications of decision support systems. We have defined a decision and its main components, such as criteria and alternatives, and the decision-making process. Furthermore, we discussed the basic architecture of a decision support system as well as implementing a new decision support system.

REVIEW EXERCISES

1. A decision is a problem faced by a decision maker in which he or she must choose from a set of criteria given a set of alternatives. True or false?
2. What is the objective performance measurement dimension of an alternative?
   a. Criterion
   b. Problem
   c. Decision
   d. All of the above
   e. None of the above
3. Which of the following is a classification of a decision according to its nature?
   a. Tactical decisions
   b. Operational decisions

   c. Unstructured decisions
   d. Strategic decisions
   e. All of the above except c
4. The alternative exclusion phase is preceded by the criteria identification phase such that all alternatives are considered. True or false?
5. Monitoring a decision would be beneficial for future replications of the decision. True or false?
6. A decision support system always makes the decision for decision makers. True or false?
7. What are the three processes that make up an ETL?
8. Data warehouse records are typically normalized for minimization of data storage. True or false?
9. Which of the following are true about the simple weighted average methodology?
   a. A weighted score is obtained for each alternative.
   b. All criteria must be on the same numeric scale.
   c. The weights are subjectively elicited from the decision maker.
   d. Only a and b are true.
   e. All of the above are true.
10. Compromise programming minimizes the distance of the alternatives to the best possible score of each criterion. True or false?
11. Genetic algorithms start with an initial solution pool instead of just a single solution. True or false?
12. Which of the following is true about AHP?
   a. AHP is applicable to weight criteria and alternatives.
   b. AHP uses a symmetric matrix for pairwise comparisons.
   c. The basic question for a pairwise comparison is "How important is Element A as compared to Element B?"
   d. All of the above are true.
   e. Only a and c are true.
13. Dashboards are composed of multiple panels that can show spreadsheets, graphs, and other graphical tools for decision making. True or false?
14. The following are valid sources of data except:
   a. Spreadsheet reports
   b. Operational company databases
   c. External surveys
   d. Flat text files
   e. All of the above are valid sources of data.
15. The frequency of data warehouse update is done during the problem and objective definition phase. True or false?
16. Data for various decision models must always come from data marts. True or false?

17. The following statements are true about implementing a decision support system except:
    a. Model evaluation is done to determine the effectiveness of the decision model.
    b. Defining the problem is the hardest part.
    c. Monitoring the performance of the system after implementation is not needed.
    d. All of the above are true.
    e. None of the above is true.

PRACTICE EXERCISES

1. Given the following decisions, classify each decision according to its nature.
    a. Rye is faced on choosing the girl to marry and live happily ever after with. He has to choose between two alternatives: Jo, a teacher, or Den, an interior designer.
    b. A plant supervisor needs to decide on which supplier to choose for providing him with office supplies.
    c. Em and Asthor are discussing where to have a dinner date later.
    d. A teacher is mulling whether to fail or pass a student of information systems, which is offered every semester.
    e. Mark is considering pursuing a PhD in Materials Engineering.
2. Given the following decisions, classify them according to their scope.
    a. The dean of the college of engineering is deciding on whether to introduce a new course program on Systems Engineering.
    b. Kit, the accountant, is considering buying the company's gift certificates at a lower cost.
    c. Apple, in 2010, considered expanding to a different market segment, specifically tablets.
    d. A staff nurse needs to schedule nurses for the holiday week.
    e. A radiologist needs to evaluate proposals for a new computer tomography (CT) scanner for the entire department.

For problems 3 and 4, consider the following decision data (higher scores mean better performance).

	Criteria (weight)		
Faculty Candidate	**Educational Attainment (50%)**	**Teaching Experience (25%)**	**Research Experience (25%)**
Ian Espiritu	7	7	2
Emmylou Esolana	6	7	4
Irish Maravillas	6	8	3
Ryan Noleal	7	8	6
Joan Salise	5	10	6
Paul Yecyec	4	6	3

3. Given the aforementioned score sheet and weights, decide on who to hire as a new faculty hire for the Industrial Engineering Department.

4. Use compromise programming to calculate for the best applicant.
5. Given the following rules, encode each applicant into a 3-bit binary chromosome.

a. Educational attainment $= \begin{cases} 1, & \text{if } \geq 7 \\ 0, & \text{if } < 7 \end{cases}$

b. Teaching experience $= \begin{cases} 1, & \text{if } \geq 9 \\ 0, & \text{if } < 9 \end{cases}$

c. Research experience $= \begin{cases} 1, & \text{if } \geq 4 \\ 0, & \text{if } < 4 \end{cases}$

6. Given the following upper pairwise comparison matrix, compute for the priorities of the criteria using the AHP methodology.

	A	B	C	D	E	F
A	1	2	1	2	3	9
B		1	3	3	3	4
C			1	3	4	5
D				1	3	4
E					1	2
F						1

7. Petek is asking her three friends for advice on which career to pursue. She is choosing from three alternatives: teaching (T), supply chain analyst (C), or a Six Sigma black-belt consultant (S). Her three friends provided the following preferences:

Sid's Preferences

	T	C	S
T	1	1/2	3/4
C		1	3
S			1

Nick's Preferences

	T	C	S
T	1	1	1
C		1	1
S			1

Mickey's Preferences

	T	C	S
T	1	2	3
C		1	2
S			1

However, it is shown that her three friends have varying preferences. She then decides to weigh their preferences and decide based on the weighted priorities. Petek's preferences for her three friends' advice are as follows:

Petek's Preferences on the Preferences of Her Friends

	Sid	Nick	Mickey
Sid	1	2	1
Nick		1	1/3
Mickey			1

Which career should Petek pursue?

Section VI

Case Studies

Section VI of this book provides some case studies to demonstrate the concepts and techniques covered in Sections II to V. Specifically, Chapter 26 covers the use of Microsoft Access and Microsoft Visual Studio to develop a Windows application and a Web application for a healthcare information system. Chapters 27 and 28 show two real-world case studies developed in the healthcare field. In Chapter 27, a quality assurance system to evaluate the efficiency of the radiology practices rooted on a large enterprise database (Oracle) is illustrated in detail. The database system is further used in a clinical application to monitor peak skin dose to improve patient safety, which is discussed in Chapter 28.

26

Development of a Healthcare Information System Using Microsoft Access and Visual Studio

Healthcare is important to the well-being of our society and economy. Healthcare involves patients, diseases, healthcare providers, insurance companies, pharmacies, pharmaceutical and medical equipment manufacturers, government agencies, and so on. To support an efficient management of healthcare and decisions about disease control, the quality of patient care, and healthcare cost control, we want to build a healthcare information system that manages data related to the quality and the cost of healthcare. The healthcare information system includes data about diseases, patients, healthcare providers, insurance companies, and pharmacies. This chapter describes the use of Microsoft Access and Visual Studio to develop this healthcare information system. This chapter covers four phases of developing a database for this healthcare information system: enterprise modeling, conceptual data modeling, logical database design, and database implementation using Microsoft Access. After the database is developed, Microsoft Visual Studio is used to build the Windows-based healthcare information system and the Web-enabled healthcare information system containing the Access database.

26.1 Enterprise Modeling

The database for this healthcare information system must include the following data:

- Patients: PatientID; social security number; name; address; phone number; gender; date of birth; employer; employment position; emergency contact with name and phone number; dependent(s) including name and date of birth; insurance information with name, address, and phone number of the insurance company and the patient's member ID and GroupID; medical record of each visit containing date of visit, doctor, symptoms, disease, prescription of medicine(s), medical service(s) performed, and recovery time
- Healthcare providers: ProviderID, name, address, phone number, gender, specialty, affiliated hospital, medical education including degree and year, language(s), schedule including appointments with patient names and times of availability
- Insurance companies: CompanyID; name; address; phone number; insurance plans offered including name, ID, and description of each insurance plan; members enrolled in each insurance plan including each member's name, memberID, GroupID, member premium, employer premium, enrollment date, and dependent(s) (including name and date of birth); claims (including ClaimID, PatientID, medical services charged, patient co-pay amount and insurance coverage for each service, balance)

- Pharmacies: PharmacyID, name, address, phone number, medicine inventory (including MedicineID, medicine name, medicine description, unit price, current inventory level, and minimum inventory level), prescriptions (including PatientID, patient name, medicine, quantity, last pickup date, and refill frequency)

Each healthcare provider maintains (i.e., adds, deletes, and updates) data of the healthcare provider and patients. Each insurance company maintains data of its health insurance plans, members, and claims. Each pharmacy maintains data of its medicine inventory and prescriptions.

Employers, patients, healthcare providers, insurance companies, and pharmacies may use the database to perform the following data queries:

- Employers
  1. List the name of insurance companies and their insurance plans whose employer premium is smaller than or equal to a given premium level (EQ1).
  2. List the name of insurance companies and their insurance plans whose member premium is smaller than or equal to a given premium level (EQ2).
  3. Present a table showing the total number of healthcare providers in each insurance plan of each insurance company (EQ3).
- Patients
  1. List the name, address, phone number, gender, affiliated hospital(s), medical degree and year, and language(s) of healthcare providers in a given health insurance plan, a given city for a given specialty (PAQ1).
  2. List the time availability of a healthcare provider with a given name on a given date (PAQ2).
  3. Give the total amounts of patient co-payments paid for a given patient and the patient's dependent(s) in a given year (PAQ3).
  4. List the name, address, and phone number of all healthcare providers who specialize in heart diseases and whose address is in Phoenix, Arizona (PAQ4).
- Healthcare providers
  1. Give the name, address, and phone number of patients whose last annual physical exam was more than 1 year ago (PRQ1).
  2. Give disease(s) treated and medicine(s) prescribed for a given patient by a given healthcare provider within the last year (PRQ2).
  3. List all medicines that have been used to treat a given disease, sorted in the decreasing order of use frequency (PRQ3).
  4. Present appointment times and patient names for a given healthcare provider on a given date (PRQ4).
  5. Give the total amount of payments that a given healthcare provider received from patient payments and insurance payments in the current year (PRQ5).

- Insurance companies
    1. Give the total number of patients, female patients, and male patients treated by each healthcare provider, sorted in alphabetical order of the provider name (IQ1).
    2. Give the total number of patients who are 65 years or older and are treated by each healthcare provider of a given insurance company in the last year (IQ2).
    3. Give the total number of members enrolled in each health insurance plan of each insurance company (IQ3).
    4. Give the total amount paid by each health insurance plan of a given insurance company in the current year, sorted in alphabetical order of the plan name (IQ4).
    5. Present the average amount paid for treating a given disease by each insurance plan of each insurance company (IQ5).
    6. Give the average recovery time of a given disease (IQ6).
    7. Give the total amount of premiums received, the total amount of insurance payments, and the total amount of profit for a given insurance company in a given year (IQ7).
    8. Increase the member premium and the employer premium of a given insurance plan by 5% (IQ8).
    9. Delete a given insurance plan (IQ9).
- Pharmacies
    1. Give the total number of patients who are treated for a given disease in a given month and the total amount of a given medicine used to treat the disease in that month (PHQ1).
    2. Give the name and ID of all medicines whose inventory level falls below the minimum level at a given pharmacy (PHQ2).
    3. Give the medicine, refill amount, last pickup date, and refill frequency for a given prescription of a patient (PHQ3).

26.2 Conceptual Data Modeling

The Entity-Relationship (E-R) model of the database for the healthcare information system is given in Figure 26.1. The Patient entity and its attributes contain data of patients. A patient may be a dependent of another patient. Hence, there is a unary relationship, DependOn, to specify who depends on whom. The insurance information of a patient is not listed as attributes of Patient because the insurance information is related not only to the Patient entity, but also to the InsurancePlan entity. The insurance information of a patient is considered as attributes of the relationship Enroll between the Patient and the InsurancePlan entities. Information in a medical record of a patient is not listed as attributes of the Patient entity because information is related to multiple entities and to the Visit relationship that links these entities. In general, we list a data item as an attribute of an entity if the data item is related to only this entity. If a data item is

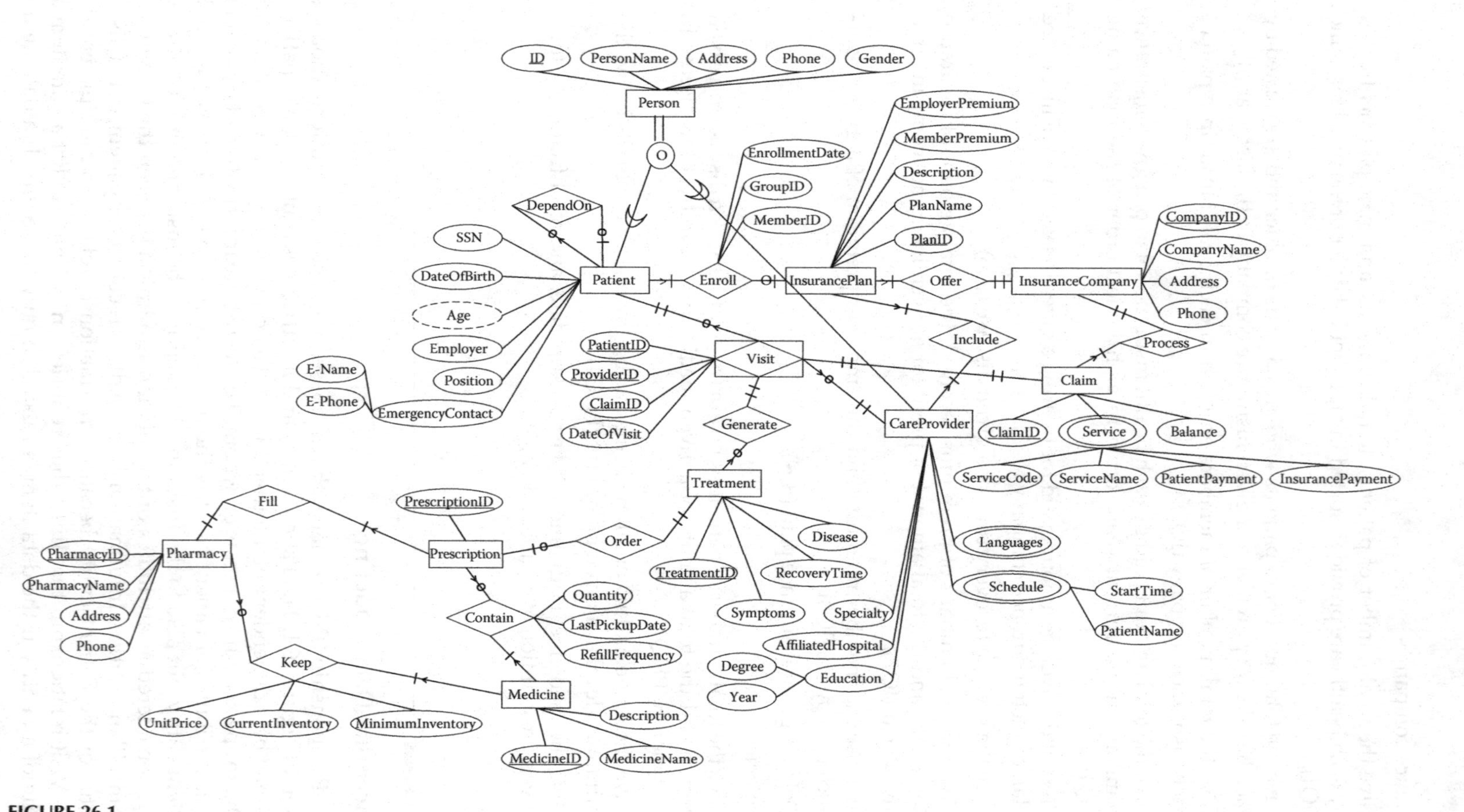

FIGURE 26.1
E-R model of the healthcare database.

related to more than one entity, we consider to include the data item as an attribute of a relationship.

The CareProvider entity and its attributes contain data of healthcare providers. The Schedule attribute of the CareProvider entity is a multi-valued attribute containing multiple time slots in a doctor's schedule. Each time slot is defined by StartTime. The length of a time slot is the time difference between two consecutive values of StartTime. The PatientName attribute has a patient name if the time slot has been assigned to a patient or has a NULL value if the time slot is available.

The InsuranceCompany entity and its attributes contain data of insurance companies. Insurance plans offered by an insurance company are placed in another entity, the InsurancePlan entity, to let the Patient entity relate to the InsurancePlan entity instead of the InsuranceCompany entity. The Patient entity is related to the InsurancePlan entity through the Enroll relationship, which keeps the insurance information of each patient. The InsurancePlan entity contains data of those insurance plans that have at least one member. Hence, an insurance plan, which is offered by an insurance company but has no enrolled member, is not contained in the InsurancePlan entity. The CareProvider entity is related to the InsurancePlan entity through the Include relationship, indicating which healthcare providers are included in which insurance plans.

Since a patient's visit to a healthcare provider results in a claim with one or more services charged, the Visit relationship links the Patient, the CareProvider, and the Claim entities. Although a patient's visit to a healthcare provider may lead to the treatment of a disease, some visits such as those for annual physical exams do not lead to a treatment. The Visit relationship is not linked to the Treatment entity because an instance of a relationship is expected to link to an instance of each entity in the relationship. The ternary Visit relationship is turned into an associate entity. The relationship of the Treatment entity with the Visit associate entity is captured through the Generate relationship. The Treatment entity contains data for the treated disease and the recovery time. A visit may result in a multiple treatments for multiple diseases (e.g., high blood pressure and diabetes). The Prescription entity only has the Precription ID attribute.

The Claim entity is related to the InsuranceCompany entity through the Process relationship because a claim is processed by a particular insurance company. Since a claim may have a remaining balance after the patient payment and the insurance payment for each service, the Claim entity has an attribute for the remaining balance.

The Medicine entity and its attributes contain data of medicines including MedicineID, name, and description. The Contain relationship between the Prescription and the Medicine entities has the attributes for the quantity, last pickup date, and refill frequency for each medicine.

The Pharmacy entity has attributes for identification, name, address, and phone number. The Pharmacy entity is related to the Medicine entity through the Keep relationship since a pharmacy keeps medicines. The attributes of the Keep relationship contain data for the unit price, current inventory level, and the minimum inventory level of each medicine that a particular pharmacy keeps. The unit price, the current inventory level, and the minimum inventory level of each medicine may vary with pharmacies.

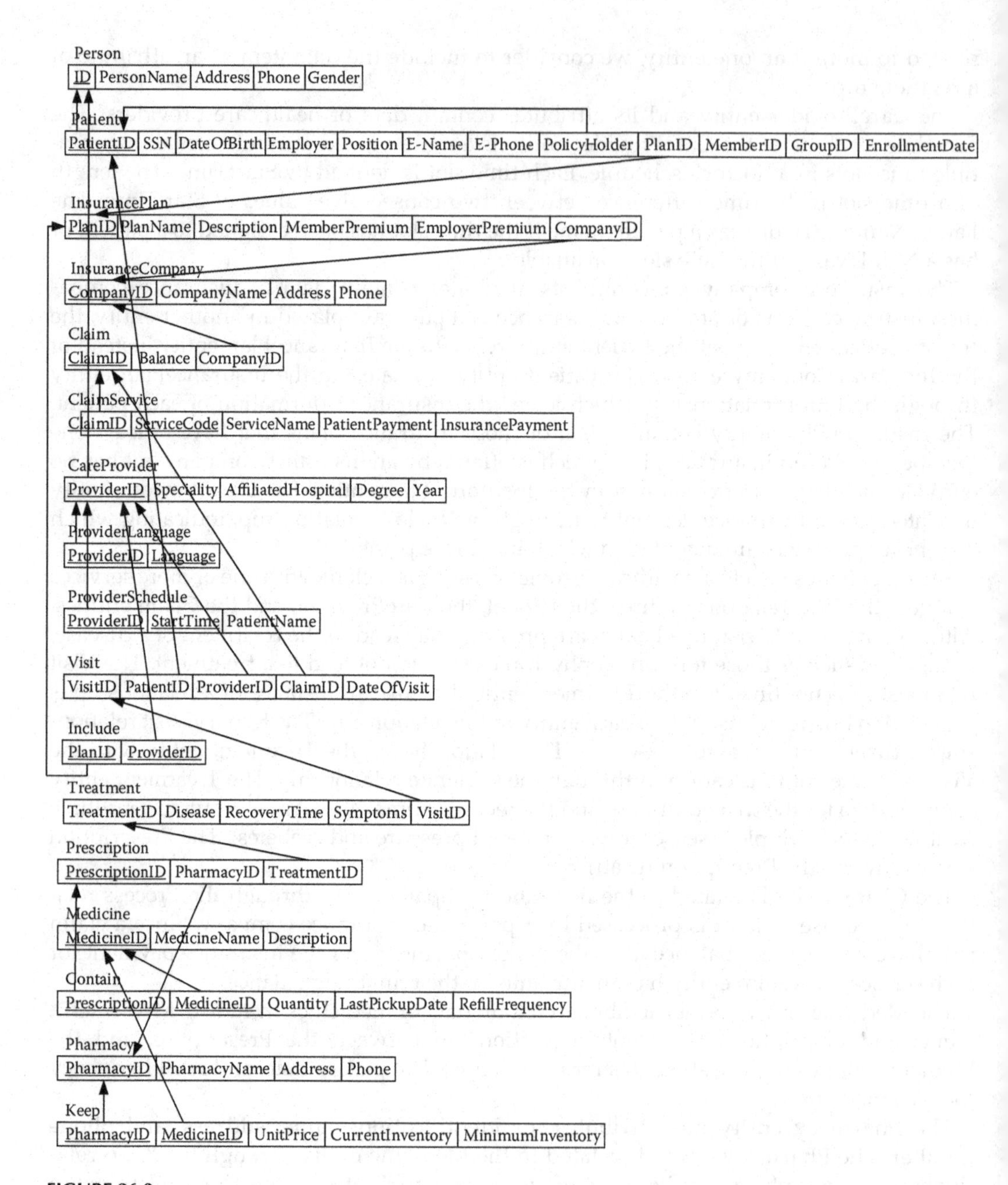

FIGURE 26.2
Relational model of the healthcare database.

26.3 Relational Modeling

The E-R model in Figure 26.1 is transformed into the relational model in Figure 26.2. Each entity in the E-R model is first represented by a relation. Each relationship in the E-R model is then represented by adding foreign key(s) to appropriate relation(s) and linking foreign key(s) to corresponding primary key(s) and specifying the degree and the minimum and maximum cardinalities of each relationship.

26.4 Database Implementation Using Microsoft Access

The relational model in Figure 26.2 is implemented in Microsoft Access. Each relationship in the relational model is implemented by a table in the Access database. For each table, the Design View is used to define data fields along with their data types and properties. For the primary keys, including PatientID, PlanID, CompanyID, ProviderID, ClaimID, VisitID, PrescriptionID, MedicineID, and PharmacyID, Short Text is used as the data type. For a data field, which is a foreign key, the Lookup data type is used to refer back to the data field of the corresponding primary key field to ensure the referential integrity constraint. When the Lookup data type is set up for a data field, which is a foreign key, a relationship between the table containing the foreign key and the table containing the corresponding primary key is established. The relationship is further edited by selecting Enforce Referential Integrity, Cascade Update Related Fields, and Cascade Delete Related Fields in the Edit Relationship window. After the design of all the tables is completed, the Datasheet View is used to add data records to each table. Figure 26.3 shows the data type of the data field in each table. Figure 26.4 shows the Relationships View of all the tables and their relationships.

Figure 26.5 shows the Design View of each query listed in Section 26.1 and implemented in the Access database for the healthcare information system. Note that IQ1 is designed as an embedded query. It is built using IQ1a, IQ1bb, and IQ1cc, where IQ1bb is built using IQ1b and IQ1cc is built using IQ1c. IQ7 is also an embedded query that is built using IQ7a and IQ7b.

26.5 Windows-Based Application and Web-Enabled Application Using Microsoft Visual Studio

Figure 26.6 gives the hierarchy of Windows forms or Web forms in the Windows-based or Web-enabled application for the healthcare information system. Figures 26.7 to 26.13 give the Design View and the Code View of each Windows form at the top and the middle levels in the hierarchy of Windows forms in Figure 26.6. Figures 26.14 and 26.15 show two examples of Windows forms at the bottom level in the hierarchy of Windows forms in Figure 26.6: FormU1 to present the data of the Person table and FormPR1 to present the query result of PRQ1. Figures 26.16 to 26.22 give the Design View of each Web form at the

(a) CareProvider

Field Name	Data Type
ProviderID	Short Text
Speciality	Short Text
AffliatedHospital	Short Text
Degree	Short Text
Year	Number

(b) Claim

Field Name	Data Type
ClaimID	Short Text
Balance	Currency
CompanyID	Short Text

(c) ClaimService

Field Name	Data Type
ClaimID	Short Text
ServiceCode	Short Text
ServiceName	Short Text
PatientPayment	Currency
InsurancePayment	Currency

(d) Contain

Field Name	Data Type
PrescriptionID	Short Text
MedicineID	Short Text
Quantity	Number

(e) Include

Field Name	Data Type
PlanID	Short Text
ProviderID	Short Text

(f) InsuranceCompany

Field Name	Data Type
CompanyID	Short Text
CompanyName	Short Text
Address	Short Text
Phone	Short Text

(g) InsurancePlan

Field Name	Data Type
PlanID	Short Text
PlanName	Short Text
Description	Long Text
CompanyID	Short Text
MemberPremium	Currency
EmployerPremium	Currency

(h) Keep

Field Name	Data Type
PharmacyID	Short Text
MedicineID	Short Text
UnitPrice	Currency
CurrentInventory	Number
MinimumInventory	Number

(i) Medicine

Field Name	Data Type
MedicineID	Short Text
MedicineName	Short Text

(j) Patient

Field Name	Data Type
PatientID	Short Text
SSN	Short Text
DateOfBirth	Date/Time
Employer	Short Text
Position	Short Text
E-Name	Short Text
E-Phone	Short Text
Responsible	Short Text
PlanID	Short Text
MemberID	Short Text
GroupID	Short Text
EnrollmentDate	Date/Time

(k) Person

Field Name	Data Type
ID	Short Text
PersonName	Short Text
Address	Short Text
Phone	Short Text
Gender	Short Text

FIGURE 26.3
Data type of each data field in the healthcare database. (a) The CareProvider table, (b) The Claim table, (c) The ClaimService table, (d) The Contain table, (e) The Include table, (f) The InsuranceCompany table, (g) The InsurancePlan table, (h) The Keep table, (i) The Medicine table, (j) The Patient table, (k) The Person table.

(l) Pharmacy

	Field Name	Data Type
🔑	PharmacyID	Short Text
	PharmacyName	Short Text
	Address	Short Text
	Phone	Short Text

(m) Prescription

	Field Name	Data Type
🔑	PrescriptionID	Short Text
	PharmacyID	Short Text
	LastPickupDate	Short Text
	RefillFrequency	Short Text
	TreatmentID	Short Text

(n) ProviderLanguage

	Field Name	Data Type
🔑	ProviderID	Short Text
🔑	Language	Short Text

(o) ProviderSchedule

	Field Name	Data Type
🔑	ProviderID	Short Text
🔑	StartTime	Date/Time
	PatientName	Short Text

(p) Treatment

	Field Name	Data Type
🔑	TreatmentID	Short Text
	Disease	Short Text
	RecoveryTime	Number
	Symptoms	Short Text
	VisitID	Short Text

(q) Visit

	Field Name	Data Type
🔑	VisitID	Short Text
	PatientID	Short Text
	ProviderID	Short Text
	ClaimID	Short Text
	DateOfVisit	Date/Time

FIGURE 26.3 (Continued)
Data type of each data field in the healthcare database. (l) The Pharmacy table, (m) The Prescription table, (n) The ProviderLanguage table, (o) The ProviderSchedule table, (p) The Treatment table, and (q) The Visit table.

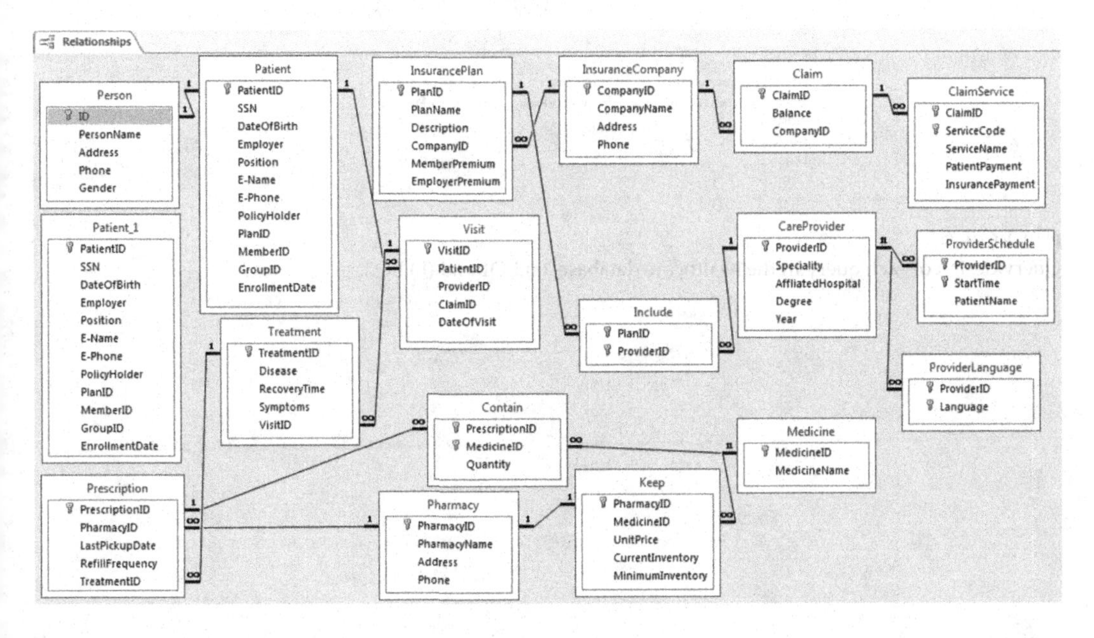

FIGURE 26.4
Relationships View of all the tables and their relationships in the database for the healthcare information system.

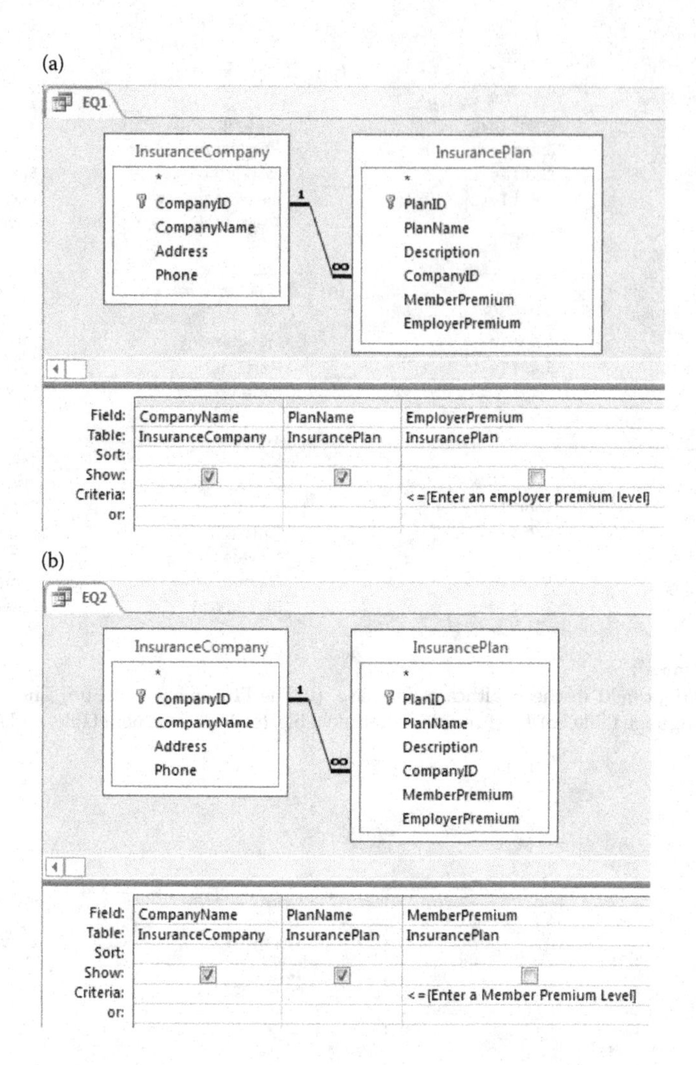

FIGURE 26.5
Query design of each query in the healthcare database, (a) EQ1 and (b) EQ2.

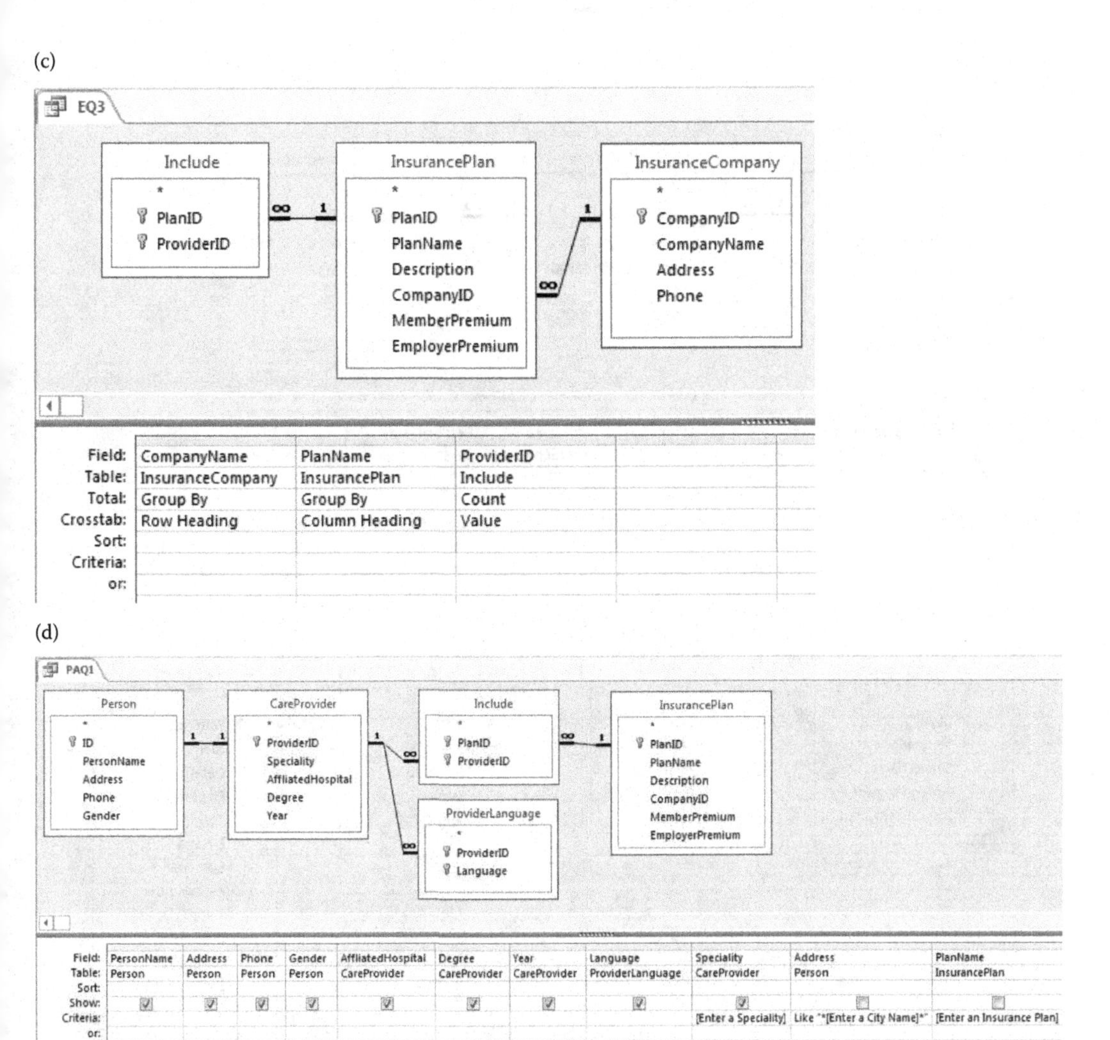

FIGURE 26.5 (Continued)
Query design of each query in the healthcare database, (c) EQ3 and (d) PAQ1.

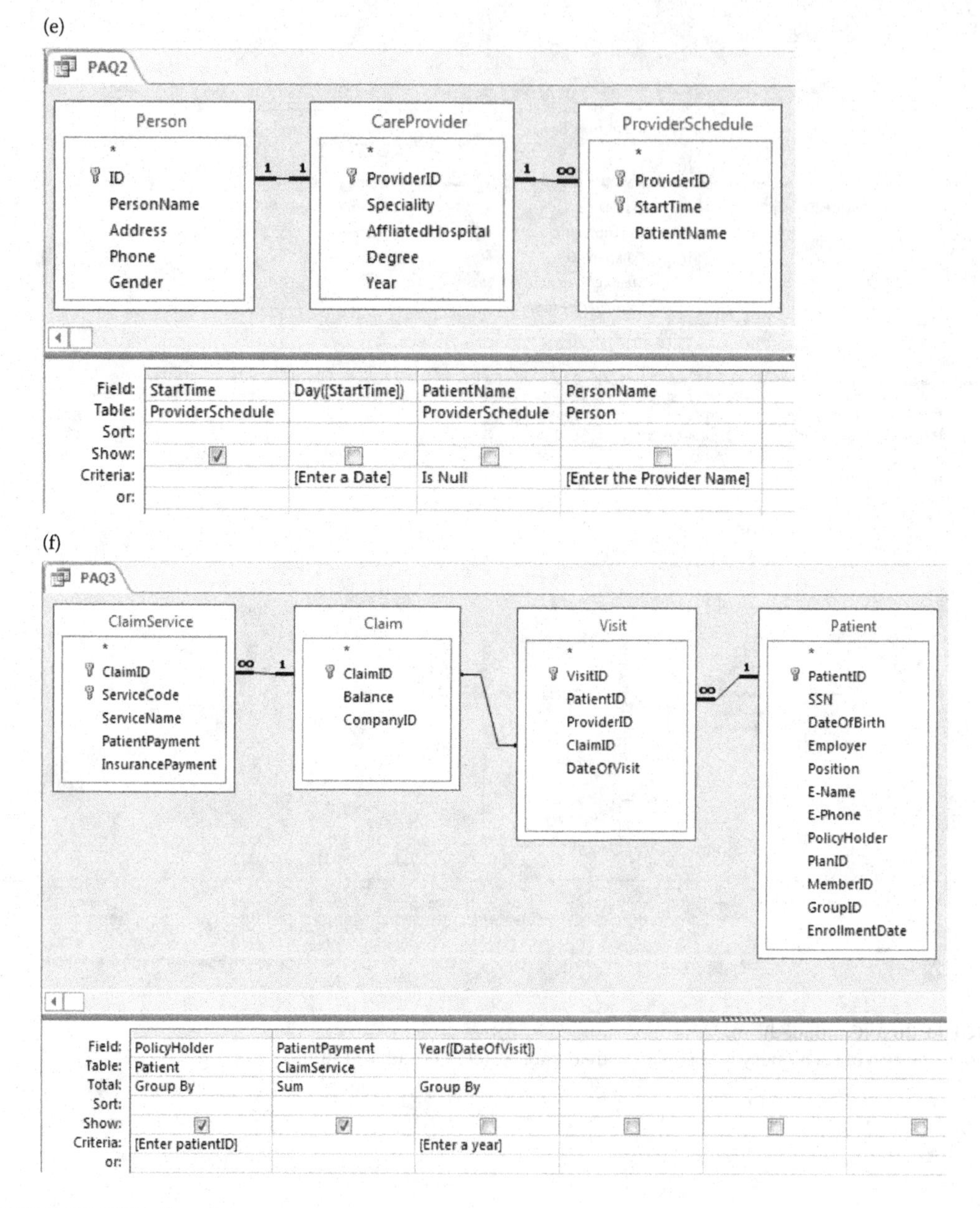

FIGURE 26.5 (Continued)
Query design of each query in the healthcare database, (e) PAQ2 and (f) PAQ3.

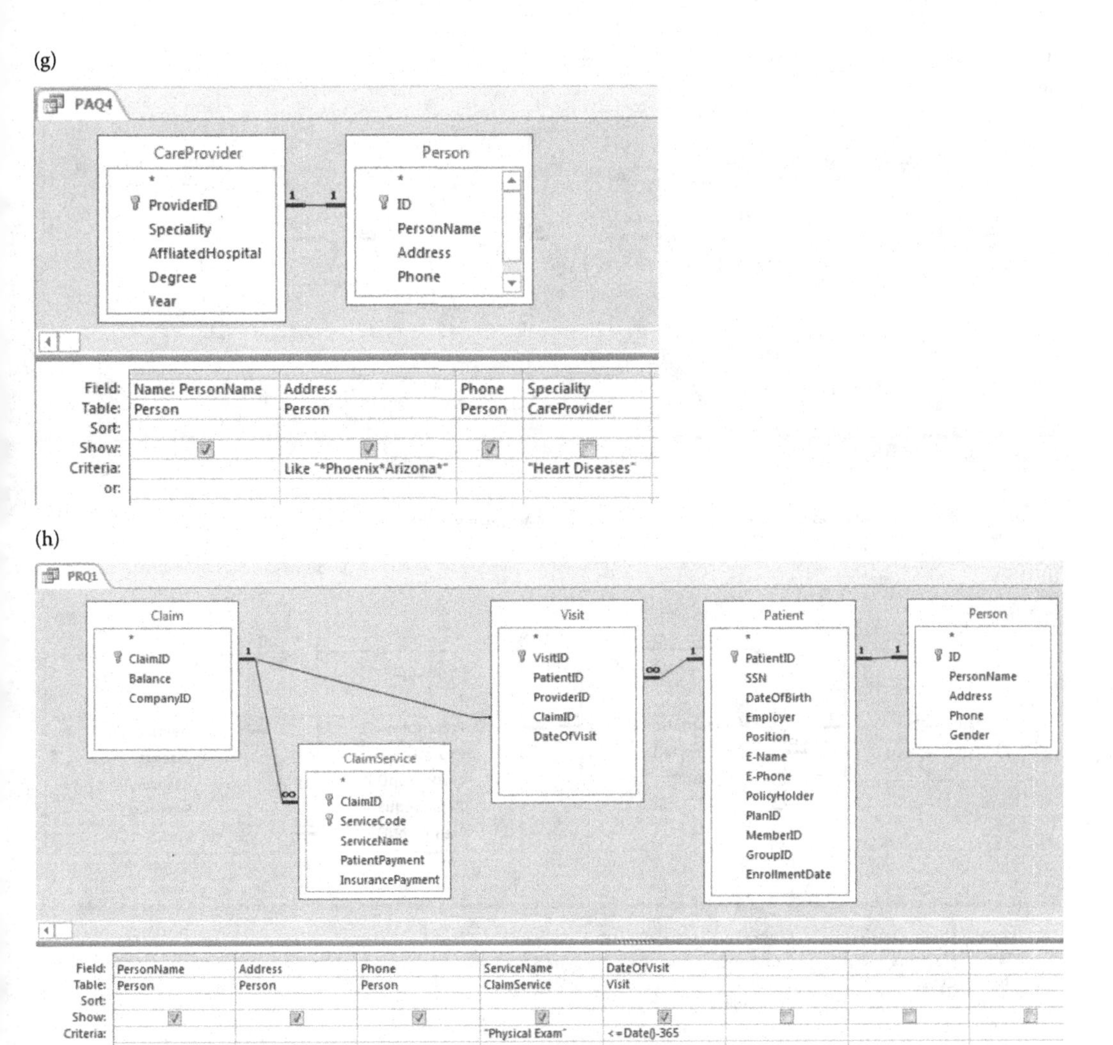

FIGURE 26.5 (Continued)
Query design of each query in the healthcare database, (g) PAQ4 and (h) PRQ1.

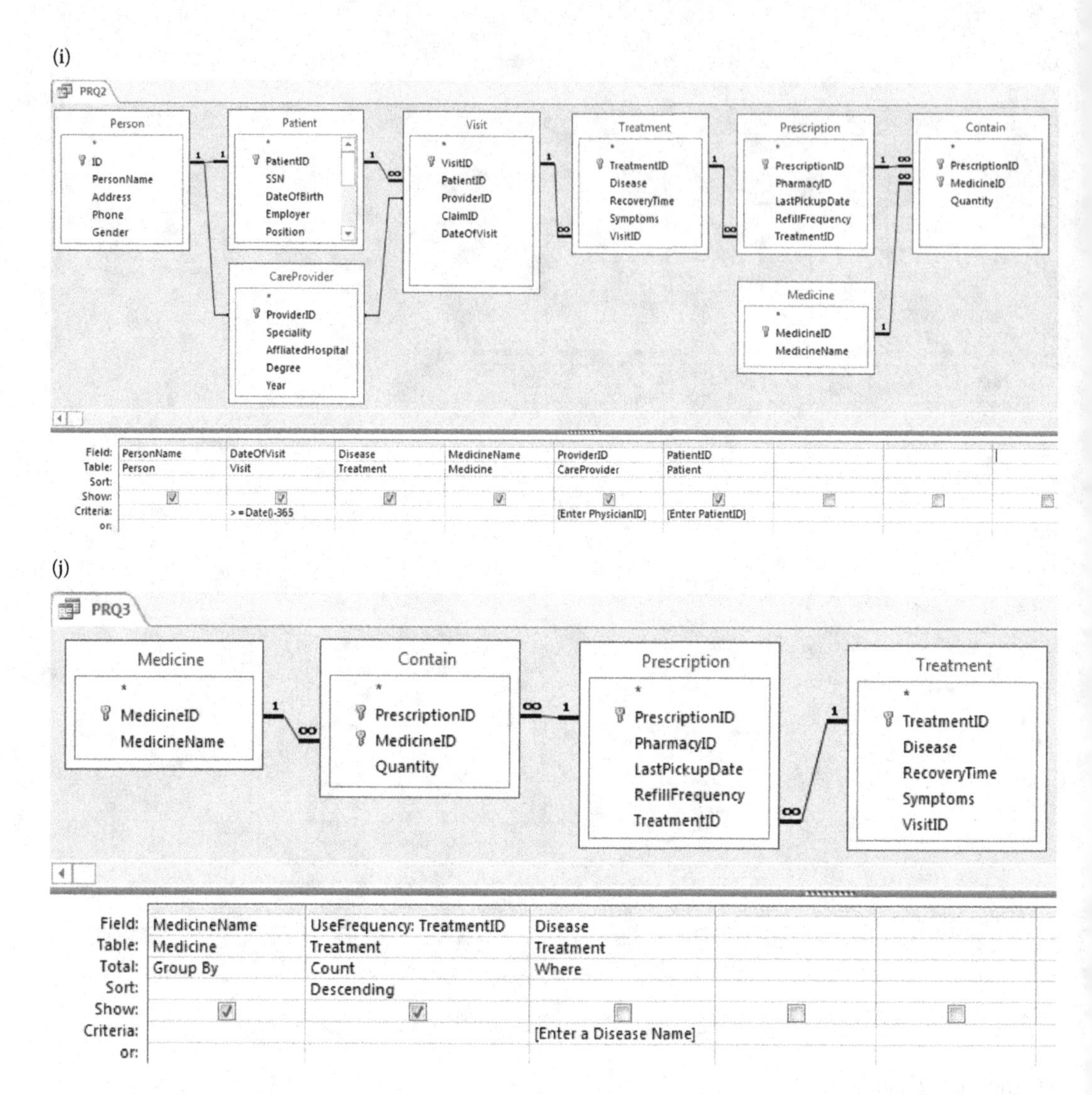

FIGURE 26.5 (Continued)
Query design of each query in the healthcare database, (i) PRQ2 and (j) PRQ3.

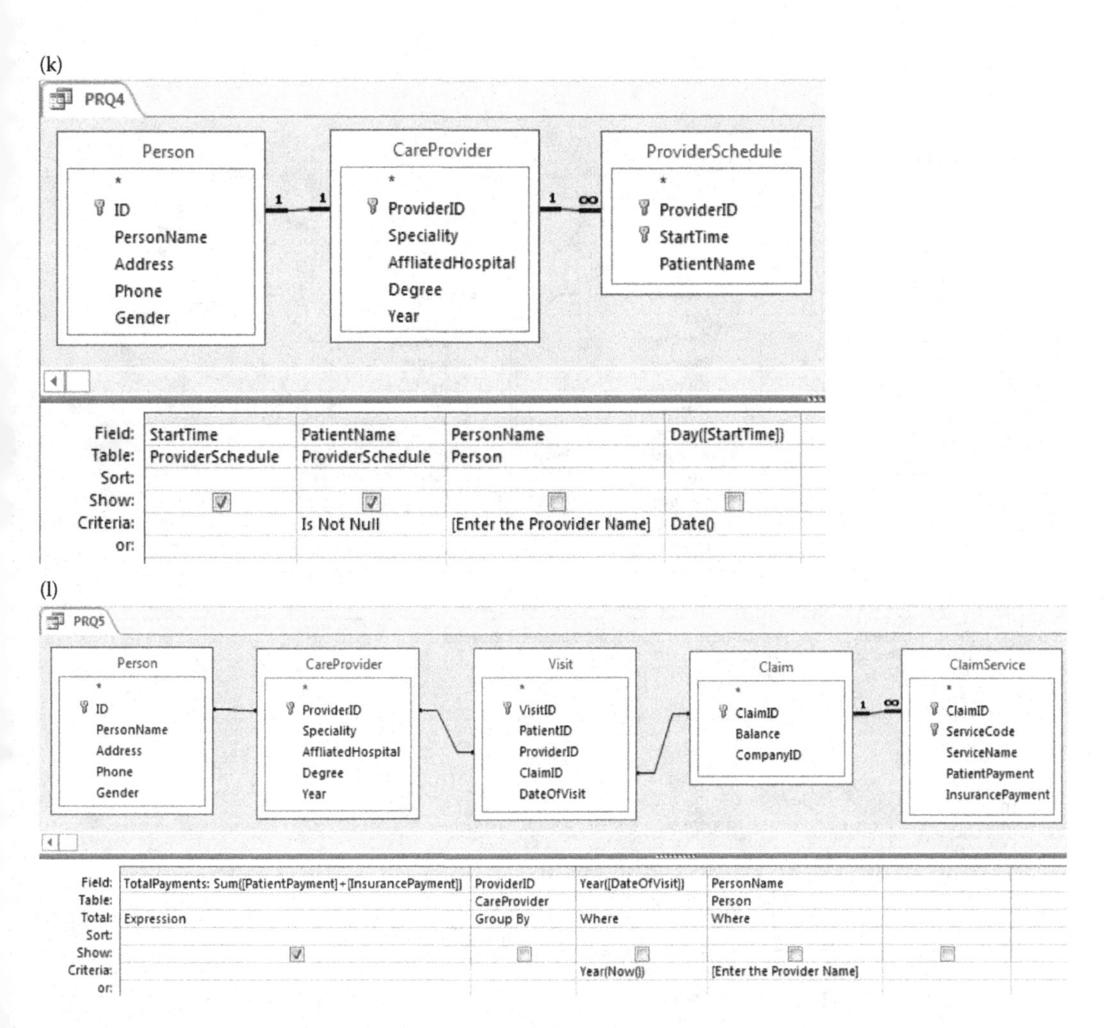

FIGURE 26.5 (Continued)
Query design of each query in the healthcare database, (k) PRQ4 and (l) PRQ5.

(m)

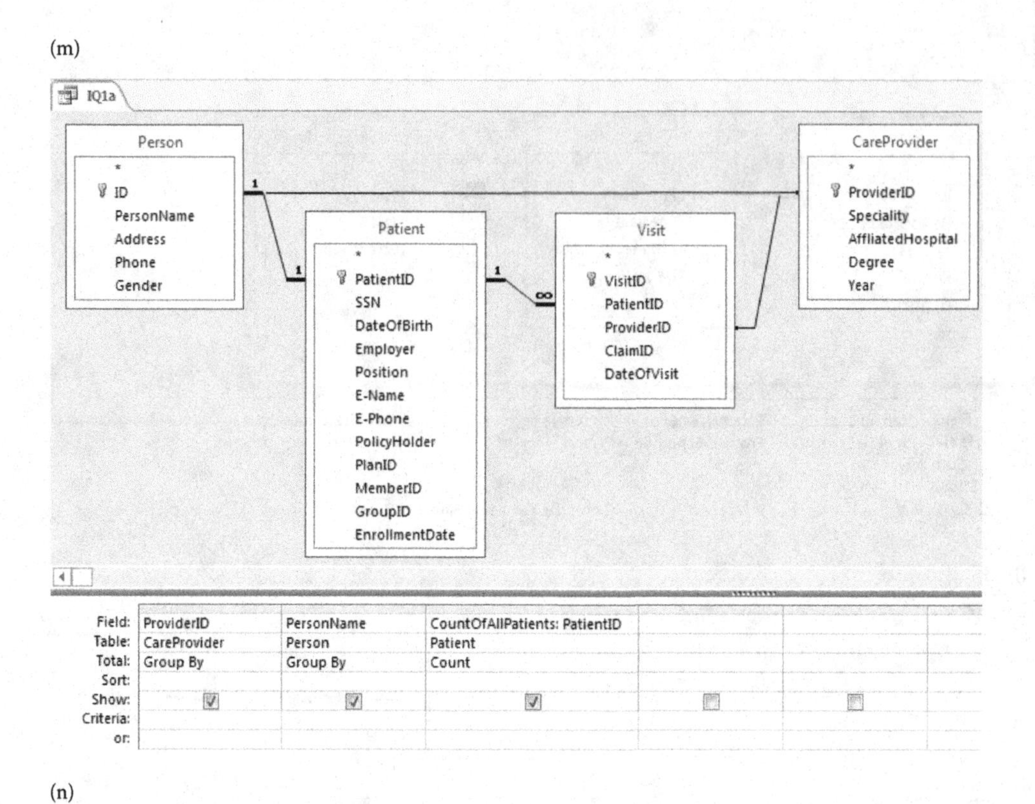

(n)

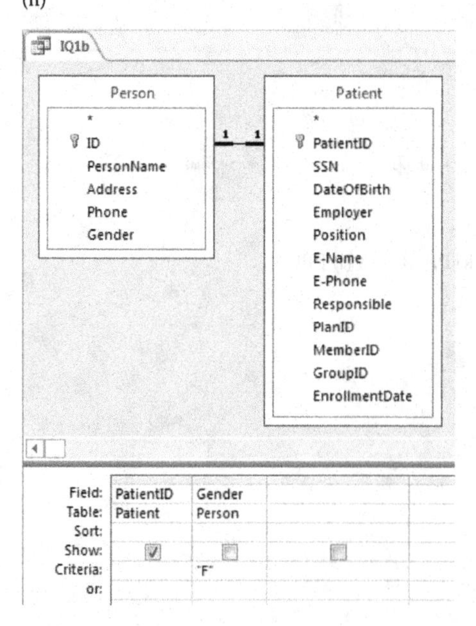

FIGURE 26.5 (Continued)
Query design of each query in the healthcare database, (m) IQ1a and (n) IQ1b.

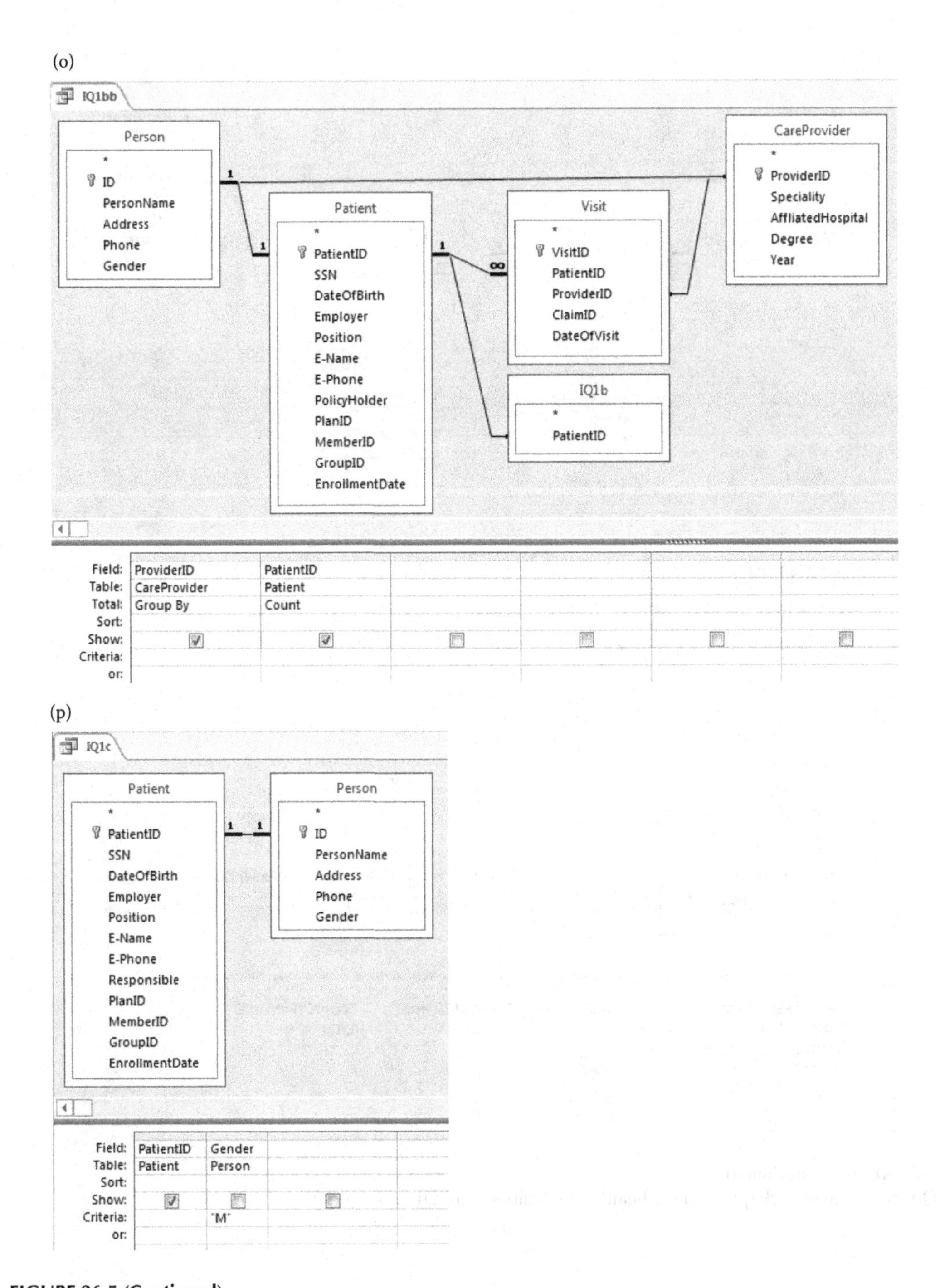

FIGURE 26.5 (Continued)
Query design of each query in the healthcare database, (o) IQ1bb and (p) IQ1c.

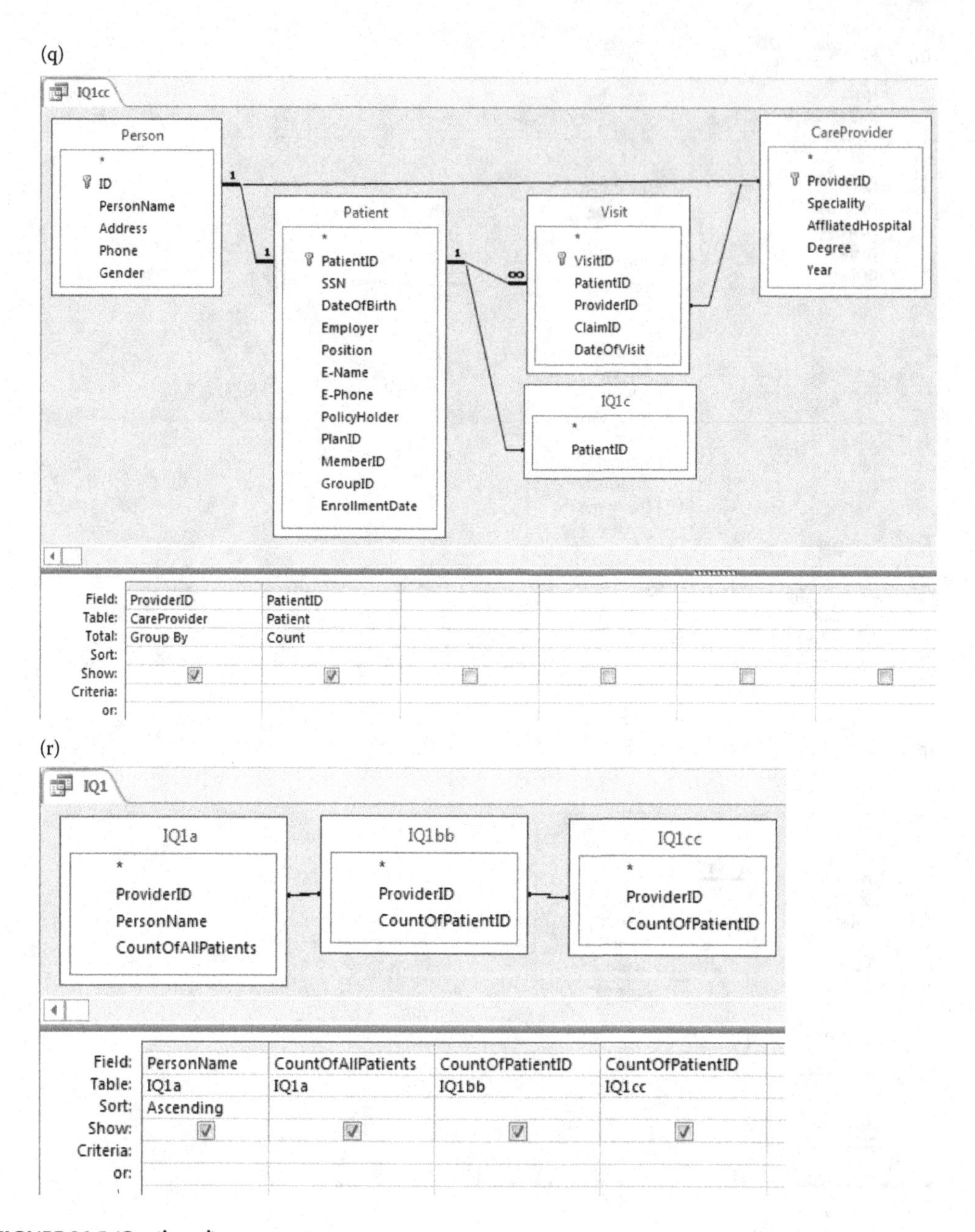

FIGURE 26.5 (Continued)
Query design of each query in the healthcare database, (q) IQ1cc and (r) IQ1.

(s)

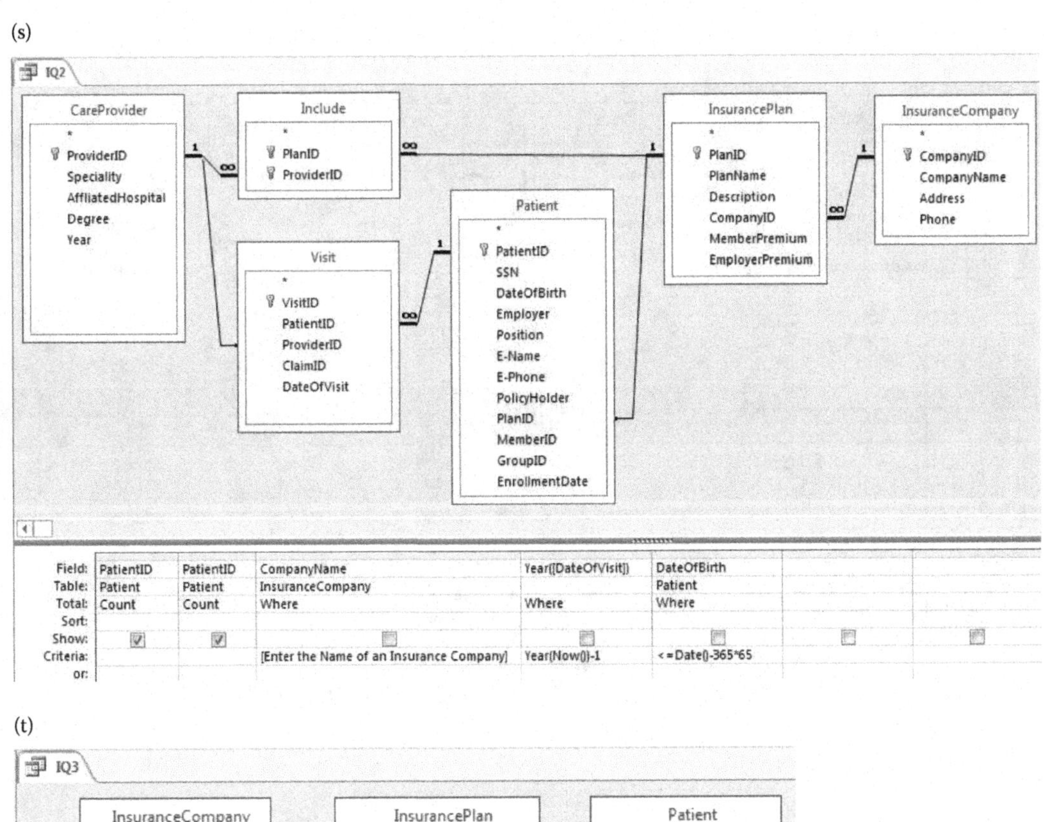

(t)

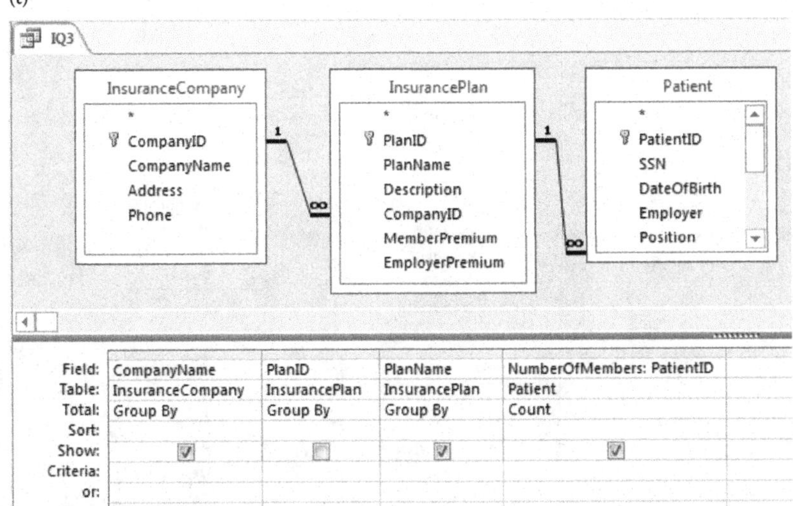

FIGURE 26.5 (Continued)
Query design of each query in the healthcare database, (s) IQ2 and (t) IQ3.

(u)

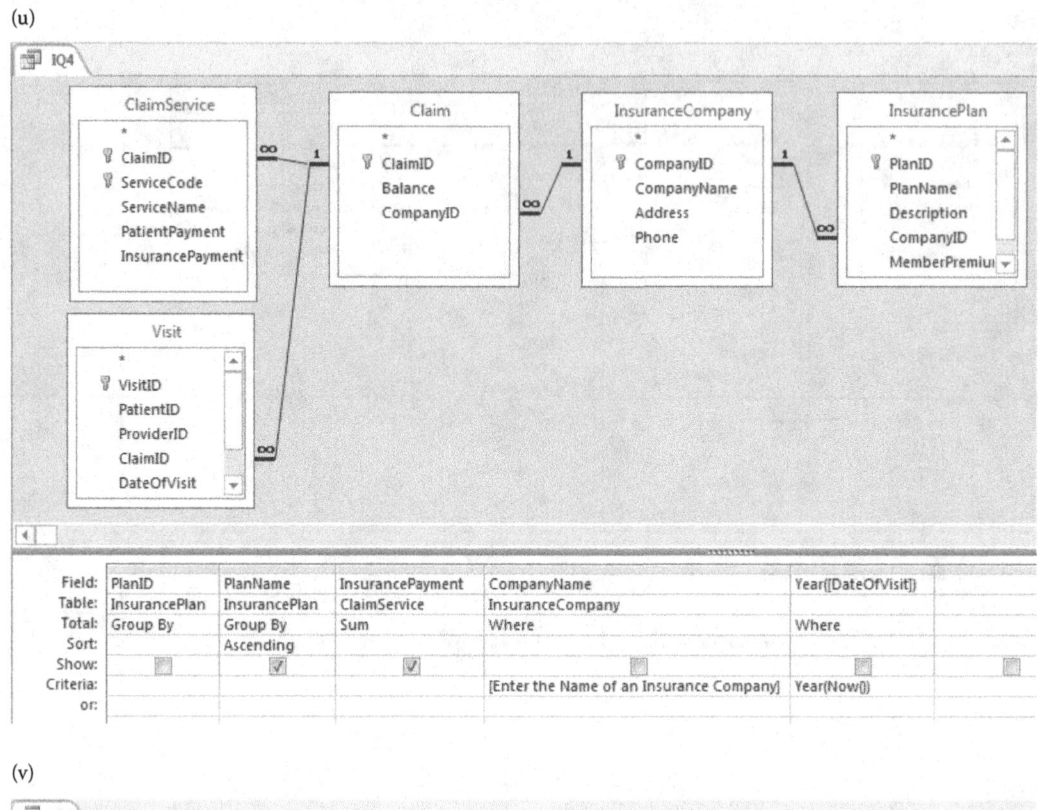

(v)

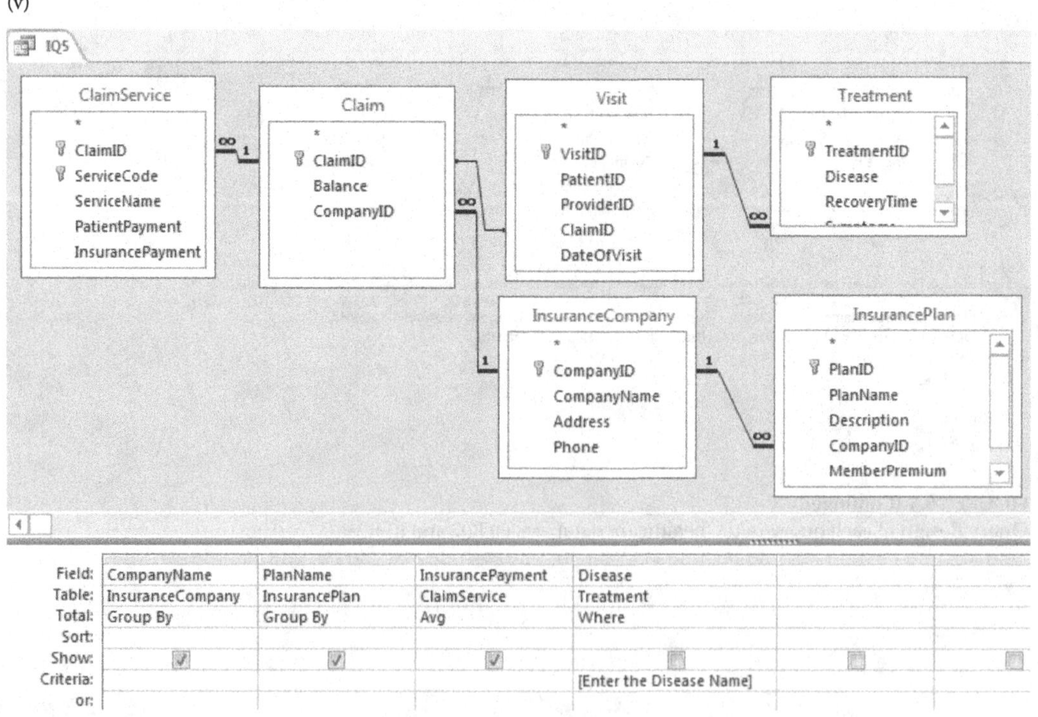

FIGURE 26.5 (Continued)
Query design of each query in the healthcare database, (u) IQ4 and (v) IQ5.

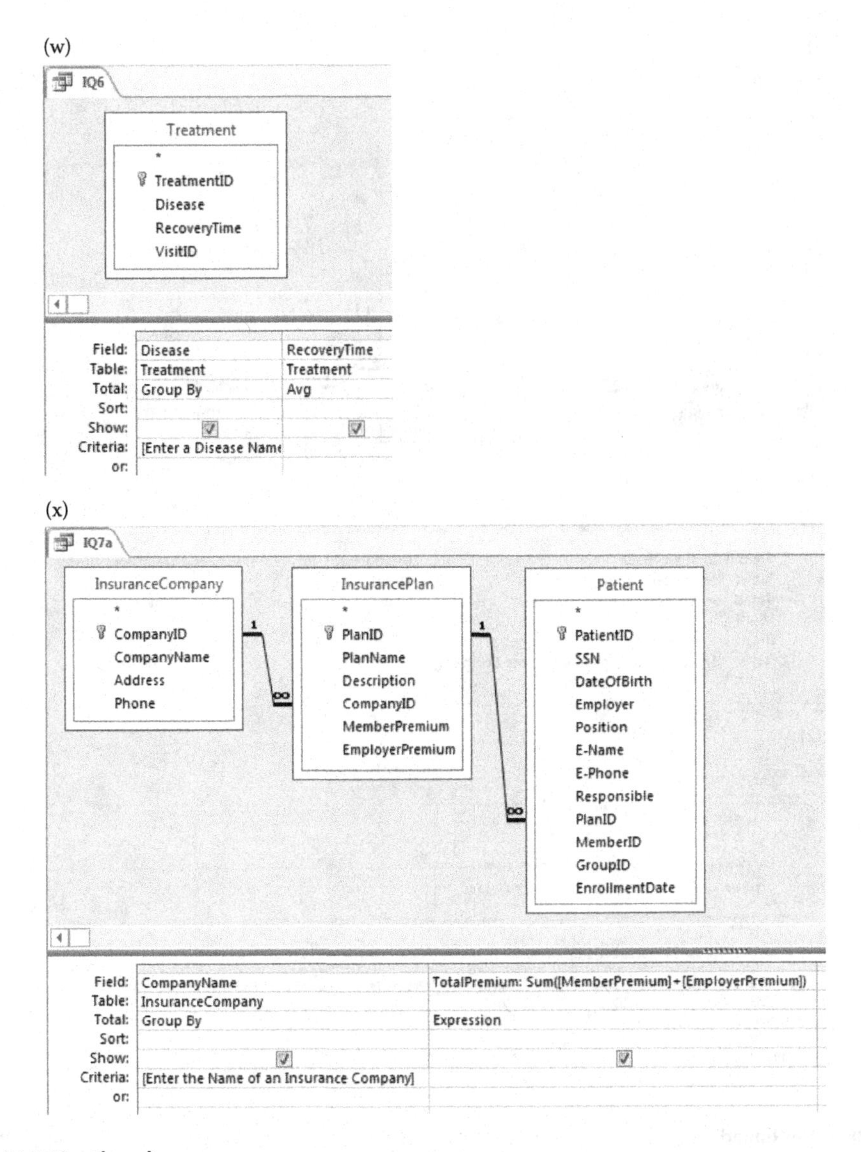

FIGURE 26.5 (Continued)
Query design of each query in the healthcare database, (w) IQ6 and (x) IQ7a.

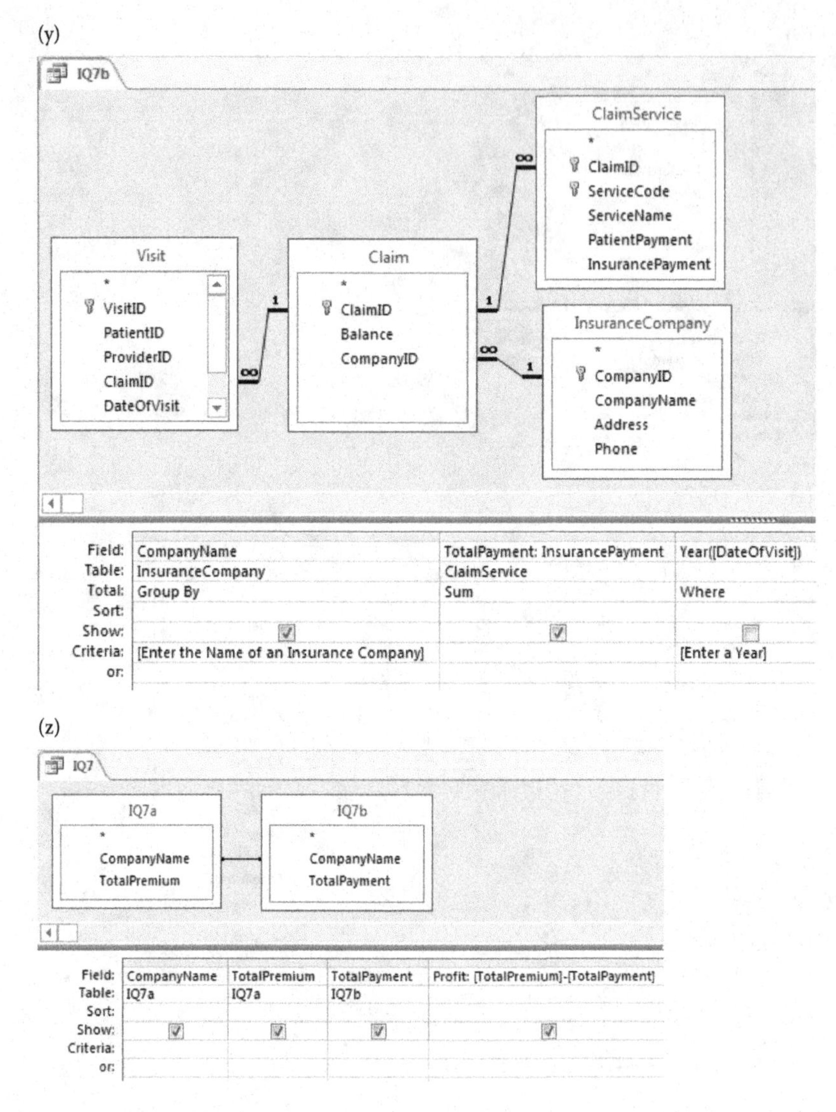

FIGURE 26.5 (Continued)
Query design of each query in the healthcare database, (y) IQ7b and (z) IQ7.

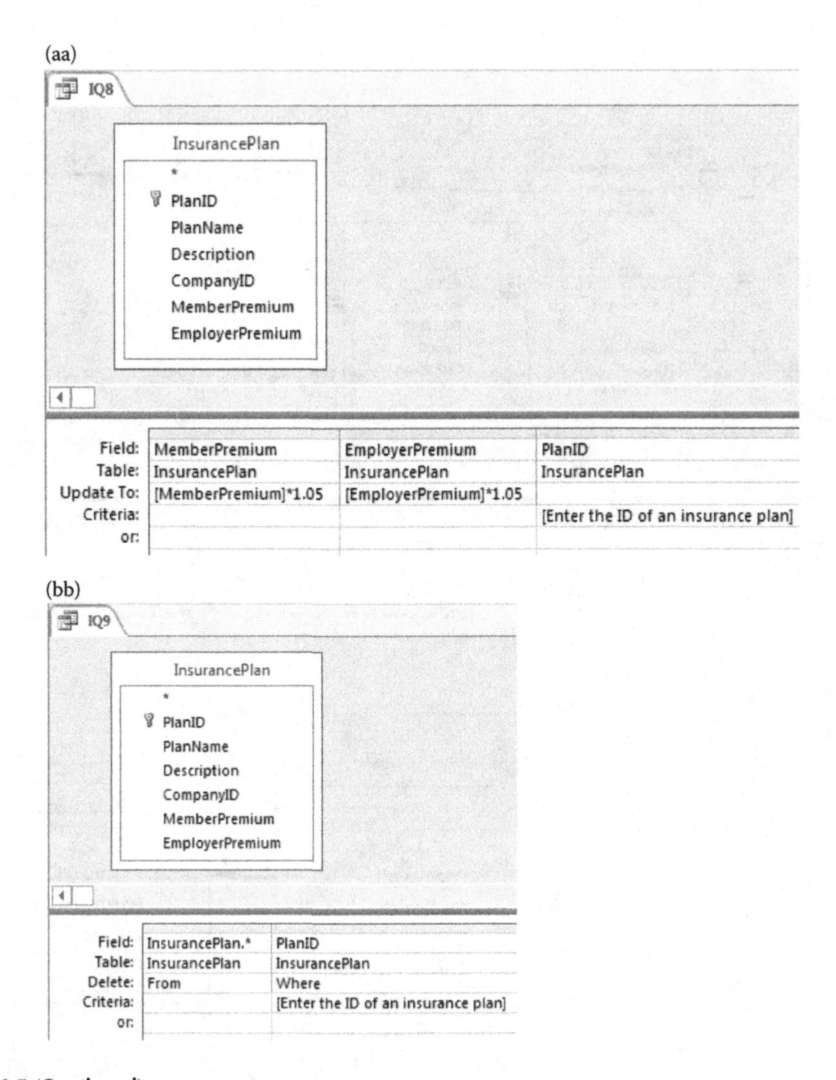

FIGURE 26.5 (Continued)
Query design of each query in the healthcare database, (aa) IQ8 and (bb) IQ9.

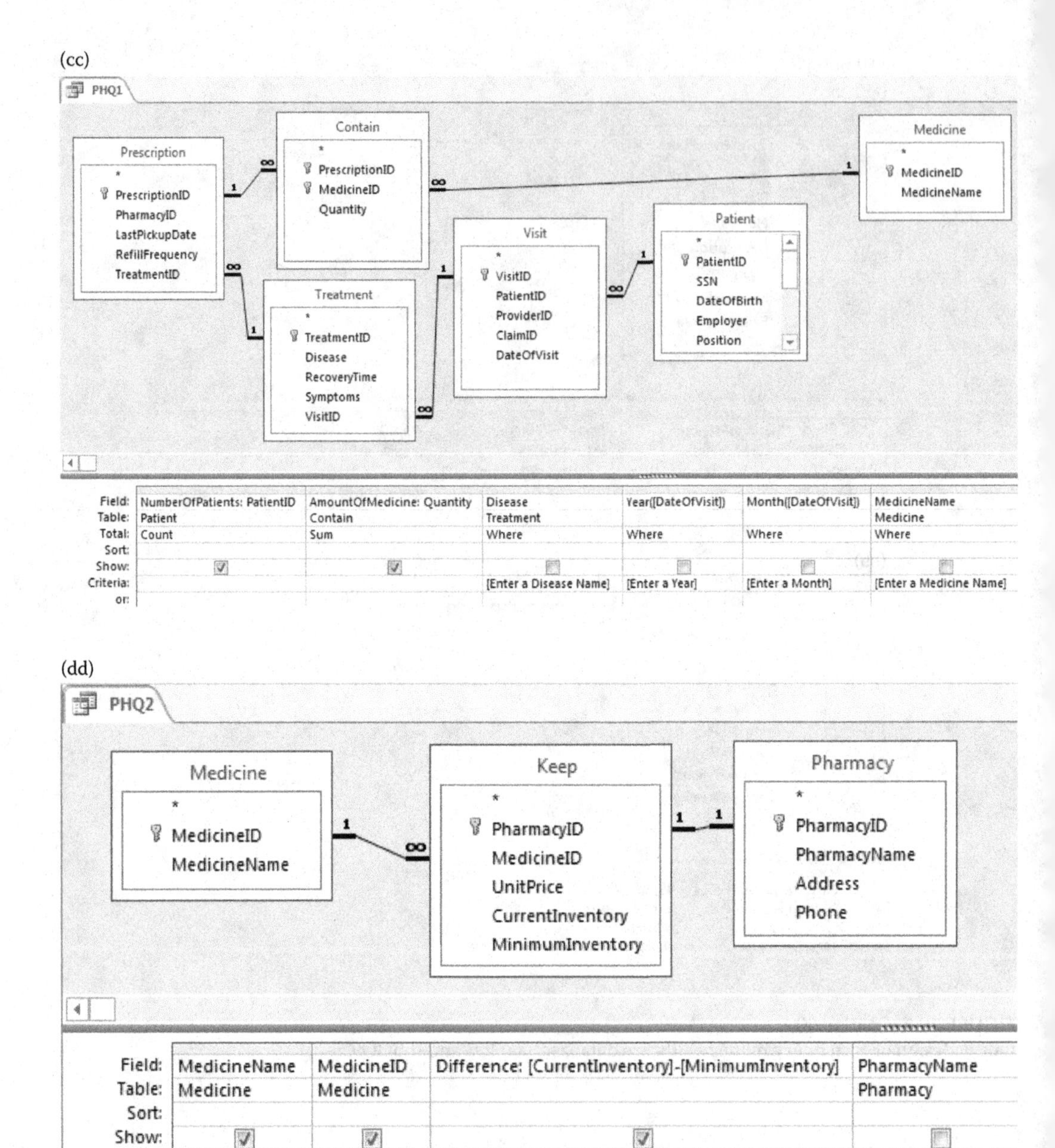

FIGURE 26.5 (Continued)
Query design of each query in the healthcare database, (cc) PHQ1 and (dd) PHQ2.

(ee)

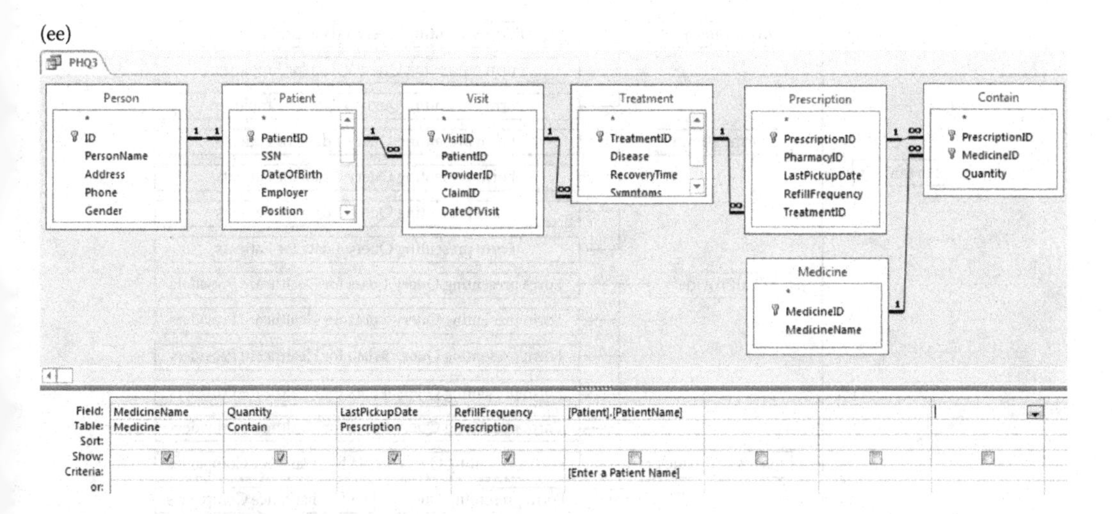

FIGURE 26.5 (Continued)
Query design of each query in the healthcare database, (ee) PHQ3.

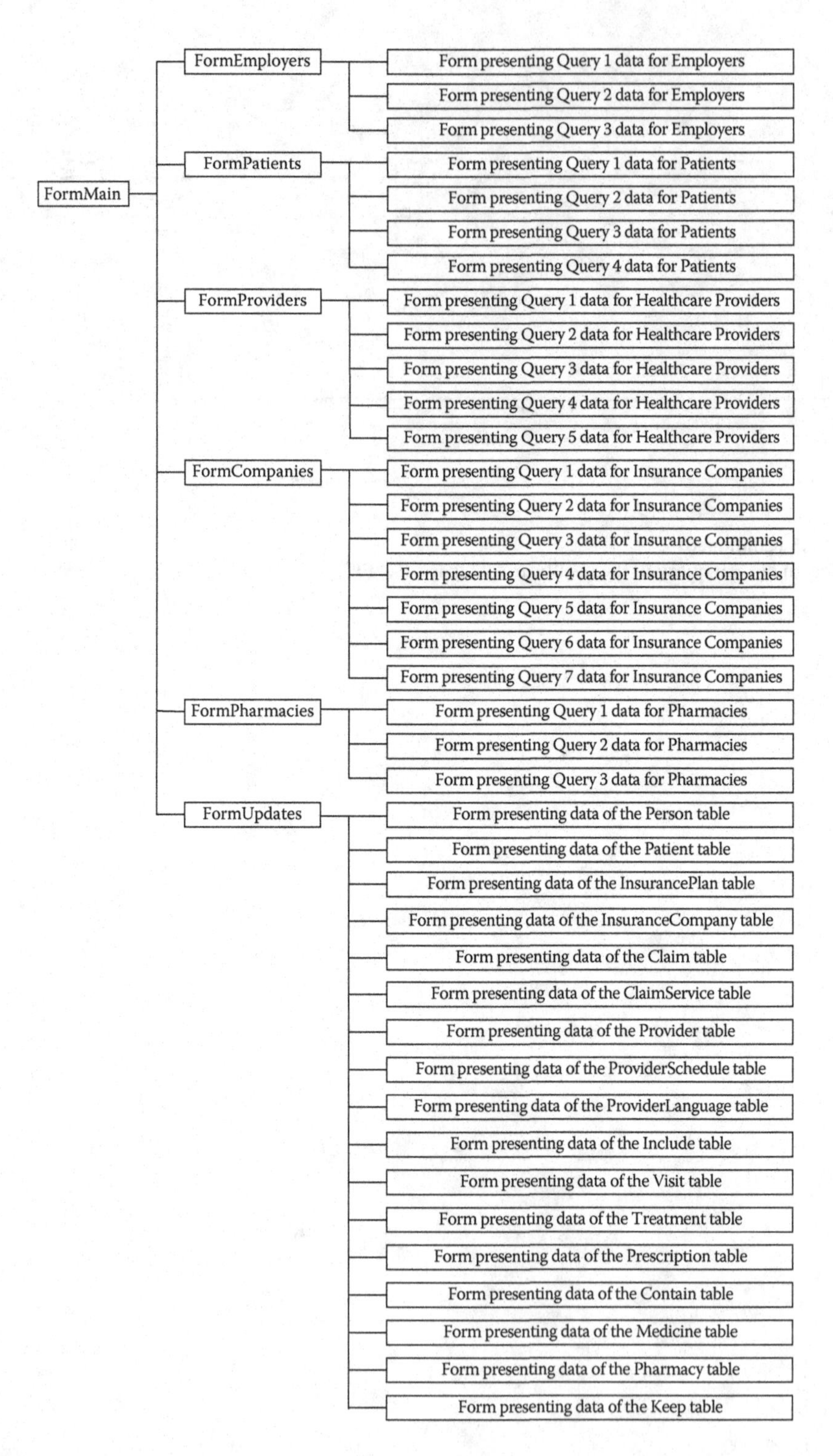

FIGURE 26.6
Hierarchy of Windows forms or Web forms in the Windows-based or Web-enabled application for the healthcare information system.

(a)

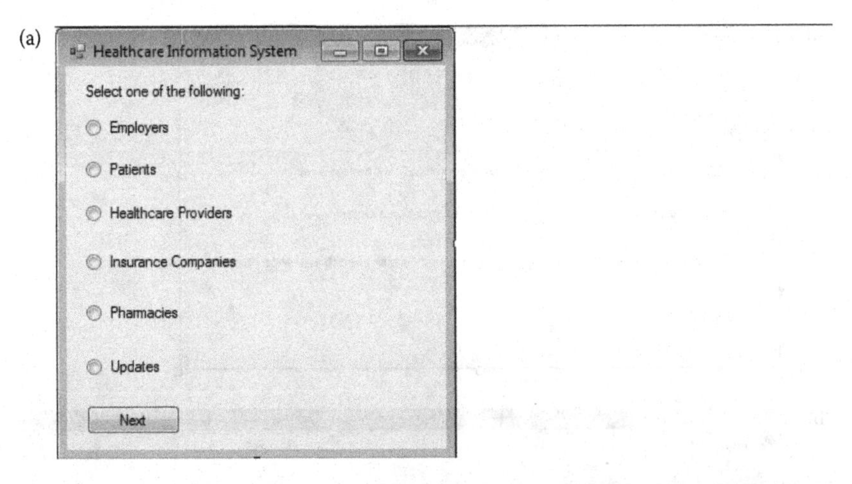

(b)

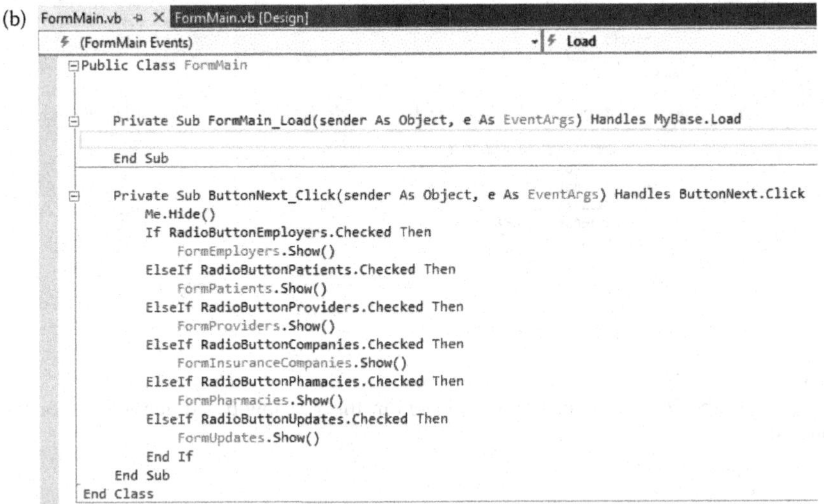

FIGURE 26.7
(a) Design View and (b) Code View of FormMain in the Window-based application for the healthcare information system.

(a)

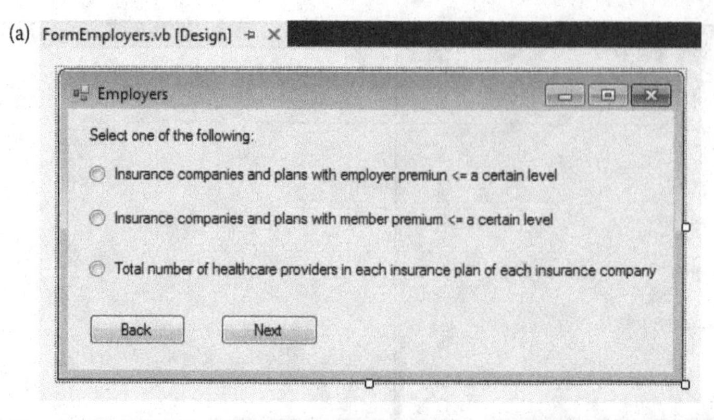

(b)

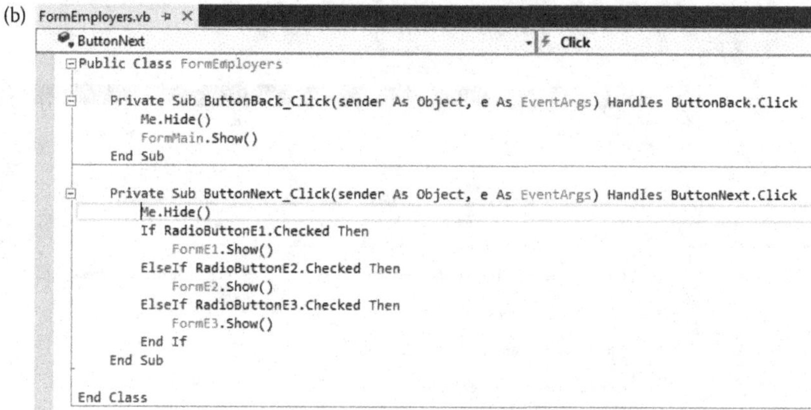

FIGURE 26.8
(a) Design View and (b) Code View of FormEmployers in the Window-based application for the healthcare information system.

(a)

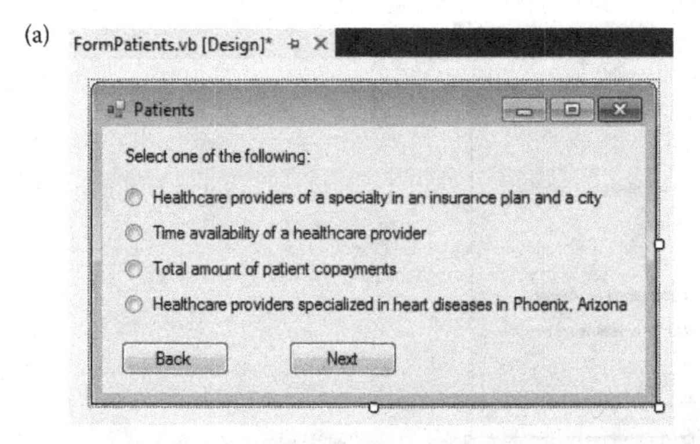

(b)

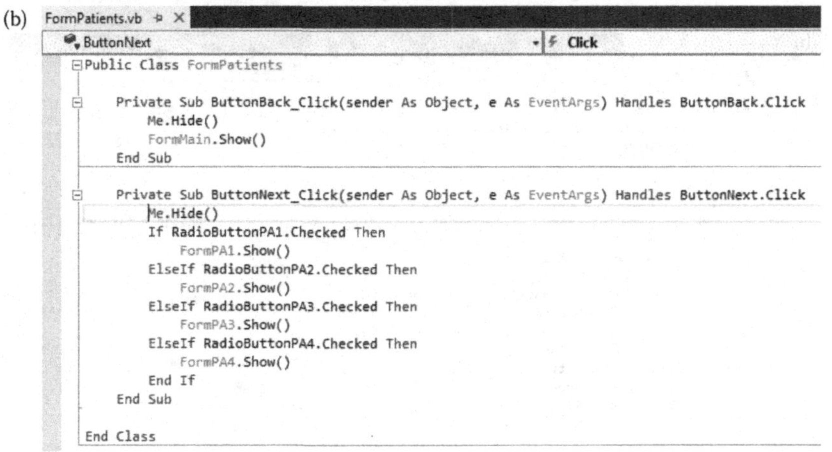

FIGURE 26.9
(a) Design View and (b) Code View of FormPatients in the Window-based application for the healthcare information system.

(a)

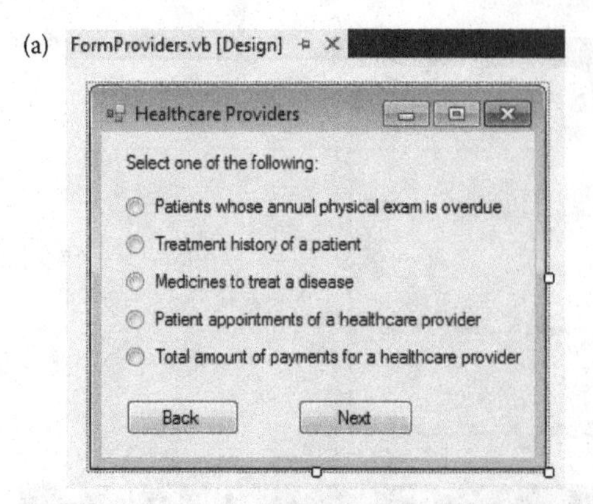

(b)

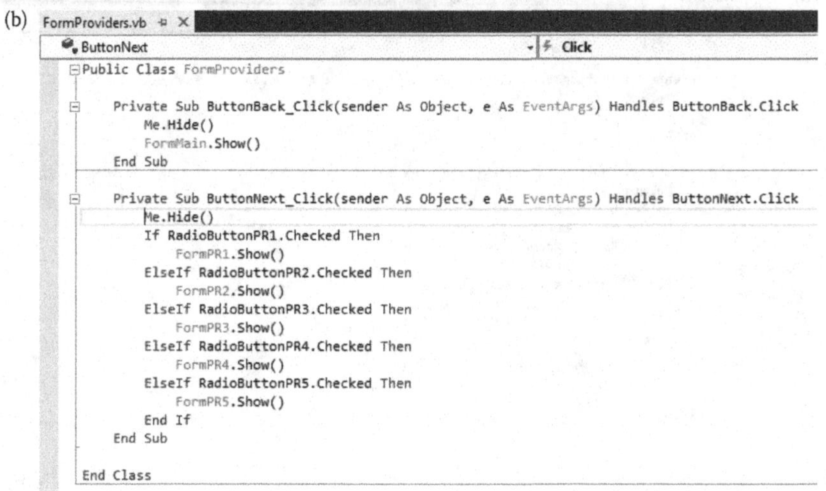


```
Public Class FormProviders

    Private Sub ButtonBack_Click(sender As Object, e As EventArgs) Handles ButtonBack.Click
        Me.Hide()
        FormMain.Show()
    End Sub

    Private Sub ButtonNext_Click(sender As Object, e As EventArgs) Handles ButtonNext.Click
        Me.Hide()
        If RadioButtonPR1.Checked Then
            FormPR1.Show()
        ElseIf RadioButtonPR2.Checked Then
            FormPR2.Show()
        ElseIf RadioButtonPR3.Checked Then
            FormPR3.Show()
        ElseIf RadioButtonPR4.Checked Then
            FormPR4.Show()
        ElseIf RadioButtonPR5.Checked Then
            FormPR5.Show()
        End If
    End Sub

End Class
```

FIGURE 26.10
(a) Design View and (b) Code View of FormProviders in the Window-based application for the healthcare information system.

(a)

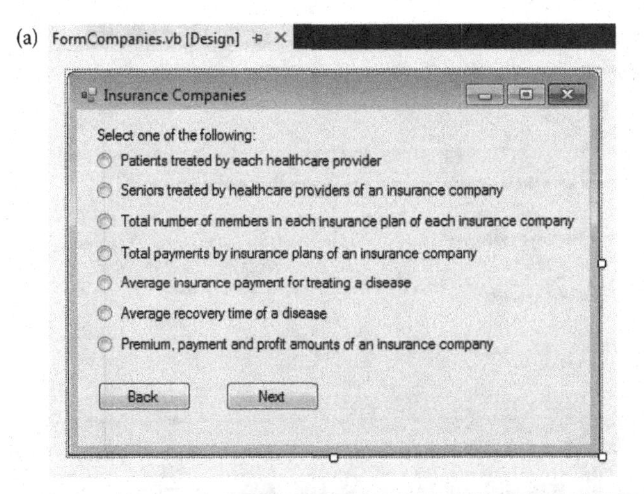

(b)

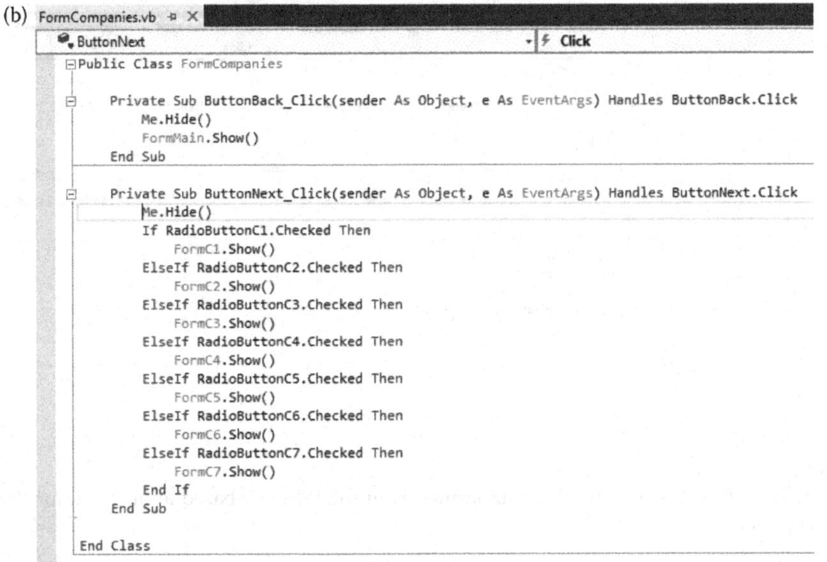


```
Public Class FormCompanies

    Private Sub ButtonBack_Click(sender As Object, e As EventArgs) Handles ButtonBack.Click
        Me.Hide()
        FormMain.Show()
    End Sub

    Private Sub ButtonNext_Click(sender As Object, e As EventArgs) Handles ButtonNext.Click
        Me.Hide()
        If RadioButtonC1.Checked Then
            FormC1.Show()
        ElseIf RadioButtonC2.Checked Then
            FormC2.Show()
        ElseIf RadioButtonC3.Checked Then
            FormC3.Show()
        ElseIf RadioButtonC4.Checked Then
            FormC4.Show()
        ElseIf RadioButtonC5.Checked Then
            FormC5.Show()
        ElseIf RadioButtonC6.Checked Then
            FormC6.Show()
        ElseIf RadioButtonC7.Checked Then
            FormC7.Show()
        End If
    End Sub

End Class
```

FIGURE 26.11
(a) Design View and (b) Code View of FormCompanies in the Window-based application for the healthcare information system.

(a)

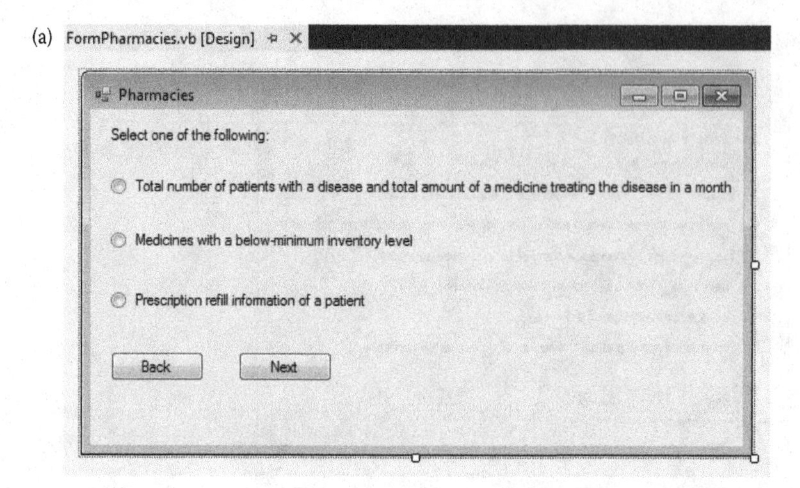

(b)

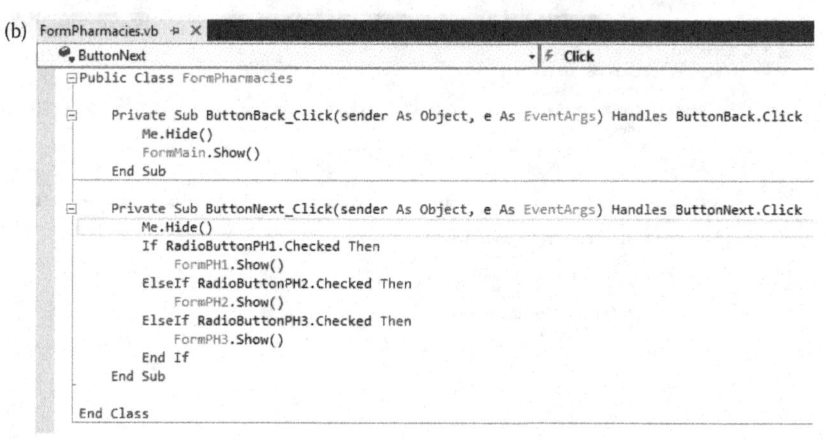

FormPharmacies.vb

ButtonNext | Click

```
Public Class FormPharmacies

    Private Sub ButtonBack_Click(sender As Object, e As EventArgs) Handles ButtonBack.Click
        Me.Hide()
        FormMain.Show()
    End Sub

    Private Sub ButtonNext_Click(sender As Object, e As EventArgs) Handles ButtonNext.Click
        Me.Hide()
        If RadioButtonPH1.Checked Then
            FormPH1.Show()
        ElseIf RadioButtonPH2.Checked Then
            FormPH2.Show()
        ElseIf RadioButtonPH3.Checked Then
            FormPH3.Show()
        End If
    End Sub

End Class
```

FIGURE 26.12
(a) Design View and (b) Code View of FormPharmacies in the Window-based application for the healthcare information system.

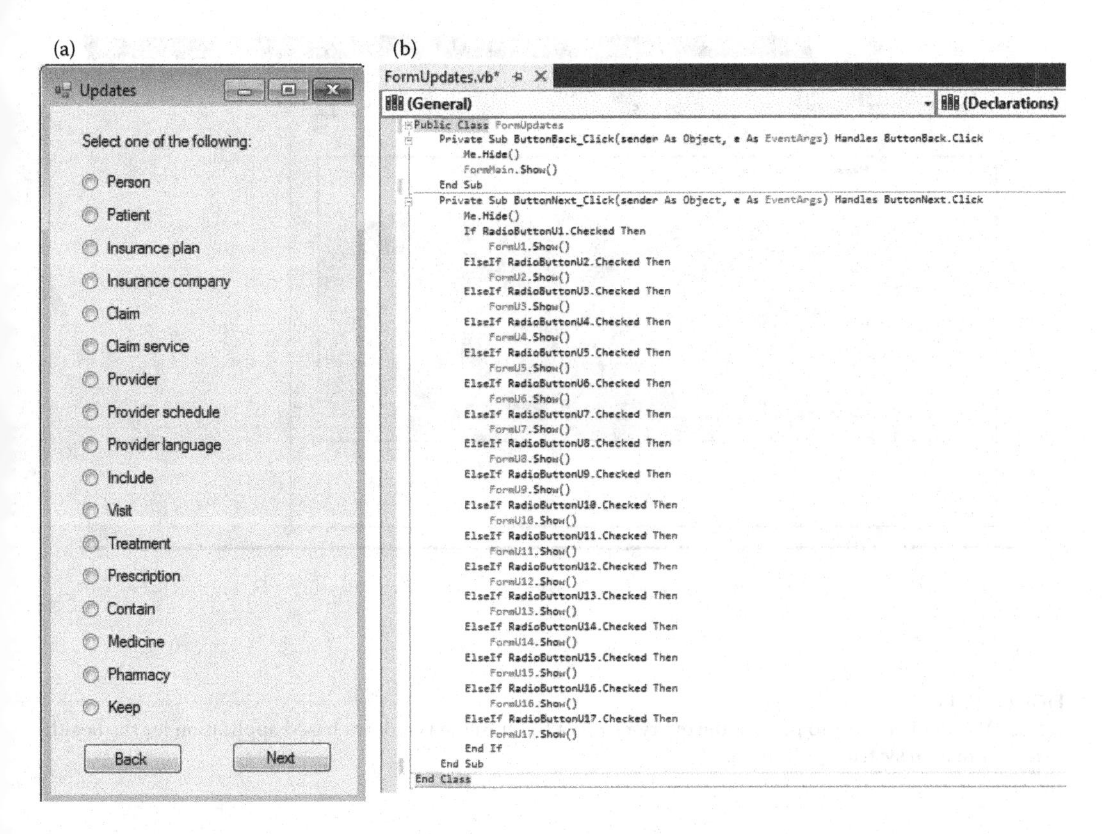

FIGURE 26.13
(a) Design View and (b) Code View of FormUpdates in the Window-based application for the healthcare information system.

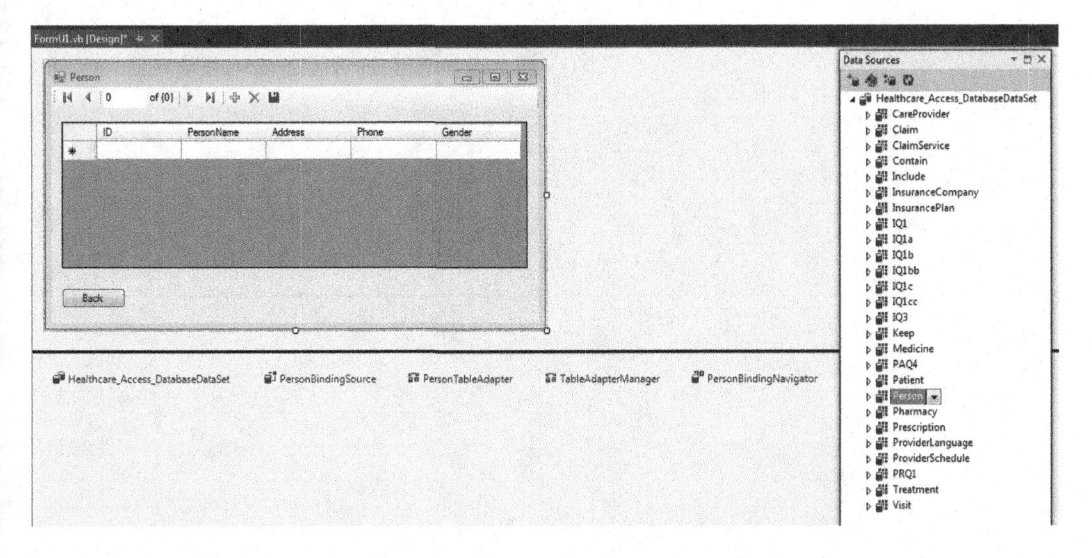

FIGURE 26.14
Design View of FormU1 to present data in the Person table in the Windows-based application for the healthcare information system application.

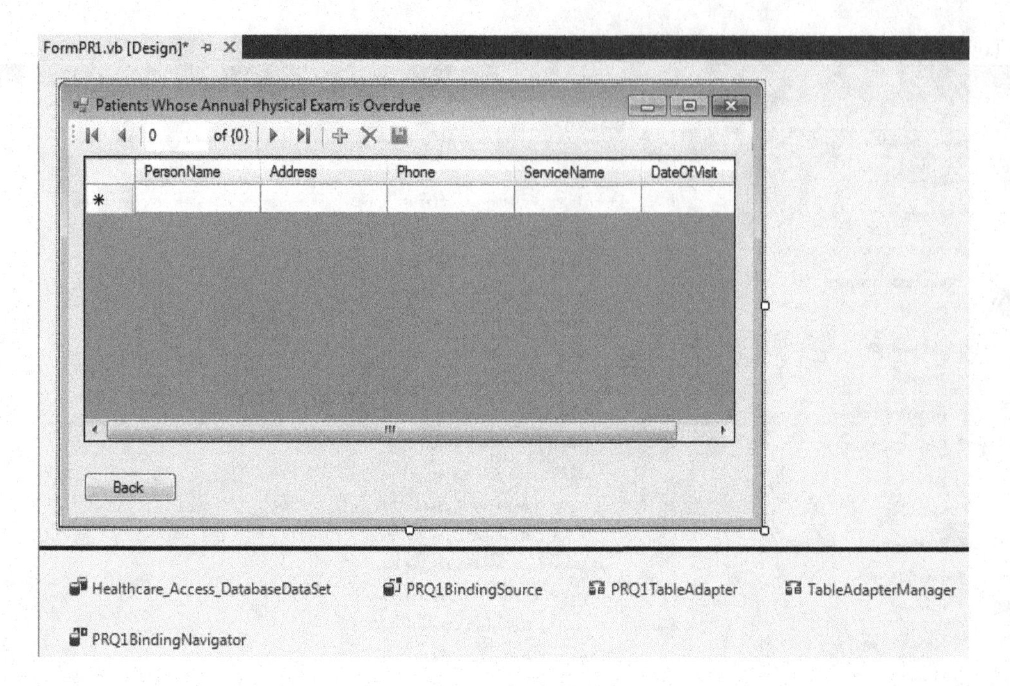

FIGURE 26.15
Design View of FormPR1 to present the query result of PRQ1 in the Windows-based application for the healthcare information system application.

(a)

(b)

```
<%@ Page Language="VB" AutoEventWireup="false" CodeFile="FormMain.aspx.vb" Inherits="Main" %>

<!DOCTYPE html>

<html xmlns="http://www.w3.org/1999/xhtml">
<head runat="server">
    <title>Healthcare Information System</title>
</head>
<body>
    <form id="form1" runat="server">
    <div>

        <asp:Label ID="Label1" runat="server" Text="Select one of the following:"></asp:Label>

    </div>
        <asp:RadioButtonList ID="RadioButtonList1" runat="server">
            <asp:ListItem>Employers</asp:ListItem>
            <asp:ListItem>Patients</asp:ListItem>
            <asp:ListItem>Healthcare Providers</asp:ListItem>
            <asp:ListItem>Insurance Companies</asp:ListItem>
            <asp:ListItem>Phamarcies</asp:ListItem>
            <asp:ListItem>Updates</asp:ListItem>
        </asp:RadioButtonList>
        <asp:Button ID="ButtonNext" runat="server" Text="Next" />
    </form>
</body>
</html>
```

(c)

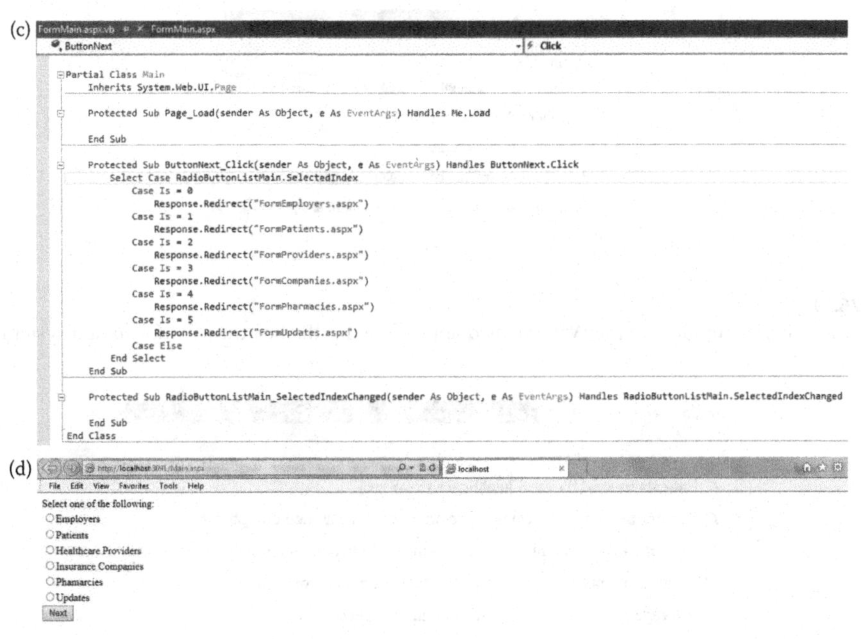

```
Partial Class Main
    Inherits System.Web.UI.Page

    Protected Sub Page_Load(sender As Object, e As EventArgs) Handles Me.Load

    End Sub

    Protected Sub ButtonNext_Click(sender As Object, e As EventArgs) Handles ButtonNext.Click
        Select Case RadioButtonListMain.SelectedIndex
            Case Is = 0
                Response.Redirect("FormEmployers.aspx")
            Case Is = 1
                Response.Redirect("FormPatients.aspx")
            Case Is = 2
                Response.Redirect("FormProviders.aspx")
            Case Is = 3
                Response.Redirect("FormCompanies.aspx")
            Case Is = 4
                Response.Redirect("FormPharmacies.aspx")
            Case Is = 5
                Response.Redirect("FormUpdates.aspx")
            Case Else
        End Select
    End Sub

    Protected Sub RadioButtonListMain_SelectedIndexChanged(sender As Object, e As EventArgs) Handles RadioButtonListMain.SelectedIndexChanged

    End Sub
End Class
```

(d)

FIGURE 26.16
(a) Design View of FormMain, (b) HTML Source View of FormMain, (c) Code View of FormMain, and (d) FormMain when the Web application is executed.

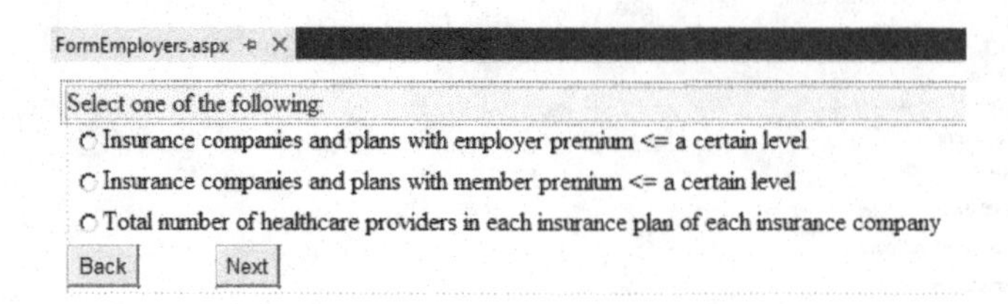

FIGURE 26.17
Design View of FormEmployers in the Web-enabled application for the healthcare information system.

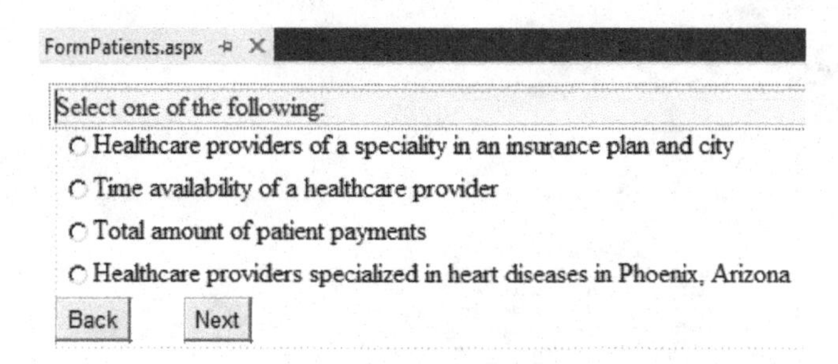

FIGURE 26.18
Design View of FormPatients in the Web-enabled application for the healthcare information system.

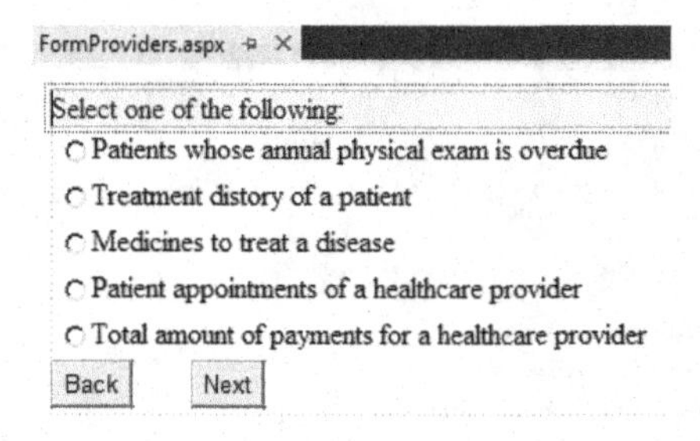

FIGURE 26.19
Design View of FormProviders in the Web-enabled application for the healthcare information system.

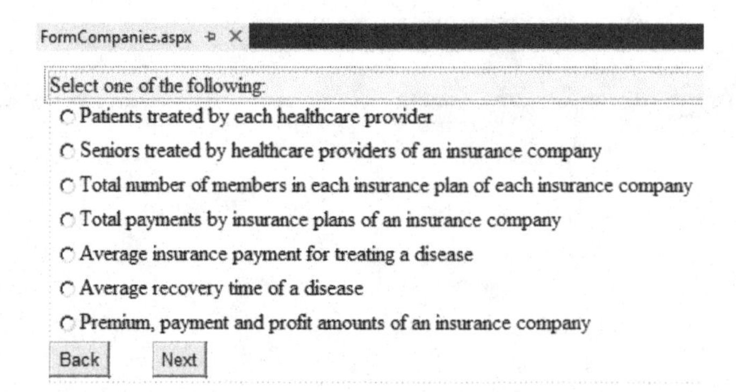

FIGURE 26.20
Design View of FormCompanies in the Web-enabled application for the healthcare information system.

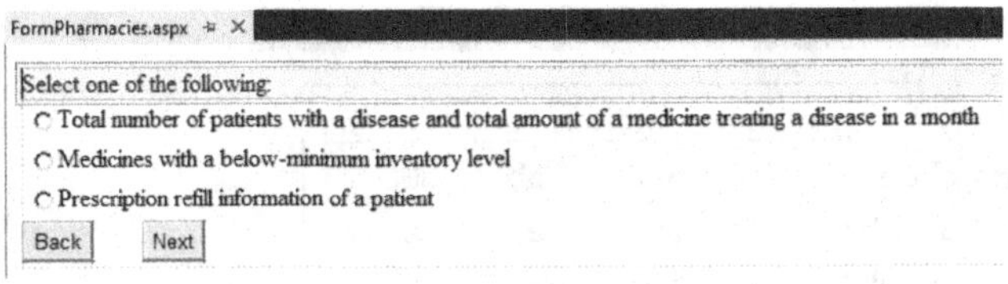

FormPharmacies.aspx
Select one of the following:
Total number of patients with a disease and total amount of a medicine treating a disease in a month
Medicines with a below-minimum inventory level
Prescription refill information of a patient
Back
Next

FIGURE 26.21
Design View of FormPharmacies in the Web-enabled application for the healthcare information system.

FormUpdates.aspx
Select one of the following:
Person
Patient
Insurance plan
Insuarance company
Claim
Claim Service
Provider
Provider schedule
Provider language
Include
Visit
Treatment
Prescription
Contain
Medicine
Pharmacy
Keep
Back
Next

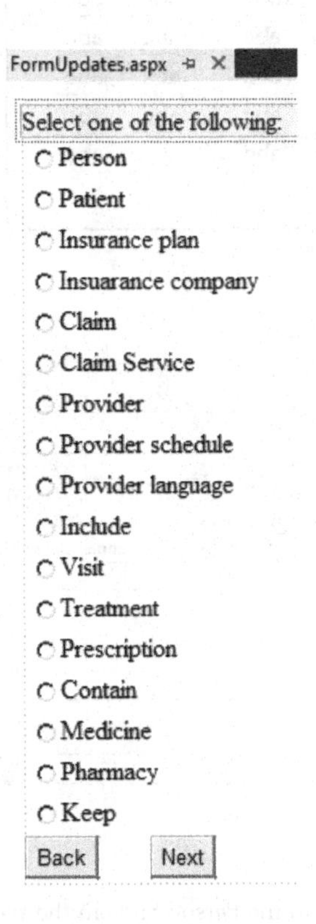

FIGURE 26.22
Design View of FormUpdates in the Web-enabled application for the healthcare information system.

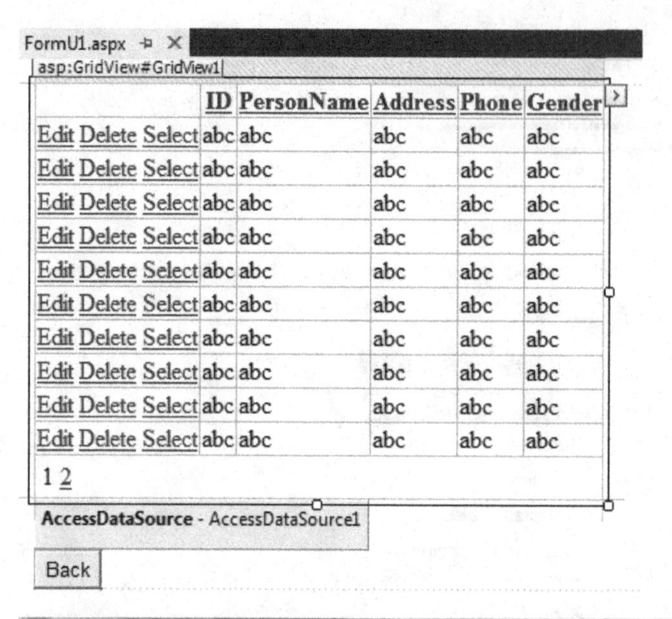

	ID	PersonName	Address	Phone	Gender
Edit Delete Select	abc	abc	abc	abc	abc
Edit Delete Select	abc	abc	abc	abc	abc
Edit Delete Select	abc	abc	abc	abc	abc
Edit Delete Select	abc	abc	abc	abc	abc
Edit Delete Select	abc	abc	abc	abc	abc
Edit Delete Select	abc	abc	abc	abc	abc
Edit Delete Select	abc	abc	abc	abc	abc
Edit Delete Select	abc	abc	abc	abc	abc
Edit Delete Select	abc	abc	abc	abc	abc
Edit Delete Select	abc	abc	abc	abc	abc
1 2					

http://localhost:3091/FormU1.aspx

Healthcare Information Sys...

File Edit View Favorites Tools Help

	ID	PersonName	Address	Phone	Gender
Edit Delete Select	PR1	John Smith	1 Main Street, Phoenix, Arizona	6021111111	
Edit Delete Select	PR2	Mary Fowler	1 Mill Avenue, Tempe, Arizona	4802222222	
Edit Delete Select	PA1	Mike Smith	1 University Street, Tempe, Arizona	4803333333	
Edit Delete Select	PR3				
Edit Delete Select	PR4				
Edit Delete Select	PR5				
Edit Delete Select	PR6				
Edit Delete Select	PR7				
Edit Delete Select	PR8				
Edit Delete Select	PR9				
1 2					

Back

FIGURE 26.23
Design View of FormU1 to present data in the Person table in the Web-enabled application for the healthcare information system application.

top and the middle levels in the hierarchy of Web forms in Figure 26.6. Figure 26.23 gives an example of Web forms at the bottom level in the hierarchy of Web forms in Figure 26.6: FormU1 to present the data of the Person table.

26.6 Summary

This chapter illustrates the entire process of developing the healthcare information system. The process includes the specification of system requirements; the E-R modeling of the healthcare database; the relational modeling of the healthcare database; the implementation of the healthcare database in Microsoft Access; the development of the Windows-based application for the healthcare information system, with the healthcare database embedded; and the development of the Web-enabled application for the healthcare information system, with the healthcare database embedded.

EXERCISES

1. Complete all the other Windows forms at the bottom level of the hierarchy of the Windows forms in Figure 26.6 for the Windows-based application for the healthcare information system, and give the Design View and the Code View of each Windows form.
2. Complete all the other Web forms at the bottom level of the hierarchy of the Web forms in Figure 26.6 for the Web-enabled application for the healthcare information system, and give the Design View, the HTML Source View, and the Code View of each Web form.

27

Development of an Online System for Imaging Device Productivity Evaluation in Radiology Practices

This chapter demonstrates a real-world case study in radiology practices using a back-end relational database and a front-end ASP.NET solution to evaluate imaging exam equipment utilization.

27.1 Introduction

Health information technology (HIT) consists of a diverse set of technologies for securely transmitting and managing health information for use by consumers, providers, the government, and quality entities and insurers. It includes the capture, storage, use, and transmission of health information through electronic processes and has increasingly become the most promising tool for improving the overall quality, safety, and efficiency of the health delivery system (Wu et al., 2006). A typical HIT system is composed of Patient Registries, Accounting/Practice Management System, Computerized Physician Order Entry with Clinical Decision Support, ePrescribing, Electronic Health Records, Patient Health Records, Appointment Scheduling, Telemedicine, Radiology Information System (RIS), and Picture Archiving and Communication System (PACS), to name just a few. Various sets of standards are necessary to support communication among the components, such as Health Level Seven for electronic health information exchange, integration, sharing, and retrieval, and Digital Imaging and Communications in Medicine (DICOM) for imaging storage, transmission, and retrieval.

This chapter will present a case study on developing an online Radiology Exam Efficiency System (REES) to monitor medical imaging equipment utilization in radiology practices. REES interfaces with RIS, a networked software suite used for managing medical imagery and associated data, and PACS, a system for managing workflow and billing, and is implemented based on DICOM standards.

27.2 Imaging Equipment Utilization

Around 80% of patients being seen will have radiology exams for diagnosis, treatment, and disease progression monitoring. The importance of measuring the productivity of expensive imaging equipment has been brought to full public discussion by the current healthcare debate. In particular, the use of imaging equipment such as magnetic resonance imaging (MR), computed tomography (CT), and positron emission tomography (PET) is mentioned as partly contributing to the rising costs of healthcare (McGlynn et al., 2008; Brice, 2009). The Centers for Medicare and Medicaid Services (CMS) has proposed that imaging devices costing greater than $1,000,000 be amortized for replacement based on a 90% service utilization. This computes to a lower per-patient reimbursement from CMS. Note that CMS proposed an increase in utilization (from the currently used 50% utilization), which was based on survey data sampled for a 1-week period at six locations (Brice, 2009). Efficiency evaluations of imaging equipment productivity are human intensive, so it is common to rely on summary statistics such as the number of patients imaged for a sampled period of time and the analysis of patient tracking data (Barnes, 1966; MacEwan, 1982; Hanwell and Conway, 1996; Huang, 2008). What is needed are formally defined metrics for benchmarking efficiency at the imaging device level, which describes the details of actual equipment utilization in the health service organization. However, current RIS and PACS do not provide this information. A new tool, termed "REES," is developed to be fully integrated into a DICOM-compliant infrastructure, with the ability to capture and maintain radiology-specific exam measures of interest for all diagnostic and procedural uses of imaging modality. DICOM is an internationally accepted standard and all-imaging equipment, and is currently provided with DICOM. REES leverages DICOM's rigorous and uniform set of standardized electronic descriptions and provides immediate transferability of a set of structurally defined productivity benchmarks.

27.3 REES Design

As seen in Figure 27.1, REES is designed as a multi-tier implementation (Wang et al., 2011). The system consists of an imaging application programming interface (API) layer accepting and parsing imaging files from PACS; a database layer within which (1) a patient-centric Imaging Exam database is designed to track imaging exam details applied to each patient episode and (2) a DICOM Knowledge database is designed to map the standard and proprietary DICOM tags of interest; an application layer generating imaging equipment utilizaton; and a client layer interfacing with users via the Web. Each of the components is described in the following sections.

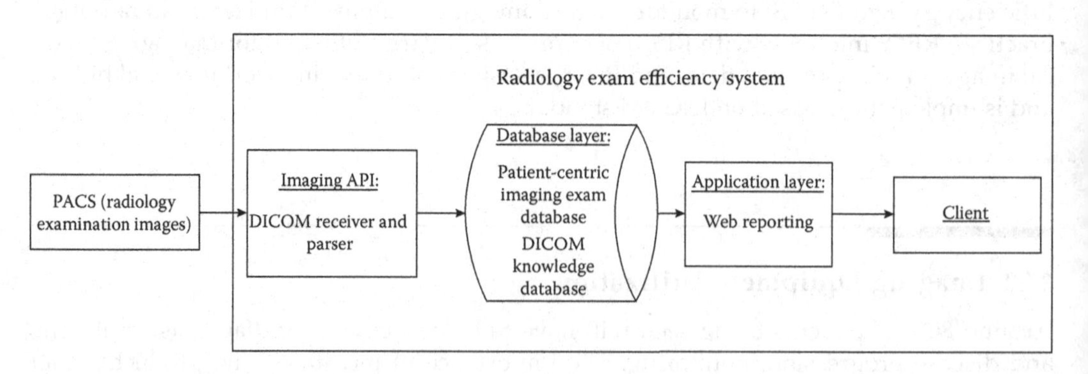

FIGURE 27.1
REES architecture.

27.3.1 Imaging API Design: DICOM Receiver and Parser

The DICOM receiver and parser module receives images from DICOM-compliant devices/systems, extracts the tag information, and stores them into the Imaging Exam database. The receiver takes advantage of the DICOM transport protocol, which is the *de facto* standard in medical imaging transfer fully supported by the majority of device manufacturers and PACS vendors. REES is able to receive and process medical images from various locations, modalities, and equipment. Following the receipt of images, the DICOM header information is parsed and transformed to extensible markup language (XML) format, and then desposited to the tables in the database. Note that REES does not store image data (it can accept the DICOM secondary capture) to enhance performance and reduce storage needs.

27.3.2 Database Design

27.3.2.1 DICOM Knowledge Database

DICOM tags commonly differ by modality, vendor, and software version even for the same vendor and modality. The continuously evolving DICOM standard often results in a given commercially available product version being left behind by newer DICOM committee standards. Nonetheless, even the base standard provides extensive information about the imaging exam, the exact medical examination protocols, and the imaging modality acquisition settings. Every DICOM-compliant image file contains modality-/exam-specific attributes, including patient age, gender, examination ordered, protocol and series descriptors, location of equipment, equipment software version, exam time, etc. Because of the variation in supplied DICOM tags from different scanners, the DICOM Knowledge database design (Figure 27.2) is included. Its schema identifies known scanners (defined as those having their hardware and software specifications as well as the standard and proprietary dose-related tags present with this equipment) as being fully known by the database. Specifically, each known scanner (termed "Known_scanners") belongs to one group of specific tags (termed "Grp_specific_tags"). In Known_scanners, the Grp_ID defines the group of proprietary data elements that will be harvested (chosen once, during new equipment installation or with a new software version) from the DICOM header (in addition to the standard elements, e.g., image Unique IDentifier [UID], accession). The advantage of such schema design is that it provides the system extensibility. For example, when a new scanner is added, or when there is an update of the software version for an existing scanner, one record is added to Known_scanners, and its list of dose-related tags are determined and appended into Grp_specific_tags. Importantly, this extension does not interfere with the

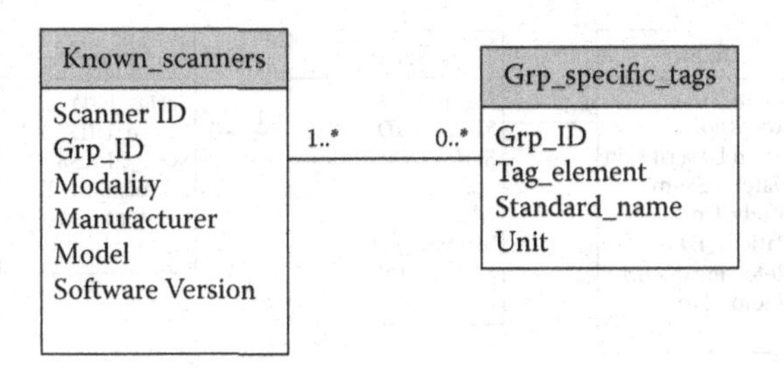

FIGURE 27.2
DICOM Knowledge database schema.

processing of existing images, and it does not require any modification of programming with new images. Overall, the cost of adding an unknown scanner software version to the Knowledge database is minimal.

Since the Knowledge database serves as the foundation of the REES, its comprehensive and accurate content is essential for a successful implementation. The list of vendor modality-specific tags was initially created as a team effort: radiology specialists from each modality, DICOM engineering staff, quality assurance staff, and the system developer.

27.3.2.2 Patient-Centric Imaging Exam Database

The Imaging Exam database is designed to store all harvested tags while avoiding redundancy, maintaining a balanced storage to optimize data access efficiency. The resultant schema contains five parts: patient information, exam information (i.e., protocol), series information, image information, and vendor modality-specific related information (i.e., dose). When images are sent from varied imaging devices into REES, the DICOM header is extracted; the data elements are parsed out and populated into the database. The specified elements are placed in database tables appropriate to the level of information.

In Figure 27.3, the Patient table has one record for each patient, with information such as Patient_Local_ID, Gender, and Date Of Birth (DOB). The table named "Exam" contains the basic information pertinent to each exam common with each modality. This includes exam information (exam description, referring doctor and radiologist), exam time, location, etc. The Series table stores the information related to series, such as Series_UID and Series Description, and the Images table keeps the information at the image level, such as Image_UID and image acquisition time. The Dose_Related_Info table contains the data elements appropriate for the modality, device, and software. The schema is in a serial arrangement, as shown in Figure 27.3, where the relationships between two neighboring tables are all one to many. That is, one Patient may have one or more exams, one Exam record contains one or more series, one Series usually have one or more Image records, and one Image record may be related to one or more Dose_Related_Info records. In this manner, the hierarchical structure of patient, exam, series, images, and dose information is reflected in the database design.

The Imaging Exam database connects with the Knowledge database to populate Dose_Related_Info. When each exam reaches REES, tags for modality, manufacturer, model, and software version will be used to match the image to a known scanner in the Knowledge base. If the match is successful, REES will fetch the list of specific desired tags from the Knowledge database table, Grp_specific_tags, and populate these tags (taken from the image header) into the Imaging Exam database table, Dose_Related_Info. If the match fails (the scanner/software is unknown), the database will automatically alert the database

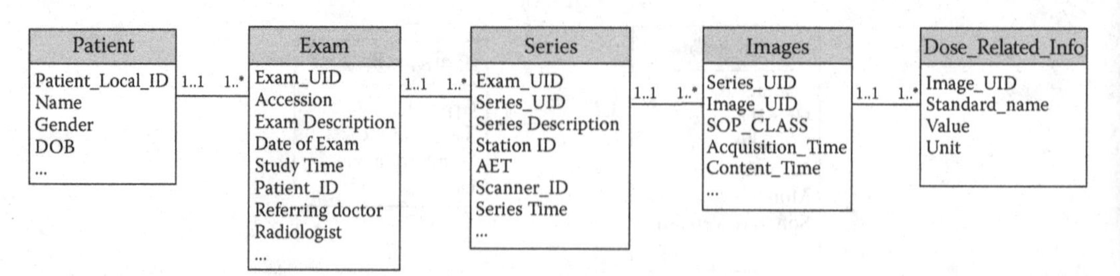

FIGURE 27.3
Patient-centric Imaging Exam database schema.

manager or the medical physicist of the need to exactly specify DICOM tag dose-related fields for future archiving.

Such database design minimizes data redundancy and accelerates a database search. The lower-level tables (left side of the table in Figure 27.3) have a greater number of records and, thus, result in a significant computational cost to conduct the query as compared to the higher-level (right side of the table in Figure 27.3) tables. For example, to list the exams of specific patients, a query on the Patient and Exam tables is sufficient without involving the Series, Images, and Dose_Related_Info tables. Additionally, the design of the Dose_Related_Info table gives improved flexibilty for querying any desired information since the number of dose-related data elements, and their meaning and units are not restricted by the table structure. In this chapter, we will mainly focus on the use of Exam and Series information to evaluate equipment utilization, whereas in Chapter 28, we will demonstrate the use of Dose_Related_Info to calculate the dose exposure for the patient to address quality assurance issues.

27.3.3 Application Layer

27.3.3.1 Benchmark Metric for Imaging Exam Efficiency Assessment

Figure 27.4 depicts five proposed benchmark efficiency metrics. In Figure 27.4, Patient A and Patient B undertake consecutive exams, each shown as having a number of series. DICOM tags (Table 27.1) are inspected for these image data and provide the basis for computing the five proposed benchmark values. (Note that DICOM uses the term "Study," but we use the more common term "Exam" in this case study.) The acquisition time (tag 0008,0032) of the first image of the first series of Patient A denotes the start times of the exam and the first series, respectively. The exam and series start times are image acquisition instances (time stamps), not time images that are transferred to PACS. In this case study, we limit the examples to the MR service, although all DICOM imaging (and treatment) devices can be included. With MR, it is necessary to determine the proper radio-frequency pulse

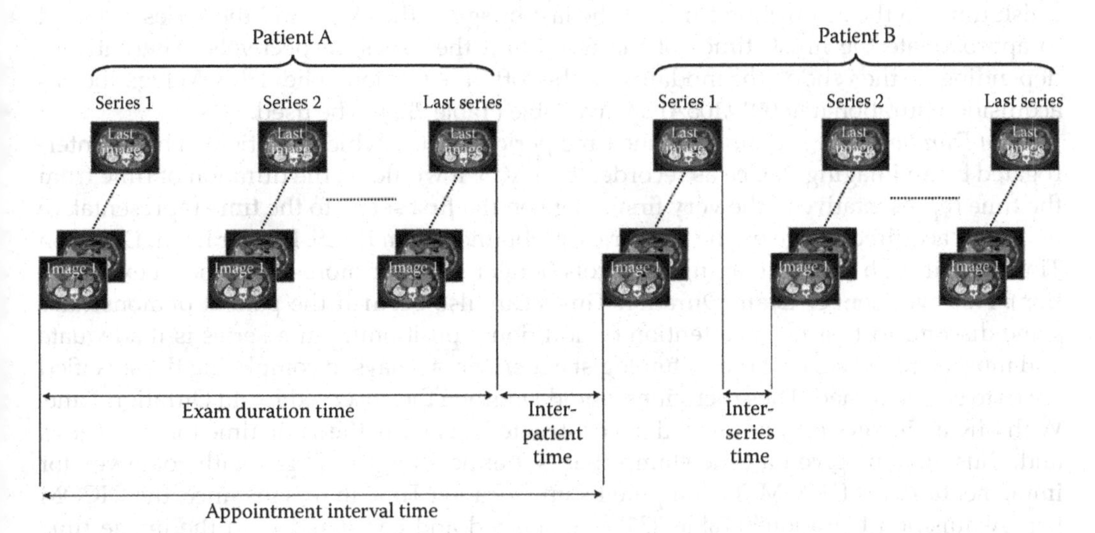

FIGURE 27.4
Definitions of device productivity metrics are shown. Exam Duration, Inter-Patient Time, Appointment Interval Time, and Inter-Series Time are derived from the Exam, Series, and Image time stamps as provided with DICOM.

TABLE 27.1
Partial Listing of Captured DICOM Tags Used by the Imaging Exam Time Monitor

Name	DICOM Tag	Description
Modality	(0008, 0060)	Equipment type
Station Name	(0008, 1010)	Equipment identifier
Patient ID	(0010, 0020)	Patient hospital identification number
Study Instance UID	(0020, 000D)	Study identifier
Study Date	(0008, 0020)	Date that the study started
Study Description	(0008, 1030)	Exam description (from RIS)
Protocol Name	(0018, 1030)	Protocol description (from scanner)
Series Instance UID	(0020, 000E)	Series identifier
Series Description	(0008, 103E)	Series description
Service Object Pair (SOP) Instance UID	(0008, 0018)	Image identifier
Acquisition Time	(0008, 0032)	Image start time
Image Type	(0008, 0008)	Original or derived
Acquisition Duration	(0019, 105A)	Length of a Series with one vendor's MR
X-ray On Time	(0043, 104E)	Length of a Series with one vendor's CT

length or amplitude prior to starting the scan. This initial (tens of seconds) interrogation of the patient may (or may not) be sent to PACS and included in these data even though the patient is on the table. However, as with X-ray CT, an anatomical localizing image is initially obtained and sent to PACS and would be included in the calculation of efficiency metrics. Importantly, images that are derived (added reconstructions, reformatted images, etc.) would not be included in the efficiency metrics. At some time, it might be useful to clarify impediments in the workflow in moving the completed exam into PACS by observing these time stamps. Since the technologist can initiate an exam on another patient while derived images from a prior patient are being created, derived images are not appropriate for inclusion in benchmarking device efficiency. No tag in DICOM actually records a finish time, so the acquisition times of the last image of the exam and the series are used to approximate the finish times of the exam and the series, respectively. Alternatively, depending on the vendor, the modality, or the software version, other DICOM tags such as acquisition duration (tag 0019,105A) are available (Table 27.1) to be used.

Exam Duration Time is defined as the time period during which a patient is being interrogated by an imaging device as recorded by PACS. It would be the duration of time from the time representative of the very first image on the first series to the time representative of the last acquired image on the last series as obtained from PACS. Longer Exam Duration Times occur with longer imaging protocols (longer series or more series for an examination). However, longer Exam Duration Times can also occur if the patient demonstrates some discomfort that needs attention or additional positioning, if a series is inadequate and must be repeated, or if the technologist experiences delays in completing the specified series to be performed. These activities would lengthen the observed Exam Duration Time. With MR, all images may be created (reconstructed) at nearly the same time for each series and, thus, may be given a time stamp that is identical for all images with some vendor implementation of DICOM. To compute Exam Duration Time in this instance, the DICOM tag "Acquisition Duration" (Table 27.1) is inspected and compared with the image time stamps to estimate the total Exam Duration Time. The Exam Description that populates DICOM (tag 0008,1030) commonly originates in the Radiology Information System (RIS) via a Hospital Information System referral order. The indication for the exam is reviewed, and a protocol is selected as directed by the imaging physician. In this manner, the very same Exam Description can have appropriately varied Exam Duration Times. Many protocols exist for commonly performed variations of exam types. Generating reports using the DICOM tag (0018, 1030) "Protocol Name" provides data of the specific collection of series comprising a protocol that was performed for the ordered exam.

Table Utilization is defined as the percentage of time that the device or scanner is actually used for imaging:

$$\text{Table Utilization}(\%) = \frac{\sum \text{Exam Duration Time over a work period}}{\sum \text{Scheduled Working Time over a work period}}$$

Its computation requires the measurement of Exam Duration Time coupled with the specification (by the user) of the normally scheduled imaging service (device) working hours for a period. The numerator is the sum total of Exam Durations for patients for a given period of time, that is, a workweek, divided by the site-specific scheduled working hours for that period. In this manner, the various service operations of out-patient and hospital-based imaging services are accommodated. For example, should the scheduled working hours on weekdays for the equipment be 7:00 A.M. to 5:00 P.M., then the Table Utilization efficiency for the equipment having a summed total of Exam Duration Time of 5 hours on a weekday is 5/10 or 50%.

Inter-Patient Time is defined as the time that a device is not being used for obtaining images of patients. In Figure 27.4, this time is determined as the difference between the first acquired image on Patient B and the last acquired image on Patient A as recorded in PACS. It roughly equates to empty table time. Its value is affected not only by many necessary activities, including room preparation, patient egress assistance, device setup, and patient positioning, but also by patient availability and health status, availability of nursing and medical staff, and equipment-related duties. A long Inter-Patient Time may be an indicator that the equipment needs to be used more efficiently and is a key productivity metric.

Appointment Interval Time is defined as the time allotted and expected to be used for actually performing the examination. It is equal to the sum of the Inter-Patient Time and the Exam Duration Time. Appointment Interval Time equals the measured starting time of an exam with successive patients, as shown in Figure 27.4. Its usefulness as a productivity tool is primarily related to out-patient scheduling services that allocate patient appointment time (slots) available with an imaging resource (device). Lengthy appointment times that are necessary indicate examinations that are device intensive. An analysis of this metric can assist with the optimal scheduling of patients and the potential for reducing wait times and improving patient flow. Since Appointment Interval Time is dependent on Exam Duration Time, the allotted time for an exam can be (and are) variable since individual scanner protocols vary in length dependent on the specifics of the number and the type of series.

Inter-Series Time is defined as the time between successive series for the same patient. Its value can be affected by the equipment software and the operators' selection of acquisition protocols. Some MR scanners (and also CT) have software that automatically moves to the next series in a protocol to minimize this time. Instances in which protracted Inter-Series Time may occur can be necessary as with routine contrast administration, coil placement, repositioning of the patient, and discussions with radiology staff regarding the refinement of the imaging protocol. An analysis of protracted Inter-Series Time may give an insight to the improvements of operational efficiency when viewed using the detailed listing of the imaging protocol.

27.3.3.2 Web Interface

Access to the database is accomplished via Web-based reporting modules through any Web browser. Secure login for the Web user interface is via institutional password protection. Different users are assigned different permission levels so the users can only select those modules that they would need to access. Once the users are authorized, the request is sent to the ASP.NET-based Web server, which executes the request by invoking a connection to the database server (queries to the database are via ADO.NET). All "standard" efficiency reports are dynamically available using the report configuration tool and are displayed to the users via ASP.NET Web pages.

27.4 Use Case

Figure 27.5 shows the Web screen using the Report module. The user can choose reports from Exam Duration, Table Utilization, Inter-Patient Time, Appointment Interval and Inter-Series Time, First Patient Start Time, and Patient High-Frequency Summary.

Using Table Utilization as an example (Figure 27.6), the user selects the date range and the type of report, the modality, the station ID, and the time base for analysis. Exam Descriptions may be specified or all inclusive. Data may optionally be retrieved in a tabular format and exported to Excel for further analysis. An example result is shown in Figure 27.7.

This case study demonstrates that a wealth of information is readily available within the DICOM image header and can be queried to generate standardized benchmark metrics of productivity at the imaging scanner level. This approach to productivity is narrowly focused. It does not address basic measures such as the appropriateness of the

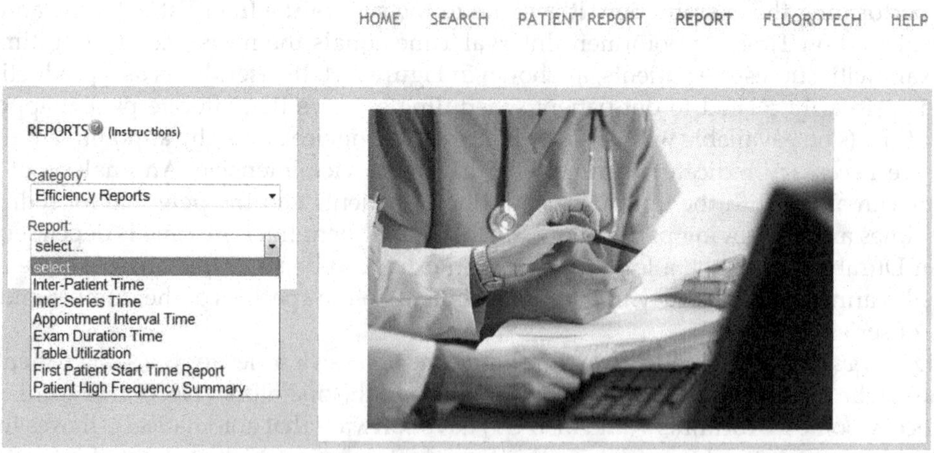

FIGURE 27.5
The Report module Web interface.

REPORTS (Instructions)

Category:
Efficiency Reports

Report:
Table Utilization
select...
Inter-Patient Time
Inter-Series Time
Appointment Interval Time
Exam Duration Time
Table Utilization
First Patient Start Time Report
Patient High Frequency Summary

Table Utilization

This report summarizes the statistics of table utilization.

Please define your scope conditions below then click button **GenerateReport**.

Start Date:
1/1/2013
The first date that the report will include.

End Date:
3/21/2013
The last date that the report will include.

Modality:
MR

Mayo Campus:
PX
Campus naming convention is used. (PX is Phoenix)

Exam Description:
All
Exam Description

Period:
Month
The selection of time period depends on the number of days the user inputs (total days determined by Start and End Dates entered above). If the number of days is 1 you can only choose "Hour"; if the number of days is 2-7, you can only choose "Day"; if the number of days is 8-30, you can choose "Day" or "Week"; if the number of days is 31-90, you can choos e "Week" or "Month"; if the number of days is 91-365, you can choose "Month" or "Quarter"; if the number of the days exceeds 366, you can choose "Quarter" or "Year"

GenerateReport

FIGURE 27.6
Generating Table Utilization Report.

order, reporting of critical results, or the delivery time of the final interpretative report to the ordering physician. Still, this tool records highly accurate and consistent measurement of exactly when patient image acquisitions actually occurred in the context of a specific device. These times are usually calibrated to the enterprise clock and the medical record, which can be coupled with other enterprise data systems. It is unlikely that,

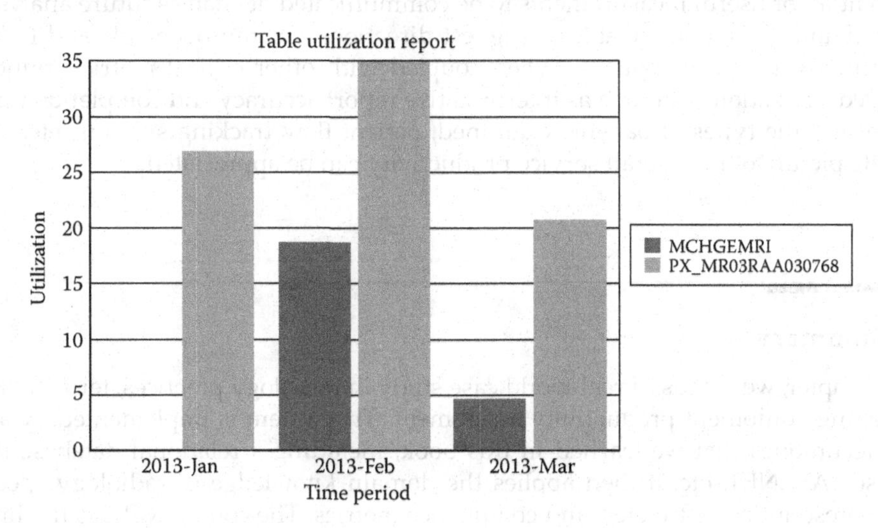

FIGURE 27.7
Example Table Utilization Report.

by using this data collection method, the method itself would cause bias in the results. Productivity measures obtained this year should differ in a meaningful way from data that are obtained next year, only due to efforts directed for improvement in service and not due to the method of data collection. As re-measurement is a necessary requirement of any continuous improvement initiative, it is critical that follow-up measurements are using consistent metrics (Abujudeh et al., 2010).

REES immediately provides nationally (and internationally) standardized definitions of productivity measurement. It is able to do this simply because DICOM is the common standard for medical imaging device communication and is available to all. Since all data are directed to PACS, collecting a copy for productivity analysis is readily available. Re-assessments of productivity to be analyzed following staffing changes, appointment calendar changes, etc., can be performed with directly comparable metrics and little added expense. Business operations staff are provided with tools that can assist their decision making.

CMS uses modality-specific efficiency measures in their review of medical imaging device utilization. Device utilization, in turn, is coupled with reimbursement via the expectation that replacement costs for expensive medical imaging equipment (amortization) are a necessary component of a payment fee schedule. An informatics-based approach that includes actual instances of patient exam details and device utilization aligns this one CMS-driven external review criteria with information for an improved understanding of actual needs.

27.5 Conclusion

In the context of the general Six Sigma Design, Measure, Analyze, Improve, and Control process (Brue, 2007), a non-arbitrary set of efficiency metrics that is fully standardized across facilities is defined and proposed. Agreed-upon efficiency standards (benchmarks) are essential for useful measurements to be communicated. It enables future analysis that is factual and, thus, is more able to suggest directions for improvements and follow-up monitoring of practical changes. When coupled with other critical metrics found with RIS-based utilization data such as interpretative report accuracy and completion time, the number and the types of patients examined, patient flow tracking systems, etc., a more complete picture of the overall service productivity can be appreciated.

27.6 Summary

In this chapter, we discuss a real-world case study in radiology practices, termed "REES," for imaging equipment productivity assessment. The system is implemented by a number of techniques that we learned in this book, including a relational database (Oracle database), ASP.NET, etc. It then applies the domain knowledge of radiology specialists in dose prescription, estimates, and compliance metrics. The core of REES is the Imaging Exam database, which is rich in imaging information. The database forms the foundation of equipment utilization reports as well as a clinical tool used for evidence-based quality assurance practices. In the next chapter, we will discuss the use of the database in conjunction with Visual Basic for Applications (VBA) for a specific medical quality assurance problem, specifically skin dose monitoring.

EXERCISES

1. Imaging equipment acquisition is costly, and as such, equipment utilization must be minimized to prolong its life. True or false?
2. The REES is used to fully integrate any DICOM-compliant machine and compute utilization performance metrics. True or false?
3. What is the use of the imaging API component of REES? Explain its purpose.
4. List the two databases in the database component of REES.
5. ADO.NET is used as a front-end Web-based GUI solution of REES. True or false?
6. Consider an imaging device that is scheduled to be operational from 8 A.M. to 5 P.M. today. Three patients were treated today, and the imaging device took 30 min each to complete. Compute the device's table utilization for the day.
7. The performance metric ______________ is the time measured between the last image acquired from Patient A to the first image acquired from Patient B.

28

Development of a Radiology Skin Dose Simulation Tool Using VBA and Database

This chapter discusses the use of the database in conjunction with visual basic for applications (VBA) to monitor skin dose exposure in medical radiology practices.

28.1 Quality Assurance in Radiology Exam

In 2009, Sodickson et al. reviewed 22-year records of patients having computed tomography (CT) examinations and concluded that "cumulative CT exposures added incrementally to the baseline cancer risk" and that "a subgroup (of patients) is potentially at a higher risk due to recurring CT imaging." As Brenner and Hall (2007) pointed out, approximately 62 million CT examinations were performed in the United States in 2006, and it is reasonable to assume that the use of CT is increasing due to a variety of factors. A strong need exists for an information system to track the radiation dose exposure for each specific patient.

Quality assurance (QA) monitoring of radiation use with certain diagnostic and procedural imaging is required by state regulatory bodies and by The Joint Commission (2007) in the United States. Of the eight U.S. National Quality Measures Clearinghouse metrics specific to radiology (Agency for Healthcare Research and Quality, 2010; National Quality Measures Clearinghouse, 2010), two deal with radiation exposure: reducing CT dose and documentation of X-ray fluoroscopy duration. Since these requirements are enterprise-wide, radiologists, cardiologists, vascular surgeons, urologists, and other permitted users of X-ray and nuclear equipment are required to be monitored to satisfy national goals. Recently, a public discussion in the U.S. national news has expressed concerns regarding the use of CT, including its expanding use and instances of erythema that have occurred following certain CT examinations (Brenner and Hall, 2007; Opreanu and Kepros, 2009). These concerns drive medical facilities to provide an enhanced monitoring of radiation use (Amis et al., 2007).

While Picture Archiving and Communication Systems (PACS) can provide some insights from archived examinations, they suffer from several drawbacks:

- PACS are typically department-level devices and do not offer insight into a patient's exposure across multiple departments or sites (Langer, 2009).
- Commercial PACS, for performance reasons, generally do not permit performing a query on a live production system.
- Radiation dose–relevant Digital Imaging and COMmunication (DICOM) tag values can be missing, inconsistent, or inappropiate.

At this time, records of radiation use are commonly hand recorded in paper logbooks typically at the operator's control console or are manually retrieved from PACS. However, fetching the whole exam to determine relevant information (e.g., dose, exam duration time) is obviously cost inefficient, as is the common approach of manually reviewing a large number of exams to generate user- or protocol-specific radiation quality reports. A standardized toolkit is needed to automatically capture dose information and accurately calculate patient-specific dose exposure in a timely fashion.

28.2 Peak Skin Exposure

Numerous episodes of severe skin damage and harm to patients have occurred due to elevated X-ray exposures during fluoroscopically guided interventional procedures (Food and Drug Administration, 1994; Wagner et al., 1999). Patient skin doses from the use of X-rays in interventional X-ray angiography (XA) procedures continue to exceed the threshold doses for deterministic effects such as erythema and epilation (Miller et al., 2002; Balter and Miller, 2007; Ukisu et al., 2009). Since 1994, the Food and Drug Administration and, more recently, the Joint Commission (2006) have specified that skin exposures from fluoroscopy will be routinely monitored as a QA metric. High skin doses that cause erythema must be identified and the patient must be informed. Appropriate medical care must be provided should this occur. An X-ray exposure to a localized region of the skin (i.e., peak skin dose) that exceeds 15 gray (Gy) is now included as a sentinel event, and it is mandated to be reported to the Joint Commission (2006).

Nonetheless, peak skin dose itself is rarely, if ever, monitored because of the current inability to compute and localize the skin exposure of the patient. Today, monitoring is commonly accomplished via paper or electronic recording at the end of examination. The result of predicting an episode of erythema or epilation is known to be inaccurate estimates of peak skin dose (Miller et al., 2003; Morrell and Rogers, 2006).

The Imaging Exam database we discussed in Chapter 27 is able to provide some meaningful dose information. Thus, we will discuss a case study on the use of the database, in conjunction with a set of modular simulation tools implemented in VBA and Excel to enhance the routine QA monitoring of peak skin dose exposures.

28.3 Simulation Tool for Peak Skin Dose Monitoring

The complete system is comprised of simulated male and female patients, simulated and configurable C-arms, the Imaging Exam database we discussed in Chapter 27, and a VBA graphical user interface (GUI) to collect some system parameters necessary for peak skin dose estimation. A schematic of the dataflow is summarized, as shown in Figure 28.1.

28.3.1 Patient Phantom

SolidWorks (×64 Edition SP 5.0, 2009) simulation software was used to construct standardized male and female phantoms using biometric values using NASA Man-Systems Integration Standards (1995). While multiple phantom options exist (Kerr et al., 1976; Eckerman, 1987), mathematical models enable the straightforward location of any point

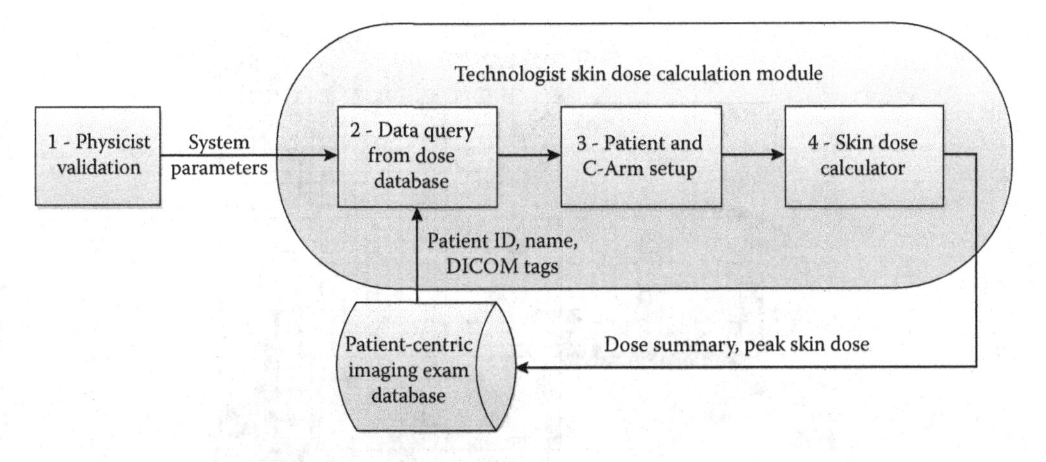

FIGURE 28.1
Skin dose simulation tool. (1) A physicist validates and enters system parameters once per system. (2) The technologist enters the modality, the station ID, and the date to retrieve DICOM data from the Imaging Exam database after each patient exam is finished. (3) The technologist checks the patient's position on the table and the C-arm setup at the end of the exam. (4) Skin Dose Calculator performs the calculation and archives a patient skin dose record to the Dose database.

on the surface of the model because the geometry, shape, and dimensions are known. This was considered advantageous in this application. The height of the standard adult male phantom is approximately 1.86 m and the weight is approximately 90 kg, while the height for the adult female phantom is approximately 1.67 m and the weight is 72 kg. While the arms (Figure 28.2) are included for completeness, they are not used in the calculation of skin dose. In clinical practice, the arms are carefully avoided during any XA procedure and would be extended away from the region of X-ray interest. The female phantom breast is included in this model, and the skin dose to the breast is computed.

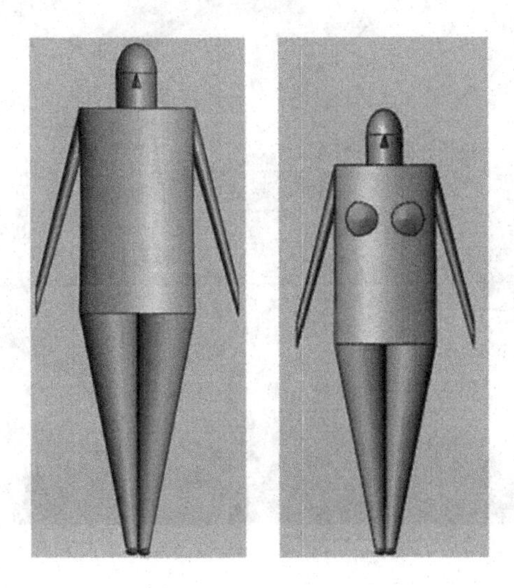

FIGURE 28.2
Standard male and female phantoms. Sizes are based on NASA Man-Systems Integration Standards (1995).

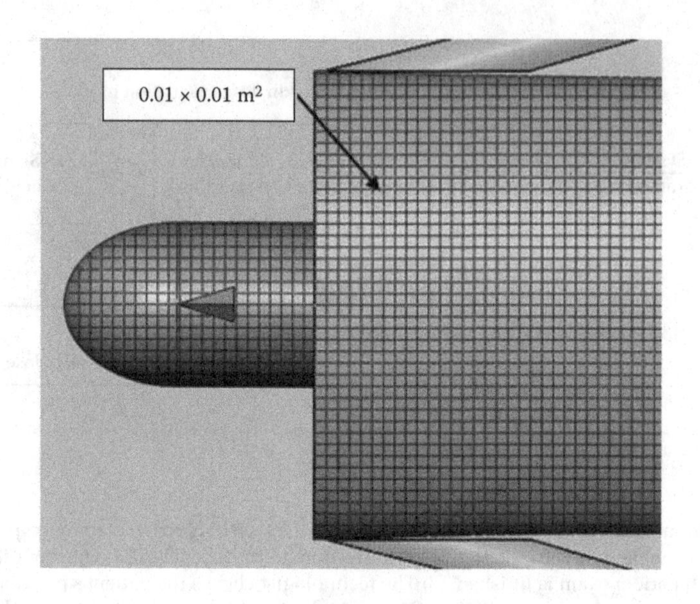

FIGURE 28.3
Standard male phantom. The surface area of the model (except for the arms) is localized to each 0.01×0.01 m^2. These regions are referenced to the origin (top center of the head). (From Khodadadegen, Y. et al., *Journal of Digital Imaging*, vol. 24, no. 4, 626–639, 2011.)

The surface of the phantom is reduced to regions of 0.01×0.01 m^2 (Figure 28.3). Because these three-dimensional mathematical male and female models are constructed from defined elliptical cylinders and cones, the coordinates of the surface of the body are derived using geometric computations, which can be used for skin dose calculation.

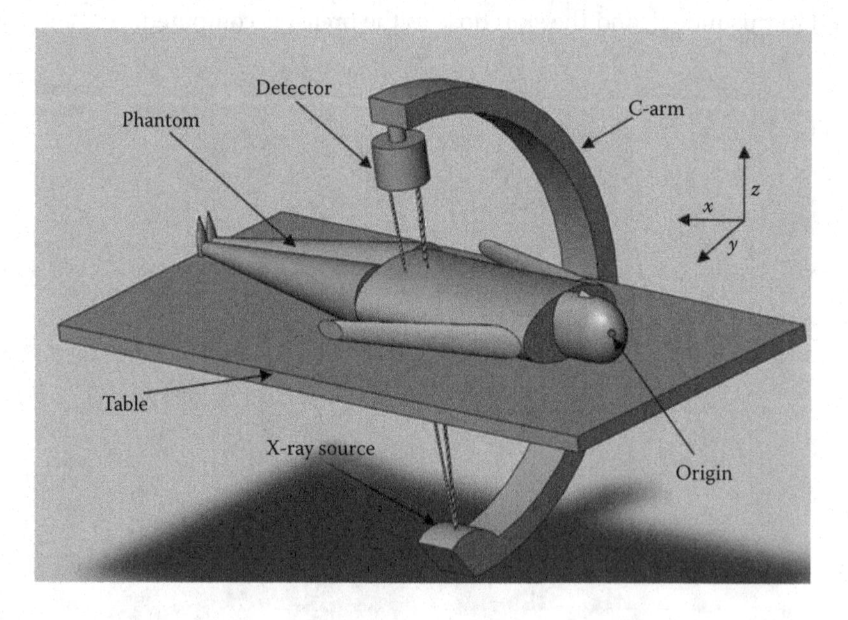

FIGURE 28.4
Simulated fluoroscopic X-ray equipment configuration and system coordination. (From Khodadadegen, Y. et al., *Journal of Digital Imaging*, vol. 24, no. 4, 626–639, 2011.)

28.3.2 Virtual Equipment Configuration and Geometry

Solidworks simulation software was used to depict conventionally available angiographic C-arm equipment, including X-ray image intensification and solid-state (flat-panel digital) X-ray receptors. The patient can be positioned on the table with the patient's head to the right (Figure 28.4) as with head first or with the patient's feet to the right as with feet first. The patient can also be positioned as prone, supine, or facing the left or the right side. The software provides a visual confirmation of each episode positioning for technologist and physicist validation. Figure 28.4 is a typical single C-arm digital detector configuration with the patient positioned head first and supine. The direction of the $+x$ axis is the longitude (moving cranial–caudal), while the direction of the $+y$ axis is to the patient's left side. The $+z$ axis directs to the front of the patient. The software simulation image display

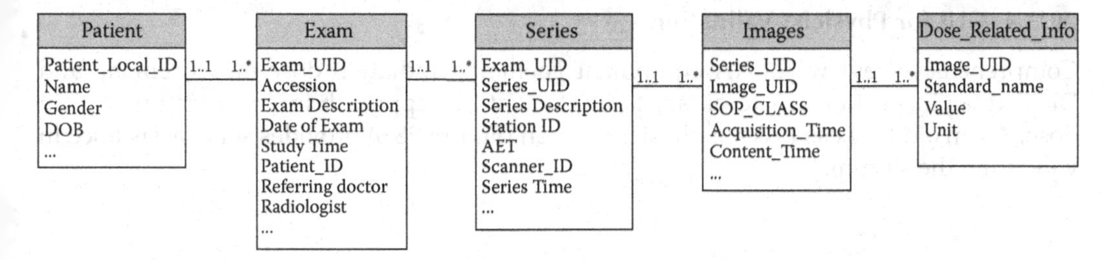

FIGURE 28.5
Patient-Centric Imaging Exam database schema.

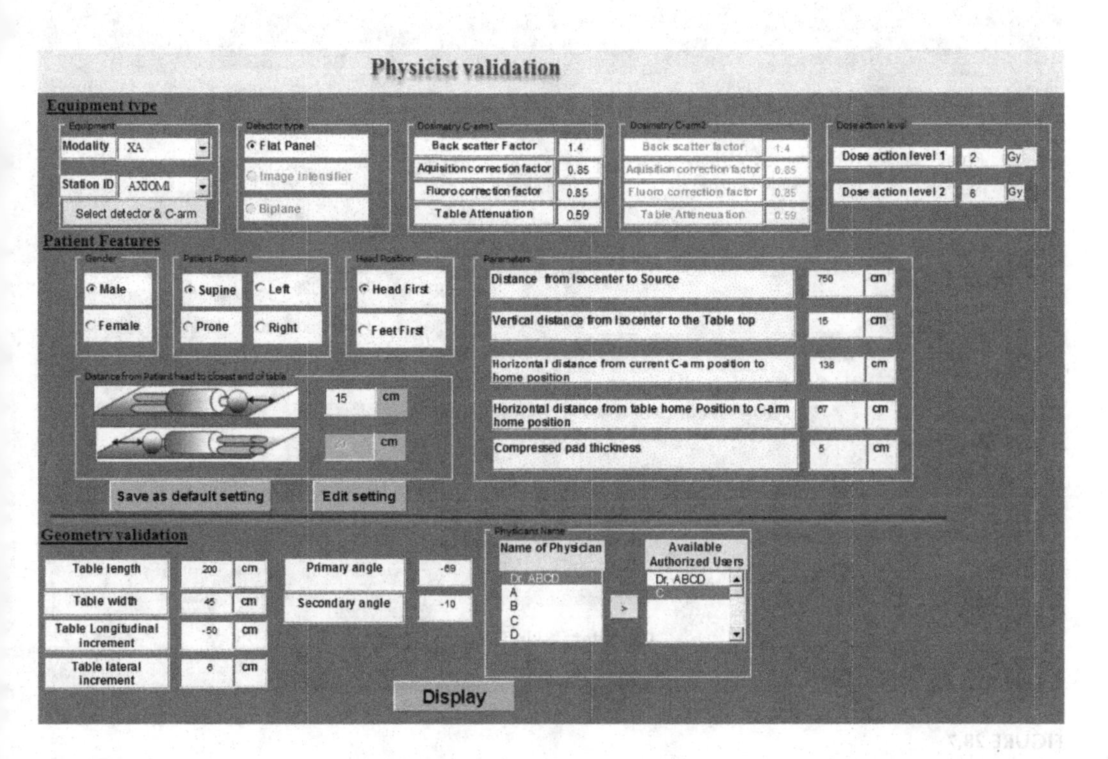

FIGURE 28.6
Physicist validation interface once per system. (From Khodadadegen, Y. et al., *Journal of Digital Imaging*, vol. 24, no. 4, 626–639, 2011.)

actually rotates the displayed gantry following keyboard data entry (to visualize in comparison with the actual gantry).

28.3.3 Imaging Exam Database

The Imaging Exam database (Figure 28.5) is queried to retrieve 11 elements for the dose calculation. These include two C-arm positioning angles, two table positioning information, two distance measures beween the radiation source to the table, four dose parameters given by the equipment, and one patient orientation information. Most of the information is contained in the Series relation—it can be jointly queried given the patient ID and the exam ID.

28.3.4 GUI for Physicist Validation

Comprehensive reviews of XA equipment geometry, radiation dose measurement, and DICOM tag inspection are necessary to implement this approach for localizing peak skin dose. As an aid to these tasks, a physicist screen (Figure 28.6) provides some assistance in validating the system.

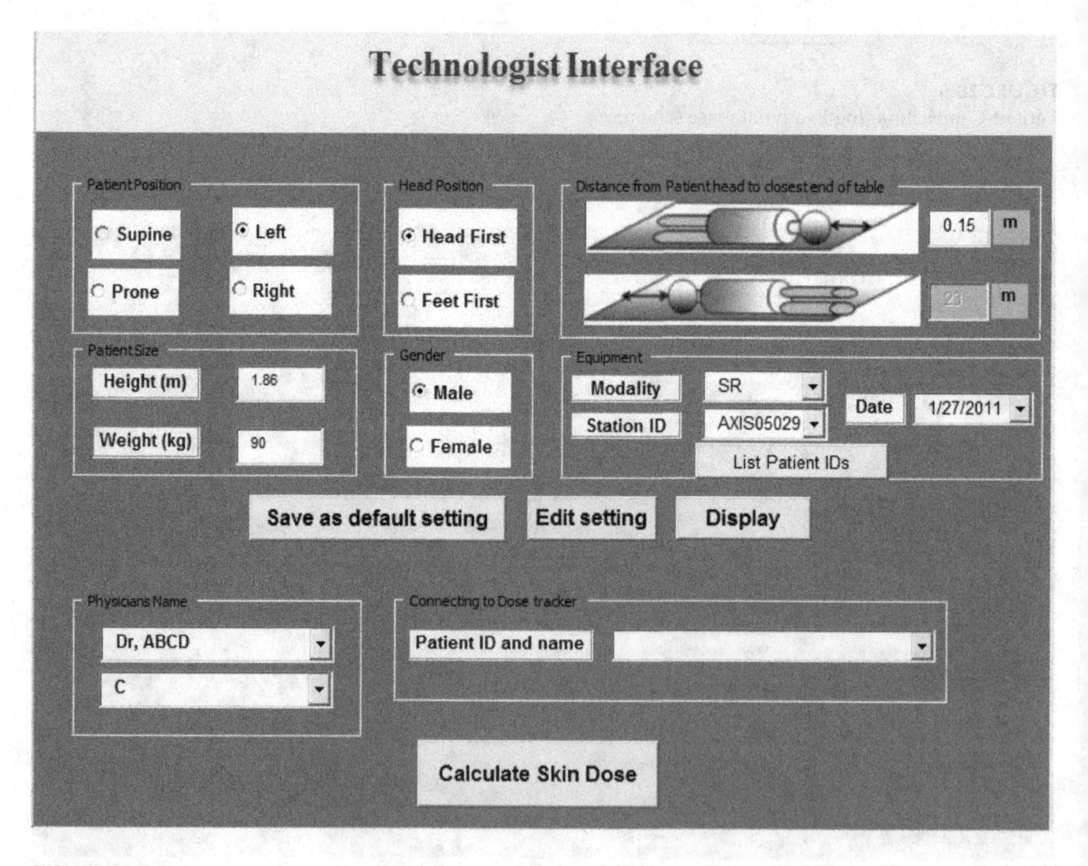

FIGURE 28.7
Technologist Screen validates the patient phantom size and positioning. It automatically connects to the Imaging Exam database for calculating and recording patient dose once per procedure. (From Khodadadegen, Y. et al., *Journal of Digital Imaging*, vol. 24, no. 4, 626–639, 2011.)

As seen in Figure 28.6, this validation is implemented on an Excel spreadsheet with various controls (textbox, list box, label, command button, and option button). This GUI allows the physicists to review and validate the exam setup. In addition, the physicists may input the system parameters that are not available in the Imaging Exam database, for example, compressed pad thickness. The equipment chosen is displayed, and by clicking Display, the positional information on the C-arm image is updated so that the physical gantry, the virtual C-arm, and the tag data are co-registered.

28.3.5 GUI for Technologist Validation

Upon the completion of each procedure, the technologist enters the gender, height, and weight estimates, along with the patient's orientation on the table (supine/prone, feet or head first, left or right lateral), and a corresponding phantom is automatically loaded. This is done by using the GUI, as shown in Figure 28.7. Similar to the physicist validation GUI, this is implemented on an Excel spreadsheet with various controls. The technologist selects the gender, height, and weight of the patient to establish the match for the phantom. Physician operators are chosen from a pre-determined listing of users. Patient positioning is noted, including the distance from the top of the patient's head to the end of the

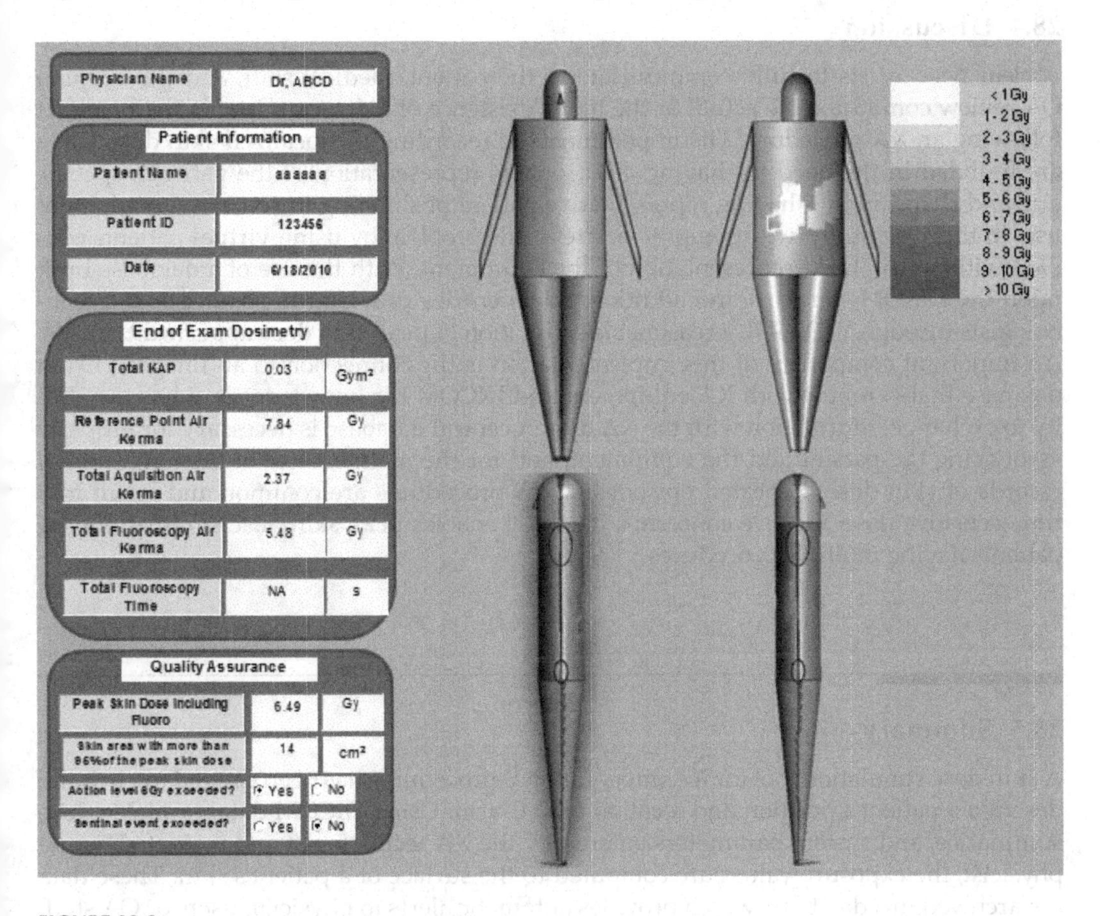

FIGURE 28.8
Patient Summary Dose Report. (From Khodadadegen, Y. et al., *Journal of Digital Imaging*, vol. 24, no. 4, 626–639, 2011.)

table. The equipment selection confirms the imaging laboratory being used and is used to retrieve the patient-specific acquisition and fluoroscopy information for calculation at the end of the procedure. Clicking Calculate Skin Dose results in a patient-specific dose screen (Figure 28.8).

28.3.6 Peak Skin Dose Calculation Engine

A point within a $0.01 \times 0.01\text{-m}^2$ area is assigned the computed value. Each representing point on the discretized surface has the coordinate $(m \pm 0.5, n \pm 0.5, f(m \pm 0.5, n \pm 0.5))$, assuming m and n as integers. To expedite computation, the irradiated regions (projected X-ray field on the surface of the patient) are initially determined using the positioning of the C-arm. Since the purpose of this chapter is to demonstrate the application of VBA and Excel and the database for QA in medical practices, for details of the calculation engine, please refer to Khodadadegen et al. (2011).

28.4 Discussion

A clear need exists to fully communicate to the patient, medical staff, and the hospital QA review committee the actual or the likely existence of a deterministic radiation effect following an XA procedure. The impediments to recording the localized skin dose value are founded in the need for having a reasonable representation of the patient's specific size and shape and for having representative patient positioning in actual XA equipment use. In this case study, we attempt to address this problem by using virtual patients coupled with a visualized representation of the equipment. With the use of a database built upon the DICOM standard and additional data entries provided to physicists and technologists through VBA GUI, a reasonable estimation of peak skin dose appears attainable. An important component of this implementation is the connection to an Imaging Exam database that is robust with XA equipment and DICOM tag knowledge and has the ability to exchange information with the XA tool. A central database is necessary for properly identifying the patient and the equipment, and for the maintenance of the longitudinal records of skin dose. Repeated episodes of XA procedures are common and result in a time-sensitive dose additive concern; a database enables peak skin dose summations for patients having multiple procedures.

28.5 Summary

A skin dose simulation tool for XA interventional procedures has been created to virtually describe a patient's position and location on a C-arm. Using DICOM data, printable dose summaries, and system parameters entered by the XA technologist and validated by the physicist, the exposure values are computed to the surface of a patient's skin. These data are archived in a database, which provides automatic alerts to physician users or QA staff. An increase in the number of patients having two, three, or more interventional procedures is expected. Furthermore, a skin dose simulation tool would reasonably include an accounting of skin dose received via other modalities. For example, computed tomography brain perfusion commonly results in 1 to 2 Gy to a localized portion of the skin, which would be additive to patients having brain XA procedures.

EXERCISES

1. QA monitoring of radiation dosage is a serious issue and, as such, needs to be monitored. True or false?
2. One of the barriers of radiation dose monitoring is that doses are recorded in electronic logbooks of different communication media. True or false?
3. It is easy to predict an episode of erythema due to deterministic estimates of peak skin radiation dose. True or false?
4. The GUI for technologist validation is created using what programming language?
5. Which of the following factors affect localized skin dose value?
   a. Patient's size.
   b. Patient's position.
   c. Patient's shape.
   d. All of the above are true.
   e. None of the above is true.

References

Abujudeh, H. H., R. Kaewlai, B. A. Asfaw, and J. H. Thrall. 2010. Key performance indicators for measuring and improving radiology department performance. *RadioGraphics*, v. 30, pp. 571–583.

Agency for Healthcare Research and Quality. 2010. Available at http://www.ahrq.gov/.

Amis, E. S., P. F. Butler, K. E. Applegate, S. B. Birnbaum, L. F. Brateman, J. M. Hevezi, F. A. Mettler, R. L. Morin, M. J. Pentecost, G. G. Smith, K. J. Strauss, and R. K. Zeman. 2007. American College of Radiology white paper on radiation dose in medicine. *Journal of American College Radiology*, v. 4, pp. 272–284.

Balter, S. and D. Miller. 2007. The New Joint Commission Sentinel Event pertaining to prolonged fluoroscopy. *Journal of American College Radiology*, v. 4, no. 7, pp. 497–500.

Barnes, R. M. 1966. *Motion and Time Study: Design and Measurement of Work*. New York: Wiley.

Breiman, L., J. H. Friedman, R. A. Olshen, and C. J. Stone. 1984. *Classification and Regression Trees*. Boca Raton, FL: CRC Press.

Brenner, D. J. and E. J. Hall. 2007. Computed tomography: An increasing source of radiation exposure. *New England Journal of Medicine*, v. 357, pp. 2277–2284.

Brice, J. 2009. New payment formula proposes Medicare cuts for high-tech imaging. *Diagnostic Imaging*. Available at: http://www.diagnosticimaging.com/ct/content/article/113619/1385012. Accessed July 1, 2014.

Brue, G. 2007. *Six Sigma for Managers*, 1st edn. Oakbrook Terrace, IL: McGraw-Hill, pp. 11–12.

Buchanan, B. G. and E. H. Shortliffe. 1984. *Rule Based Expert Systems: The MYCIN Experiments of the Standford Heuristic Programming Project*. Reading, MA: Addison-Wesley.

Cattell, R. G. 1991. *Object Data Management—Object-Oriented and Extended Relational Database Systems*. Reading, MA: Addison-Wesley.

Codd, E. F. 1971. A database sublanguage founded on the relational calculus. In *Proceedings of the 1971 ACM SIGFIDET Workshop on Data Description, Access and Control*, pp. 35–68.

Codd, E. F. 1990. *The Relational Model for Database Management*. Boston: Addison-Wesley.

Cormen, T. H., C. E. Leiserson, R. L. Rivest, and C. Stein. 2009. *Introduction to Algorithms*, 3rd edn. Cambridge, MA: MIT Press.

Deutsch, A., M. Fernandez, and D. Suciu. 1999. Storing semistructured data in relations. In *Workshop on Query Processing for Semistructured Data and Non-Standard Data Formats*, Jerusalem, January 1999.

Eckerman, C. M. 1987. Specific absorbed dose fractions of energy at various ages from internal photon sources. Appendix A: Description of the mathematical phantoms. *ORNL/TM-8381/VI*. Oak Ridge, TN: Oak Ridge National Laboratory.

Food and Drug Administration. 1994. Public Health Advisory: Avoidance of Serious X-Ray-Induced Skin Injuries to Patients during Fluoroscopically-Guided Procedures, September 30.

Frank, A. and A. Asuncion. 2010. *UCI Machine Learning Repository*. Irvine, California: School of Information and Computer Science, University of California. Available at http://www.archive.ics.uci.edu/ml.

Hanwell, L. L. and J. M. Conway. 1996. *Utilization of Imaging Staff: Measuring Productivity*. Sudbury, MA: American Healthcare Radiology Administrators, pp. 1–30.

Heckerman, D. and E. Shortliffe. 1992. From certainty factors to belief networks. *Artificial Intelligence in Medicine*, v. 4, no. 1, pp. 35–52.

Hinchcliffe, D. 2011. The "Big Five" IT trends of the next half decade: Mobile, social, cloud, consumerization, and big data. *ZDNet*. Oct. 2.

Hofmann, P. 2011. The big five IT megatrends. Available at http://www.slideshare.net/paulhofmann/the-big-give-it-mega-trends.

Huang, H. K. 2008. Utilization of medical imaging informatics and biometrics technologies in healthcare delivery. *International Journal of Computer Assisted Radiology and Surgery*, v. 3, no. 1–2, pp. 27–39.

Joint Commission. Radiation Overdose as a Reviewable Sentinel Event. 3/7/2006. Available at http://www.jointcommission.org/NR/rdonlyres/10A599B4-832D-40C1-8A5B-5929E9E0B09D/0/Radiation_Overdose.pdf.

Kerr, G. D., J. M. L. Hwang, and R. M. Jones. 1976. A mathematical model of a phantom developed for use in calculations of radiation dose to the body and major internal organs of a Japanese adult. *Journal of Radiation Research*, v. 17, pp. 211–229.

Khodadadegen, Y., M. Zhang, P. Pavlicek, P. Robert, B. Chong, B. Schueler, K. Fetterly, S. Langer, and T. Wu. 2011. Automatic monitoring of localized skin dose with fluoroscopic and interventional procedure. *Journal of Digital Imaging*, v. 24, no. 4, pp. 626–639.

Langer, S. 2009. Issues surrounding PACS archiving to external, third-party DICOM archives. *Journal of Digital Imaging*, v. 22, no. 1, pp. 48–52.

MacEwan, D. W. 1982. Radiology workload system for diagnostic radiology: Productivity enhancement studies. *Journal of the Canadian Association of Radiologists*, v. 33, no. 3, pp. 183–196.

Man-Systems Integration Standards. 1995. Revision B, NASA-STD-3000, vol. 1, July 1995.

McGlynn, E. A., P. G. Shekelle, S. Chen, D. P. Goldman, J. A. Romley, P. S. Hussey, H. Vries, M. C. Wang, M. J. Timmer, J. Carter, C. Tringale, and R. M. Shanman. 2008. Identifying, categorizing, and evaluating healthcare efficiency measures. Rand Corporation Final Report to Agency for Healthcare Research and Quality, HHS. Contract 282-00-0005-21, April.

Miller, D. L., S. Balter, P. T. Noonan, and J. D. Georgia. 2002. Minimizing radiation-induced skin injury in interventional radiology procedures. *Radiology*, v. 225, no. 2, pp. 329–336.

Miller, D. L., S. Balter, P. E. Cole, H. T. Lu, A. Berenstein, R. Albert, B. A. Schueler, J. D. Georgia, P. T. Noonan, E. J. Russell, T. W. Malisch, R. L. Vogelzand, M. Geisinger, J. F. Cardella, J. S. George, G. L. Miller, and J. Anderson. 2003. Radiation doses in interventional radiology procedures: The RAD-IR study. Part II: Skin dose. *Journal of Vascular and Interventional Radiology*, v. 14, no. 8, pp. 977–990.

Morrell, R. E. and A. T. Rogers. 2006. A mathematical model for patient skin dose assessment in cardiac catheterization procedures. *British Journal of Radiology*, v. 79, pp. 756–761.

Munoz, D. F., M. D. Lascurain, O. Romero-Hernandez, F. Solis, L. Santos, A. Palacios-Brun, F. J. Herreia, and J. Villasenor. 2011. INDEVAL Develops a New Operating and Settlement System Using Operation Research. *Interface*, January–February, p. 8.

Murphy, C. 2011. Create. *Information Week*, March 14, 2011, p. 23.

National Quality Measures Clearinghouse. 2010. Available at http://www.guideline.gov/browse/xrefnqmc.aspx.

Null, L. and J. Lobur. 2006. *The Essentials of Computer Organization and Architecture*. Sudbury, MA: Jones and Bartlett Publishers.

Opreanu, R. C. and J. P. Kepros. 2009. Radiation doses associated with cardiac computed tomography angiography. *JAMA*, v. 301, no. 22, pp. 2324–2325.

Pfleeger, C. P. 1997. *Security in Computing*. Upper Saddle River, NJ: Prentice Hall PTR.

Quinlan, J. R. 1986. Induction of decision trees. *Machine Learning*, v. 1, pp. 81–106.

Ravishankar, M., S. L. Pan, and D. E. Leidner. 2011. Examining the strategic alignment and implementation success of a KMS. *Information Systems Research*, p. 39.

Skoudis, E. 2002. *Counter Hack*. Upper Saddle River, NJ: Prentice Hall PTR.

Sodickson, A., P. F. Baeyens, K. P. Andriole, L. Prevedello, R. D. Nawfel, R. Hanson, and R. Khorasani. 2009. Recurrent CT, cumulative radiation exposure, and associated radiation-induced cancer risks from CT of adults. *Radiology*, v. 251, no. 1, pp. 175–184.

Stair, R. and G. Reynolds. 2013. *Fundamentals of Information Systems, Course Technology*, 7th edn.

Stevens, W. R. 1994. *TCP/IP Illustrated*, vol. 1. Boston: Addison-Wesley.

The Joint Commission. 2007. Oakbrook Terrace, IL: The Joint Commission Sentinel Event Policy and Procedures.

TrendMicro. 2011. *Consumerization of IT*. Available at http://www.trendmicro.com/cloud-content/use/pdfs/business/reports/rpt_consumerization-of-it.pdf.

Ukisu, R., T. Kushihashi, and I. Soh. 2009. Skin injuries caused by fluoroscopically guided interventional procedures: Case-based review and self-assessment module. *American Journal of Radiology*, v. 193, pp. 59–69.

Wagner, L. K., M. D. McNeese, M. V. Marx, and E. L. Siegel. 1999. Severe skin reactions from interventional fluoroscopy: Case report and review of the literature. *Radiology*, v. 213, pp. 773–776.

Wang, S., W. Pavlicek, C. Roberts, S. Langer, M. Zhang, M. Hu, R. Mornin, B. Schueler, C. Wellnitz, and T. Wu. 2011. An automated DICOM database capable of arbitrary data mining (including Radiation Dose indicators) for quality monitoring. *Journal of Digital Imaging*, v. 24, no. 2, pp. 223–233.

Wickens, C. D., S. E. Gordon, and Y. Liu. 1998. *An Introduction to Human Factors Engineering*. New York: Longman.

Wu, S., B. Chaudhry, J. Wang, M. Maglione, W. Mojica, E. Roth, S. C. Morton, and P. G. Shekelle. 2006. Systematic review: Impact of health information technology on quality, efficiency, and costs of medical care. *Annals of Internal Medicine*, v. 144, no. 10, pp. 742–752.

Ye, N. 2003. *The Handbook of Data Mining*. Mahwah, NJ: Lawrence Erlbaum Associates.

Ye, N. 2008. *Secure Computer and Network Systems: Modeling, Analysis and Design*. John Wiley & Sons.

Ye, N. 2013. *Data Mining: Theories, Algorithms, and Examples*. Boca Raton, FL: CRC Press.

Ye, N., B. Zhao, and G. Salvendy. 1993. Neural-networks-aided fault diagnosis in supervisory control of advanced manufacturing systems. *International Journal of Advanced Manufacturing Technology*, v. 8, pp. 200–209.

Index

Page numbers followed by f and t indicate figures and tables, respectively.

K

L

M

Printed in the United States
by Baker & Taylor Publisher Services

Printed in the United States
by Baker & Taylor Publisher Services